S0-ALL-459

THE DECLINE OF NATURE

THE DECLINE OF NATURE

ENVIRONMENTAL HISTORY AND THE WESTERN WORLDVIEW

GILBERT F. LAFRENIERE

ACADEMICA PRESS
BETHESDA, DUBLIN, LONDON

Library of Congress Cataloging-in-Publication Data

LaFreniere, Gilbert F.
 The decline of nature : environmental history and the Western worldview /
Gilbert F. LaFreniere.
     p. cm.
 Includes bibliographical references and index.
 ISBN-13: 978-1-933146-40-9
 ISBN-10: 1-933146-40-0

 1. Nature—Effect of human beings on—Western countries. 2. Environmental
degradation—Western countries. 3. Environmental quality—Western countries.
4. Western countries—Environmental conditions. I. Title.

 GF75.L34 2007
 304.2—dc22
                          2007021825

Copyright 2007 by Gilbert F. LaFreniere

All rights reserved. Printed in the United States of America. No part of this book
may be used or reproduced in any manner whatsoever without written permission
except in the case of brief quotations embodied in critical articles and reviews.

Editorial Inquiries:
Academica Press, LLC
7831 Woodmont Avenue, #381
Bethesda, MD 20814
Website: www.academicapress.com
To order: (650) 329-0685 phone and fax

For my wife, Pamela, and son, Sean

TABLE OF CONTENTS

FOREWORD

> ...The difference in mind between man and the higher animals, great as it is, is certainly one of degree and not of kind.... If it be maintained that certain powers, such as self-consciousness, abstraction etc., are peculiar to man, it may well be that these are the incidental results of other highly-advanced intellectual faculties; and these again are mainly the result of the continued use of a highly developed language. — Charles Darwin, *The Descent of Man*

Whatever the differences between species, the evolutionary paradigm makes clear that our kind is distinctive in being a language animal. Still, we differ genetically from our closest kin, the chimpanzee, only by a slight margin. But that small difference makes all the difference. We have become storytelling culture-dwellers. *The Decline of Nature* is a sustained examination of the foundational narratives that overdetermine our lives.

It is possible for a culture to engage these stories critically. It is also enormously difficult to do so. The stark reality for a culture-dwelling species is simply this: we either come to own our stories. Or else the established stories own us, relentlessly perpetuating themselves regardless of the consequences. In the case of the environment the issue is simple. Own our stories. Or else.

Easily said. But owning our cultural stories is an enormous challenge. In some cases the consequences of the unexamined life are relatively benign. In the case of dysfunctional relations between our cultural systems and the evolved planetary bio- and ecosystems, the unexamined life will be catastrophic. Clearly, the dominant stories are destroying planetary life-support systems. And just as clearly, if those systems collapse, so too will civilization.

No doubt, there are many powerful members of the US Congress and the administration, as well as business and educational leaders, and millions of ordinary citizens who continue to deny the scientific reality of a worsening environmental crisis. Ironically, the very denial of ecocrisis confirms the hold of the unexamined life on the cultural system. However, the powerful few and the multitudinous millions who are in denial will find a clear exegesis of the stories that own them herein. LaFreniere exposes the narratives that ground the symbolic compulsion that leads to denial.

The Decline of Nature is a masterful critique of the stories that own us. LaFreniere's analytical effort is a veritable tour de force. The intellectual resources he draws upon are current, scientifically solid, and virtually irrefragable. And, remarkably, he pulls these diverse materials into a seamless and compelling narrative. The dogmatically religious will be taken aback, as there are no sacred cows herein. LaFreniere's critique on occasion pinches even his staunch supporters, such as myself, at one point or another. Which means that, above all else, *The Decline of Nature* is a *must read book*, even among the already committed.

Readers outside the fold of environmentalism and the academy will find Professor LaFreniere's narrative conceptually challenging. As the Norwegian deep ecologist Arne Naess often remarks, thinking is painful. Make no mistake, this book demands that readers engage their neocortex, however disconcerting that might be. We live in a time when conceptual "fluff and puff" books dominate the talk shows and the shelves of mass-market book sellers.

Most of these hyped books are not worth the candle. In contrast, regardless of the demands on the reader, *The Decline of Nature* should be read by every adult American – whether butcher, baker, or candlestick maker. Society is a collectivity that shares political rights and obligations. One of these obligations is to participate in the civic discourse which is essential to resolving the environmental crisis. Reading this book will enhance the possibilities for responsible citizenship.

What the citizen will come to discover is that unless the underlying assumptions, the mythopoeic fundamentals of Western culture, are brought to the surface for public deliberation, then environmental policy will continue in the same channel as the last twenty years or so: much ado about little that's fundamental, a shifting of the chairs on the deck of the Titanic while the band plays on. I repeat myself. Own your stories. Or they own you.

Academics such as myself and LaFreniere are sometimes characterized as "hothouse orchids." Meaning that we could not survive in the "real world" outside universities and colleges. I find that notion implausible. Above all else, *The Decline of Nature* is a book firmly in the neo-Darwinian fold. Cultural systems have always been and will always be put to the sternest of tests. This, then, is a book about survival in the real world.

Max Oelschlaeger, Northern Arizona University

PREFACE

The world is in some mysterious way numinous. It possesses almost inscrutable characteristics of what humans have described as transcendent value or "spirit." This virtually indefinable quality has manifested itself to humans in various ways: as paranormal phenomena; as personal visions; and in the religions which form the core of belief systems underlying the world's great civilizations. Whatever "spirit" may be and however it relates to value in nature may be intellectually irresolvable, but one thing is certain. Once a given culture establishes its own "truth," relatively fixed, rigid religious traditions permeate the minds of the individual members of the culture, including those "higher" cultures which we call civilizations. In pre-modern civilizations the unquestioning followers of a religion were what Eric Hoffer called "true believers." In modern Western or westernized civilizations true believers have also embraced cryptoreligions, the worldly ideologies of communism, fascism, and capitalism, and have often defended their beliefs to the death.

Americans living in the early twenty-first century have become highly sensitive to the varieties of true belief held in the United States, where a resurgence of Christian fundamentalism and the pseudo-theory of "intelligent design" have entered the political arena much to the amazement of predominantly secularized western Europeans. What has all this to do with environmental history as it relates to the Western worldview?

During the past several decades environmental history and environmental philosophy have emerged as an outgrowth of the Environmental Movement. Much of the research associated with their development has focused upon Western attitudes toward nature, thereby eliciting a critique of the Western worldview and its interpretation of the linked concepts of God, man, nature, history, and society. Within a given worldview a conception of history (some explanation of its pattern and meaning) tells the stories which reassure the culture's masses of true believers. In Western European civilization the Judeo-Christian story of creation as the beginning of history, leading through several millennia to the end of history with the Apocalypse and Last Judgment, is our foundational historical myth. This mythopoeic conception of history, originating in the Judaic civilization before Christ, has been central to the development of Western European civilization since the Early Middle Ages, and it is still taught to the majority of American citizens in homes and in churches. No contrary interpretation of the meaning of history arose to challenge this philosophy of history until the Age of Enlightenment, beginning in the late seventeenth century.

The idea of history as progress grew out of Western scientific and technological developments during the sixteenth and seventeenth centuries. Nevertheless the Christian idea of history, the idea of providence, held fast among the majority of Europeans and Americans until the time of Darwin's *Origin of Species*, but scientists and historians of the early modern and nineteenth centuries had inadvertently laid the foundations for challenging the Judeo-Christian interpretation of history. Nevertheless, although largely eradicated in Europe, the idea of providence, combined with an idea of progress understood as endless technological innovation and economic growth, dominates the American worldview to the present day.

Since the Renaissance, a third alternative conception of history with ancient roots, the idea of cycles, has competed with the ideas of providence and progress, culminating in the great syntheses of Oswald Spengler and Arnold Toynbee in the twentieth century. Toynbee combined cycles with the ideas of providence and progress, whereas Spengler eschewed the latter two philosophies of history. Moreover, since the rise of environmental history in the late 1960s the discredited cycles described by Spengler in *The Decline of the West* have taken on new meaning. The rise, flourishing, decline, and collapse or devolution of civilizations has been linked to environmental degradation by historians such as J. Donald Hughes and Clive Ponting, and scholars such as Brian Fagan and Jared Diamond in related disciplines. Thus, the cyclical philosophy of history has been given new life by studies in environmental history, archeology, and anthropology. However, recent and contemporary scholars have failed to recognize the renascence of the cyclical philosophy of history (as an explanation of the rise and fall of civilizations caused in part by the degradation of regional ecologies) which I describe in this book.

The foregoing developments and an article I published twenty years ago entitled "World Views and Environmental Ethics" have been a major part of the inspiration to write a book which attempts to probe to the heart of our Western worldview, with all of its kaleidoscopic changes through the medieval and modern centuries. Admittedly this is a daunting task, but much of the work had already been done for me by environmental historians, environmental philosophers, and others, and merely required integration into a single narrative.

To sum up, the purpose of this book is to explain historical changes in the Western worldview regarding Western humanity's relationship to nature and history. Nature has been regarded for too long as little more than the stage upon which the human historical drama has been played. Our conceptions of history have to a great extent determined our understanding of nature. In the modern progressive worldview nature is generally perceived as merely the material resources sustaining improvement in the material conditions of human life, and this idea is often alloyed with the providential interpretation of history. In this book I also attempt to explain how our understanding of nature would be changed from the perspective of a cyclical philosophy of history such as the one Oswald Spengler put forth in *The Decline of the West*, but with modifications based upon recent scholarship in environmental history and archeology. Also, as the links between our speculative philosophies of history and our perceptions of nature are explained in the context of an account of the environmental history of Western civilization, other aspects of the Western worldview such as our conceptions of God, "man," and society are further illuminated.

Thus, the plan of the book is to lay the foundations for an environmental history of Western civilization (focusing on the history of ideas) in the first two chapters, followed by eight chapters explicating this history, and three final chapters emphasizing the significance of this account for the present "environmental crisis" and the so-called "death of environmentalism." To my mind, the providential and progressive philosophies of history which are stamped upon the Western mind as functioning "truths" explain to a great extent why Western civilization and its imitators around the globe are incapable of coming to terms with the draconian changes needed to reverse the ecologically devastating consequences of fulfilling the "American dream" of the abundant material life. Worse, the reversion to religious fundamentalism and its rejection of evolutionary theory in the United States only further narrows our constricted view of long-term historical reality and stiffens our inertial resistance to a change in worldview. Only *metanoia*, a profound change in worldview which might supersede traditional religious beliefs or integrate them into a worldview that reconciles our recent ecological and historical knowledge of nature with our ways of making a living from it, can save the planet from destruction of the evolutionary ecosystems which gave rise to us and all other species.

The Decline of Nature is an unusual book because it combines an analysis of Western interpretations of the meaning of history (i.e., patterns of providence, progress and cycles) with an environmental history of the Western world (European civilization and its colonies) responsible for the modernization and attendant ecological degradation of other civilizations around the globe. It is an important book for a time when confusion over the meaning of both nature and history undergirds fierce social and political disagreements which divide the United States while the Environmental Movement has fallen into the background of American cultural debate.

ACKNOWLEDGMENTS

I wish to thank Willamette University, Salem, Oregon for the 2001 fall sabbatical during which I initiated research on this book. In the early stages of research I consulted with J. Donald Hughes and Carolyn Merchant, whose own work is so important to the thesis developed in *The Decline of Nature*. I owe a debt of gratitude to Mary Plank, Willamette University Administrative Assistant for her assistance with typing the first draft of the manuscript and to Editor Robert Redfern-West and technical assistant Ginger McNally of Academica Press for their support in refining and formatting my manuscript into a finished product. I am also indebted to the journal *Environmental History* for permission to publish material from an article ("Rousseau and the European Roots of Environmentalism") published in *Environmental History Review*, Volume 14, Number 4, Winter, 1990.

The Decline of Nature synthesizes ideas which grew out of decades of teaching environmental ethics and environmental history at Willamette University, and owes a great deal to the students who read and discussed my own articles and other books and papers at length. Also, discussions with Willamette University colleagues Peter Eilers, Lane McGaughey, and Richard Lord influenced my thinking about environmental history and religion, and Professor William Smaldone suggested the title of Chapter Ten, "The Twentieth Century Road to Ruin." The manuscript was read in whole or in part by Professor Klaus P. Fischer, intellectual historian, Hancock College, Santa Maria, California and J. Donald Hughes, environmental historian, University of Denver, and the book was improved significantly from their suggestions. Max Oelschlaeger, Professor of Philosophy and Religion, Northern Arizona University, was for all purposes my editor in refining the first draft of the manuscript into a much improved and expanded draft. I cannot thank him enough for his thorough and incisive reading of *The Decline of Nature*. His encouragement and support was paramount to my completing the book.

Finally, I am grateful for the contributions of my wife, Pamela and son Sean, regarding both the substantive content of the manuscript and technical assistance with the computer.

INTRODUCTION

The View from Fairmount Hill

    It is less than two miles from downtown Salem, Oregon to the top of Fairmount Hill, the northern terminus of the South Salem Hills, a broad ridge of basalt which rises as much as 1000 feet above the northward flowing Willamette River, which joins the Columbia River about seventy miles to the north. The capitol of Oregon is still a relatively quiet place compared to bustling Portland an hour's drive to the north. Living on the eastern slope of Fairmount Hill, we have a good view of the Cascade Mountains and its foothills, and one can make out the speeding semi trucks on the I-5 highway three miles east of our house.

    For twenty years I have walked this hill daily to stay in condition for hikes and mountain climbs. Halfway through this daily ritual there is a spectacular viewpoint which looks westward over the Willamette Valley to the Coast Range. In early May, this year's snow has melted off of the Coast Range, but it is still visible on the Cascades above 3000 feet. The valley and hills in the foreground are cloaked in the fresh green colors of spring vegetation, and below the distant horizon the Coast Range is a collage of grey and purplish shadows beneath cumulus clouds blowing in from the west. One's first impression is of an idyllic pastoral landscape framed by a wild and rugged mountain backdrop.

    Gradually my reverie at the viewpoint gives way to the intrusion of multiple mechanical sounds. A lawnmower or two and a power-driven edger are distinctive enough to separate from the drone of a larger machine in the background. A small plane heads westward from the Salem airport, and as I look skyward I notice the north-south alignment of contrails from the jets heading towards San Francisco and Los Angeles.

    Refocusing on the landscape before me, the Willamette River is bounded by a broad band of trees that snakes through prime farmlands on the river's floodplain. I am reminded of a recent article in The Oregonian, Portland's only major newspaper, describing the impact of industrial effluent, urban runoff, and pesticides and fertilizers from intensive mechanized agriculture upon the water quality and ecology of the river, recently ranked as one of the most polluted rivers in the United States. Looking beyond the valley to the Coast Range, I can barely discern the square patches of clear-cut forest now that the snow is gone. My pastoral and wild mountain landscapes have been mentally deconstructed into an intensively utilized, humanized patchwork of farmland and timberland whose wild inhabitants, particularly the carnivores, have been displaced or decimated in numbers. With the grizzlies, wolves, and most of the black bear gone, deer have proliferated and this has helped the state's cougar population to rebound after private hunting of them with dogs was voted down some years ago. A state population of less then 200 cougars in the 1960s has exploded to approximately 5,000 and state fish and game hunters have been given the task of reducing them to about 3,000. Everything on the Oregon landscape is controlled and managed, from one corner of the state to another.

Walking back to the east side of Fairmount Hill, the clear-cuts at the higher elevations of the Cascades are white with snow. Driving the highway over Santiam Pass towards Bend was enjoyable for many years, although the forest near the summit of the pass had become infested with wood-boring beetles. Several years ago both trees and parasites were removed, however, by a devastating wildfire that destroyed hundreds of square miles of national forest surrounding the high peaks of the Cascades where the wilderness areas are located. Heavy human use in the wilderness of the Willamette National Forest has polluted streams to the point that a water purification kit must now be packed in to avoid illness from *Giardia* and other pathogens.

Humans have been transforming the Willamette Valley for thousands of years. The Kalapuya Indians girdled and burned the trees, especially Douglas fir, to open up grasslands that would be inviting to deer and elk to complement their acorn, camas, and fish-based diet. The forested mountains of the region appear to have been much less affected by Native Americans than the valleys and foothills.

Leaving Fairmount Hill behind us, a flight from Portland to New York displays an even greater level of human impact upon the landscape from coast to coast. Lowland and valley areas are farmed or heavily grazed, national forests and private timberlands are generally subjected to clear- cutting and planting of desired commercial species, and the skies above are laced with contrails reflecting the networks of freeways, highways, and secondary roads below. Although scattered wild areas remain, wilderness today exists in name only. Widespread air pollution, acid rain, and global warming is the consequence of more than two centuries of industrial and technological revolution. Humanity dominates the planet, and species diversity and ecosystems diminish more rapidly than ever before. More than half of Earth's land surface is utilized by humans and much of the remaining land is in deserts, high mountains, and glaciers, all relatively barren ecologically. We also use 55 percent of the world's fresh water, and 40 percent of the total photosynthetic productivity of the sun. [1] Although this book tells the story of Western civilization's impact upon nature since the Middle Ages, I begin here with a brief discussion of global warming because, although it may well deliver the *coup de grace* to many species and ecosystems, I wish to make the point at the very beginning of this account that even without global warming we have done a fairly thorough job of eradicating wilderness and numerous species of plants and animals. Global warming, however, will not only cause great harm to what is left of the biosphere, it will also present the greatest challenge to the human species since the 20th century proliferation of nuclear warheads, the other sword of Damocles which hangs precariously over human and non-human life.

Glaciation and Natural Cycles

When J.B. Bury, the great early 20th century classical scholar and author of *The Idea of Progress*, asked himself, rhetorically, the hypothetical question of how long into the future *Homo sapiens* would have to go on improving the condition of the species for progress to have occurred historically, he answered "perhaps 60,000 years, or ten times the duration of known human history."[2] This was in 1925, when the first edition of *The Idea of Progress* was published. As a secular humanist he did not hesitate to think in terms of a duration ten times greater than the apocalyptical Christian vision of all history, from creation to apocalypse, enduring a mere 6,000 years. Bury's conception of time was informed by modern geologic knowledge and its description of Earth history over hundreds of millions and even billions of years since the origin of the solar system approximately 4.6 billion years ago.

In the course of Earth's history there have been multiple periods of glaciations which can be traced, through the evidence of glacial deposits, back into Pre-Cambrian time. Glaciation is anything but random, and dependent upon the complex interaction of several variables. These variables include, in order of importance, the positioning of continents relative to the polar regions, three interacting cycles which determine the amount of solar energy our planet receives (the Malenkovitch hypothesis), and secondary factors such as the overall planetary albedo (the reflection of energy into space by light surfaces such as snow and ice) and the distribution of ocean currents and air masses in the atmosphere.

All of these processes are in some way cyclical, beginning with the plate tectonic cycle. Ice ages can occur only when adequate land masses are present at or near the Polar regions. The plate tectonic cycle involves the creation of new basaltic crust at oceanic ridges, the accumulation of differentiated granitic crust to form island arcs and continents near subduction zones, and the horizontal movement of both types of crust along transform faults (see Chapter Eight). Over long periods of time the granitic continents accumulate, bump together, and pull apart, at times being positioned at or near the poles. During the Mesozoic era, the Age of Dinosaurs, of 260 million until 65 million years ago, the continents were not positioned in a manner favorable to glaciation. However, during the Cenozoic era of the last 65 million years they have moved into positions increasingly favorable to glaciation. In the process of shifting, they have also created mountain ranges and plateaus where ice accumulates in far northern and southern regions. These Cenozoic uplifts have also greatly influenced ocean currents and atmospheric circulation.

Plate tectonic cycles have set the stage for the glacial and interglacial cycles of the Pleistocene Ice Ages which are dependent upon the cyclical nature of variations in Earth's astronomical relationship to the sun, which determines incoming solar energy.

Milutin Milankovich was a civil engineer in the Austro-Hungarian Empire who became fascinated with the causes of periodic glaciation. His research resulted in the discovery of the importance of three related cycles which caused alternating glacial and interglacial periods during the Pleistocene epoch. The cycles include the periodic shift in Earth's orbit about the sun from circular to elliptical and back (100,000 years), the amount which Earth is tilted on its axis as it revolves around the sun (42,000 years from maximum to minimum tilt and back), and the top-like wobble of the Earth on its axis (22,000 years). Milankovich's correlation of glacial and interglacial cycles with the orbital, tilt, and wobble cycles has been corroborated by geologic evidence during the years since publication of his discoveries in English in 1969.[3] Another cyclical variable, sunspots, also appears to affect Earth's climate, but its role in climate change is not yet fully understood.[4]

The growing realization that anthropogenic global warming is now occurring and the fact that it is scientifically indisputable recently led *Time* magazine to devote a lengthy cover story to the future consequences of human-caused climate change. Photographs of stranded polar bears on melting Arctic ice flows, drought-ridden, desiccated land in Ethiopia, monsoon-flooded lowlands in India, forest fires in the Yukon, and a gallery of endangered or threatened species capture the casual reader's attention and lead them towards a grim conclusion. "In a solar system crowded with sister worlds that either emerged stillborn like Mercury and Venus or died in infancy like Mars, we're finally coming to appreciate the knife-blade margins within which life can thrive. For more than a century we're been monkeying with those margins. It's long past time we set them right."[5]

Almost simultaneously with *Time*'s special issue on global warming, Elizabeth Kolbert's *Field Notes from a Catastrophe: Man, Nature, and Climate Change* appeared, touted as the most important book about life on Earth since the appearance of Rachel Carson's *Silent Spring* in 1962. Documenting the evidence for global warming by accompanying climate scientists to the Greenland Ice Sheet, questioning native American inhabitants in Alaska and interviewing scientists at the Goddard Institute for Space Studies in New York, Kolbert argues convincingly that global warming is unequivocally anthropogenic with the potential to generate temperatures in the twenty-first century that are likely to be hotter than at any time in the two million year history of the Pleistocene. The 11,000 years since the Pleistocene glaciers have been rapidly melting away (known to geologists as the Holocene) has given way to what some climate scientists have christened the "anthropocene," an essentially new epoch dominated by human impacts upon natural processes on a geological scale since the Industrial Revolution.[6]

A more thorough analysis of the consequences of anthropogenic climate[7] change is to be found in Tim Flannery's *The Weather Makers: How Man is Changing the Climate and What It Means for Life on Earth* (2005). Some of the most devastating ecological and biological changes in the twenty-first century and beyond will occur in the landscapes and seascapes of the Arctic and Antarctic regions. Like Kolbert, Flannery describes the existing evidence of higher temperatures, melting ice, and diminished populations of species in the polar oceans. Flannery's conclusions regarding devastating ecological impacts in both the Arctic and Antarctic are particularly ominous.

In Antarctica, the relationship between the extent of shelf ice, plankton, krill, and the mammals that feed on krill is fragile enough that even a moderate temperature increase of a few degrees could turn the South polar seas into an ecological desert. The presence of sea ice promotes the growth of microscopic plankton that provides the base of the South polar food chain. Krill, tiny shrimp organisms, feed upon the plankton and in turn provide the food that supports penguins, seals, and whales. Flannery shows that krill have been in sharp decline since the 1970s, reducing the krill to fifty percent of their 1926 populations. As temperatures have risen, Antarctic sea ice and krill have rapidly diminished.[8]

The Arctic has fared no better under global warming. Winters there today are 4 to 5 degrees Fahrenheit warmer than they were only thirty years ago, with profound ecological consequences on both land and sea. The polar bear is immediately at risk as Arctic ice retreats because it is dependent upon pack ice for hunting the ringed seal that lives and breeds there. Already, the great white bears are slowly starving due to the earlier break-up of pack ice and increasing distances between floating ice, causing some of them to drown in recent years. Documenting the negative impacts of Arctic warming upon several other species, Flannery concludes that the imminent loss of the polar bear "may mark the beginning of the collapse of the entire Arctic ecosystem." [9]

Arctic warming has also contributed to the devastation of northern forests and their inhabitants. Two insect species indigenous to more Southerly latitudes, the spruce bark beetle and the spruce bud worm, have killed tens of millions of trees in southern Alaska, and they continue to rage out of control, along with the wildfires that thrive on unhealthy, damaged forests. Flannery also foresees the demise of tundra-dependent species such as caribou, lemmings, and bird populations as warming drives the spread of northern forests over the tundra all the way to the Arctic Ocean.[10]

Elizabeth Kolbert spent some time accompanying a permafrost expert in Alaska, where she learned that the frozen ground has warmed by three to six degrees since the early 1980s. The "active" layer above the permafrost thaws over the summer and allows plant growth. Gradually, with winter freezing, plant matter is pushed into the permafrost. As the permafrost melts with increasing global temperatures the plant remains release large quantities of methane. Given that almost a quarter of all the land in the Northern Hemisphere is underlain by permafrost, warming will be further increased by large quantities of methane released into the atmosphere, adding to the warming caused by the decrease in Earth's albedo by the loss of pack ice in the polar regions. These negative feedback processes have the potential to outstrip human attempts to reverse the warming process.[11] There is no longer any reasonable doubt regarding the overwhelming evidence that anthropogenic global warming is already well under way. It is not unreasonable, therefore, to conclude that Oswald Spengler's warning at the beginning of the twentieth century (see end of Chapter One), that modern, industrialized humanity has intensified and accelerated the destruction of the natural world, has been validated by the factual knowledge of the environmental sciences and environmental history. However, Spengler could not have imagined that humans would actually break the established pattern of the Pleistocene glacial-interglacial climate cycles of the past two million years!

The shift from Pleistocene glacial-interglacial cycles to a prolonged and un-naturally hot global climate under projected temperature increases from global warming is well documented. Jim Hansen (Director of the NASA Goddard Institute for Space Studies and adjunct Professor of Earth and Environmental Sciences at Columbia University's Earth Institute) explains that under a business-as-usual scenario of increased production of carbon dioxide, methane, and other greenhouse gases, global warming during this century is likely to increase global temperatures by an average of about five degrees Fahrenheit. The Earth has not been this warm since three million years ago, when sea level was about eighty feet higher than it is today.[12] Hansen poses the question whether global warming today will cause a comparable sea-level rise and postulates the time it would take. "A rise in sea level, necessarily, begins slowly. Massive ice sheets must be softened and weakened before rapid disintegration and melting occurs and the sea level rises. It may require as much as a few centuries to produce most of the long-term response.[13]

An alternative scenario of strong controls on greenhouse gas emissions would involve the leveling off of CO_2 emissions over the next decade, slow decline for a few decades and a rapid decrease in emissions by mid-century. Even under this (probably unlikely) scenario there would be an increase of slightly less than two degrees in global temperatures during the same period of time. The warmest interglacial periods during the last three million years were only two degrees Fahrenheit warmer than today's temperatures! Sea level was about sixteen feet higher than present levels during these interglacial periods. Thus the only certainty is that sea level will rise. Uncertainty resides in answering how much and how soon that rise will be. A long period of slow melting will accompany the softening and weakening of the Greenland and Antarctic ice caps before they disintegrate and move seaward.[14]

Mass extinctions of species during the geological past were triggered either by asteroid collisions (see Chapter Nine) or excessive increases in heat typically related to extensive volcanism and the production of massive volumes of CO_2 and other green house gases such as CH_4 (methane). The release of "frozen methane," for example, may have been the major cause of the ten-degree Fahrenheit temperature increase responsible for the mass extinction marking the Paleocene-Eocene boundary about 55 million years ago.[15] According to Flannery, the extreme warming occurred over decades or centuries, when basaltic magma in the north and central Atlantic ignited vast deposits of clathrates (icy, methane-rich solids) contained in oceanic sediments.

Enormous amounts of methane combined with oxygen to form CO_2, which caused atmospheric concentrations to rise from 500 parts per million to about 2,000 parts per million. Norwegian scientists have recently discovered several crater-like structures dating from 55 million years ago when the great volumes of CO_2 were released. The CO_2 acidified the oceans simultaneously with extreme atmospheric warming, causing extinctions so widespread that they define the Paleocene-Eocene epochs.[16] Thus, evidence from the geologic past defines and clarifies the range of ecological degradation to be expected from various degrees of global warming induced by humanity. The human contribution to climate change requires an understanding of environmental history as well as geology, however.

The unique accomplishment of William F. Ruddiman in *Plows, Plagues and Petroleum* (2005) was to link global climate change to major aspects of environmental history, such as human environmental impacts upon ecosystems resulting from the discovery of agriculture, grazing, and other aspects of the Neolithic Revolution. Farming and grazing in the early stages of agriculture and the development of ancient civilizations occurred in regions of Eurasia which were 90 percent forested. Gradual deforestation over a period of approximately 8,000 years produced a carbon release of about 300 billion tons, forming enough carbon dioxide to offset the decline of atmospheric CO_2 since the end of the Wisconsin Ice Age approximately 12,000 years ago. Inadvertently, deforestation by humans also may have prevented the onset of a new ice age.[17]

Around 8,000 years ago, following a natural CO_2 peak of about 270 ppm ca. 10,500 years ago, instead of continuing to decrease as it had after previous shifts from glacial to interglacial periods, the CO_2 concentration began a slow rise from about 260 ppm to more than 280 ppm by the start of the Industrial Revolution late in the 18th century. "Natural processes caused the atmospheric CO_2 peak nearly 10,500 years ago and the subsequent decrease until 8,000 years ago, but humans have caused the anomalous CO_2 increase since that time."[18]

Ruddiman has demonstrated convincingly that these differences in CO_2 concentrations over the past 8,000 years are chiefly the result of deforestation, while the increases in methane concentrations over the past 5,000 years resulted mainly from rice irrigation in China and Southeast Asia. Other factors such as human waste, livestock emissions, and biomass burning would have provided only minor increases in methane because human population growth was relatively slow during these millennia. Ruddiman was also able to scientifically correlate short-term decreases in CO_2 accumulation with plague events in early civilizations.[19] Large declines in human population caused by bubonic and other plagues temporarily slowed the cutting of forests in Eurasia and even allowed them to grow back temporarily, thereby temporarily reversing the overall long-term increase in atmospheric CO_2. These anomalies show up as bumps or "wiggles" on the CO_2 curve for the last 8,000 years. From nearly a millennium ago to the present climate scientists have observed a decrease in global temperatures which is known as the Little Ice Age (ca.1300-1850), followed by a sudden rise in temperatures caused by the enormous CO_2 and CH_4 output resulting from the Industrial Revolution and rapid human population growth.

William Ruddiman is not an environmentalist, but, rather, considers himself an objective scientist who chides environmentalists for their "high-minded, preachy appeals to Jean Jacques Rousseau's notion of the 'noble savage,' the concept of a primitive but wise people who once lived lightly on the land and in complete harmony with the environment." [20] Thus, he sees environmentalists as mistakenly portraying the Industrial Revolution as the only serious human assault on pristine nature. Ruddiman has learned from environmental history, in the course of researching the causes of global climate change, that humans have been degrading Earth's ecosystems at least since the Neolithic Revolution around 8,000-9,000 years ago. Consequently, he recognizes that global climate change may not be the greatest problem facing humanity, even though it may well be the most devastating. "In the short term, many other environmental concerns are already more worrisome, especially major ecological changes. Over the longer term, humanity's concerns will probably shift to the gradual depletion of the 'gifts' that Earth has freely provided, including fossil fuels, groundwater, and topsoil."[21]

John Lovelock is even more pessimistic than Ruddiman about the seriousness of the threat of global warming to humanity and nature. Originator of the Gaia Hypothesis, Lovelock postulates that the consequences of human-induced global warming will be far greater than Ruddiman and other climate scientists can yet imagine. The Gaia Hypothesis proposes that the near-surface and surficial area of the Earth, including the solid rock lithosphere, hydrosphere, atmosphere, and biosphere, all interact in such a manner as to sustain habitability for the changing life forms that are its inhabitants.[22] However, when Lovelock referred to the whole system as functioning as a single living superorganism, most other scientists condemned the idea in scathing terms. Subsequently, it has been an uphill battle for Lovelock's hypothesis to gain respectability within the communities of the earth and life sciences. Over time, Lovelock's description of Gaia as a "superorganism" has been recognized as a metaphor for a planetary system whose components interact to maintain homeostasis or equilibrium, making the idea more acceptable to contemporary scientists.

Like Ruddiman, Lovelock recognizes the degree of damage other than global warming which humans have inflicted upon Earth. He is well aware that we "need to recognize that the Earth's natural ecosystems regulate the climate and the chemistry of the Earth and are not there merely to supply us with food and raw materials. Our attempts to replace these ecosystems with farmland or forestry plantations have led in recent years, in Indonesia and elsewhere in the tropics, to disaster both regional and global in scale."[23] Extensive clearance of forest for agriculture and the draining of peat bogs in South East Asia have caused massive wildfires that temporarily produced the equivalent of 40 percent of worldwide total carbon dioxide from fossil fuel combustion.

Lovelock believes that modern environmentalism cannot succeed without a widespread acceptance by humanity of the Gaia concept of a living planet which "counters the persistent belief that Earth is a property, an estate, there to be exploited for the benefit of humankind."[24] He also retains some hope that humanity may effect a change in worldview by exposing children to greater knowledge of the natural world, explaining how Gaia operates and showing how human beings belong to it. Unfortunately, we are mired in the beliefs of traditional religions instead. Lovelock anticipates the main arguments which I put forth in *The Decline of Nature* when he writes: "The humanist concept of sustainable development and the Christian concept of stewardship are flawed by unconscious hubris. We have neither the knowledge nor the capacity to achieve them. We are no more qualified to be the stewards or the developers of the Earth than are goats to be gardeners."[25]

The foregoing discussion of the latest knowledge and insights of leading climate scientists and commentators on their work is intended to emphasize the point that the history of environmental degradation and the findings of contemporary climate scientists are closely connected. Ruddiman needed to learn environmental history in order to understand graphic patterns of changes in CO_2 and CH_4 levels. The global warming crisis is the culmination of thousands of years of human manipulation and transformation of Earth's ecosystems, with the spike of extreme development and environmental impact coming over the last two centuries since the beginning of the Industrial Revolution. Until the 19th and 20th centuries regional environmental degradation was associated with the rise and fall of individual civilizations, but the recent population explosion and massive global development have raised the degree of environmental impact to the global scale, thereby creating the crisis of abnormally rapid global warming which will affect all of the world's terrestrial and marine ecosystems, compounding already severe regional ecological damage.

Thus, it is necessary to tell the story of the human attitudes and practices which have caused extensive ecological degradation (for many millennia since the beginnings of the Neolithic Revolution) within the contemporary context of growing panic over global warming in the 21st century. The same Western beliefs in the ideas of providence and progress which have led us to the present climatic impasse not only prevail in exaggerated form in the United States today, but have given rise to the spread Western economic ideology and practices globally to the point that nations such as India and China are now practicing the same kinds of ecologically destructive development that the United States and Western Europe used to transform entire continental regions into almost totally humanized and ecologically degraded environments. Without having learned to respect nature, i.e., relatively undamaged ecosystems and species diversity, for its own intrinsic value, we are not likely to change the way we have treated ecosystems in the past.

China and India, like Western Europe and to a lesser extent North America, were severely affected by human action long before the Industrial Revolution and its coal and oil-powered machines accelerated the rate of transforming ecosystems into humanized environments. Humans with spears before the Neolithic Revolution, and picks and shovels after agriculture was introduced, and cattle, goats, and sheep were sufficient for transforming biomes (terrestrial ecosystems) into farmlands, grazing lands, towns and cities wherever organized communities grew up. Kirkpatrick Sale claims that only when *Homo sapiens* developed the tools (weapons) for hunting the mammoths and other now extinct great mammals of the late Pleistocene did humans distance themselves sufficiently from the natural world that the practices of controlling nature and dominating it in an unprecedented manner could begin. With the invention of agriculture, nature would be subdued unrelentingly. Subsequently, every civilized culture would develop its own set of rationalizations for dominating nature. Western civilization appears to have developed a particularly destructive worldview in this regard.

The question which this book attempts to answer is as follows: To what extent do our beliefs about the nature of reality, our worldview, affect the degree or intensity of our manipulation of nature? In other words, do our ideas about man and nature have major environmental consequences? Sale might argue that no matter what worldview a civilization possesses, it inadvertently destroys its regional ecology because humans are genetically and culturally programmed to behave differently from every other species on Earth, to the point that it has become broadly accepted around the world that humans are the superior and dominant species of the planet. Implicitly, world historian Oswald Spengler would have agreed with this viewpoint with the exception of Western, Faustian civilization, which, he claimed, possesses a far more destructive set of attitudes toward nature then any other known civilization. Sale, in fact, supports this position when he writes that Western civilization is "a culture that prides itself on its distance from the natural world and the natural cycles and rhythms, that regards its mission as needing (in Francis Bacon's words) 'to conquer and subdue nature' with its indomitable technology...."[26] In the case of modern Western civilization we have seen arguments in defense of nature intensify over the past two centuries. However, the older, traditional destructive attitudes, rooted in the medieval idea of providence and the modern idea of progress, persist and dominate our institutions and behavior to the present. These aspects of our worldview can perhaps be modified if we dare to expose them to the light of historical analysis in academia, and debate in the public square. How likely is this to occur? Perhaps increasing awareness of global warming and its connection to environmental history will lead us towards metanoia, a change in worldview. If not, we very likely face a coming "dark age."

NOTES

[1] Kirkpatrick Sale, *After Eden: The Evolution of Human Domination* (Durham and London: Duke University Press, 2006), pp. 2-3.
[2] J. B. Bury, *The Idea of Progress* (New York: Dover Publications, Inc., 1987), pp. 5-6. Bury also wrote on p. 5: "If there were good cause for believing that the earth would be uninhabitable in A.D. 2000 or 2100 the doctrine of Progress would lose its meaning and would automatically

disappear." We seem to be approximating such pessimism about Earth's future today, at least in the minds of some intellectuals.

[3] Tim Flannery, *The Weathermakers: How Man is Changing the Climate and What It Means for Life on Earth* (New York: Atlantic Monthly Press, 2005), pp. 41-43.

[4] Ibid, p.44.

[5] Jeffrey Kluger, "By Any Measure, Earth Is at the Turning Point," *Time*, Vol. 167, No. 14, April 3, 2006, p.42.

[6] Elizabeth Kolbert, *Field Notes from a Catastrophe: Man, Nature, and Climate Change* (New York: Bloomsbury Publishing, 2006), p. 181.

7 Tim Flannery, *The WeatherMakers*. See Chapter 22 for Flannery's scenario of the future effects of global warming upon civilization. On page 209, he writes: "We have seen that human health, water and food security are now under threat from the modest amount of climate change that has already occurred. If humans pursue a business-as-usual course for the first half of the century, I believe the collapse of civilization due to climate change becomes inevitable."

[8] Tim Flannery, *The Weather Makers*, pp. 100-102.

[9] Ibid., p.102.

[10] Ibid., p. 98-100.

[11] Elizabeth Kolbert, *Field Notes from a Catastrophe*, pp. 13-22.

[12] Jim Hansen, "The Threat to the Planet," *New York Review of Books*, Volume 53, No. 12 (July 13, 2006), pp. 12-16, p. 13.

[13] Ibid.

[14] Ibid.

[15] Ibid., p. 14.

[16] Tim Flannery, *The Weather Makers*, pp. 50-53.

[17] William F. Ruddiman, *Plows, Plagues and Petroleum* (Princeton and Oxford: Princeton University Press, 2005), pp. 87-88.

[18] Ibid., p. 87.

[19] Ibid., pp. 119-126.

[20] Ibid., p. 184.

[21] Ibid., p. 178.

[22] James Lovelock, *The Revenge of Gaia: Earth's Climate Crisis and the Fate of Humanity* (New York: Basic Books, 2006), pp. 22-25.

[23] Ibid., pp. 131-132.

[24] Ibid., p. 135.

[25] Ibid., p. 137.

[26] Kirkpatrick Sale, *After Eden*, p. 127.

CHAPTER ONE: THE CYCLE IN NATURE AND HISTORY

Introduction

During the last thirty or so years environmental history has grown into a major category of historical study, in the United States most frequently taught by historians specializing in American history, and typified in its early years by Roderick Nash's classic *Wilderness and the American Mind* (1967) and Joseph Petulla's *American Environmental History* (1977). My route into this new discipline was totally idiosyncratic and circuitous, leading from a general interest in natural history to geology and biology, then swerving unexpectedly to European history and the philosophy of history, ultimately integrating these varied precedents into the teaching of and writing about environmental history. I believe that this book is unique in approaching environmental history, particularly European environmental history, from the perspective of the speculative philosophy of history, the study of scholarly endeavors to recognize patterns of flux, cycles, providence, and progress in history. I have attempted to present an intellectual history of these ideas within a framework of facts and descriptive environmental history which assist the reader in understanding Western philosophies of history in the new light of contemporary environmental history and vice versa. Before further developing this idea, I think that it will enhance the reader's understanding of the relationship between the history of ideas and the history of humanity's interaction with Earth's ecosystems if I briefly explain my unusual pathway to environmental history.

I spent the first eight years of my life growing up in the purely artifactual environment of Manhattan Island. A few summer trips to the lakes of northern New Jersey and regular visits to my grandparents on Staten Island were the only reminders of the world of nature beyond the concrete and steel of Manhattan during those years. However, with the onset of World War II my father took a job in the steelyards of Staten Island as a ship-fitter, and the drab city blocks of our Third Avenue and Thirty Third Street neighborhood were suddenly replaced by an old neighborhood of two-story wooden houses along tree-lined streets. There were even open fields and a stream nearby where endless World War II "battles" were re-enacted by platoons of pre-teenagers. Having arrived on Staten Island at the age of eight, the next several years involved multiple probings into the mysteries of science, as I acquired telescope, microscope, chemistry set, and dozens of books dealing with animal stories and wildlife, especially the wonders of African wildlife revealed by Martin and Osa Johnson and other popularizers of the wild. Above all, I remember the integrating "summa naturala" of all this frenetic activity, Vinson Brown's *The Amateur Naturalist's Handbook*, with its keys to North American animals, birds, minerals, and rocks. Trips to the American Museum of Natural History in Manhattan became increasingly frequent, and an ardent fishing friend, as well as trips on my father's small fishing boat for commercial parties, drew me further and further from Port Richmond to distant corners of Staten Island.

It was my second year in high school when our biology teacher showed our class a bubbly, black, shiny, fibrous mass of goethite iron ore from the old Civil War mines above upper Clove Lakes. My curiosity about the old iron mines stimulated a shift in interest from biology to mineralogy and geology, and I traced the geology of Staten Island from a small scale map and then plotted the locations of the old limonite and goethite mines onto topographic maps. A friend and I then tracked them down weekend after weekend and began to build our mineral and rock collections. A year later my parents bought a twenty-four acre farm in the hills of western Massachusetts near Northampton, and my mineral collecting and amateur geologizing was further energized by my discovery of the great variety of mines and quarries in the area.

About the same time that I was collecting minerals and rocks, I came upon and read a book which profoundly influenced my thinking, unconsciously as well as overtly, and which prefigured my interest in environmental history decades later. Fairfield Osborn's *Our Plundered Planet* was first published in 1948, and I can still recall the appearance of the dust-jacket at the time of reading it through. Suddenly, as if a trap-door in the cosmos had been sprung, my growing veneration of the marvelous species of African, Asian, and other mammals was undercut and compromised by Osborn's chilling revelation of the destruction of Earth's ecosystems, species diversity, and resources at alarming rates. *Our Plundered Planet* was a passionate plea for moderation of modernity's assault on nature combined with a brief global environmental history.[1] Osborn was a zoologist with the New York Zoological Society, and his manuscript was reviewed and enriched by such notable "conservationists" (i.e., both conservationists and pre-environmentalists) of his time as William Beebe, Hugh Hammond Bennett, Walter C. Loudermilk, William Vogt, and Aldo Leopold. Although his book was sensationalistic in its alarmist cries portending planetary disaster, it was not the work of a hack journalist, but rather of an esteemed biological scientist. Why, then, did it not light the fire of what we today call "environmentalism," and its retrospective academic ally, environmental history? To light a fire requires fuel, and the fuel of concern for nature simply did not exist in Osborn's time. Only when many individuals felt threatened by air and water pollution in the 1960's, their fears enhanced by Rachel Carson's *Silent Spring* (1962) and the implicit warning of the National Wilderness Act of 1964, did they turn to political action and proselytization leading to the Environmental Movement (1962-1980 ?), legislation of the National Environmental Policy Act (NEPA, 1969), and the subsequent environmental legislation demanded by a public mandate.

However, in 1948, buoyed up by the Allied victory in World War II, and gearing up for the peaceful application of growing industrial power, the American public was generally deaf to the supplications in defense of nature presented by Fairfield Osborn. Some of us, like me, were strongly influenced, however, at a time when the only magazines available about nature on the news-stands were *Field and Stream* and *Sports Afield*.

Given my growing interest in mineralogy, it was natural to be thinking about a geology major when I entered the University of Massachusetts in 1955. Six years and two academic degrees later, I left Dartmouth College with the intention of working as a geologist for the rest of my life. However, during those early years of collecting and learning, I made some observations that would only make sense in the context of environmental history much later on. Walking the woodlands of the "Berkshires," as the hills of western Massachusetts were often called (even though, technically, the Berkshires referred to the hills of the far western part of the state), I quickly became aware of how populated this now semi-wild region had once been on the basis of the surprising number of old stone walls I encountered in the middle of extensive century-old forests. Deer and bobcats roamed over a network of abandoned farms which enclosed the lead, manganese, lithium, emery, and other mines that I systematically explored. I will return to colonial environmental history in a later chapter, but I mention it here as my first inkling that something like what we now call environmental history was needed to explain those abandoned farms and the return of wildness to western Massachusetts.

A trip to Europe in 1959 was the deciding factor that gradually shifted my interest towards history, although I have never lost interest in geology. In short, I returned to graduate school in 1966 to study European history, having read Oswald Spengler's *Decline of the West* and Arnold Toynbee's *A Study of History* following my first European experience. My Ph.D. Dissertation on *Jean-Jacques Rousseau and the Idea of Progress* reflected a developing interest in the philosophy of history, assimilated from Spengler and Toynbee, within the context of traditional European intellectual history. The remaining factors which consolidated my eventual interest in environmental history were, of course, the Environmental Movement, but also my teaching of environmental geology and, especially environmental ethics (environmental philosophy) at Willamette University in Salem, Oregon since 1979. It was the development of this accidental eclecticism over many years that allowed me to understand environmental history from a unique perspective. Certainly an unusual amalgamation of disciplines is characteristic of environmental history itself, which also raises the question of "What is environmental history?"

Environmental History and the Speculative Philosophy of History

Deceptively, at first glance this might appear to be a simple question. Isn't environmental history the new branch of historical study that focuses upon the interaction of mankind and his surrounding environment, including the impact of humans upon the environment over time as well as the environmental factors influencing human utilization of nature as resources? Some variation on this definition is given in most environmental history books. However, because humankind is the lead player in the drama of environmental history as it is in traditional history, it should follow that human evolution and pre-history are also essential to the study of environmental history. This aspect of environmental history is related to pre-modern attempts, from classical antiquity through the Middle Ages and Renaissance in Western civilization, to probe into human nature in order to answer the question, "What is man?" In the course of seeking an answer to this question, the ancient Greeks and Romans emphasized the political and social aspects of human nature, and Christians of the medieval period emphasized the idea of the "creation" and the overwhelming importance of the eternality of the human soul. In making these judgments concerning human nature, pre-modern civilization in the West closed down the possibility of understanding humanity within the context of nature and natural time, although this is less true of the ancients than of medieval thinkers. For example, the observations of nature made by Plato, Aristotle, and other Greek philosophers suggested to them dynamic changes occurring over long periods of time, as in Plato's observations of major changes in the Athenian environment since early human settlement, as well as his cyclical interpretation of history involving multiples of tens of thousands of years. However, only in the modern period, particularly in the Enlightenment, did the extension of scientific method to the study of earth processes involving geology, climate, soil formation, and even the biological transformation of mankind over time allow scientists and philosophers to begin to visualize the long-term interaction of mankind and his surrounding environment. The contributions of Michel de Montaigne (1533-1592) on climate, Le Comte de Buffon (1707- 1788) on earth origins and the history of the earth, and Buffon, Pierre Maupertuis (1698-1759) , Denis Diderot (1713-1784), and Jean-Jacques Rousseau (1712-1778) on the transformation or evolution of mankind were necessary prerequisites to the possibility of writing anything we would call "environmental history" today. One recent author makes a strong case that Rousseau's *Discourse on the Origins of Inequality Among Men* was a kind of proto-green history, for example, thereby beginning this genre of historical study long before the term "environmental history" was made popular during the Environmental Movement.[2]

In addition to knowledge of traditional anthropocentric history, the environmental historian must be familiar with the physical-biological environment, including, "from the ground up," geology, geography, hydrology, the broader aspects of biology such as plant communities and ecosystems, and the spatial and temporal aspects of meteorology and climatology. Above all in the biological realm, the "*ecosystem*" concept is central to the study of environmental history, particularly as definitions and meanings of "ecosystem" have changed radically in recent decades as a consequence of what history, archeology, and anthropology have taught us about human impact upon both terrestrial and aquatic ecosystems. "Pristine nature" virtually has been defined out of existence, leaving us with a preponderance of anthropogenically altered environments quite different from what the original concept of ecosystems as *natural* systems intended. Before considering the additional realms of importance for environmental historians, such as socioeconomic institutions, considerations of race, gender, and class, and the impact of the multicultural and postmodern academic revolutions on historical research and writing, it would be useful to analyze the ecosystem concept in more detail.

Older definitions of "ecosystem" either openly stated or implied that the biological systems described were "natural" in the sense that they were not much influenced by human activity but were rather the product of natural physical and biological processes. In *Nature's Economy: a History of Ecological Ideas*, environmental historian Donald Worster pointed out the fact that the term "ecosystem" was coined in 1935 by biologist Arthur Tansley to take the place of the more value-laden term, "community."[3] In *A History of the Ecosystem Concept in Ecology* (1993), Frank Golley wrote: "*Ecosystem* referred to a holistic and integrative concept that combined living organisms and the physical environment into a system."[4] As recently as 1992, my preferred definition of an ecosystem was a "community of organisms living in a specified locale, along with the nonbiological factors in the environment- air, water, rock, and so on- that support them, as well as the ensemble of interactions among all these components."[5] Earlier in the twentieth century, ecologist Frederic Clements had applied the term "climax community" to systems of vegetation that dominated particular geographic regions and that culminated the sequences of plant cover that followed disturbance by fire, storms, humans and other causes of ecological disequilibrium. Eschewing the anthropomorphic connotations of "community," contemporary ecologists have attempted to expunge the term from the vocabulary of ecological science. Thus, recognizing this shift, Worster defines "ecosystem" as follows: "the collective entity of plants and animals interacting with one another and the nonliving (abiotic) environment in a given place, some ecosystems are fairly small and easily demarcated, like a single pond in New England, while others are sprawling and ill defined, as hugely ambiguous as the Amazonian rain forest or the Serengeti plain."[6]   Worster also emphasizes the importance to modern ecologists of energy flow and the cycling of nutrients in ecosystems.

Many ecologists prefer to apply the term "ecosystem" more discretely, classifying the earth's ecosystems into typical biomes (terrestrial ecosystems) such as steppe, taiga, northern coniferous forest, mid-latitude deciduous forest, Mediterranean scrub or chaparral, and so forth. Thus, the term is stretched or shrunk to meet particular descriptive needs, thereby creating considerable contradiction and ambiguity in its application. This multiplicity of meanings has opened the door to a persistent controversy over whether or not cities, as anthropogenic artifacts for the most part, should be recognized as ecosystems. A few geographers and biologists have made such arguments. My own position regarding this controversy is that to think of cities as a kind of ecosystem is both illogical and potentially insidious and damaging to attempts to preserve natural environments. On the other hand, to make an *analogy* between ecosystems and cities is a useful way of pointing out their differences. Cities are not ecosystems *per se*, however, because: (1) There is no evidence that early users of the ecosystem concept, long before the word existed, intended to include human artifacts such as cities; (2) the ultimate criterion upon which the classification of ecosystems (as biomes) is based is the vegetation or "plant community" because they are the *producers* upon which all other organisms (herbivores, carnivores, omnivores, and decomposers) depend. No such extensive plant communities are located in modern cities, into which most food is transported, and out of which waste is exported; and (3) cities, especially megalopolitan cities, cause massive destruction of ecosystems outside of their locale in order to exist (e.g., plains converted to monocultural agriculture, forest cutting, mining, damming rivers for water, overfishing). Even ancient agrarian-based cities were not self-sufficient, but depleted and degraded surrounding ecosystems as well as humanized agrarian systems. The ancient city was a population center sustained by humanized agrarian systems, but not an ecosystem in itself. At best, a city can be considered as an ecosystem by analogy or metaphorically, rather than in an objective scientific sense. The difference between these two meanings appears to have eluded some biologists and technocrats. This is a potentially dangerous ambiguity which could be used to rationalize the development of dwindling ecosystems into humanized environments considered to be of a higher order economically and aesthetically.

Recent knowledge which recognizes a greater degree of human impact upon ecosystems which were once considered to be "natural" further blurs the distinction between natural ecosystems and anthropogenically altered ecosystems. If already altered by humans, why should the earth's biomes not be transformed into something preferable for human use? As we shall see in Chapter Two, virtually the entire Mediterranean basin, including its maquis (stunted forest), garrigue (low brush) and steppe (grasslands), is essentially a human artifact produced over thousands of years of selective hunting and gathering, herding, burning, and farming. The classical Greco-Roman, Mesopotamian, and ancient Mayan civilizations all caused massive environmental degradation and typically caused permanent changes to the original ecosystems where they became established. Even hunter-gatherers such as most Native American Indian tribes caused massive changes to ecosystems through burning, selective hunting, and other practices.

Although biomes play an important role in most environmental histories, other aspects of the physical and biological parameters of ecosystems may be emphasized. Plate tectonics and its role in determining the present distribution of ecosystems and animal and plant species is central to Alfred Crosby's classic, *Ecological Imperialism: The Biological Expansion of Europe, 900-1900* (1986), whereas J. Donald Hughes, in *Pan's Travail: Environmental Problems of the*

Ancient Greeks and Romans (1994) focuses on the relationship between the geography of the Mediterranean basin and human impact upon its ecosystems over time. Authors outside of the discipline of history also contribute to the field of environmental history by emphasizing the role of soils or climate in history, for example, archeologist Charles L. Redman in his *Human Impact on Ancient Environments* (1999) and archeologist Brian Fagan in *Floods, Famines and Emperors: El Niño and the Fate of Civilizations* (1999).

Having considered the scientific, ecological side of environmental history, let us turn to the historical side of the field, along with aspects of social and economic institutions and the progeny of the American "cultural revolution" of the 1960's, including multiculturalism, issues of gender, race, and class, and the postmodern turn in philosophy and history. An excellent general introduction to these topics is appended to Donald Worster's edited collection of environmental history essays, *The Ends of the Earth* (1988). In his "Appendix: Doing Environmental History" Worster situates environmental history within the context of the twentieth century reformation in historical writing (historiography) in which scholars began to unearth long submerged "...layers of class, gender, race, and caste." The environmental historians are the latest of the reformers, probing down into the earth itself to give history a broader, more profound meaning than what it had traditionally produced in the national and political histories of the nineteenth and twentieth centuries.[7] Worster points out the roots of environmental history to be found in the work of such historians of the American West as Frederick Jackson Turner and members of the French *Annales* School. He categorizes environmental history into three types dealing with different aspects of the field. Applying categories from Marxist thought, they can be described as substructural, superstructural, and ideological. The substructural category deals chiefly with the study of nature itself and is well represented by Worster's own *Nature's Economy*. Superstructural environmental history links the socioeconomic realm to ecology and emphasizes modes of production used to gain a living from the realm of nature. A classic in this category is Carolyn Merchant's *Ecological Revolutions: Nature, Gender, and Science in New England* (1989). Finally, ideological environmental history, the uppermost level of abstraction, is a kind of intellectual history, a history of ideas concerning human attitudes toward nature and the ethics and laws established by humans in the course of their encounters with nature. Roderick Nash's *Wilderness and the American Mind (1967)* initiated this branch of environmental history, and Carolyn Merchant's *The Death of Nature* (1980) applied it to European beliefs and attitudes toward nature.[8]

The veracity of history in general has long been questioned. Voltaire wrote that "History is a pack of tricks the living play on the dead," and Napoleon surmised that "History is a pack of lies agreed upon." In the twentieth century Henry Ford reiterated this skepticism in his famous remark, "History is bunk!" Nevertheless, historians since classical antiquity have set themselves standards for truth and accuracy in recounting the events of the past, as in nineteenth century political historians' goal of writing about historical events "wie es eigentlich gewesen" (as they actually were). Yet every college history major knows, or should have learned, that historical writing always entails powerful subjective elements. The prejudices and worldview of any given historian, the historical era and the nation in which he writes, the climate of opinion at the time of writing, the author's personality and his or her position regarding gender, race, class, politics and other variables, all contribute to his or her philosophy of history, selection of facts, and interpretation of any given events. In writing environmental history the author's environmentalist, anti-environmentalist or relatively neutral perspective is usually not difficult to recognize. To make the challenge of writing truthful history even more imposing, historians like Carl Becker and Charles Beard, writing in the 1930s, were already well aware that history faced serious epistemological and methodological problems, as expressed in the title of Becker's *Every Man His Own Historian*.

Several decades after Becker and Beard expressed their concerns over historical relativism, historical writing in the United States was further challenged, first by the spread of poststructural and postmodern thought which critiqued the fundamental assumptions of modernity. Both of these developments were to have a profound influence on the writing of environmental history. By fracturing traditional histories of the scientific, technological, military and economic triumphs of the white European male into the competing strands of race, gender, and class, social history exposed the seamy underside of the great American success story. In the process of challenging the sexist and white supremicist racist narratives of the nineteenth and twentieth centuries, the new social histories, beginning in the 1960's, reinforced the relativism of Becker and Beard, thereby helping to clear the way for the spread of Continental critical theory. Structuralism was the first wave of the French-dominated intellectual fashions which followed existentialism. It was rooted in the linguistic analysis of the French-Swiss scholar Ferdinand de- Saussure and the anthropological methodology of Claude Lévi-Strauss. Saussure maintained that language is a self-contained system of signs within which the relationship between the word and the concept represented; i.e., between the *signifier* and the *signified*, was more or less arbitrary and subject to ambiguity and obfuscation. As a consequence, what we think of as "objective reality" is, to varying degrees, veiled and obscured from us by language. Lévi-Strauss, a Belgian-French social anthropologist, viewed culture as analogous to a language, and recognized in its systems of communication the key to identifying and understanding the universal structures of the human mind. He attempted to accomplish this through comparative cultural studies of the symbols, myths, and social organization of primitive as well as civilized cultures, focusing in particular upon dichotomies or "binary oppositions" of male/female, public/private, etc. as indications of the subliminal or unconscious logic of a given culture.

The poststructuralists, although they built upon the linguistic and cultural analyses of de Saussure and Lévi-Strauss, denied that any ultimate knowledge could be gained from these methodologies because "texts could be interpreted in multiple, if not infinite, ways because signifiers had no essential connection to what they signified."[9] The ambiguous relationship of signifier to signified turns language into a quagmire of distortion and misrepresentation. Thus, Jacques Derrida, the foremost postmodern philosopher of "deconstructionism," maintained that texts "repressed as much as they expressed in order to maintain the fundamental Western conceit of 'logocentrism,' the (erroneous) idea that words expressed the truth of reality."[10] Within the discipline of environmental history itself the once self-explanatory meanings of such words as "nature," "wilderness," and "ecosystem" have already been compromised by postmodern critical analysis, each of these terms being deconstructed into multiple and ambiguous meanings. Also, the rejection of the Western intellectual "canon" and such received truths promulgated by a white, European elite as "man," "reason," and "progress" (the modernist values bequeathed to us by the Enlightenment) are now morally and intellectually challenged, along with their implications for the natural world.

> Disenchanted eyes are now cast onto the West's long history of ruthless expansionism and exploitation- the rapacity of its elites from ancient times to modern, its systematic thriving at the expense of others, its colonialism and imperialism, its slavery and genocide, its anti-Semitism, its oppression of women, people of color, minorities, homosexuals, the working classes, the poor, its destruction of indigenous societies throughout the world, its arrogant insensitivity to other cultural traditions and values, its cruel abuse of other forms of life, its blind ravaging of virtually the entire planet.[11]

In this passage, intellectual historian Richard Tarnas gives the reader a sense of the seething resentments which have accumulated in opposition to the progressive ideals of modernity and their implementation, resentments which underlie the postmodern critique of Western thought and ideals, and which link it to social history. The latter has changed the face of environmental history in its leading journal in the United States, *Environmental History* (formerly *Environmental History Review*, and before that *Environmental Review* back until the journal was founded in 1976), and the same trend is evident throughout the literature of the field. During the 1970s and 1980s the connection between environmentalism and environmental history was quite strong, and many if not most of the articles published reflected a strong bias towards environmental values, as one might expect. During the last decade this proselytizing tendency has been expunged or muted in *Environmental History*, and the social issues of class, race, gender, and environmental justice have flourished under the editorship of Hal Rothman, who expressed this shift in perspective as follows: "Environmental history's promise has always been a redefinition of the boundaries of the discipline of history, an inclusion of the environment in the study of the past. We are not advocates of the environment; instead we are advocates of including the environment in the story of human evolution on the planet."[12] Although Rothman may have adhered to this objectivist ideal, many environmental historians, including myself, are also ardent environmentalists, and it requires a studied effort to maintain a "scientific," essentially value-free position in writing environmental history, although many scholars have questioned "scientific objectivity" as possessing its own implicit values, i.e., the values of early modern science and Enlightenment rationalism. I am not so sure either that the substitution of emphasis upon social history and multicultural values for environmentalist values is a step forward, as the world fragments at the very time that a consensus of environmental valuing is needed to prevent destruction of the global environment.

Two established branches of historical study take a broad approach to history that should prove useful to twenty-first century environmental history. These are world history and the philosophy of history, which are related through sharing a global perspective, and some world historians, such as Arnold Toynbee (1889-1975) and Oswald Spengler (1880-1936), are better known for their philosophies of history, i.e., their interpretations of meaningful patterns in history, than they are for their accounts of world history. Typically, world historians have not paid much attention to ecology or the environment, although Spengler made some profound and thought-provoking observations concerning modern man and the environment at the end of his *The Decline of the West,* and William McNeill, in *The Rise of the West*, took the role of nature and epidemiology more seriously than any other world historian, and also attempted to counter Spengler's historical pessimism. Environmental historian J. Donald Hughes, best known for his studies of environmental history in the realm of classical antiquity, has criticized twentieth century world history for emphasizing "development" as the organizing principle of world history. By "development" Hughes means "primarily economic growth and technological progress...."[13] Hughes maintains that this development-oriented framework of world history has ignored or denied the importance of ecological factors in world history. For example, J.M. Roberts's highly acclaimed *History of the World* (1993) does not contain the words "ecology" or "nature" in the index, and "environment" appears only by page 818, near the end of the book. In Roberts's discussion of the causes of Rome's fall, no reference is made to environmental causes or the ecological problems associated with Roman agricultural and grazing practices which are discussed in detail in Chapter Two of this book. As an alternative to "development" as the leading theme of world history, Hughes proposes ecological process as its organizing principle.[14] Not only is the emphasis on the principle of development wrong, asserts Hughes, but it is also pragmatically devastating. "Giving attention to development while ignoring ecological process gives implicit approval to the environmentally destructive course of the modern world economy."[15] One year after making these criticisms Hughes took the bull by the horns and published *An*

Environmental History of the World: Humankind's Changing Role in the Community of Life (2001).

A decade before Hughes' world environmental history, British historian Clive Ponting published *A Green History of the World: The Environment and the Collapse of Great Civilizations* (1991). Ponting's book has done much to increase awareness of the fact that the world's great civilizations have exacted a heavy toll upon ecosystems as the price of their development, which, in the case of several civilizations, such as Mesopotamian, Greco-Roman, and Mayan, caused their extinction as cultures.   As we shall see later on in this chapter, Ponting's ecological explanations of civilizational decline have answered questions raised by the great twentieth century philosophers of history, Spengler and Toynbee, regarding the rise and fall of civilizations.   The term *philosophy of history* is applied to two very different categories involving the search for ultimate meaning in history. On the one hand, *critical* philosophy of history is concerned with the analysis of historical method in order to determine the degree to which historians are capable of writing histories which approximate the truth of what actually happened in history. R.G. Collingwood's *The Idea of History* (1946) and Edward Hallett Carr's *What is History?* (1967) are representative of this genre.[16] *Speculative* philosophy of history entails theoretical inquiry into the meaning of history as understood in terms of patterns of historical change such as cycles, flux, providence, and progress.   Instead of simply giving an account of historical change it typically claims to have discovered the underlying mechanism or driving force of history dictated by supernatural or secular causes.   Critical or analytical philosophers of history are generally skeptical regarding the claims of speculative philosophers of history to have discovered the true pattern or meaning of history.

Although most modern writers of world histories do not claim to have discovered patterns in history the way Oswald Spengler and Arnold Toynbee did, and also claim to follow the canons of historical objectivity prescribed by the critical philosophers of history, they nevertheless give a shape or pattern to their histories which are identifiable as progressive or developmental. In other words, the modern idea of progress overtly or subliminally structures the ordering of facts in their works. As J. Donald Hughes observes:

> The optimism expressed in world history texts is an extension of the momentum of the organizing principle of development. The wars and depressions of the twentieth century rendered more difficult the maintenance of a belief in "progress," but it has been replaced with development as a desideratum. Perhaps the two are not all that different; development is certainly less questioned, but it might be suspected that it is old progress writ large.[17]

During the first half of the twentieth century modern Western mankind's persistent faith in general human progress received a staggering blow from

Oswald Spengler's *The Decline of the West* when its first volume appeared at the end of World War I (1918). The ancient Greek idea of historical cycles had been resurrected by Spengler as perhaps a more meaningful explanation of world history than the West's smug belief that, despite occasional setbacks, things had gradually improved over time since the beginning of civilization. Coming at the end of World War I, the *Decline* was received by many intellectuals as a shocking, sensationalistic revelation. In *The Outline of History* (1920) H.G. Wells, optimist that he was, also warned of the impending potential destruction of civilization if the human species failed to develop some form of global government to control warfare. Arnold Toynbee's *A Study of History* (1933, first three volumes) also presented his world history as speculative philosophy of history which challenged the Western idea of progress. So it was that traditional writing of world histories, implicitly or explicitly ordered by the pattern of progressive development, was challenged by a few brilliant historians explaining world history as ordered by alternative or complementary patterns of cycles (Spengler) and providence (Toynbee). Toynbee actually wove patterns of cycles, progress, and providence together in his massive eleven volume work. These world historians, by openly professing to have discovered new meaningful patterns of history, are known as speculative philosophers of history.[18] Although other metahistories (another term for speculative philosophies of history) were to follow, including works from anthropologist Alfred Kroeber and sociologist Pitirim Sorokin, Spengler and Toynbee were the dominating figures in the field. The most serious challenge to their ideas was William McNeill's The *Rise of the West* (1963), which rejected Spengler and Toynbee's historical models and substituted a metahistory or macrohistory which has been described as "ecological mythistory."[19]

The title of McNeill's book is directly aimed at Spengler's *The Decline of the West* and, along with later works by McNeill, was designed "to clear the air of metaphysical obscurantism, cyclical determinism, and the organicism of self-contained civilizations that pursued predictable life courses."[20] Although McNeill had been a friend and associate of Arnold Toynbee beginning in the mid-twentieth century,[21] his personal agnosticism led him to a secular approach to macrohistory which eschewed the providentialism of Toynbee along with the cultural cycles which both Spengler and Toynbee embraced. Central to McNeill's approach to world history is the belief that "an intelligible world history might be expected to diminish the lethality of group encounters by cultivating a sense of individual identification with the triumphs and tribulations of humanity as a whole."[22] The broad term for his work applied by Paul Costello, "ecological mythistory," is linked to the above quotation by two key ideas around which *The Rise of the West* is structured. Foremost of these is "diffusionism," the notion that even though civilizations may rise and fall, the technologies and intellectual heritage created by each of them are passed on by cultural diffusion in the course of world history, so that later civilizations begin with major advantages over their predecessors. In this sense there is long-term "progress." However, progressive accumulation of knowledge and technology is compromised by the fact that technological development increases the potential of humankind to destroy itself more efficiently militarily. Therefore "progress" is threatened and uncertain today despite the impressive aggregation of knowledge and technology up until our own time. The second fundamental idea in *The Rise of the West* is the use of the concept of microparasitism, which describes the endemic diseases of humanity, to build the analogy of macroparasitism, which describes the parasitical relationship between conquering peoples and the conquered, which abets cultural diffusion but creates a master-slave relationship between conqueror and conquered.

Although McNeill is not an environmental historian himself, he anticipated important ideas developed by environmental historians during the 1970s and 1980s. One of these, dubbed "McNeill's Law" by Alfred Crosby, is that endemic disease works in favor of the conqueror to decimate and weaken an invaded population, i.e., microparasitism favors conquests leading to macroparasitism, as in the European colonization of the Americas. This idea was fully developed within the context of impacts upon global ecosystems by Alfred Crosby in his classic, *Ecological Imperialism.* The use of the term "ecological" in reference to McNeill's work does not refer to ecosystems, however, but rather to human ecology, particularly the role of epidemiology in world history. On the other hand, in his later work, McNeill broadened his "ecological" perspective and understood "that with each civilized adaptation we move away from individual ecological self-sufficiency and we up the ante of the potential for human catastrophes through our abilities to destroy one another or from other ecological disasters."[23] Paul Costello, whose *World Historians and Their Goals* does an excellent job of summarizing and critiquing the works of Spengler, Toynbee, McNeill and other great twentieth century world historians (most of whom are also speculative philosophers of history), draws the conclusion that the mechanization of man, "along with the sense of crisis now surrounding environmental issues on a world scale and the need for global environmental education make world ecohistory an imperative for the profession."[24] It is a step forward that, even as the traditional historical profession as a whole continues to ignore this advice, Clive Ponting in Great Britain and J. Donald Hughes in the United States have given us the two world environmental histories mentioned earlier in this chapter. It remains to be seen whether traditional historians will respond to their wake-up call.

Although McNeill rejected cyclical explanations of the rise and fall of civilizations, it is because environmental history has provided an ecological explanation of cyclical patterns that I will discuss Oswald Spengler's philosophy of history in some detail in the following section of this chapter. Spengler's cyclical civilizations left the puzzled reader facing the Sphinx. Why did these civilizations rise and fall? Was the cause of their cyclical patterns discernable? Spengler did not answer these questions very well at all. However, to some degree environmental history gives rational, scientifically-founded answers to the questions that he raised. Given the presuppositions of environmental history that humans are regarded as animals with an evolutionary genealogy, and that supernatural explanations of natural phenomena are unacceptable (i.e., God does not interfere with natural laws), Spengler's work is preferable to Toynbee's for study within an ecological framework. Toynbee's providentialism is inconsistent with the assumptions of environmental history. As he once wrote: "When we are investigating the relations between the facts of history, we are trying to see God through History with our Intellects."[25] Toynbee did make a move towards integrating the biosphere into his world history in *Mankind and Mother Earth* (1976), but he died before it was completed, and his reflections on the importance of respect for nature are not integrated with the main body of the text. As Hughes writes: "Despite a promising title and a prefatory section that takes ecology seriously, it remains for the most part a conventional political-cultural narrative that repeats observations made in Toynbee's earlier works."[26] On the other hand, Paul Costello observes that "Toynbee became an evangelist heralding the arrival of the World State, as the only hope of human survival in the modern age of technology."[27] For Toynbee, much like H.G. Wells and William McNeill, the spread of nuclear armaments and population explosion made the World State an ecological imperative.

Oswald Spengler and Human Destruction of Nature

Oswald Spengler's *The Decline of the West* was one of the great books of the twentieth century, written and published shortly before, during, and through the end of World War I. Despite its faults, the *Decline* is of great symbolic and critical significance for the end of the Century of Progress (1815-1914) and the growing critique of the very idea of human progress during the twentieth century. Other great philosophers of history, i.e., seekers of a pattern or meaning in the world-historical process, such as Arnold Toynbee and William McNeill, had to understand and grapple with Spengler's ideas as a prelude to their own interpretations of the larger, overall meaning of history.

In the course of Western history, three major interpretations or philosophies of history have been elaborated by the intellectuals of Classical Greco-Roman and Western European civilizations. These include the ideas of history as cyclical, providential, and progressive. The idea of cycles dominated classical history, although elements of the idea of progress and perhaps even a weak, minority idea of progress may have had a limited following in Classical antiquity.[28] Human history was generally understood by the ancients, however, as a manifestation of cosmic cycles which determined the rise, flowering, stagnation, and ultimate decline of all human societies. Plato even quantified the historical cycle (possibly based upon Pythagorean numerology but more likely founded upon Babylonian speculations concerning cosmic cycles based upon astronomical observations) into alternating 36,000 year periods of pre-civilized stability and 36,000 year periods of civilizational rise, development, and decline into chaos before repeating the next 72,000 years (See Chapter Thirteen for further discussion of Plato's cycle).[29] The idea of Providence, born in association with Judeo-Christian and antecedent religious traditions, recognized the Hand of God in the pattern of world history, generally positing a beginning, the Creation, and an end of history in the form of the Apocalypse and Last Judgment. For Western Christians especially, both the history of the Cosmos and of Man was short and selective, ending in eternity with God's chosen on his right hand and the damned on his left in hellfire. Finally, the modern idea of progress has applied the Christian, providential notion of linear history, with a recognizable beginning and end, to the secular development of humanity over the course of time. The beginning of human progress is generally identified with the invention of agriculture and the rise of cities based upon food surpluses and specialization of labor. The creation myth is also founded upon the agrarian-based rise of cities and the accompanying invention of writing, and compresses all of pre-history into the last 6,000 years. Historical thinkers of the modern, post-medieval world have gradually developed ideas of progress since the late seventeenth century, and a general belief in the probability or certainty of future human progress, spurred by scientific and technological development, persists among the general populace of Western and westernized nations to the present day.

The appearance of the first volume of *The Decline of the West* in 1918 initiated a controversy over the nature and meaning of history which flourished in Europe and the United States until about the time of the beginnings of multicultural and environmental history in the 1960s and early 1970s. However, these very different approaches to understanding the past have rarely crossed paths, and the speculative philosophy of history, including Spengler's cyclical perspective, generally has been ignored by American intellectuals in recent decades.[30] Environmental history itself has become more parochial and multicultural in its perspectives, and it has generally ignored the subject of philosophy of history as seemingly irrelevant to its own objectives. However, over the past decade an increasing number of studies by anthropologists, environmental historians, and philosophers have focused on the connections between cultural decline or extinction and long-term regional environmental degradation, the most recent being Jared Diamond's *Collapse: How Societies Choose to Fail or Succeed.*[31] In the course of this broader approach to environmental history, new answers to old questions regarding the decline of civilizations have been forthcoming, along with an increased awareness of the potential significance of the philosophy of history for environmental history and the history of Western civilization. One of the Sphinx-like riddles that has been answered by the environmental history of regional ecological degradation is the question raised by Spengler's explanation of the typical duration and ultimate decline of the great cultures, such as ancient Mesopotamia and Egypt, Classical Greece and Rome, and the ancient Mayan civilization. Before reconsidering Spengler from this macroscopic environmental perspective, we need first to summarize his philosophy of history as well as his own view of the significance of the environment for historical change.

Ultimately, the roots of Spengler's cyclical conception of history extend to the ancient civilization of antiquity, in which a variety of speculations regarding cyclical patterns flourished. To what extent Spengler built his cyclical speculations upon the writings of Plato and other ancient philosophers it is difficult to say. However, the influence of Goethe's philosophy of nature and Nietzsche's reiteration of the ancient myth of the eternal return upon Spengler is well known.

Johann Wolfgang von Goethe (1749-1832) exercised the greatest influence on Spengler. Through the study of nature, Goethe developed a philosophy of nature emphasizing the universality of natural cycles. "The change of day and night, of the seasons, of the blossoms and fruits, and of everything confronting us from epoch to epoch …these are the essential motive forces of earthly life." [32]   Like Johann Gottfried von Herder (1744-1803) and Friedrich Wilhem Joseph Von Schelling (1775-1854), Goethe broke with the rationalist Enlightenment idea of history to expound an approach in which "history is a showplace in which the cultures of unique people, each embodying a certain spirit (geist), grow and develop like biological organisms."[33] These cultures could only be understood through the imaginative reconstruction of their unique styles in contrast to other cultures.

> To the German Romantics, the world of both man and nature was vital through and through. Far from being an inert ensemble of matter in mechanical motion, as the philosophes of the Enlightenment conceived it, nature is a gigantic organism teeming with life and vitality. The Romantics agreed with the Aristotelian doctrine that each being possessed an inbuilt plan of development already latent in its germinal form. Nature is thus replete with organic forms seeking to actualize their inner potentialities. [34]

Thus, Spengler's conception of distinct, individual cultures was rooted in Goethe's organic and cyclical philosophy of nature.[35]   In particular, Spengler utilized Goethe's conception of the metamorphosis of plants, i.e., their stages of development, as a model for cultural archetypes in history, each expressing an underlying "prime symbol," a unifying conception of the nature of space which would determine the pattern of a culture's development. Western or "Faustian" culture, for example, would manifest its conception of space as infinite in the form of Gothic cathedrals, the unique mathematics of the calculus, soaring skyscrapers, the science of an expanded cosmos and space exploration, the exploration and conquest of much of the globe, and conceptions of unlimited urban growth and boundless energy supplies allowing human progress indefinitely into the future.

Spengler also inherited ideas from Friedrich Nietzsche (1844-1900) which were central to his philosophy of history in *The Decline of the West*. These included Nietzsche's conception of classical or "Apollonian" culture, his sense of decadence and nihilism as characteristic of the modern age, the idea that the will to power is man's basic drive, and, above all, his revival of the classical myth of eternal recurrence. [36] As Mircea Eliade points out, this myth reaches back even beyond ancient Greece to Babylonian conceptions of periodic cataclysms and planetary revolutions derived from Babylonian astronomical speculations. These ideas appear to have influenced Plato's conception of the historical cycle, which subsequently influenced the Stoics, who perpetuated the idea of the periodic renewal of the world, as did the Neo-pythagoreans.[37] As a dedicated Grecophile, Nietzsche imbibed the cyclical conception of history which Spengler modified in the *Decline*.

Prior to Spengler's *Decline* there were occasional critiques of the idea of human progress. However, they were generally ignored, and popular belief in the inevitability of human progress, generally understood as owing to rapid increases in scientific and technological knowledge, was probably stronger at the beginning of the twentieth century than it is today. Spengler's major accomplishment was to cast serious doubt upon this secular faith at the beginning of the chain of events (World Wars I and II, atomic warfare, the Cold War, social alienation and challenges to democracy, and the escalation of environmental degradation to a global scale) which would make the twentieth century more of a battleground than a stepping stone to a benign future.

In the introduction to Volume I of the *Decline*, Spengler critiques or "deconstructs" the modern Western belief that a long-term pattern of progressive improvement, the basis for the idea of progress, is to be found in world history. He chides academic historians for having promulgated the tripartite division of world history, ancient-medieval-modern, as representing a global historical pattern when what it really represents is the meaning of history as perceived by the intellect of Western European man, a relatively recent arrival on the historical stage. Thus, with Western European man's parochial perspective of history, perceived through the distorting lens of the Judeo-Christian worldview, a Western European-centered, "Ptolemaic" distortion of historical reality has represented itself as the modern, informed basis for understanding the world-historical process. As Spengler wrote:

> The most appropriate designation for this current West-European scheme of history, in which the great Cultures are made to follow orbits round us as the presumed center of world-happenings, is the *Ptolemaic System* of history. The system that is put forward in this work in place of it I regard as the *Copernican discovery* in the historical sphere, in that it admits no sort of privileged position to the Classical or the Western Culture as against the cultures of India, Babylon, China, Egypt, the Arabs, Mexico as separate worlds of dynamic being which in point of mass count for just as much in the general picture of history as the Classical, while

frequently surpassing it in point of spiritual greatness and soaring power.[38]

Spengler then goes on to document the Persian and Jewish religious roots of the linear pattern of history which would later be embellished by Western Europeans to represent world history as progress. In the ancient Persian and Jewish views, their "world history" would be culminated by religious "Redemption" for their own parochial societies. In the modern version, the addition of a "modern" epoch (in the seventeenth century) to the preceding ancient and medieval ones would reflect a general improvement in the overall condition of secular humanity, which would be further realized in the temporal future rather than through divine intervention to create the Millennium.[39] In sharp contrast, Spengler recognizes neither a spiritual nor a secular progressive pattern in history, but rather a world history in which "'Mankind' ... has no aim, no idea, no plan, any more than the family of butterflies or orchids."[40]

> I see, in place of that empty figment of one linear history which can only be kept up by shutting one's eyes to the overwhelming multitude of the facts, the drama of *a number* of mighty cultures, each springing with primitive strength from the soil of a mother-region to which it remains firmly bound throughout its whole life-cycle; each stamping its material, its mankind, in its own image.... Each Culture has its own new possibilities of self-expression which arise, ripen, decay, and never return. There is not one sculpture, one painting, one mathematics, one physics, but many, each in its deepest essence different from the others, each limited in duration and self-contained, just as each species of plant has its peculiar blossom or fruit, its special type of growth and decline.... I see world-history as a picture of endless formations and transformations, of the marvelous waxing and waning of organic forms. The professional historian, on the contrary, sees it as a sort of tapeworm industriously adding on to itself one epoch after another.[41]

Having established the nature of the individual Culture (In Spengler's usage, Culture and Civilization are capitalized.) as a kind of superorganism (an idea long since demolished by later historians), Spengler explains that each great Culture possesses an essence or "soul" which is represented by a *prime symbol*, an abstraction of the particular manner in which the Culture is actualized through time in extended space. The prime symbol determines the distinctive style of a Culture through the way it perceives, and historically develops in, space. For Spengler, the Classical prime symbol emphasized the "strictly limited, self-contained Body...," whereas the Western (Faustian) Culture thought in terms of "infinitely wide and infinitely profound three-dimensional Space...."[42]

Once the prime symbol has been fully actualized by a Culture as it passes through the organic stages of development: Spring, Summer, Autumn, Winter, Spengler sees it as destined to mortify and lose its motive force when it becomes "Civilization," the half-dead "thing - become" as opposed to the vibrant "thing - becoming."[43]   Although "Civilization" arises in the Winter stage of all great Cultures, in some instances, as in ancient Egypt, the Culture may continue for many centuries or even a millennium or more in this petrified, essentially unchanging stasis before it disappears or is absorbed into another great Culture. The most outrageous and unsettling thesis in *The Decline of the West*, of course, is that Western European or Faustian Culture has passed into the Winter stage and Civilization since the end of the Enlightenment, thereby guaranteeing a steady decline in Western values and cultural vitality in modern times and in the centuries ahead.

Before further exploring Spengler's analysis of the Civilization stage of modern Western European Culture, with its Faustian striving towards the infinite, let us first consider Spengler's chronology of the world's great Cultures, with particular attention to their *duration*, their beginnings, their transformation into Civilization, and their decline and fall. The importance of Spengler's characterization of the duration of his organic Cultures will be made clear when we examine his philosophy of history from the perspective of contemporary environmental history.

Having, with a great intellectual debt to Goethe and his idiosyncratic "scientific" methodology, established to his own satisfaction that Cultures are living organisms, Spengler, through his analysis of the stages of the great Cultures, arrived at an ideal life-span of one thousand years (a millennium) for these historical entities, comparable with the ideal individual human life span of three-score years and ten.[44]   With this norm in mind, Spengler went so far as to insist that the inner structure of each great Culture was identical to all the others, and that "without exception all great creations and forms in religion, art, politics, social life, economy and science appear, fulfill themselves and die down *contemporaneously* in all the Cultures ...."[45]   Thus, a major part of the *Decline* is devoted to matching the historical phenomena of the great Cultures chronologically.

The only partially systematic presentation of the duration of the great Cultures, i.e., Babylonian (Mesopotamian), Egyptian, Chinese, Indian, Mexican (Mayan-Aztec), Magian (Arabic), Classical (Apollinian), and Western European (Faustian) in the *Decline* is included in the charts appended to Volume I. The other references to duration are scattered throughout the text. For the purpose of later comparing the durations of Spengler's culture with cultural deterioration based upon regional environmental degradation, the chronologies of the following five cultures are most useful: Babylonian; Egyptian; Mexican; Classical; Western.

Spengler considered the Babylonian culture to have been born sometime before 3000 B.C. and to have entered decline after 2500, for a rough duration of a millennium (ca. 3200-2200). Egyptian Culture is more precisely dated as having been born 2900 B.C. and becoming "Civilization" by 1680 B.C. for a duration of 1200 years (and continuing in the Winter stage into a long period of Civilization). Mayan Culture is considered to begin ca. 160 A.D. or earlier and to enter the Winter stage by 960 A.D., for a duration of 800 years plus the Winter stage. For Classical Culture, Spengler gives a birth date of ca. 1100 B.C. and claims that the Winter stage begins after 300 B.C., for a duration of 800 years plus the Winter stage. It is unclear in most cases where Winter ends as a stage and where Civilization, as an overlapping stage, begins. Finally, for our own Western Culture, the birth date is ca. 900 A.D. or the tenth century, with Winter and Civilization setting in at the end of the Enlightenment, ca. 1800, giving Faustian Culture a duration of about 900 years plus the duration of the Winter stage after the onset of Civilization.

Any question of the precision or imprecision of Spengler's cultural birth dates, beginnings of Civilization, and overall duration of his great Cultures is not of fundamental importance here. Numerous refinements of cultural chronology have been made since Spengler's time. What is important is that Spengler's observations of historical chronology led him to the conclusion that the world's great Cultures typically existed for about one millennium before falling into the stagnation of "Civilization," or extinction. Particularly during the past several decades, the same basic facts regarding the historical duration of cultures (which we more typically refer to as "civilizations" in common usage, when referring to more advanced cultures which developed cities and specialization of labor based upon agriculture) have been scrutinized by archeologists and environmental historians from a far more detached, scientific perspective than Spengler's. Generally, they have convincingly demonstrated that the cyclical pattern of cultural rise and fall is, in most cases, largely the result of gradual environmental degradation, particularly in sensitive arid, semi-arid, sub-tropical, and tropical regions. Their arguments will be presented after we evaluate Spengler's contribution to the philosophy of history and, inadvertently, to environmental history as well.

Academic historians and critical philosophers of history concerned with the methodology of historical writing, such as the British philosopher of history R. G. Collingwood, have had no great difficulty in exposing major fallacies in *The Decline of the West*. Among Spengler's worst mistakes are: his claim to possess "physiognomic tact," the intuitive perception necessary to recognize the true significance of historical facts and patterns; his entire "morphological method," which reifies a simple biological metaphor into a supposed objective reality in which cultures are *actually organisms*; and perhaps worst of all, his deterministic bending of the facts to fit the same, identical pattern of development in each of his "great cultures." And yet, despite these fallacies, his general depiction of the pattern of cycles in world history, as being preferable to the linear patterns of providence and progress in terms of comprehensively explaining the facts of world history, has stood the test of time and historical criticism, particularly from the perspective of natural science.

Writing at mid-century, H. Stuart Hughes also enumerated the weaknesses of Spengler's *Decline*, but then went on to say "that all cyclical theorists, whether or not they lay claim to scientific validity, play the role of intuitive seers. They are all doing what Spengler alone quite frankly says he is doing."[46] Furthermore, Hughes claims that considering the *Decline* as literature, Spengler's *Decline* is without equal in the field of cyclical philosophy of history because his striking figurative language and powerful images "give to his work a character of excitement, of tension, and of evocative melancholy."[47] In the *Decline*, Hughes wrote, Spengler also made his readers aware of the emergence of a new barbarism in the twentieth century.

Even so harsh a critic of Spengler's *Decline* as R. G. Collingwood recognized, like Hughes, that the cyclical interpretation of history revived by Spengler was objectively preferable to the popular ideas of providence and progress when he wrote: "The cyclical view of history is thus a function of the limitations of historical knowledge. Everyone who has any historical knowledge at all sees history in cycles; and those who do not know the cause think that history is really built thus."[48] Extrapolating from Collingwood's analysis, Hughes writes:

> Thus the cyclical interpretation - whether Spengler's or any other writer's - is subject to constant change. Each generation, each individual, will set up the cycles differently. But at least this much is definite. For the past two generations, for the early and mid-twentieth century, the cyclical approach has proved to be particularly rewarding. It has appeared to explain more things than the straight-line interpretation with which earlier generations tried to express the meaning of history. In the pragmatic test, the cyclical method has been gaining ground.[49]

What Hughes could not envision at mid-century was a new sub-discipline of history, environmental history, along with parallel research in archeology and anthropology, which would attempt to explain the pattern of historical cycles on the basis of a more objective, essentially scientific approach which would have the potential to link the explosive development of new facts to the philosophy of history.

In addition to challenging the entrenched popular beliefs in providence and progress (which are part of the Western European worldview) with a neoclassical cyclical perspective of history which set the tone for twentieth century world historians, Spengler also laid some of the ground work for the comparative history of individual cultures. Ethnology, a major branch of modern anthropology, analyzes and compares cultures and their historical development in much the same way that Spengler compared the great civilized cultures of the past. The American anthropologist Alfred Kroeber, world historian Arnold Toynbee, and sociologist Pitirim Sorokin all studied the *Decline* carefully before elaborating their own interpretations of the cyclical pattern in history. One philosopher of history, Roger W. Wescott, credits Spengler with having developed the best classification of continental or major cultures in the field of historiology, the historical equivalent of ethnology.[50]

Finally, Spengler contributed some remarkable insights into the particularly destructive character of Faustian Culture regarding the environment, the natural world upon which all of the world's cultures are ultimately dependent. Having established his "morphological method" as involving two very different ways of knowing, i.e., the distinction between understanding *nature* as opposed to understanding *history*, in the first volume of the *Decline*, Spengler devotes more attention to nature in the second volume. For example, he strongly rejects Darwinian evolutionary theory, based upon the idea of slow, gradual, accumulative changes resulting in the creation of new species, for the notion of sudden changes, i.e., mutations, as the basis for speciation over geologic time. He also applies the idea of mutations in nature to Culture when he observes that "swift and deep changes assert themselves in the history of the great Cultures, without assignable causes, influences, or purposes of any kind."[51] Similarly, he maintains "that the type of the higher Culture appeared suddenly in the field of human history. Quite possibly, indeed, it was some sudden event in the domain of earth-history that brought forth a new and different form into phenomenal existence." Here, Spengler is referring to the shift from pre-history to civilization which has been so intensively studied during the past century, but which has been subjected to an *ecological* approach only during the past few decades.

In a long footnote early in Volume II of the *Decline,* Spengler shows that he was well aware of the influence of nature, particularly in the form of climate change, upon human history, although his overall assessment of human causes of environmental degradation in the *Decline* is ambivalent. He recognizes the local displacement of nature by culture during the stage of Civilization in every Culture, but he doesn't seem to appreciate the permanent ecological damage resulting from the rise and fall of Cultures until he describes the Winter-Civilization phase of Faustian, Western European Culture in the nineteenth and twentieth Centuries:

> Another blank is the history of the countryside or landscape (i.e., of the soil, with its plant mantle and its weathering) in which man's history has been staged for five thousand years. And yet man has so painfully wrested himself from the history of the landscape, and withal is so held to it still by myriad fibres, that without it life, soul, and thought are inconceivable. So far as concerns the South-European field, from the end of the Ice Age, a hitherto rank luxuriance gradually gave place in the plant-world to poverty. In the course of the successive Egyptian, Classical, Arabian, and Western Cultures, a climatic change developed all around the Mediterranean, which resulted in the peasant's being compelled to

fight no longer *against* the plant-world, but *for* it - first against the primeval forest and then against the desert. In Hannibal's time the Sahara lay very far indeed to the south of Carthage, but today it already penetrates to northern Spain and Italy. Where was it in the days of the pyramid- builders, who depicted sylvan and hunting scenes in their reliefs? When the Spaniards expelled the Moriscos, their countryside of woods and ploughland, already only artificially maintained, lost its character altogether, and the towns become oases in the waste. In the Roman period such a result could not have ensued.[52]

In this passage Spengler anticipates the environmental determinism of Ellsworth Huntington and other scholars who began to recognize the powerful role of climate change in history during the twentieth century, leading to recent discoveries which even link specific events to El Niños, as in Brian Fagan's *Flood, Famines and Emperors: El Niño and the Fate of Civilizations*. However, as we shall see, although the passage just cited shows that Spengler was well aware of the impact of climatic change on humans, he was unaware of the profound changes in regional ecology which humans were capable of effecting from the earliest civilized cultures, such as the Babylonian (Mesopotamian) and the Mexican (Mayan). Only when he examined the environmental effects of the Winter (i.e., modern) stage of Faustian culture did he come to the realization that humankind could bring about the nearly total destruction of nature. In Vol. II of the *Decline*, and especially in the final chapter, "The Machine," Spengler gives us a grim and poetically powerful analysis of the nature-destroying tendencies of Faustian Civilization. Although Spengler saw all Culture culminating in dense, sprawling cities in the Civilization stage, the Faustian Civilization, with its prime symbol of infinite space, would expand this tendency to truly demonic proportions:

Even now the world-cities of the Western Civilization are far from having reached the peak of their development. I see, long after A.D. 2000, cities laid out for ten to twenty million inhabitants, spread over enormous areas of country-side, with buildings that will dwarf the biggest of today's and notions of traffic and communication that we should regard as fantastic to the point of madness.[53]

In Faustian Civilization, as in all higher Cultures, it is the rise of the city, in the winter stage, to *megalopolis*, the great world-city, which creates the ultimately destructive domains of money and the machine (i.e., advanced technics). Describing the rise of megalopolis, Spengler observes that it "...first defies the land, contradicts Nature in the lines of its silhouette, *denies* all Nature."[54] Within this superurban milieu the idea of money itself begins to take on a life of its own, to value things, including the soil and individual humans, with reference to itself. The intellectual power of money expressed in the upper economic stratum of megalopolitan society, now detached from soil and climate, "is no more limited in potential scope by actuality than are the quantities of the mathematical and the logical world."[55] For Spengler, megalopolis or "Cosmopolis" represents the petrification of a once ensouled Culture, the transformation of the "thing-becoming," into the stoney, purely artifactual "thing-become" which represents "the exact epoch that marks the end of organic growth and the beginning of an inorganic and therefore unrestrained process of massing without limit."[56]   In the modern world, what drives all of this countryside-devouring megalopolitan sprawl is the uniquely Faustian combination of the abstract conception of money which produced the modern corporation and credit systems, and the seemingly infinite production of machines through modern technology.

    Spengler emphasizes the uniqueness of Faustian technics as having, since the Middle Ages, possessed the intrinsic tendency to thrust "itself upon Nature, with the firm resolve to *be its master.* Here, and only here, is the connection of insight and utilization a matter of course."[57]   In other words, Spengler understood the Western striving for dominion over nature to have been present from the beginnings of Western European civilization (much as medieval historian Lynn White, Jr. did), long before Francis Bacon's rationalizations of our "right" to dominate nature. Like White, Spengler recognized the early scientific researches of Gothic monks as religiously inspired and directly linked to modern technics. However, it was the discovery of the steam-engine which raised Faustian dominion to an un-dreamed of level of control, "which upset everything and transformed economic life from the foundations up. Till then nature had rendered services, but now she was tied to the yoke as *a slave ....*"[58]

> As the horse-powers run to millions and milliards, the numbers of the population increase and increase, on a scale that no other Culture ever thought possible. This growth is a product of the Machine, which insists on being used and directed, and to that end centuples the forces of each individual. For the sake of the machine, human life becomes precious. Work becomes the great word of ethical thinking; in the eighteenth century it loses its derogatory implication in all languages. The machine works and forces the man to co-operate. The entire Culture reaches a degree of activity such that the earth trembles under it. And what now develops, in the space of hardly a century, is a drama of such greatness that the men of a future Culture, with other soul and other passions, will hardly be able to resist the conviction that "in those days" nature herself was tottering. The politics stride over cities and peoples; even

the economics, deeply as they bite into the destinies of the
plant and animal worlds, merely touch the fringe of life and
efface themselves. But this technique will leave traces of
its heyday behind it when all else is lost and forgotten. For
this Faustian passion has altered the Face of the Earth.[59]

In the final pages of the *Decline* Spengler further elaborates the frenzied
proliferation of the machine and its achievements, powered by the work of
engineer, entrepreneur, and factory worker. Simultaneously, he sees the machine
becoming ever less human as it spreads its web over the earth. "Man has felt the
machine to be devilish, and rightly. It signifies in the eyes of the believer the
deposition of God. It delivers sacred Causality over to man and by him, with a
sort of foreseeing omniscience is set in motion, silent and irresistable." Thus, he
sees Faustian man and all the world influenced by him as having become "the
slave of his creation. His number, and the arrangement of life as he lives it, have
been driven by the machine on a path where there is no standing still and no
turning back."[60] As a consequence we are obedient slaves and not masters of the
"devilish" machine, from the steamships and early skyscrapers of Spengler's own
day to the mindless proliferation of computers, jetliners, and cell phones of our
own. To Spengler's mind, nothing can derail us from this daemonic treadmill of
endless money and technique but the end of nature, for the essence of money, as a
form of thought, "*fades out as soon at it has thought its economic world to
finality*, and has no more material upon which to feed."[61] Long before any
"environmental crisis" or "environmental movement," Spengler said it all
regarding modern, Faustian man's destruction of nature. More importantly, he
claimed that the specific cause of the global destruction of nature was to be found
in the relentless striving for dominion over nature which was an intrinsic part of
the Western European worldview. With conquest and colonization, Faustian
attitudes toward nature and their accompanying techniques of domination through
money and the machine had infected the entire globe, with dire long-term
ecological consequences.

Given Spengler's astute awareness of global environmental crisis at the
beginning of the twentieth century, one might expect that he would have intuited
massive regional environmental degradation and resource depletion in earlier
cultures as they passed into the Winter stage and Civilization. Perhaps he did not
because his attention was so strongly focused on matching parallel stages of
development in his "great Cultures." Nevertheless, he was remarkably close to
having understood what we have since learned from environmental history, that,
with rare exceptions, human civilizations have followed cycles of birth, youth,
maturity, old age, and extinction or decline not because of some metaphysical
design or some unknown secular basis for the history of civilizations, but simply
because they have generally destroyed the ecology and resource base in the
regions in which they once flourished.

NOTES

[1]. Fairfield Osborn, *Our Plundered Planet* (New York: Pyramid Books: 1970). Little, Brown edition published March, 1949.

[2]. Jonathan Bate, *The Song of the Earth* (Cambridge, Massachusetts: Harvard University Press, 2000), pp.43-49.

[3]. Donald Worster, *Nature's Economy: A History of Ecological Ideas* (Cambridge: Cambridge University Press (1977, 1985) p.378.

[4]. Frank Benjamin Golley, *A History of the Ecosystem Concept in Ecology* (New Haven and London: Yale University Press, 1993), p.8. On the same page Golley quotes from Alfred George Tansley's (1871-1955) article titled "The Use and Abuse of Vegetational Concepts and Terms" in the periodical *Ecology*:

But the more fundamental conception is, as it seems to me, the whole *system* (in the sense of physics), including not only the organism-complex, but also the whole complex of physical factors forming what we call the environment of the biome--the habitat factors in the widest sense.

It is the systems so formed which, from the point of view of the ecologist, are the basic units of nature on the face of the earth.

These *ecosystems*, as we may call them, are of the most various kinds and sizes. They form one category of the multitudinous physical systems of the universe, which range from the universe as a whole down to the atom. (Tansley, 1935, 299).

[5]. William Ophuls with A. Stephen Boyan, Jr., *Ecology and the Politics of Scarcity Revisited: The Unraveling of the American Dream* (New York: W.H. Freeman and Company, 1992), p.19.

[6].Donald Worster, *The Wealth of Nature: Environmental History and the Ecological Imagination*, (New York: Oxford University Press, 1993), pp.50-57.

[7]. Donald Worster, ed., *The Ends of the Earth* (Cambridge: Cambridge University Press, 1988), p.289.

[8]. Ibid., *The Ends of the Earth*, pp.291-293.

[9]. Joyce Appleby, Lynn Hunt and Margaret Jacob, *Telling the Truth about History* (New York: W.W. Norton and Company, 1994), p.215.

[10]. Ibid.

[11]. Richard Tarnas, *The Passion of the Western Mind: Understanding the Ideas That Shaped Our World View* (New York: Ballantine Books, 1991), p.400.

[12]. Hal Rothman, "A Decade in, the Saddle: Confessions of a Recalcitrant Editor," *Environmental History*, 7:1 (January, 2002), p.20.

[13]. J. Donald Hughes, Ed., *The Face of the Earth: Environment and World History* (Armonk, New York: M.E. Sharpe, 2000), p.4.

[14]. Ibid. p.9.

[15]. Ibid, p.12.

[16]. Robin G. Collingwood, *The Idea of History* (London: Oxford University Press, 1946). Edward Hallett Carr, *What is History?* (New York: Alfred A. Knopf, 1967).

[17]. J. Donald Hughes, *The Face of the Earth*, p.7.

[18]. Oswald Spengler, *The Decline of the West* (New York: Alfred A. Knopf, Vol. 1, 1926; Vol. II, 1928). Original German publication, 1918, 1922. H.G. Wells, *The Outline of History* (Garden City: Garden City Publishing, 1920). Arnold J. Toynbee, *A Study of History* (London: Oxford University Press, 1933, 1939, 1954, 1961).

[19]. Alfred L. Kroeber, *Configurations of Culture Growth* (erkeley and Los Angeles: University of California Press, 1944) and *Style and Civilizations* (Ithaca: Cornell University Press, 1957). Pitirim A. Sorokin, *Social and Cultural Dynamics* (1937-1941; reprint, New York: Bedminster, 1962). William H. McNeill, *The Rise of the West: A History of the Human Community* (Chicago: University of Chicago Press, 1963).

[20]. Paul Costello, *World Historians and Their Goals: Twentieth Century Answers to Modernism* (DeKalb: Northern Illinois University Press, 1993), p.183.
[21]. William H. McNeill, *Mythistory and Other Essays* (Chicago: University of Chicago Press, 1986), pp.187-191.
[22]. Ibid,, p.16.
[23]. Paul Costello, *World Historians and Their Goals*, p.211.
[24]. Ibid,, p.226.
[25]. Ibid,, p.70.
[26]. Donald Hughes, ed., *The Face of the Earth*, p. 90.
[27]. Paul Costello, *World Historians and Their Goals*, pp.90-91.
[28]. See Ludwig Edelstein, *The Idea of Progress in Classical Antiquity*, (Baltimore, Maryland: The Johns Hopkins Press, 1967).
[29]. J.B. Bury, *The Idea of Progress* (New York: Dover Publications, Inc., 1955, 1932), pp. 9-10.
[30]. Oswald Spengler, *The Decline of the West*, Vol. I, p. 18.
[31]. Jared Diamond, *Collapse: How Societies Choose to Fail or Succeed* (New York: Viking, 2005).
[32]. Klaus P. Fischer, *History and Prophecy: Oswald Spengler and The Decline of the West* (New York, Bern, Frankfurt am Main, Paris: Peter Lang, 1989), p. 96. The quotation is from Goethe.
[33]. Ibid., p. 91
[34]. Ibid.
[35]. Ibid, pp. 94-97.
[36]. Ibid, p. 106.
[37]. Mircea Eliade, *The Myth of the Eternal Return* (Princeton, N.J.: Princeton University Press, 1954, 1971), pp. 121-123.
[38]. Oswald Spengler, *The Decline of the West*, Vol. I, pp.18-20.
[39]. Ibid., p.21.
[40]. Ibid., pp.21-22.
[41]. Ibid., p.174.
[42]. Ibid., p.106; Vol. II, p.99.
[43]. Ibid., Vol. I, p.110.
[44]. Ibid. p.112.
[45]. H. Stuart Hughes, *Oswald Spengler* (New York: Charles Scribner's Sons, 1962), p.162.
[46]. Ibid.
[47].Ibid., p. 158 (Hughes quoting R.G. Collingwood).
[48]. Ibid.
[49]. Roger W. Westcott, "The Enumeration of Civilizations," *History and Theory*, Vol. IX, No. 1 (1970), p.74.
[50]. Oswald Spengler, *The Decline of the West*, Vol. II, p.35.
[51]. *Ibi.d.* p.36.
[52]. Ibid., p.39.
[53]. *Ibi.d.*, p.101.
[54]. Ibid., p.94.
[55]. Ibid., p.98.
[56]. Ibid., p.100.
[57]. Ibid., p.501.
[58]. Ibid., p.502.
[59]. Ibid., p.503.
[60]. Ibid., p.504.
[61]. Ibid., p.506.   Also, on p.505: "Nature becomes exhausted, the globe sacrificed to Faustian thinking in energies."

CHAPTER TWO: ECOLOGY AND THE RISE AND FALL OF ANCIENT CIVILIZATIONS

Introduction

Modern man first became aware of ancient Mesopotamian civilization only as recently as the mid-nineteenth century,[1] and we have accumulated most of our knowledge of other ancient civilizations only very recently as well. Much of the latest scientific information related to these civilizations is concerned with global climate change and accompanying shifts in the distribution of ecosystems since the end of the Pleistocene. These climatological and ecological parameters have been of particular importance in helping to explain the greatest shift in life style in the history of the 100,000-150,000 year old human species, *Homo sapiens*, the shift from hunting and gathering societies to cultures based on agriculture and grazing known as the Neolithic Revolution. Until recent decades historians believed that humans had left the "state of nature" of hunting and gathering to develop agriculture and urban civilization because of a desire to improve, to find a better way of life, an alternative to the "primitive" state of nature in which life was, as philosopher Thomas Hobbes put it, "solitary, poor, nasty, brutish and short." As we shall see, archeologist Brian Fagan and others who have studied the Neolithic Revolution in the light of new climatological and ecological knowledge thoroughly disagree with the idea that humans sought to find a release from the "miseries" of hunting and gathering.

In *Floods, Famines, and Emperors*, Fagan summarizes the climatological changes since the last major Pleistocene glaciation, the "Wisconsin" stage (Würm in Europe) which ended about 15,000 years ago. Global warming has continued since then (with only one major interruption, the "Younger Dryas" period, the millennium of cooling from around 10,000-9,000 B.C.)[2] until about 5,000 years ago when world-wide sea level stabilized at approximately the present-day level. Reminding us that the ice-age climate shifts led to major challenges to human ingenuity (resulting in the primitive inventions that developed in the course of hunting, fishing, gathering, and the building of shelters), Fagan also points out that slowly growing human populations subjected to post-glacial aridification tended to concentrate along bodies of water, especially in valleys such as those of the Tigris-Euphrates, Jordan, and Nile rivers.

The ancestors of *Homo sapiens* appear to have evolved in tropical Africa and to have spread to Europe, Asia, and the Americas during the last (Wisconsin) ice age. At the end of the Wisconsin glaciation, the global warming which caused retreating glaciers and rising sea levels also caused a reconfiguration of ocean currents and weather patterns. The shift towards a drier climate in the Near East and Mediterranean Basin forced growing populations of hunters and gatherers to a greater dependence upon riparian environments. Fagan estimates that after tens of thousands of years of very slow population growth, the human population had reached about eight and a half million by 13000 B.C.[3] As population growth began to reach the limits or carrying capacity of the ecosystems upon which humans were dependent, *Homo sapiens* had increased its knowledge of the available variety of food resources, particularly plant foods, which it could depend upon for sustenance. However, the post-glacial climatic shift and regional dessication left some human populations at levels beyond the carrying capacity of the land. Whereas many other mammalian populations died off in these regions, humans survived by making the transition to agriculture. The encyclopedic knowledge of edible plants, gleaned by experience over the millennia of endless foraging, was the key to the gradual invention of agriculture during the Neolithic Revolution.

According to Fagan, the transition from hunter-gatherers to the earliest farmers and villagers would make humans far more vulnerable to the kinds of short-term climatic fluctuations which are controlled by El Niños and related oceanic and atmospheric phenomena. Relatively mild as the post-glacial climate was, the concentration of humans into farming communities dependent upon a regular water supply from rains and/or river flows was a formula for periodic disaster involving famine, malnutrition, disease, social and political instability, and warfare.

By 9000 B.C. in Mesopotamia, the Tigris-Euphrates alluvial plain supported dense oak forests and stands of wild cereal grasses which in turn supported both abundant gazelle herds and numerous hunter-gatherers. Following the Younger Dryas cooling, the nut-rich forests which provided an important part of the human diet shifted nearly one hundred kilometers away from the valley, forcing its inhabitants to utilize wild cereals and clovers to supplement their depleted diet. Concurrently, while game populations crashed, deliberate plantings were made to increase local food supplies.[4] Thus did the nomadic hunter-gatherers, who had already begun to settle in villages near once-abundant food supplies, begin the settled way of life as farmers dependent upon their crops for the majority of their nutritional needs. Brian Fagan concludes, therefore, that: "We humans became farmers because we had to. We changed the way we lived and interacted with one another almost overnight because we were at the mercy of distant Atlantic currents that brought rain and mild winters to our homelands. We have been at the mercy of short-term climate change ever since."[5]

Mesopotamia

The birth of Mesopotamian civilization was slightly before and almost contemporary with that of Egypt around 3000 B.C. Historian H.W.E. Saggs notes that permanent agricultural settlements came later in Egypt than in Mesopotamia not because Egypt was more backward, but because left to a free choice, humans "prefer to remain hunters and gatherers and do not settle permanently to the toil of farming until it is forced upon them."[6] Saggs also points out that centralized power came much later in Mesopotamia than in Egypt, and in a very different way. Prior to the middle of the fourth millennium, in Southern Iraq, where rainfall is inadequate for growing crops, village peasants learned to irrigate by diverting waters of the Euphrates into abandoned stream channels. After ca. 3500 B.C. the major attributes of civilization began to appear: cities (as city-states); monumental architecture; widespread use of metals; and, most importantly, the invention of writing.[7] It is beyond the scope of this book to consider the subsequent history of warfare between the city-states, the consolidation of power, and the succession of empires which followed the rise of Sumeria from 2900-2400 B.C. Rather, what interests us here are the long-term environmental consequences of Mesopotamian civilization.

An excellent summary of the environmental history of Mesopotamia is presented by Clive Ponting in *A Green History of the World: The Environment and the Collapse of Great* Civilizations (1991). Ponting points out the relative ecological vulnerability of arid regions as compared to temperate regions such as Western Europe, generalizing that arid to semi-arid regions with high population densities "began to be affected within a thousand years of the adoption of agriculture and a settled way of life."[8] Referring to central Jordan, he writes that "within about a thousand years of the emergence of settled communities, villages were being abandoned as soil erosion caused by deforestation resulted in a badly damaged landscape, declining crop yields and eventually inability to grow enough food." [9]

The barren, desolate, largely treeless landscape of southern Iraq, the site of Sumerian civilization, shocked and puzzled nineteenth century archeologists. How this once highly productive grain-growing region was turned into a desert is no mystery, however, for the combined destructive processes of waterlogging of silty, poorly drained soils on nearby flat floodplain surfaces, and high evaporation rates during hot, dry summers produced a long-term salt build up that first caused the Sumerians to shift to more salt-tolerant barley instead of wheat, and eventually prevented farming entirely as salinization of irrigated soils increased over time. The only solution to the problem, long periods of leaving the land fallow and unwatered, was ignored because of relentless demands for growing more food. Ponting concludes that it took from about 3500 B.C. until 2400 B.C. to reach the point at which salinization took an increasing toll resulting in a rapid decline in the food surplus. By 1700 B.C. no wheat at all could be grown in southern Mesopotamia, and Sumeria had " declined into insignificance as an underpopulated, impoverished backwater of empire."[10] Ponting's epitaph for Sumerian civilization is as follows:

> The artificial agricultural system that was the foundation of Sumerian civilization was very fragile and in the end brought about

its downfall. The later history of the region reinforces the point that all human interventions tend to degrade ecosystems and shows how easy it is to tip the balance towards destruction when the agricultural system is highly artificial, natural conditions are very difficult and the pressures for increased output are relentless. It also suggests that it is very difficult to redress the balance or reverse the process once it has started.[11]

Egypt

In contrast to the environmental difficulties encountered by Mesopotamian civilization, Egypt was ecologically blessed with the "gift of the Nile," a far more manageable, dependable, and generally benign source of irrigation than the Tigris-Euphrates river system. The Nile did periodically flood at destructive levels and, when rainfall thousands of miles to the south of Egypt was sparse, frequently failed to provide sufficient discharges to supply a gradually increasing population with adequate crops to avoid malnutrition and famine. However, during a majority of years overall, the Nile was bounteous in her gift to the people of the Nile Valley. This was true to such a degree that the Egyptians viewed nature as sacred and held in balance by the goddess Ma'at since the time of creation.[12] The annual Nile flood brought deposits of alluvial silt as natural fertilizer and simultaneously flushed accumulated salts from the soil beneath the rays of almost perpetual sunshine. Is it any wonder that this beneficial set of circumstances was viewed as a gift of the gods?

It was therefore almost inevitable that the Egyptian pharaohs would foster "the belief that they controlled the mysterious inundation, the very fountain of human existence."[13] This myth was undoubtedly very useful to the pharoahs of the Old Kingdom. The Old Kingdom rulers were perceived as personifying Ma'at, (cosmic order, justice, and prosperity) through their magical control of both the life-giving Nile floods and the civilized existence of the Egyptian people. According to Spengler, this mythically-sustained political equilibrium developed at the beginning of the Old Kingdom and lasted through the end of the Middle Kingdom, from about 2700 B.C. to 1800 B.C.[14] A series of disastrously low flood years led to the famines and peasant revolts that caused the Old Kingdom to collapse and eventually give rise to the Middle Kingdom.[15] The fact that Egyptian civilization survived for thousands of years thereafter led Spengler to invent the ingenious distinction between "Culture" and "Civilization" in *The Decline of the West*. Egypt was the major exception to the approximately one thousand year Spenglerian cycle and could be reconciled to his rigid system only by the concept of "Civilization," understood as the continuation of a fully developed Culture in a rigidified or petrified condition incapable of creating anything essentially new. The prime symbol or master pattern of the Culture having been fully realized in the thousand year cycle, the Culture could perpetuate its completed and ossified paradigm for centuries or even millennia after it had realized its potential.

Regarding the artistic impotence of the stage of Civilization, Spengler writes: "Instead of the steady development that the great age had pursued through the Old and Middle Kingdoms, we find *fashions* that change according to the taste of this or that dynasty."[16]

Although Spengler was generally aware of the role of a nurturing environment in the growth of his cultures, his use of the organic metaphor of cultural development probably prevented him from probing more deeply into the environmental causes of cultural rise and fall. For example, he makes no reference to the profoundly different ecological conditions which gave rise to the severely pessimistic worldview of Mesopotamian culture as compared to the brightly optimistic Egyptian worldview, i.e., the effects of the unreliable and often destructive Tigris-Euphrates river system in contrast with the relatively benign and reliable Nile River.[17]

From the perspective of contemporary environmental history, the Nile River is the classic model of natural sustainability, the ability of a culture to use nature as a source of resources in such a manner that the resources are not degraded and are available in perpetuity for future generations. The natural sustainability of the annual Nile floods and the floodplain soils which they inundated was, as "the gift of the Nile," the environmental basis for the general long-term stability and longevity of Egyptian civilization. As J. Donald Hughes writes:

> Egypt existed as an autonomous civilization from before 3000 B.C. to after 1000 B.C. , and during that period maintained a relatively consistent pattern in economy, government, religion, and ecological viewpoints and techniques. It is likely that the stability of Egyptian civilization was the result of the sustainability of Egypt's ecological relationships.[18]

Clive Ponting agrees: "The most striking example of a society establishing a sustainable balance between the natural environment and its demand for food is Egypt."[19] Ponting further points out that this sustainability continued until the introduction of Western practices of irrigation and fertilization for commercial production for export in the nineteenth century.[20] The final blow to the sustainability of the Nile came with construction of the Nasser (Aswan) Dam in the mid-twentieth century, producing a congeries of unintended environmental consequences ranging from salinization, loss of natural fertilization, channel erosion and beach erosion, to the spread of schistosomiasis as well as the loss of the Mediterranean sardine fishery due to depleted nutrients resulting from the entrapment of sediment behind Nasser Dam. As if this were not enough in terms of loss of sustainability, we can now add the threat of global warming which very probably will alter the weather and rainfall patterns of the regions in which the Blue and White Nile have received the abundant rainfall that allows the Nile to flow through the Egyptian desert as an exotic river.

The historical record tells us of the disastrous social and political consequences resulting from major fluctuations in the flood level of the Nile, including the demise of the Old Kingdom and the two centuries of anarchy that followed prior to establishment of the Middle Kingdom. According to Ponting, the overall historical trend in Nile flood levels has been downwards, "probably caused by declining rainfall in the highlands that are the source of the Nile, but with major fluctuations within that trend."[21] If this gradual dessication is magnified by human-caused global warming, then modern Egypt in the twenty-first century and beyond could find itself hard-pressed to feed itself.

Contemporary discussions of environmental sustainability typically fall into two major categories. They either refer to the ability of the human population to support or sustain itself in food and other essentials of life indefinitely into the future, or they consider the foregoing within the context of sustaining large representative portions of relatively unhumanized ecosystems along with steady-state human societies. Which of these two kinds of sustainability that we choose to pursue will have an enormous influence upon the nature of future human progress. In the case of Egypt's history, the variety of sustainability of the first kind, what we might call anthropocentric sustainability, was achieved for about 5,000 years. However, regarding the second type, which we will call ecocentric sustainability, the Egyptians ultimately failed, despite the fact that they held the natural environment in high regard and worshiped many of the native species of animals that originally co-existed with them in the Nile floodplain and the surrounding area.

J. Donald Hughes has observed that " the ecological attitudes and practices of the Egyptians were rooted in a worldview that affirmed the sacred values of all nature, and of land in particular."[22] Nevertheless, despite this deep respect for nature, the Egyptians deforested the Nile Valley, depleted grasslands through overgrazing, and gradually destroyed most animal habitat. Elephant, rhinoceros, wild camel, and giraffe were all gone or rare by the end of the Old Kingdom, and barbary sheep, lion, and leopard had become scarce. Waterfowl and marsh animals suffered the same fate.[23] Saggs concurs that this ecological destruction was brought about unwittingly, " not from overhunting, but from disturbing the ecological balance, particularly by setting under way increasing dessication away from the Nile." [24] Destruction of the producers, the plants at the base of every ecosystem, was completed over time, through loss of forest, grasslands, and wetlands, to produce an ecological wasteland where once an almost edenic natural paradise existed. How much more quickly might ecosystems by destroyed by civilizations lacking a deep respect for nature?

The Maya

The ancient Mayan civilization existed in the region extending from present-day southern Mexico through Guatemala and Belize to western Honduras. It has long been viewed as a striking example of the sudden collapse of a civilization for unknown reasons. Its origins were also shrouded in mystery, and even by the mid-twentieth century, the idea that Mayan civilization was a result of cultural diffusion from either Egypt or Southeast Asia had not been fully discredited by archeologists and historians. By the 1960s, however, there was a growing consensus that connections with Egypt and other cultures, based upon such superficial similarities as those existing between Mayan hieroglyphics and pyramids and those of Egypt, were impossible. Having rejected diffusionist theories as " the product of imagination," in 1964 archeologist Henry Stierlin wrote:

> In America evolution followed completely different paths from those of the Old World. The background to its culture in no way resembles that of the great agrarian empires of the ancient world. There is no possible comparison between the plains irrigated at regular intervals by the flooding of great rivers such as the Nile, the Tigris, the Euphrates, the Indus and the Yellow River, and the lands of the Maya, where the soil cover is relatively thin and the rains frequently abundant.[25]

Until the last thirty years very little detailed knowledge of the ecology of the Mayan region existed. By the close of the twentieth century, however, sufficient knowledge had accumulated to prove that long-term ecological deterioration, involving the destruction of soils, erosion and rapid sedimentation, and destruction of native plant and animal communities played a major role in the rapid fall of Mayan civilization into obscurity. These new developments are succinctly summarized in anthropologist Charles L. Redman's *Human Impact on Ancient Environments* (1999).

Redman begins his scientific explanation of the ecological basis for the sudden demise of the Maya by pointing out that the Mexican and Mayan regions were home to prehistoric societies for thousands of years, during which corn was transformed from a weed plant of marginal use for food to the staple food of Mesoamerica. It was first domesticated at least as early as 5000 B.C. and, through several millennia of selection by humans, greatly increased in size and quality. Given the soil and climate of the Mayan landscape, slash-and-burn agriculture (also known as swidden or milpa), which did not lend itself to extensive irrigation works, was used to grow corn along with gourds, squash, and beans.[26] Although the best soils generally were in the lowlands, as population grew, crops were planted on steeper slopes to the extent that at the zenith of Mayan civilization, 75% of their environment was cleared for agriculture and other uses.[27]

The most thorough scientific study of the relationship between Mayan civilization and environmental degradation presented by Redman is the Central Petén Historical Ecology Project, which was designed to understand historical changes in the tropical forest of northern Guatemala. Somewhat unexpectedly, however, the study showed that the Maya had been responsible for major ecological changes and damage to the region's ecosystem. The key to these effects was the relationship between the fact that 75% of the nutrients in tropical forests are held in the vegetative cover rather than the soil (whereas in temperate regions most of the nutrients are held in the soil) and the fact that extensive slash-and-burn agriculture over time gradually depletes the soil of its nutrients.[28] Burning of the tropical vegetative cover releases abundant nutrients adequate for two years of crops, but the impoverished lateritic soil must then be left fallow for at least three to six years before repeating the slash-and-burn procedure. With more intensive use, tropical soils lose their ability to support crops. Unfortunately, over the centuries Mayan agriculture became both more extensive and more intensive to support a maximum population estimated at between 5 to 10 million people.[29] The ecological consequences of these agricultural practices were discovered by the Central Petén Historical Ecology Project on the basis of their study of archeological settlement patterns, the pollen record, erosion and deposition of sediment, and losses of chemicals from soils corroborated by chemical changes in lake sediments.

The conclusions of the Petén study were that: (1) increased deposition of phosphorus and silica during the period of Mayan occupation was evident in sediment cores taken from lake bottoms and was the result of burning of vegetative cover, increased erosion, construction of stone buildings, and other human activities; (2) high rates of silica deposition in lakes was indicative of deforestation, cultivation, and settlement construction by the Maya; and (3) the preceding evidence suggests maximum deforestation between 1000 and 2000 years ago, culminating farming and settlement activities of the Maya beginning 3000 to 4000 years ago to produce "an anthropogenic ecosystem through much of the Holocene." Similar studies in other Mayan regions such as Copán corroborate the results of the Petén project, and suggest that high demand for wood products for fuel, for the production of lime plaster for houses and monuments, and for the construction of dwellings added to agricultural demands as causes of deforestation.[30]

Observing that "the drain on the land of dense population, intensive agricultural manipulation, and construction of massive settlements increased to the point where the system was no longer sustainable... by the end of the tenth century A.D.," Redman concludes that "this collapse was primarily due to the extended period of intense human exploitation, albeit aided by microclimate variability." This climatic variability refers to a relatively dry period that appears to coincide with the ninth and tenth century decline in Mayan civilization.[31]

Fagan's *Floods, Famines and Emperors* sheds further light on this climatic connection by considering the ecological cause of the collapse of Mayan civilization in the context of both long-term climatic fluctuation and the ultimately destructive nature of Mayan society. Citing recent scientific data for the past eight thousand years of climate change, Fagan tells us of persistent wet conditions until about 1000 B.C., after which increasingly dry conditions continued for nearly two millennia, "peaking between about A.D. 800 and 1000, at the very time of the Maya collapse. The drought cycle of these two centuries was the driest period of the last eight thousand years."[32] These droughts appear to be the result of little-understood atmosphere-ocean interactions such as El Niño events, decadal shifts in the North Atlantic Oscillation, and several decade cycles of rainfall variation. Fagan observes that despite these short-term climate swings and the fragility of the soils in the Mayan region, classic Maya civilization flourished for more than eight centuries before the harsh realities of overpopulation and environmental stress toppled its proud leaders. That it survived so long in such a demanding environment is a tribute to the skill of Maya farmers."[33]

Concerning the social factor leading towards the sudden collapse of the Maya, Fagan explains how the perception of early Mayanists, such as Sylvanas Morley, of the Maya as " a civilization without cities, ruled by peaceful religious leaders with a passion for calendars and the movements of heavenly bodies," was overturned in the second half of the twentieth century by new knowledge showing the Maya as capable of supporting large urban populations as well as high population densities in the intensively farmed countryside, but eventually under the rule of warring lords whose military ambitions, ancestor worship, and pyramid and temple construction were related to increases in population and intensive farming that were ecologically destructive. Fagan tells us that recently improved translations of the Maya script reveal Maya society as:

> a patchwork of competing city-states ruled by bloodthirsty lords, obsessed not with calendars and rituals but with genealogy and military conquest. Past masters of diplomacy and the manipulation of prestige, the ruthless Maya lords nurtured powerful ambitions that destroyed their environment and brought down their great cities.[34]

Recognizing the ecological and sociopolitical factors as acting in the long-term to overextend Mayan civilization until severe drought caused its sudden collapse, Fagan documents the same environmental factors that Redman has explained with recent scientific evidence: exhausted soils, deforestation, and extensive sheet erosion, concluding that the collapse of the Maya " is a cautionary tale in the dangers of using technology and people to expand the carrying capacity of tropical environments."[35] Following severe drought, starvation, and rebellion, survivors of the ninth and tenth century debacle dispersed into small, self-sustaining villages as a small remnant population.

Historian Clive Ponting's account of the rise and fall of Mayan civilization closely parallels that of Fagan, with the major difference being Ponting's exclusion of the climatic factor, probably because his *A Green History of the World* (1991) was published nearly a decade before *Floods, Famines and Emperors*, when less scientific information was available to show the influence of climate on the fate of the Maya. Thus, documenting the increasingly intensive agricultural practices of the Maya, including hillside terraces and the construction of raised fields in swampy areas, Ponting recognizes deforestation, soil deterioration and declining crop yields, population pressures pushing fields and terraces onto more marginal soils, and increasing siltation of rivers as combining to create an ecologically-based agricultural crisis which led to declining food production.[36] He then documents evidence for malnutrition in the period before 800 A.D., "when the skeletons from burials of the period show higher infant and female mortality and increasing levels of deficiency diseases brought about by falling nutritional standards."

Major social consequences resulting from the decline in food supplies would have included increased warfare between cities over declining resources, very high death rates, and a catastrophic fall in population, "making it impossible to sustain the elaborate superstructure the Maya had built upon their limited environmental base."[37] Add to Ponting's argument the subsequently discovered factor of climatic shift to severe drought mentioned by Redman and scientifically corroborated by Fagan, and Ponting's ecological argument is made complete. Ecological systems are designed by natural processes to withstand long-term variations in climate. The anthropogenically altered ecosystem of the Maya was not.

Geographer Jared Diamond's *Collapse: How Societies Choose to Fail or Succeed* (2005) includes the most recent account of the rise and fall of Mayan civilization. With the exception of his account of the Maya and speculations regarding the future of Western civilization and the global network of societies today, Diamond does not give a comparative account of the world's great civilizations (Mesopotamian, Egyptian, Chinese, Indian, Greco-roman, Byzantine, Islamic, and Western European). Rather, he focuses for the most part on the causes of collapse of relatively primitive, isolated or island civilizations. As he writes himself, his reason for devoting a chapter to the Maya " is to provide an antidote to our other chapters on past societies, which consist disproportionately of small societies in somewhat fragile and geographically isolated environments, and behind the cutting edge of contemporary technology and culture." [38] Also, Diamond focuses largely upon the relationship between substructure and superstructure (i.e., the relationship between nature as a resource base and the socioeconomic practices of the societies utilizing those resources), with less stress upon the ideologies (worldviews) of individual societies. Nevertheless, his research strongly supports the major thesis of this book, which is that, in general, the world's civilizations have overutilized nature-as-resources, and thereby contributed to the decline and complete or partial collapse of great civilizations, the ultimate cause stemming, to varying degrees, from the attitudes toward nature found within the worldviews of different civilizations. Thus, the ideological component (i.e., the history of ideas and religious belief systems) in this book complements Diamond's study of the more direct and immediate causes of the collapse of societies.

Throughout *Collapse* Diamond makes it clear that human-caused (anthropogenic) ecological degradation predisposes societies to collapse when external changes, especially climate change, cause extended or severe drought, subjecting societies to extreme stress. The history of the Maya exemplifies this paradigm dramatically. Thus, the "mysteries" of the rise, decline, and fall of civilizations, which once baffled philosophers of history such as Spengler and Toynbee, can now be explained scientifically as a result of recent ecological methods like palynology (the study of accumulations of pollen) and dendochronology (tree-ring dating) combined with C-14 and other dating methods. Diamond has summarized this new evidence much as archeologists Redman and Fagan, and historians Hughes and Ponting have. Environmental history and associated disciplines have accounted for the historical rise and fall of civilizations in this manner for the last two decades, but, surprisingly, Diamond gives no credit to these efforts (with the exception of British environmental historian Clive Ponting). This has caused him to "re-invent the wheel" regarding the Maya. Nevertheless, it is important that he has done so because the role of ecology in the history of civilization has been so woefully ignored by the American public, and also because his analysis lends further support to my own thesis that *all* great civilizations have severely damaged their regional ecology, in some cases causing their collapse.

In the case of the Maya, Diamond states that: "They did damage to their environment, especially by deforestation and erosion. Climate changes (droughts) did contribute to the Maya collapse, probably repeatedly. Hostilities among the Maya themselves did play a large role. Finally, political/cultural factors, especially the competition among kings and nobles that led to a chronic emphasis on war and erecting monuments rather than on solving underlying problems, also contributed."[39] Diamond explains the effects of population growth which caused the expansion of Mayan agriculture from the fertile soils of the valleys to less fertile upland soils which were acidic and low in phosphates. These soils on hillslopes quickly lost their productivity and suffered rapid erosion, causing rapid sedimentation of valley bottoms. The pine forests that once protected the hillsides were all eventually cleared for fuel, construction, and for making plaster. Diamond adds " that deforestation may have begun to cause a 'man-made drought' in the valley bottom because forests play a major role in water cycling, such that massive deforestation tends to result in lowered rainfall." [40] Thus, even the drought which brought about the collapse of Maya civilization was very likely at least partially induced by human ecological degradation.

Oswald Spengler's classification of civilizations accommodated the Maya and Aztec cultures into a single civilization which he called the "Mexican" Culture. Although his rigid explanation of cultural development generally limits the possibilities of cultural diffusion, he clearly recognized cultural diffusion between the Maya and the Aztecs in order to merge them into a single cultural entity. Linking the two cultures through their similarities in architecture, agriculture, religion, etc., Spengler perceives the springtime of Mexican civilization in the architecture of Copan, Tikal, Chichen Itza, and other cities of the period from 160-450 A.D. Its equivalent of our own Western Late Gothic and Renaissance he finds in the "full glory of Palenque and Piedras Negras" around 450-600, and the analogue of the Early Modern West around 600-960. His final winter stage he places at 960-1165 (equivalent to 1800-2000 in Western European civilization). Thereafter, in the petrified "Civilization" phase, the Aztecs dominate, with their megalopolitan capital at Tenochtitlan, which "grew enormous and housed a cosmopolitan population speaking every tongue of this world-empire."[41]

Spengler appears to have had no idea of the Maya collapse as understood in ecological and climatological terms. He does recognize a political and military shift of power to the League of Mayapan and "great wars and repeated revolutions" from 960-1165, after which the Aztecs controlled the Mexican world, but these events are cited as if the landscape and climate had remained static throughout the history of Mexican civilization. In this regard, Spengler's account of the ancient Maya serves as a useful index of how far ecological, archeological, anthropological, climatological, and historical research have come since early in the twentieth century. History can never again be portrayed as a grand human drama played on the static stage of Earth. Only when he described Western, Faustian culture in its Winter stage did Spengler begin to grasp this shift in historical perspective.

Greco-Roman Civilization and the Fall of the Roman Empire

Ever since Edward Gibbon's monumental book entitled *The Decline and Fall of the Roman Empire*, published between 1776 and 1787, controversy has surrounded the probable causes of the fall of the Western Roman Empire. The Byzantine civilization which survived a thousand years, as a continuation of the Eastern Roman Empire, seems to fascinate most readers of history far less than "the fall of Rome". Many historians argue that there was no "fall" at all, but instead a gradual transition during the third and fourth centuries A.D. which allowed the Christianized Roman Empire to contract to a more defensible center at Constantinople after the barbarian hordes had overrun the West. Thus, with this perspective in mind, two centuries of modern historians have set forth their explanations of Rome's fall, with varying emphases on the social, economic, political, military, religious, demographic, and other factors which led to the official end of the Roman Empire in 476 A.D. when the last emperor died and Odoacer established a Gothic kingdom in Italy.

In the course of centuries and the publication of hundreds of books and monographs on Rome since Gibbon, the dispute over the causes of Rome's fall has occurred as though history, the story of civilized mankind on planet Earth, occurred on the earth as a platform or stage upon which the human drama is played. The earth itself, with its mountains and valleys, rivers and soils, forests and deserts, generally has been treated as a kind of static backdrop over which armies tread and upon which kingdoms rise and fall, as though the earth itself had no history of its own. There is an obvious reason for this anthropocentric historiography which depicts a dynamic humanity on a static earth. The reason is that, even as recently as the nineteenth century, history was written within the framework of a Western worldview much beholden to Christian theology and to the idea that the earth was created by the deity about 6000 years ago. In contemporary terms, this was about the time of the beginning of civilization, the origin of which we have learned from archeology only since the nineteenth century. Contemporaneously, in the nineteenth century, modern geology and Darwinian evolutionary biology underscored the great antiquity of the earth, thereby undercutting the chronological foundations of anthropocentric historical writing. Nevertheless, further anthropocentrically girded by humanism since the Renaissance, historians to the present day have remained essentially anthropocentric, and uninterested in historical causes of a natural kind. Mankind as a whole today is not far removed chronologically from explaining natural events, especially catastrophes such as earthquakes, floods, and famines, in terms of divine intervention or "acts of God." If this is so, what of more subtle and gradual earth processes such as climate change, deforestation, erosion, siltation, and the degradation of soil fertility? Until the 1970s and 1980s these aspects of the human environment have been ignored by historians ubiquitously, even though historical geographers had begun to uncover the links between human history and earth history at an earlier date. On the other hand, we have seen that Spengler was aware of this kind of inquiry nearly a century ago.

Looking back at classical antiquity, one is immediately struck by the "modern" sounding observations of Plato's *Critias* in the *Dialogues*. Reflecting upon the physical environment of ancient Athens in the fourth century B.C., Plato notes that the soil of the region around Athens was once excellent and its production copious, and that even what is now left is a match for any in the world. However, what was left in the region in his own time is likened to " the skeleton of a body wasted by disease; the rich, soft soil has been carried off and only the bare framework of the district left... and there was abundant timber on the mountains, of which traces may still be seen."[42] Plato adds that the yearly "water from Zeus" which once benefitted the soil as underground water, springs, and rivers, now runs off a barren ground into a deep sea.[43]

Plato does not go on in the *Critias* to explain the causes of the ecological degradation which he has described (poor soil management, deforestation, and overgrazing leading to erosion, flooding, and siltation of stream beds), but he gives us a picture of changes in the earth over time initiated by human industry, and with the implication that the Athenian farmlands, forests, and water resources of former times have become the much degraded resources available to Athenians in his own time. Elsewhere in his writings Plato describes a cyclical pattern in history in which the human environment deteriorates from a Golden Age of relative simplicity to an overcivilized society which ultimately disintegrates into chaos.[44] Taken together, these ideas appear to indicate a sophisticated awareness on Plato's part of the connection between ecological degradation and civilizational decline.

Reflecting upon Plato's observations, political philosopher William Ophuls concludes that Athens, having used up most of its own resources even before Plato's time, was forced to engage in trade, from which imperialism naturally followed. Thus, Ophuls sees Athens as the paradigm for both ecological self-destruction and its extension through the development of colonies. As Ophuls writes:

> Modern scholarship has demonstrated that the rise and fall of all ancient civilizations were deeply affected, if not determined, by such ecological factors as deforestation, salinization, and desertification: the ancient Mesopotamians, Egyptians, Greeks, and Romans, as well as the Mayans of the New World and successive dynasties in India and China, were all deeply implicated in a sometimes fatal process of ecological degradation and destruction. And the destruction did not stop at home but was extended abroad through a process of imperialistic expansion and exploitation that brought Nubian gold to Egypt, the cedars of Lebanon to Carthage, the lions of Africa to Rome, and slaves from everywhere to all of ancient civilization. Rome, in particular, ruthlessly mined captured provinces for their human and ecological wealth.[45]

The brutal conquests, exploitation, and slavery associated with the ancient Mediterranean empires are clearly rooted, according to Ophuls, in the abuse of nature related to the growth of civilizations, which he calls "predatory development." All civilizations, including our own, have practiced predatory development, which leads inexorably to "the four great ills of civilization": (1) overexploitation of nature; (2) organized violence against outsiders; (3) political and/or religious tyranny over insiders; and (4) extreme socioeconomic inequality or slavery.[46] In Ophuls's view, the cycle of civilization, throughout recorded history, has been inextricably linked to the gradual ecological degradation of each region inhabited by a great civilization, and Greece and Rome were among the worst offenders. If Ophuls is correct, as he appears to be from the perspective of this chapter, how can it be that even down to the present day historians of the Greco-Roman world have almost entirely neglected the ecological factor in explaining the fall of the Roman Empire?

I attempted, above, to explain this failing on the part of modern and contemporary classical historians, but I should add that the isolation of historical scholarship from scientific, and especially ecological, knowledge has been compounded in recent decades by the increasing specialization and narrowness evident in all the scholarly disciplines. Fortunately, however, a few innovative scholars have dared to cross a rickety bridge between the sciences and humanities to allow reinterpretations of history in the light of the relatively recent field of ecology. J. Donald Hughes's *Pan's Travail: Environmental Problems of the Ancient Greeks and Romans* (1994) and his earlier *Ecology in Ancient Civilizations* (1975)[47] have accomplished the ground-breaking work of historical reinterpretation leading to innovative cultural insights such as Ophuls' concept of predatory development. Before summarizing Hughes's contribution towards illuminating the causes of Rome's fall from the perspective of ecology, however, we should consider the explanations of Rome's fall offered by traditional (anthropocentric) historians.

Edward Gibbon, who is generally perceived as having largely invented the entire ongoing controversy regarding the causes of the fall of Rome, was, as a man of the Enlightenment, strongly influenced by the *philosophes*,[48] popularizers of modern science and its social and philosophical implications, particularly the Baron de Montesquieu and his classic *The Spirit of the Laws* (1750). Building upon Renaissance and seventeenth century historical scholarship, Montesquieu's *Causes of the Greatness and Decadence of the Romans* (1734) also surely gave impetus to Gibbon to write his own masterpiece. Above all, Montesquieu provided Gibbon with a general awareness of the influence of climate upon human nature and society, but possibly with insight into the human modification of nature as well. Montesquieu typically thought of such human-induced changes as benign, as in the case of terracing and other practices designed to improve and extend agriculture.[49] This influence may be reflected in Gibbon's *Decline and Fall of the Roman Empire*, but it is hard to tell because Gibbon remained focused almost exclusively upon the human causes of Rome's fall.

Toward the end of chapter thirty-six of the *Decline and Fall*, which covers events from the sack of Rome by the Vandals in 455 to the deposition of the final Roman emperor in 476, Gibbon writes: "Since the age of Tiberius, the decay of agriculture had been felt in Italy; and it was a just subject of complaint that the life of the Roman people depended on the accidents of the wind and the waves."[50] Unfortunately, Gibbon does not explain whether the "decay of agriculture" is due to degradation of the land or to other human causes related to " the irretrievable losses of war, famine, and pestilence." He also mentions the depopulation of regions such as Aemelia, Tuscany, and the adjoining provinces, which suggests that he perceived the decay of agriculture in terms of the combination of the factors of war, famine, and disease as impairing the ability of farmers to work the land rather than as a consequence of degradation of the land itself. We shall see that Gibbon was treading remarkably close to the environmental causes of Rome's fall in these speculations.

The end of chapter thirty-eight in the *Decline and Fall* includes a section entitled "General Observations on the Fall of the Roman Empire in the West," the closest thing to a focused explanation of Rome's fall to be found in Gibbon's history. There is no evidence in this discussion that Gibbon believed in any kind of environmental basis for the fall of Rome. However, in addition to pointing out the growing disparity between the imperial military power and the power of government, the ignorance of the Romans regarding the numbers of their enemies beyond their frontiers, and the general enfeeblement of the urban population, Gibbon dared to place a major proportion of blame for the fall on Christianity. He wrote:

> The clergy successfully preached the doctrines of patience and pusillanimity; the active virtues of society were discouraged; and the last remains of the military spirit were buried in the cloister; a large portion of public and private wealth was consecrated to the specious demands of charity and devotion; and the soldier's pay was lavished on the useless multitudes of both sexes who could only plead the merits of abstinence and chastity.[51]

Gibbon goes on to castigate the theological discord, religious faction, the distraction of the attention of emperors from military frontier to church synod, and persecution of non-believers resulting from the conversion of Constantine in 313 A.D. Nevertheless, despite his obvious disgust with the debilitating effects of Christianity upon the late Roman Empire, Gibbon concedes that the new state religion "broke the violence of the fall, and mollified the ferocious temper of the conquerors."[52]

Three quarters of a century passed between the completion of Gibbon's *Decline and Fall* in 1787 and the appearance of a book in 1864, *Man and Nature: Or, Physical Geography as Modified by Human Action.* The author, George Perkins Marsh, was born in Woodstock, Vermont in 1801, his father having been a successful lawyer and politician. Like his father, Marsh became a lawyer and politician, and eventually became U.S. Minister to Turkey in 1848-54 and minister to Italy from 1861 until his death in 1882. Marsh was a man of many practical as well as scholarly talents, but he is remembered today chiefly by ecologists and environmentalists as a forerunner of awareness of human impact upon the natural environment. As the editor of a Harvard centennial edition of *Man and Nature* writes concerning the cutting of the forest and the tilling of the soil with axe and hoe, "before Marsh no one had assessed the cumulative effect of all axes and hoes. For him the conclusion was inescapable. Man depends upon soil, water, plants, and animals. But in obtaining them he unwittingly destroys the supporting fabric of nature."[53]

In sharp contrast to Edward Gibbon's humanistic assessment of the causes of the decline and fall of the Roman Empire, Marsh begins *Man and Nature* with a chapter that attributes the fall of Rome to a combination of environmentally destructive human practices and " the brutal and exhausting despotism which Rome herself exercised over her conquered kingdoms, and even over her Italian territory...."[54] Although Marsh explicitly acknowledges the political, economic and military factors contributing to decline, he is unequivocal about the ecological consequences of Roman imperialism:

Vast forests have disappeared from mountain spurs and ridges; the vegetable earth accumulated beneath the trees by the decay of leaves and fallen trunks, the soil of the alpine pastures which skirted and indented the woods, and the mould of the upland fields, are washed away; meadows, once fertilized by irrigation, are waste and unproductive, because the cisterns and reservoirs that supplied the ancient canals are broken, or the springs that fed them dried up; rivers famous in history and song have shrunk to humble brooklets.[55]

Implied by this description are the processes of deforestation, erosion, flooding, and siltation, the environmental effects of which Marsh describes as the sedimentation and destruction of harbors, the blocking of stream channels by sandbars where they enter the sea, and the accumulation of sediment which has " converted thousands of leagues of shallow sea into unproductive and miasmatic morasses."[56] Aware of the testimony of history and the extensive ruins of ancient cities which he observed in the eastern Mediterranean region, Marsh concludes that much of the Roman Empire which was most favored in its original soil and climate " is now completely exhausted of its fertility, or so diminished in productiveness, as, with the exception of a few favored oases that have escaped the general ruin, to be no longer capable of affording sustenance to civilized man."[57]

During the interval of 1787-1864 which separates the completion of Gibbon's *Decline and Fall* from *Man and Nature* a quiet but momentous revolution in the sciences had occurred. First of all, beginning with the formulation of the basic principles of geology, especially uniformitarianism (slow, gradual changes in the earth over immense periods of time), by the Scotsman James Hutton in the 1780s, and culminating in the first systematic integration of the work of Hutton and continental European geologists in Charles Lyell's *Principles of Geology* (1830-33), the processes and chronology of the earth were understood without reference to the supernatural. Building upon these achievements, modern evolutionary biology followed Darwin's naturalistic explanation of the fossil record in *The Origin of Species* (1859), and modern archeology applied the stratigraphic methodology of geology to human culture. Thus, the great disparity between the perspectives on Rome's fall held by Gibbon and Marsh is the product of scientific enlightenment and its application to history. Unfortunately, Marsh's integrated scientific-historical approach to history was to be the exception rather than the rule for historians down to the present day, with the exception of environmental historians who developed their own sub-discipline outside the mainstream of academic history.

The Greeks and the Romans are often treated by historians as distinct, separate civilizations. On the other hand, they are also considered together as Greco-Roman or Classical civilization because of their interaction over a long period of time. One of the greatest modern historians of antiquity, Michael Grant, writes that "Greece and Rome were inextricably intermingled .... Theirs was a single world, in which the Romans took up the inheritance of the Greeks, and adapted it to their own language and national traditions."[58]   Grant believes that it is essential, if Classical civilization is to be given its full, significant meaning, to consider Greece and Rome in conjunction. An important related point made by Grant is that the conquest of the Hellenistic Greek world by Rome did not mean the end of Greek society, but rather the maintenance and perpetuation of Greek culture within the Roman Empire.[59]   Much of the greatness of Rome was owed to the originality of Greek philosophy, art, and architecture. Unfortunately, Roman treatment of nature was also foreshadowed by ecologically destructive practices in Greece.

Grant's explanation of Rome's fall is given entirely in human terms, chiefly involving internal political, economic, and social disintegration on the one hand, and external pressures from barbarians on the other. The innumerable civil wars of the late Empire undermined the imperial succession, enforcement of conscription into the army gradually weakened, German troops and officers came to dominate the army, and civilian populations resented and resisted the gigantic taxes needed to pay for the soldiers. Simultaneously, more and more people essentially renounced their citizenship either by joining gangs of marauding bandits, seeking the protection of powerful landlords, or joining the ranks of increasing numbers of monks who payed homage to the City of God rather than the crumbling City of Man. The extent to which the Christian worldview undercut the authority of the Roman state must have been considerable, much as Edward Gibbon had surmised more than two hundred years earlier.[60]

Despite the rise of environmental history as a renegade interdisciplinary field in the 1970s, its influence upon traditional anthropocentric historical writing has been negligible. Here and there, lip service is occasionally paid to the ecological awareness that grew out of the environmental movement, as when C. Warren Hollister wrote in 1982, regarding Rome's decline and fall and the demise of Greco-Roman civilization: "Many reasons have been alleged: climatic changes, diseases, bad ecological habits, sexual orgies, slavery, Christianity--even lead poisoning. None of them makes much sense."[61]   Although Hollister rejects environmental causes of decline without explaining why, and selects as significant causes " the failure of the Roman economy to change or expand, and the parasitical character of the western cities" along with the barbarization of the Roman armies, nevertheless he does recognize the possibility that ecological factors might have contributed to the fall of Rome. A decade later, however, in a subsequent edition, ecology is still ignored as a significant factor contributing to the collapse of Greco-Roman civilization.

There is some irony in the fact that modern historians of classical antiquity throughout the twentieth century were unable to recognize or acknowledge the importance of ecology in the Roman world when the Romans themselves appear to have been well aware of environmental degradation, just as Plato had been. Historian David J. Herlihy surveyed the ecologically pessimistic perceptions of the ancient Romans in an essay on historical ecology published in 1980.[62] Selecting passages from Tertullian, St. Cyprian, St. Augustine, and the Epicurean poet Lucretius, Herlihy makes a strong case that the intelligentsia of Imperial Rome was distinctly aware of soil degradation, deforestation, loss of species, and the decline of agriculture in their time.

The early Christian father, Tertullian (ca. 155-220 A.D.), describes a crowded Roman Empire around 200 A.D. as follows:

> Everything has been visited, everything known, everything exploited. Now pleasant estates obliterate the famous wilderness areas of the past. Plowed fields have suppressed the forests; domesticated animals have dispersed wildlife.... Everywhere there are buildings, everywhere people, everywhere communities.... We weigh upon the world; its resources hardly suffice to support us. As our needs grow larger, so do our protests, that already nature does not sustain us. In truth, plague, famine, wars and earthquakes must be regarded as a blessing to civilization, since they prune away the luxuriant growth of the human race.[63]

Even two centuries earlier, the pagan Lucretius, writing before Christianity in his poetic essay *On the Nature of Things* (based in part upon the atomistic philosophy of Democritas and Leucippus) could write of once fertile fields, vineyards and pastures:

> which now can scarce grow anything, for all our toil. And we exhaust our oxen and our farmer's strength,we wear out plowshares in the fields that barely feed us ... . And now the aged plowman shakes his head and sighs again and again, to see his labors come to naught, and he compares the present age to days gone by ... when ... life was easily supported on smaller farms .... The gloomy grower of the old and withered vines sadly curses the times he lives in, and wearies heaven, not realizing that all is gradually decaying, nearing the end, worn out by the long span of years.[64]

The theme of the advancing senility of nature and the world advanced by Lucretius was also expounded by the African Roman Christians St. Cyprian (ca. 200-258 A.D.) and St. Augustine (354-430 A.D.) to reinforce their religious conviction that the earthly world was nearing its end as the Apocalypse and Second Coming of Christ approached. Thus, their perceptions of environmental decline may be somewhat distorted by their millennial expectations. St. Cyprian observed that the climate of North Africa was becoming less favorable for agricultural production, and that the mines and quarries were almost completely exhausted. By the early fifth century St. Augustine repeated the themes of

Tertullian and Lucretius that the world is crowded and overpopulated, that nature has grown old and weary, and that the end of human history is not far off. Augustine also argued vehemently for population control through abstinence (which he had not practiced earlier in his own life). If this led to the extinction of the human race, he argued, it must be because the number of souls predestined for sainthood in the eternal City of God was close to being reached.

Whereas some of those who lived during the centuries of the Roman Empire have left a record of contemporary awareness of environmental deterioration, modern historians such as J. Donald Hughes have relied upon more precise scientific and historical facts in addition to primary sources as a basis for making judgments concerning the role of ecology in the decline of Greco-Roman civilization. Dating back to George Perkins Marsh in the mid-nineteenth century, several astute observers of the Mediterranean landscape have speculated that ancient civilizations declined due to such problems as deforestation, erosion, and depleted soils.[65] In *Pan's Travail* Hughes has collated our knowledge of these aspects of the ancient world of the Mediterranean basin, and has drawn the conclusion that:

> .... environmental factors were significant causes of the decay of Greco-roman economy and society, though not the only causes, and that the most important of these factors were anthropogenic. The ancients were unable to adapt their economies to the environment in harmonious ways, placed too great a demand on the available natural resources, and depleted them. Thus they failed to maintain a balance with nature that is necessary to the prosperity of any human community.[66]

Devoting the final chapter of *Pan's Travail* to the decline of Greco-Roman civilization, Hughes separates the environmental reasons for the decline of classical civilization into those with natural causes and those caused by human activities, i.e., non-anthropogenic and anthropogenic causes. Surveying the non-anthropogenic factors of climate change and epidemic diseases, Hughes concludes that climatic change was not a major contributing cause in the decline of Classical civilization, but that epidemics were significant. He notes the increase of plague in the Mediterranean region following the conquests of Alexander and expanding Roman trade. Of particular importance, he observes that plagues increased in severity after the reign of Augustus, when Roman merchants traveled regularly to India and even reached China, thereby adding to native Mediterranean pathogens. The worst plague, introduced by troops who had served on the Euphrates River, killed at least a third of the population of Rome in 164 A.D.[67] Population losses due to plague and other introduced diseases were further compounded by the gradual spread of malaria within the Empire. Probably entering Greece in the fourth century B.C. and Italy in the second century B.C.,[68] malaria not only caused population losses, but also left its survivors in a chronically weakened condition, thereby diminishing worker productivity.[69]

While Hughes classifies malaria as a non-anthropogenic factor, he qualifies this by reminding us that the spread of malaria " is facilitated by human interference in the landscape through actions such as deforestation, with the erosion and deposition that follows."[70]   The ancients did not understand that malaria was transmitted by mosquitoes, although they were aware that the disease was associated with swamps and marshes. As we shall see, such wetlands, both fresh and salt water, began to spread during the period from the late Roman Republic to the late Empire. By the fourth century A.D., malaria had spread to the Pomptine Marshes near Rome.[71]

Recent scientific evidence in the form of DNA testing of bones has provided proof of a major epidemic of the most virulent form of malaria in Rome around 450 A.D., adding to historical accounts of malaria epidemics based upon descriptions of "sweats and chills," typical malarial symptoms. In addition to decimating the population of the Late Empire, the *Plasmodium falciparum* species of malaria may have caused Attila the Hun to turn back short of sacking Rome out of fear of the disease.[72]   Given this proof of Hughes's speculations regarding the debilitating as well as the fatal effects of malaria, if malaria arrived at the time of the Late Republic when, as Marsh points out, swamps were proliferating along the shoreline as well as along river valleys, the Roman labor force, whether free or enslaved, may well have been diminished in it efficiency for centuries, and this would apply to the military as well. Although Hughes classifies epidemics as a non-anthropogenic factor contributing to Rome's decline and fall, malaria, the most persistent threat over long periods of time, would not have been so widespread without the massive ecological degradation over the centuries described by Hughes.

First and foremost, deforestation and attendant erosion and siltation were the most devastating ecological impacts inflicted by Greco-Roman civilization upon the Mediterranean landscape. The vulnerability of exposed soil to erosion was further compounded by extensive overgrazing. Soil and rock material eroded from bare or poorly vegetated land was subsequently deposited in stream valleys, in lakes, and along shorelines to form malarial swamps.[73]   Hughes's research essentially corroborates the observations made by Plato 2,400 years ago and by Marsh in the nineteenth century. Closely related to deforestation, overgrazing, erosion, flooding, and siltation, was degradation of the remaining soil by salinization, leaching, and exhaustion of nutrients. Quoting the Russian historian Rostovtzeff that "exhaustion of the soil" happened because "men failed to support nature ,"[74] Hughes cites the political, economic, and military pressures which prevented farmers from applying preventative measures that they were well aware of (much as our contemporary profit-driven economy destroys our own soil by supporting pesticide-fertilizer-machine intensive agribusiness agriculture).

Together, excessive taxation, the collapse of small farmers, overintensive use without adequate fallow periods, the replacement of farming by grazing (subject to cycles of epidemic ungulate diseases), and the shift to monocultures on large farms or latifundia worked by slaves, all contributed to a downward spiral of soil productivity. Declining population exacerbated the problem, since fewer farm workers diminished both conservation practices and production, while their efficiency was also diminished by the debilitating effects of malaria.

Other environmental problems in the Roman Empire documented by Hughes include: species extinctions and simplification of ecosystems; impacts from extensive mining and quarrying and demands for wood for smelting of ores and firing of ceramics; pollution from lead associated with silver ore, along with arsenic and mercury which poisoned air, water, and food; urban pollution which attracted vermin and spread disease; and steady immigration from rural areas to replace members of the urban population who died from pollution and epidemics, thereby weakening agricultural productivity.[75]

Recent geologic studies combined with archeological evidence show definitively that periods of massive sedimentation during the last five or six thousand years are caused chiefly by human changes in land use rather than climatic change. Prior to studies done in the last two decades of the twentieth century, it was thought that both the Older Fill of alluvial sediment (late Pleistocene or ca. 50,000-10,000 B.P.) and the Younger Fill (ancient to early modern times or ca. 5,000-400 B.P.) were the result of pluvials or wet periods. Subsequently, studies in Greece, Italy, and Spain have shown that the Younger Fills do not correlate over large areas, and that they are the product of anthropogenic soil disturbance followed by episodes of erosion and sedimentation over limited areas.[76]

In Greece, for example, following an earlier episode related to population decrease around 2000 B.C., there are two episodes related to Greco-Roman civilization, one in the Hellenistic era (the three centuries prior to Christ) and the other in late Roman times or the early Dark Ages (ca. seventh century A.D.). Both of these events can be interpreted historically " as periods of settlement contraction following eras of high population and political controls."[77] The episodes of rapid erosion are thought to represent agrarian breakdown and the retreat of farmers to their best soils, leaving the poorer fields to pasturage. Gully erosion of these unmaintained lands then deposited sediment in the valley bottoms to form the Younger Fills. Studies of several regions of Greece also show that erosional episodes do not result from initial land clearance for agriculture, but rather that they occur " only after substantial periods of farming." It appears that people farmed their lands for long periods of time prior to deposition of these nonsynchronous fills, preventing erosion through the use of labor intensive soil practices such as terracing, especially on steeper slopes. Redman adds that "after long periods of dense occupation, upland regions appear to suffer serious erosional episodes, that lead to their depopulation and the aggregation of farmers in the still favorable settings. This in turn puts excessive pressure on these locations."[78]

Related studies in Adriatic Italy and Spain confirm the correlation between intensive human use and increased sedimentation rates in the valleys. In Italy the earliest episodes of major erosion are directly related to expanding rural settlement and agriculture during pre-Roman and Roman times, from about 500 B.C. to 500 A.D., approximately the duration of Greco-Roman civilization. Redman concludes that these "oscillations in settlement and productivity revealed that after each period of intense human activity, there was a commensurate 'environmental crisis.'   The subsequent reoccupation of the area required a change in exploitation techniques."[79] New techniques generally required investments of capitol and labor as well as greater dependency on socioeconomic systems removed from the region.

The demand for wood during the period of Greco-Roman civilization was almost insatiable. In addition to destruction of forests by clearing for agriculture and grazing, often by extensive burning, the greatest use of wood was for fuel. Hughes estimates that perhaps 90 percent of the demand for wood was used for fuel in metal refining, pottery and glass manufacturing, heating, and cooking. In addition, large quantities of wood were used in mining, shipbuilding, and war machinery.[80]

The process of deforestation in ancient times was not very well understood until the science of palynology arose in the second half of the twentieth century. Palynology involves the scientific study of grains of pollen found in deposits of soil and mud. Deposits found in lake sediments, marshes, or caves can tell us the relative abundance of trees and other plants, and also can be dated by radiocarbon. Applying the findings of paleobotanists, Hughes explains that forests in northern Greece, for example, tended to flourish more in times of stability and settlement than in times of warfare and invasion, when peasants would move up into the mountains to clear and plant crops. Under peacetime conditions peasants returned to the richer soils of lowland areas, allowing mountain forests to return. [81]

John Perlin, in *A Forest Journey: The Role of Wood in the Development of Civilization* (1989), makes the argument that deforestation in Spain, where the Roman imperial silver mines were located, indirectly contributed to the decline of the Empire through destruction of the timber needed to smelt silver. Under the emperor Vespasian (69-79 A.D.) wood used to heat bathhouses in the Spanish silver district was limited in favor of the demands of the mines "to smelt the amount of silver the Roman government needed to continue financing its growth." Following the extravagant demands of Caligula and Nero upon the imperial treasury, even conservation laws could not prevent wood shortages, and by the end of the second century A.D. silver production declined not for lack of ore, but for lack of timber for fuel to process the ore. The shortage of silver led to debasement of the coinage, beginning with the emperor Commodus (177-192 A.D.) and reducing 98 percent of its silver content to base metal by the end of the third century A.D. As a consequence, trading in commodities and services became institutionalized early in the fourth century A.D. Compounding this problem was the need to purchase grain from Africa and to support the military. Perlin documents soil depletion in Italy during the late Empire as the basis for dependency upon North African wheat. He attributes the failure of Rome to at least partly feed itself to a combination of "human causes, deforestation and the inability to take remedial action [to prevent soil depletion] as suggested by Columella."[82]

Adding to the accumulating scientific and archeological evidence supporting Hughes' analysis of environmental degradation during Classical civilization, Stephen J. Pyne's study of the environmental history of fire, *Vestal Fire* (1997), demonstrates that fire-generated environmental impacts, largely the result of burning by pastoralists and farmers, have transformed the original natural forest ecosystems of the Mediterranean into " an utterly anthropogenic landscape. Its fire-sculpted biota is an artifact of human-wrought disturbances. ... as fully a relic as the ruins at Leptis Magna...."[83] This anthropogenic vegetation is a result of 10,000 years of human disturbance in the eastern Mediterranean lands and 4000 years in the West. Relentless burning over centuries and millennia transformed the post-Pleistocene Mediterranean forests into pyrophytes, fire-loving and fire-resistant plants that comprise the typical maquis (large shrubs), garrigue (small shrubs), and steppe (grasslands) which have dominated much of the Mediterranean basin for millennia. However, Pyne does not seem to acknowledge how extensive forests were in Italy during the Roman Republic, or in Spain, Africa and other parts of the growing Roman Empire, or that much of the transformation of forest to pyrophytes occurred during the period of Classical civilization rather than before, which Hughes demonstrates by historical documentation. He does recognize the environmental degradation and the fact that relentless manipulation by humans has " worn down the biota and locally wasted away, along with its soils, its capacity for incidental recovery."[84] Thus, we are impressed by Pyne's account of the pervasiveness of herders and fire working together to modify vegetative patterns and change rates of erosion and deposition, but little light is shed on how much of this transformation took place during the thousand years of Greco-Roman civilization as opposed to an earlier period. On the other hand, there is nothing in *Vestal Fire* which refutes the weight of documentary evidence indicating that massive ecological degradation was a hallmark of Classical civilization.

British geographer I. G. Simmons writes that the agroecosystems of Ancient Greece caused excessive damage to the environment in the course of producing its meat, leather, wool, grain, fruit and wine because of overgrazing, double cropping and other unsustainable practices. He argues that the Romans did essentially the same thing. "The pre-existing cover of Italy was forest which largely disappeared during the flourishing of the Roman Empire. This loss of woodland was the most widespread effect of Roman occupation of the land, proceeding more slowly in the north and west." [85] Following some limited earlier forest destruction by pastoralists, as suggested by Pyne, Simmons supports Hughes by remarking on the great demand for wood in Classical civilization for fuel, shipbuilding and construction in addition to the effects of clearing for agriculture and grazing, with the same consequences for the Romans as for the Greeks:

> ... bare hillsides, rivers choked with silt, floods, and coastal marshlands infested with malarial mosquitoes. This syndrome was not of course confined to Italy: the Empire spread its practices along with its politics, and deposits 10 metres thick of the products of the erosion which followed deforestation occupy some valleys in the former Roman provinces of Syria and north Africa.[86]

Simmons also observes that because the Romans were heavier meat-eaters than the Greeks, the hillsides of Italy and the Empire would have been severely overgrazed, with attendant burning of forest and scrub by shepherds further increasing rates of erosion.

The evidence presented in the foregoing pages indicates a general consensus of opinion, based upon recent factual geological, historical, and palynological knowledge, which strongly favors the interpretation that environmental degradation played an important role in the decline and eventual collapse of the Roman Empire. Adding weight to this interpretation is the recent body of knowledge concerning the ecological factors involved in the rise and fall of other great civilizations, such as the Mesopotamian and the Mayan.

Empirical geologic studies of normal as opposed to accelerated erosion rates compare quantities of sediment derived from cultivated or pasture land with sediment from forested or reforested areas. These studies, conducted by the U.S. Department of Agriculture, show that erosion on cultivated lands is hundreds of times greater than on forest land, and that erosion rates on pasture lands are about sixteen to several tens of times greater.[87]   These studies fully corroborate the evidence of massive erosion of cultivated and grazing lands during the course of Greco-Roman civilization.   Deposition of the rapidly eroded sediment in flood plains, lakes, and along shallow coastal areas created a man-made environment of swamps and marshes which provided ideal breeding habitat for malaria-infected mosquitoes. The arrival of malaria in Italy around the second century B.C. coincided with the creation of new swamp habitat by rapid erosion and deposition which worsened in the following centuries. For example, Ostia, the military post and early port of Rome constructed where the Tiber River entered the Mediterranean, was built in the fourth century B.C. By the second century A.D., Ostia was sufficiently isolated from the sea by deposition of Tiber River sediments that it was supplanted by a new harbor, the Portus Romae.[88] Today, Ostia lies several miles inland from the sea, attesting to further rapid erosion in the Tiber drainage basin down to the present day.

Taken together, the degradation of farmlands and grazing lands by soil erosion, the creation of extensive breeding grounds for malarial mosquitoes by rapid deposition of eroded soil and rock, the loss of forest and mineral resources (especially silver), the gradual depopulation of Italian farmlands resulting from impoverished soils and diminished production, the population decline and physical debilitation resulting from widespread malaria and other diseases both in the city of Rome and in the countryside, and population losses in Rome related to urban pollution from lead, arsenic, and mercury all took their toll on the size and vigor of the Italian population at the heart of the Roman Empire. Similar ecological problems in Africa and the other Roman provinces compounded this weakness at the center by diminishing colonial supplies of wheat, wood, and the metal (especially the silver needed for coinage) essential to maintenance of a secure and healthy population in the capitol city. Gradual displacement of Italians and colonials by barbarians in the military forces was promulgated by deeper, long-term environmental causes which have been overlooked by the majority of historians of Greco-Roman civilization. Furthermore, despite the consensus of modern historians, the Romans themselves were clearly aware of the environmental degradation and resource shortages which ultimately led to the decline and fall of the Roman Empire and the transformation of the Classical Greco-Roman civilization in the West into a residual society dominated by political and military chaos, Christianity, and monasticism.

　　Environmental degradation as a consequence of agricultural and urban development associated with the shift from hunting and gathering to civilization and larger populations is inevitable and ubiquitous. I have described such ecological impacts in Mesopotamia, Egypt, Maya civilization, and ancient Greece and Rome, with the qualification that, in general, ecological degradation tends to be more severe in arid and tropical regions than in temperate zones such as western Europe and much of North America. On the other side of the globe, where do ancient and modern China and Japan stand in comparison to occidental regions?

　　During the early stages of the Environmental Movement, in the 1960s and 1970s, environmental philosophers and historians typically chided Western attitudes toward nature in favor of Chinese ethical thought rooted in Buddhist and Taoist traditions which see humans as living harmoniously within nature, as a part of it rather than separate from it and dominating it. Unfortunately, the ancient tradition of forest care in China grew up as an afterthought *in response* to damage inflicted upon the environment by humans. Ultimately, as in the case of the ancient Greeks and Romans, pragmatic considerations in Asian civilizations overrode the ancient aesthetic and religious ideals held up by modern environmentalists as guidelines for contemporary behavior. Animistic beliefs and Taoist nature philosophy might have produced some form of ecotopian community if China had not been a civilization with a growing population, a vast bureaucracy, and the militarism required for imperialistic expansion. Along with these ills characteristic of *all* civilizations came severe overexploitation of the Chinese environment. Generally the demands of growing cities, industries for trade as well as regional needs, and the elaborate temples of the religious elite all contributed to ecological degradation through deforestation, erosion, siltation, flooding, local climatic change, and habitat loss. By 1000 A.D. clearing of vast areas in northern China had caused severe ecological impacts. "The archeological evidence is only now being collected, but it is likely that we will find substantial evidence of deforestation and soil erosion in the northern provinces as early as the Anyang dynasty some 3500 years ago."[89]

There is a growing consensus among archeologists, sociologists, and historians that, like the civilizations of Mesopotamia, ancient Greece and Rome, Mesoamerica, and the Indus Valley, that "rapid deforestation in early China for agricultural production, irrigation, and the production of metallic implements and commodities generated the conditions of ecological crisis."[90]   In southern China population increases from the seventh to the eleventh century A.D. were propelled by breakthroughs in agricultural technology contemporary with similar breakthroughs in Western Europe. These included the plow in China and moldboard plow in Western Europe, new seed varieties, and double cropping, all of which led to further deforestation. Deforestation in China falls into three main phases: (1) cumulative deforestation from 2700 B.C. to 1100 B.C.; (2) a period of active conservation from 1100 B.C. to 250 B.C.; and (3) from 250 B.C. to the twentieth century A.D., severe deforestation due to agricultural development, warfare, and industrial transformation.[91]

Increased deforestation over more than two millennia, punctuated by attempts to conserve the forest as a resource, resulted in a disastrous increase in the frequency of floods over the long term, from few to none at the beginning of the first century A.D. to 360 each year by 1900. Nevertheless, Chinese civilization has persisted to the present day, displaying a resilience comparable to that of ancient Egypt.[92] This raises the question: why do some civilizations survive massive environmental degradation while others collapse or decline more gradually into non-existence? Spengler answered that those which survived long beyond their creative life cycles of about a millennium persisted in a rigid, static condition of cultural petrification which he termed "Civilization" in contrast to an actively creative and self-transforming civilized culture. More recently, the duration of great cultures or civilizations in the traditional sense has been explained as a function of the long-term exhaustion of ecological resources such as forests, soils, and stable ecosystems, including hydrologic systems. Donald J. Hughes' example of ancient Greece and Rome lends this hypothesis strong support, and more recently, Jared Diamond, Sing C. Chew, Charles L. Redman, Brian Fagan, and others have further substantiated the validity of this approach to the history of the world's great civilizations in general. A number of the world's great civilizations have declined or collapsed in large part due to the destruction and overutilization of ecological resources, while others have adjusted to resource scarcity through trade and technological innovation. It is no longer possible for world historians to ignore or rule out ecology as a major factor in world history.

Sing C. Chew's *The Recurring Dark Ages: Ecological Stress, Climate Changes, and System Transformation* (2007) goes beyond the fundamental generalization that all civilizations severely degrade nature in the long run to propose that environmental degradation and climate change in Eurasia has led to recurring "Dark Ages." Extending across the old world from China to Spain, Dark Ages occurred from 2200 B.C to 1700 B.C. (from India to Greece), from 1200 B.C. to 700 B.C. (from India to the eastern Mediterranean, central Europe, and Scandinavia) and A.D. 300-400 to A.D. 900 (from Japan and China to western Europe and North Africa). In each instance Chew perceives the Eurasian Dark Age as a "system crisis involving such trends as economic slowdown, the disruption of trade, political unrest, breakdown of social hierarchies, de-urbanization, increased migration, and losses in population.[93]

Chew explains these Dark Ages as periods of economic and sociopolitical devolution in individual civilizations brought on by the contingencies of ecological degradation and climate change. During these periods, civilized cultures are forced to transform themselves or risk extinction. Dark Ages are also times of ecological recovery as agriculture and other environmentally destructive land-use practices diminish.

Chew's theoretical view of historical change might be termed a "McNeillian" view of history in that it emphasizes William McNeill's explanation of world history in *The Rise of the West* and later works. McNeill sought to modify the idea of distinct, individual civilizations described by Oswald Spengler, Arnold Toynbee, Pitirim Sorokin and Alfred Kroeber during the first half of the 20th century with a diffusionist approach which showed that these distinct civilizations were surprisingly interconnected, especially in Eurasia, by such factors as trade, technology exchange, conquest, and the spread of disease. As a consequence, Chew has shifted the world historical focus from the long-term fate of individual civilizations and their eventual devolution or demise to a view of converging Eurasian problems related to long-term ("la longue duree" of Fernand Braudel) ecological degradation and climate change. In any case, whether from the Spenglarian perspective of individual civilizations, or from Chew's Neo-McNeillian diffusionist point of view, the gradual destruction of a region's soils, watersheds, and ecology ultimately leads to civilizational crisis. Chew's Dark Ages may or may not be acceptable to world historians in the long run, but his contribution to environmental history from a sociological-archeological perspective is invaluable. The importance of the ecological factor in the history of the world's civilizations can no longer be questioned.

NOTES

[1]. H.W.F. Saggs, *Civilization before Greece and Rome* (New Haven: Yale University Press, 1989), p.21.

[2]. Approximately 12,000 years ago a massive influx of glacial meltwater from North America flowed into the Arctic Ocean, shutting down the downwelling of saltwater into the deep ocean and the conveyor belt of warm water (the Gulf Stream) responsible for natural global warming in northerly latitudes. See Brian Fagan, *The Little Ice Age* (New York: Basic Books, 2000), pp.46, 214-215.

[3]. Brian Fagan, *Floods, Famines, and Emperors* (New York: Basic Books, 1999), p.79.

[4]. Ibid., p.84.

[5]. Ibid., p.90.

[6]. H.W.F. Saggs, *Civilization before Greece and Rome*, p.21.

[7]. Ibid., p.31.

[8]. Clive Ponting, *A Green History of the World* (New York: Penguin Books, 1991), p.68.

[9]. Ibid., p.69.

[10]. Ibid ., p.72.

[11]. Ibid.

[12]. J. Donald Hughes, *Pans Travail: Environmental Problems in Ancient Greece and Rome* (Baltimore: Johns Hopkins University Press, 1994), p.37.

[13]. Brian Fagan, *Floods, Famines, and Emperors*, p.105.

[14]. Oswald Spengler, *The Decline of the West* (New York: Alfred A. Knopf, Publisher, Vol. I, 1926, Vol. II, 1928), Vol. II, p.79.

[15]. Brian Fagan, *Floods, Famines, and Emperors*, p. 110.

[16]. Oswald Spengler, *The Decline of the West*, Vol. I, p.295. Spengler also cites the eclecticism of the art and architecture of the post-Middle Kingdom, the copying and parodying of Old Kingdom art, and even the practice of Rameses the Great of cutting out the names of his predecessors on buildings and replacing them with his own as evidence of the creative bankruptcy of the post-Cultural period: "Civilization" as sterility. Vol. I, p.294.

[17]. C. Warren Hollister, *Roots of the Western Tradition: A Short History of the Ancient World* (New York: McGraw-Hill Inc., 1982), pp.15-16, 26-30.

[18]. J. Donald Hughes, *Pan's Travail*, p.35.

[19]. Clive Ponting, *A Green History of the World*, p.83.

[20]. Ibid., p.84.

[21]. Ibid., p.85.

[22]. J. Donald Hughes, *Pan's Travail*, p.36.

[23]. Ibid., p.42.

[24]. H.W.F. Saggs, *Civilization before Greece and Rome*, p.22.

[25]. Stierlen, Henri, *Living Architecture: Mayan* (New York: Grosset and Dunlap, 1964), p.47.

[26]. Charles L. Redman, *Human Impact on Ancient Environments* (Tucson: The University of Arizona Press, 1999), pp.139-140.

[27]. Ibid., p.198.

[28]. Ibid., p.142.

[29]. Ibid., p.140; Clive Ponting, p.81.

[30]. Ibid., p.143-145.

[31]. Ibid., p.145.

[32]. Brian Fagan, *Floods, Famines, and Emperors*, pp.143-144.

[33]. Ibid., p.145.

[34]. Ibid., p.151.

[35]. Ibid.,pp.155-157.

[36]. Clive Ponting, *A Green History of the World*, pp.81-83.

[37]. Ibid., p.83.

[38]. Jared Diamond, *Collapse: How Societies Choose to Fail or Succeed* (New York: Viking, 2005), p.159.

[39]. Ibid., pp.159-160.

[40]. Ibid., pp.169-170.

[41]. Oswald Spengler, *The Decline of the West*, Vol. II, p.45.

[42]. Plato, *Dialogues*, Edith Hamilton and Huntington Cairns, eds., translated by A.E. Taylor (Princeton, New Jersey: Princeton University Press for Bollingen Foundation, 1961), p.1216.

[43]. Ibid., pp.1216-1217.

[44]. J.B. Bury, *The Idea of Progress* (New York: Dover Publications, Inc., 1932, 1955), pp.9-10.

[45]. William Ophuls, *Requiem for Modern Politics: The Tragedy of the Enlightenment and the Challenge of the New Millennium* (Boulder, Colorado: Westview Press, 1997), p.96.

[46]. Ibid., pp.94-97.

[47]. J. Donald Hughes, *Pan's Travail* (1994) and *Ecology in Ancient Civilizations* (Albuquerque: University of New Mexico Press, 1975).

[48]. Clarence J. Glacken, *Traces on the Rhodian Shore* (Berkeley: University of California Press, 1967), p.577.

[49]. Edward Gibbon, *The Decline and Fall of the Roman Empire* (Chicago: Encyclopedia Brittanica, Inc., 1952), Vol. I, p.592.

[50]. Ibid., p.631.

[51]. Ibid., p.632.

[52]. George Perkins Marsh, *Man and Nature*, David Lowenthal, ed. (Cambridge, Massachusetts: The Belknap Press of Harvard University Press, 1985), p.xxvii.

[52]. Ibid., p.11.

[54]. Ibid., p.9.

[55]. Ibid.

[56]. Ibid., p.10.

[57]. Michael Grant, *The Founders of the Western World: A History of Greece and Rome* (New York: Charles Scribner's Sons, 1991).

[58]. Ibid., p.137.

[59]. Ibid., pp.215-216.

[60]. C. Warren Hollister, *Roots of the Western Tradition*, p.218 in 1982 edition, p.225 in 1991 fourth edition.

[61]. Herlihy, David J., "Attitudes toward the Environment in Medieval Society" in Lester J. Bilsky, ed., *Historical Ecology: Essays on Environment and Social Change* (Port Washington, New York: Kennikat Press (National University Publications), 1980).

[62]. Ibid., p.103.

[63]. Ibid., p.105.

[64]. J. Donald Hughes, *Pan's Travail*, pp.1-3.

[65]. Ibid., p.182.

[66]. Ibid., p.187.

[67]. Ibid., p.85.

[68]. Ibid., pp.188-189.

[69]. Ibid., p.189.

[70]. Ibid., p.188.

[71]. *The Oregonian*, "Bones Yield Signs of Malaria Hastening the Fall of Rome" (Portland, Oregon), Feb. 21, 2001.

[72]. J. Donald Hughes, *Pan's Travail*, p.190.

[73]. Ibid., pp.190-191.

[74]. Ibid., pp.191-192.

[75]. Charles L. Redman, *Human Impact on Ancient Environments*, pp.110-112.

[76]. Ibid., p.114.

[77]. Ibid., p.115.

[78]. Ibid., p.117.

[79]. J. Donald Hughes, *Pan's Travail*, pp.73-77.

[80]. Ibid., p.81.

[81]. Perlin, John, *A Forest Journey: The Role of Wood in the Development of Civilization* (Cambridge, Massachusetts: Harvard University Press. 1989), pp.126-128.

[82]. Pyne, Stephen J., *Vestal Fire: An Environmental History, Told Through Fire, of Europe and Europe's Encounter with the World* (Seattle: University of Washington Press, 1997), p.82.

[83]. Ibid., p.95.

[84]. Simmons, I.G., *Changing the Face of the Earth: Culture, Environment, History* (Oxford: Basil Blackwell Ltd., 1989), p.116.

85. Ibid.

86. Ibid.

87. Arthur N. Strahler and Alan H. Strahler, *Geography and Man's Environment* (New York: John Wiley and Sons, 1977), p.217.

88. Anna Maria Liberati and Fabio Bourbon, *Ancient Rome: History of a Civilization that Ruled the World* (New York: Steward, Tabori and Chang, 1996), p.162.

89. Charles L. Redman, *Human Impact on Ancient Environments*, p.24.

90. Sing C. Chew, *World Ecological Degradation* (Walnut Creek, California: Rowman & Littlefield Publishers, Inc., 2001), p.109.

91. Ibid., p.110.

92. Ibid., Regarding the history of ecological degradation in ancient Japan and Southeast Asia, which followed trajectories similar to that of China, see pp.114-115, and 105-106.

93. Sing C. Chew, *The Recurring Dark Ages: Ecological Stress, Climate Changes, and System Transformation* (Lanham, MD: Altamira Press, A division of Rowman & Littlefield Publishers Inc., 2007), p.10. Regarding the ecological factor in the fall of the western Roman Empire, Chew writes: "The incessant socioeconomic and political forces that underlay this system expansion meant that the natural system continued to be under great assault. The resiliency of Nature to such anthropogenic acts of violence finally gave way about A.D. 300/400 with the onset of another Dark Age. By this period as well, the weather had also started to change," p. 112. A detailed discussion of the evidence for this generalization follows on pp.112-137. Chew's study further corroborates my own conclusion that ecological degradation was an important factor contributing to the demise of the western Roman Empire.

CHAPTER THREE: WORLDVIEWS AND THE CLASSICAL-MEDIEVAL METANOIA

Every human culture which has existed in the course of human history and pre-history has possessed a commonly held set of beliefs about the nature of reality, the structure of the cosmos, and humanity's relationship to the perceived parameters of the surrounding world. These cultural belief systems are usually referred to in our own time as "worldviews." Worldviews are implicit belief systems specific to particular cultures, and are constructed around a core of religious or cryptoreligious (ideological) beliefs. Most people would never think of challenging their own worldview and usually assume that other worldviews are wrong or misguided. Contemporary historians of ideas generally agree that a worldview should explain the broad categories of God, nature, man (humanity), history, and society within an integrated model or paradigm of reality.[1] Historians and anthropologists alike consider worldviews to be relative to time and place, providing a unique "cultural lens" through which the perception of nature and history is distorted to fit the needs of a particular belief system.

The conviction held by the members of a given culture that they possess certain, absolute knowledge has been characterized by classical historian J.B. Bury as "the illusion of finality,"[2] the belief, characteristic of philosophical systems as well as worldviews, that one's own age has arrived at final, absolute truth, as in the venerable philosophical systems of Aristotle, Aquinas, and Descartes. The illusion of finality is built upon the inability to recognize the relativity of human truth to time, place, and culture. In the modern world, positivism and religious fundamentalism, Marxism and Capitalism, and the Western belief in inevitable historical progress all display the illusion of finality. Periodically throughout history, however, illusions of finality have crumbled when the weight of contrary evidence challenging an existing belief system has become too great for the existing belief system to bear. When a cultural worldview is involved, an extended period of change or transformation from the existing worldview to a new worldview occurs, as in the shift from the ancient classical worldview to the medieval worldview, or the transformation of the Western medieval worldview into the modern. In the 1870s the word "metanoia," derived from the Greek word meaning change of mind or repentance, was revived as meaning "a profound, usually spiritual, transformation."[3] The term has been applied to cultural as well as personal, individual transformations in recent years.[4]

Several pithy definitions of "worldview" provide further insight into its meaning, from "some profound cosmological outlook, implicitly accepted" and "the presuppositions of thought in given historical epochs" to "a conception of the nature of cosmic and human reality that discloses the meaning of life."[5] Oswald Spengler, in *The Decline of the West*, explained all worldviews of higher cultures (civilizations) as being organized around idiosyncratic conceptions of space held by each individual culture and manifested in all aspects of the culture -- art, science, literature, philosophy, politics, and architecture. Rooted in a culture's experience of nature in its particular geographic region, this spatial conception gave rise to the "prime symbol," the basis for the distinctive outlook or weltanschaüng (worldview) of a civilization.

In the mid-twentieth century, Henri Frankfort, an archeological expert on ancient Mesopotamian and Egyptian civilization, criticized Spengler and Arnold Toynbee for studying civilizations from a perspective preoccupied with their decay. Although Frankfort brushes aside Spengler's "prime symbol" of Ancient Egypt as being at variance with the archeological evidence, he nevertheless credits him for his attempt to penetrate to the essence of cultural worldviews in a manner consistent, up to a point, with his own concept of "form."[6] Frankfort defines the "form" of a civilization as "a certain cultural 'style' which shapes its political and its judicial institutions, its art as well as its literature, its religion as well as its morals."[7] This coherence in its various manifestations " is never destroyed although it changes in the course of time." The causes of change in form over time are twofold: (1) development or changes resulting from inherent factors; and (2) historical incidents (i.e., barbaric invasions or natural disasters such as earthquakes) resulting from external forces. Taken together, Frankfort calls these changes the "dynamics" of a civilization, and the interaction of form and dynamics constitutes the basis for the history of a civilization.

The worldview of a given civilization manifests itself distinctively in the artifacts of that civilization to the point that any able archeologist, any Indiana Jones, readily recognizes not only the characteristic form of Mesopotamian, Egyptian or Mayan civilization, but also the subtle or striking differences in art, sculpture, or architecture which distinguish a particular stage of development that expresses the effects of the dynamics of the culture. As we consider the worldviews of Classical Greco-Roman and Christian medieval civilization in the following sections, we will become more aware of the utility of Henri Frankfort's distinction between form and dynamics, particularly as the complex worldview of Greco-Roman civilization changes kaleidoscopically over time.

The Worldview of Classical Antiquity

Although the worldview of classical Greco-Roman civilization was extraordinarily complex, when one compares the attitudes of a citizen of ancient Greece or Rome with those of a Christian serf or noble during the Middle Ages, a striking difference in orientation towards human existence in the secular world is discernible. Medieval humans appear, by comparison with the ancients, to be obsessively concerned with life in a spiritual or transcendental realm beyond the confines of ordinary secular existence, and all of the architecture, social institutions, art and politics of medieval European civilization reflect this preoccupation. Certainly the ancients had their religion, and for many of them numerous cults offered the consolation of life in a world beyond this life, but overall their civilization paid homage to the values of life in this world, the secular world of nature and culture, of war and peace, flood and drought, famine and disease, wine and song, and all the pleasures and pains of earthly human existence. Ancient religion was indeed important, but it was also emphatically pragmatic and closely tied to patriotism and the power of the political state, whether Greek polis or Roman Republic and Empire.

Viewed from the perspective of modern scientific positivism, in which humans are seen as biological organisms unprivileged over other species, religion, the linchpin of most worldviews, is a social phenomenon arising out of the needs of particular cultures, especially as a moderating force beneficial to the health and success of a culture as a whole. For example, cultural geographer Warren Johnson writes of the rise of many of the world's major religions:

> The sixth century B.C. was a remarkable century in retrospect; it was a period of time that produced a large percentage of the world's great religious figures: Buddha, Confucius, Lao-tze, Zoroaster, and the last of the major prophets of Israel. It is this very confluence that suggests that after a long struggle, the tide had finally begun to turn against exploitation. For to restrain rulers, religion not only had to threaten them with the punishment of hell, or something similar, it also had to achieve its effects over large areas *at roughly the same time* so as to maintain political and military equilibrium.[8]

If the antique religions of the older civilizations of the eastern Mediterranean and Tigris-Euphrates valleys, i.e., of Egypt and Mesopotamia, were moribund by this time, what was the state of religious belief in ancient Greece and Rome in the sixth century B.C.? This question might be impossible to answer in limited space were it not for the classifying and systemizing tendencies of a few twentieth century scholars of classical religion. Cutting through the forest of encyclopedic detail on the subject, Gilbert Murray did much to clarify our understanding of Greco-Roman religion in his classic *Five Stages of Greek Religion*. Michael Grant and Luther H. Martin have provided subsequent integrative analyses on Roman and Hellenistic religion, respectively.

Murray identifies a sequence of overlapping stages that include: (1) primitive fertility rites and related religious cults; (2) the Olympic pantheon of anthropomorphic gods (e.g., Zeus, Poseidon, Pluto, etc.); (3) the Great Schools of the Cynics, Stoics, and Epicureans which approached the idea of a human relationship to God philosophically; (4) the "Failure of Nerve" represented by Christianity and the competing mystery cults of the Roman Empire; and, finally,

(5) the pagan revival associated with the emperor Julian the Apostate.[9] The primitive fertility cults that grew out of the Neolithic Revolution and the invention of agriculture are looked upon by Murray as a universal foundation for the development of religion in all cultures. Demeter and Dionysus, goddess and god of grain and wine, gave rise to sophisticated mystery religions which persisted with the changing stages of Greek religion and returned with renewed vigor during the "Failure of Nerve." The Homeric "Olympian Gods," although obscure in their origins, appear to represent a mix of indigenous goddesses deriving from the early chthonic (pertaining to deities and spirits dwelling beneath the earth) or fertility cults and associated with protection of particular locations of cities, as Athena is protectress of Athens, combined with Zeus and other male gods brought from Europe during the Dorian invasions ca. 1100 B.C.

The great value of Murray's work during the first half of the twentieth century was to have shown that the religion of classical antiquity was anything but a stagnant worship of Zeus and the Olympic pantheon of gods and goddesses, but a dynamically shifting variety of religious beliefs and practices. Although the inspiration of sculptural representations and the value of the Greek gods for civic religion carried Frankfort's "form" of classical religion and pagan gods into the late Roman Empire, the "dynamics" of Greco-Roman civilization (including both the internal intellectual challenges to the authenticity of the Olympic pantheon which led to their rejection, and the external factor of the importation of exotic gods such as Isis from Egypt) led to the brilliant chain of thought which extended from the Ionian nature philosophers to Aristotle. Greek religion, in other words, became highly intellectualized among the wealthy leisure classes at first. Over time, however, the ideas and beliefs characterized by Murray's third stage of Greek religion, "the Great Schools," spread to a larger population. Meanwhile, the rites of the old fertility cults continued to be practiced in the countryside and the Olympic pantheon continued to be worshiped patriotically as protectors of the Greek polis. Later, in their Latinized versions, these gods supported the morale and civic virtue of the Roman Republic and Empire.

Murray's Great Schools are roughly contemporaneous with the Hellenistic era of Greco-Roman civilization, a period extending from the death of Alexander the Great in 323 B.C. to the annexation of Egypt into the Roman Empire under Emperor Augustus in 30 B.C. While the Great Schools continued the philosophical tradition which began with the probing of the Ionian Greeks into the rational order of nature in the sixth century B.C., they are nevertheless treated as a stage of Greek religion by Murray because of the problems they attempt to solve regarding humanity's relationship to the divine and to the cosmos.

Two profoundly disturbing events in Greek history, the decline of the polis or city-state resulting from the Peloponnesian Wars between Athens and Sparta (431-404 B.C.) and the rise and fall of the empire of Alexander the Great (336-323 B.C.), led to a period of disenchantment with politics and a questioning of values by Greek intellectuals. As Murray wrote: "The city state, the Polis, had concentrated upon itself almost all the loyalty and the aspirations of the Greek mind. It gave security to life. It gave meaning to religion."[10] In the aftermath of the replacement of the polis by military monarchies, the best minds had turned away from politics to search for new foundations for a sense of meaning in life. Following Plato's quest for an ideal state in *The Republic* and the *Laws*, the founders of the Great Schools in the fourth century B.C. sought to provide themselves and ordinary citizens with some kind of psychological or spiritual armor against the vicissitudes of Hellenistic life and the pessimism engendered by its fatalistic worldview. Murray goes on to describe the heroic quest for virtue of Diogenes the Cynic, the acceptance of the order of the Cosmos preached by Zeno and other Stoics, and the Epicurean revelation that humans should strive for virtue, but with the ultimate objective of personal happiness rather than virtue, goodness, or resignation for their own sakes. These proselytizing philosophies can be understood more fully in the context of how the Hellenistic worldview changed during the centuries following the death of Alexander the Great.

Emphasizing the point that Alexander's empire irreversibly altered the sociopolitical world of the Greeks by substituting an internationalizing vision of the entire world as polis or city-state (with its local world protected by Olympian gods), Luther H. Martin argues that Hellenistic religions were an interrelated system of responses to a major revisioning of the cosmological order. This reconceptualization of the structure of the cosmos replaced the Hellenic tripartite cosmic image of sky, earth, and underworld with an image of the earth enveloped by concentric planetary spheres. The latter cosmology evolved out of the Ionian philosophical quest for order in nature and was established in the fourth century B.C. by Eudoxus, a student of Plato, long before Ptolemy finalized the system in the second century A.D. However, the seeming perfection of the Hellenistic cosmic model was demolished by the astronomer and astrologist Hipparchus in the second century B.C. when he discovered movement of the fixed stars themselves. Thus:

> If the stars themselves were transitory, it followed that the cosmos in its totality must be ruled by a relative rather than an absolute natural order, and finally by chance. Whether understood as external forces of nature that ordered but entombed man in an inescapably determined condition, or as dramatic personae united in a dance of cosmic chaos, the cosmic spheres increasingly were understood to represent a tyranny of this-worldly existence.[11]

Gradually during the Hellenistic era, the Hellenic conception of *moira*, the assumption of a natural or cosmic order of things, was transformed into an oppressive conception of fate or *heimarmenē*. "These transformations and revaluations in the grammar of fate provided systemic basis for the religions of the Hellenistic world and for the coherent structure underlying and shaping their sociohistorical expressions."[12] This dichotomy between benign and maleficent aspects of fate was associated with the division of the Ptolemaic cosmos into sublunar (beneath the moon) and superlunar realms inhabited by lower and higher powers. The lower sublunar powers were associated with demons and chaotic forces linked to the unreliable and destructive aspects of fortune represented by *heimarmenē*. Martin enumerates several alternative religious or philosophical responses to this perceived structure of the cosmos in the form of: (1) rational inquiry into the assumed natural order, as exemplified by Plutarch, who reasoned that humans are, unlike animals, not ruled by fate because they possess the power to think and reason. Stoicism claimed knowledge of the natural order controlling fate, and advocated a philosophy of sympathetic harmony with the natural order; (2) astrology and divination shared the Stoic assumption of order in the cosmos which was accessible to human knowledge; and (3) the Hellenistic Mysteries, rites of Demeter, Isis, etc., responded to the challenge of terrestrial fate and its disorder through soteriological initiation which revealed the celestial knowledge of the cult deity.[13]

All three strategies of philosophy, divination, and mystery were developed upon a shared positive view of a cosmos accessible to human knowing. However, during the period of the Roman Empire the cosmos was increasingly perceived as negative and unknowable, stemming somehow from its largely feminine attributes, leading, according to Martin, to a masculine otherwordly alternative to fate as the arbiter of human destiny:

> Rejection of this-worldly existence in favor of a vertical relation to otherwordly and redemptive origins radically reconceived the distribution of order wholly in terms of the Ptolemaic hierarchical architecture of existence. As opposed to the mystery reintegration of the individual into some sort of ideal spiritual society, gnosis was a completely inner or personal strategy of existence, rejecting even individual bodily requirements. In the myth of revelatory ascent, the masculine redeemers of Mithraism, Christianity, and gnostic myth charted a "Way" through the celestial spheres of this-worldly determinism for the faithful to follow in their own transcendence of *heimarmenē* and final reunion with the otherwordly Father Deity.[14]

In the sixth century B.C. Pythagoras had proposed that human beings were the animal bodies that encased the spirits of fallen souls in the temporal world. He established a monastic order and rituals of abstinence, meditation, and even specialized diet to aid the release of the eternal divinity or soul from the body, its ephemeral tomb. A thousand years later, a century before the official fall of Rome, a similar spirit of otherworldliness pervaded the Greco-Roman world. Through centuries of fertility cults, gods of the hearth, Olympian deities, speculative philosophy and primitive science, and above all, the capricious role of fate, the relentless warfare, slavery, and suffering of empire- building had taken its toll, and even the soils of Rome and its provinces appeared weary and worn out with relentless use. We are led back, after a millennium, from the transcendentalism of Pythagoras to the plaintive, world-weary musings of Tertullian and Augustine. Contemporaneously, the old Greco-Roman architecture was feebly resuscitated by Emperor Constantine by re-assembling sculptures made under Hadrian and other earlier emperors. The "form" of Greco-Roman civilization persisted a bit longer, but a millennium of its "dynamics" had produced a grinding, heartless, and barbarous Mediterranean civilization about to collapse in a heap of human despair. To what extent was the triumphant Christian transcendentalism both cause and effect of Rome's fall?

The idea of God, which, along with the interrelated categories of man, nature, history, and society, raises questions that are answered explicitly or implicitly in the worldview of any culture or civilization. We have seen that in Greco-Roman civilization, although any archeologist can recognize its "form," its belief system or worldview was malleable enough to have allowed for a surprising range of religious belief over time. Recognition of these various attitudes toward God led Gilbert Murray to his five stages of Greek religion culminating in Christianity as its fourth stage and the final pagan revival as its fifth. As the idea of God changed, so did human understanding of the nature of man, nature, history, and society, reflecting the "dynamics" of a developing civilization. At the same time, historians have characterized these other attitudes which make up a worldview with somewhat less difficulty than religion or the idea of God, but our awareness of the range of Greco-Roman religious belief should be a caveat against narrow, rigid representations of the various aspects of worldviews. Nevertheless, there are profound differences between the classical worldview and the Christian, medieval worldview which superseded it during the last two centuries of the Roman Empire. Having explored classical religion and its transformations, ultimately to a dominant Christianity after Emperor Constantine's conversion of the empire to Christianity in 313 A.D. [15] and establishment of the official dogma of the medieval (Catholic) church at the Council of Nicaea in 325 A.D., there are other major differences between classical and early medieval civilization regarding the five worldview categories of God, man, nature, history, and society.

Greek religion was, like other ancient religions, closely related to the problem of explaining the forces of nature. Both the fertility gods and the Olympian gods reflect attempts to gain control over nature through appropriate sacrificial rites. As we have seen above, increasing awareness over time of the ordered structure of the cosmos led the Greeks and Romans towards acceptance of conceptions of fate and Providence based upon this order. These ideas linked Greco-Roman religion to the cyclical view of history which we shall consider below. The Greeks answered the question of "what is man?" with Aristotle's "a political animal," more specifically as a member of the polis under special protection of the Olympian gods, and also as "the measure of all things," as Protagoras put it, a self-confidence in human ability to fashion technics and civilization out of the surrounding environment, which if carried to extremes, could lead to excessive pride or *hybris*. Nature, in the classical worldview, was the sphere of activity of the gods or *theoi*, the anthropomorphized forces of nature which could be dealt with on a human level through religious ceremony. In the long run this ritual access to the gods of nature was undermined by the rationalistic search for order in nature which began with Thales, Heraclitus and Pythagoras in the sixth century B.C. and culminated in the biological science of Aristotle and Theophrastus, and the astronomical achievements which resulted in the Ptolemaic model of the physical universe. The idea of history in the classical worldview shifted from the flux or chaos of Heraclitus to the cycles of Plato and the Stoics, which were alloyed with the idea of Providence, the notion of care and guidance over the creatures of the earth by God or nature, an omniscient and beneficent God who directs the universe and the affairs of humanity. The ideal society in the early Greco-Roman worldview was seen to be the *polis* or city-state, but gradual commercialization of the *polis* related to trade and imperialism undermined the *polis* until its collapse and integration into autocratic empires under Alexander the Great and, after a period of military struggles, the Roman Empire.

The Triumph of Christianity

In contrast to the classical worldview, the essence of the Christian one which would replace it and lay the foundations of medieval civilization is its virtually obsessive preoccupation with an otherworldly, transcendental reality which it considers to be inordinately superior to the things of the secular world. The Christian religion places all of its belief and hope in the guarantee of salvation of the individual soul if only the individual accepts Christ as savior and the Christian God as the only true God, rejecting all other gods and idols. This powerful transcendentalism had deep roots both in the Judaic tradition and in the undercurrent of belief in religions or cults offering personal salvation throughout the Hellenistic and Imperial Roman periods. Further consolidating the Christian belief system was the development of Greek philosophical conceptions of a created universe, of a single, all-powerful God who designed and managed the cosmos, and of the superiority of the transcendental realm over the secular. Plato alone, building upon the otherworldly philosophies of Pythagoras and others, had given enormous impetus towards the "spiritual metamorphosis" [16] or metanoia which transformed the Greco-Roman worldview during the centuries after Christ. Within the Christian worldview, developed more fully by Saint Augustine in the early fifth century A.D., man is a fallen creature following the original sin of the Biblical Eve, and can only achieve redemption and salvation through belief in and worship of Christ. Nature within this framework of ideas is God's creation, from the celestial bodies of an earth-centered cosmos to the soul-less animals and ensouled humans who were created out of clay to inhabit Earth, the unique home created by the Judeo-Christian God for humankind. From the perspective of Augustine's *City of God*, Earth is essentially a stage upholding the cosmic drama of Christianity upon which the souls of the saved belonging to the City of God are sorted from the damned souls of the City of Man consigned to eternal hellfire. History, then, is brief and finite, beginning with the Creation of Adam and Eve, and ending with the Apocalypse, Last Judgment, and the Millennium, the return of Christ when all souls have finally been divided into the two cities for all eternity. Society within this framework is hierarchical, with the most important responsibility, the salvation of souls, in the hands of the Church under the guidance of Pope, archbishops, bishops and priests while monarchs, feudal lords, and serfs go about the mundane business of ruling, fighting wars, and farming upon the ruins of the disintegrated Roman Empire.

Profound as the metanoia or spiritual metamorphosis was during the last centuries of the Roman Empire, as we have seen, there was a powerful undercurrent of transcendentalism and searching for individual salvation or immortality throughout the classical period. The question often raised, as to why Christianity was triumphant rather than Mithraism or another mystery religion, is answered in similar fashion by several historians of Christianity in terms of the universal appeal to individuals of all classes, races, and sexes which Christianity offered in contrast to the narrower audiences of competing mystery religions.[17] C. Warren Hollister argues that the doctrine of the Trinity "gave Christians the unique advantage of a single, infinite, philosophically respectable God who could be worshiped and adored in the person of the charismatic, lovable, tragic Jesus."[18] To an inveterate classicist like Gilbert Murray, however, the triumph of Christianity was seen darkly as a morbid turn away from the values of life in this world, which he characterized as "the failure of nerve," the fourth stage of Greek religion:

> Anyone who turns from the great writers of classical Athens, say Sophocles or Aristotle, to those of the Christian era must be conscious of a great difference in tone. There is a change in the whole relation of the writer to the world about him. The new quality is not specifically Christian: it is just as marked in the Gnostics and Mithras worshipers as in the Gospels and the Apocalypse, in Julian and Plotinus as in Gregory and Jerome. It is hard to describe. It is a rise of asceticism, of mysticism, in a sense, of pessimism; a loss of self-confidence, of hope in this life and of faith in normal human effort; a despair of patient inquiry, a cry for infallible revelation; an indifference to the welfare of the state, a conversion of the soul to God.[19]

As a twentieth century secular humanist and classical scholar, Murray clearly found this spiritual transformation unpalatable even though he was well aware of its development over the course of Greco-Roman civilization. Perhaps if he had reflected at greater length over the gradual deterioration of the Roman Empire physically, morally, militarily, and politically, he would have been more accepting of the failure of nerve and the turn towards supernaturalism. Certainly, the harsh conditions of life for slaves, soldiers conscripted to fight on the frontiers, farmers tied to a losing struggle against gradually diminished productivity, and a growing awareness of the limitations of imperial power generally, all contributed towards a turn away from the material world combined with hope for something better in the transcendental realm. The Italian historian Aldo Schiavone has analyzed the destructive circle of classical antiquity which linked the spread of slavery, the rejection of labor by citizens as demeaning, and the absence of labor-saving machines with the flight from a mechanical and quantitative vision of nature (which led to an otherwordly quest for salvation as the only acceptable alternative vision of reality). He also describes a process of transformation of labor relations from the fourth to the ninth centuries A.D. which changed slaves to serfs in a way that would attach the concept of slavery to the ethnic differences of slaves, unlike Roman slavery, for which ethnic difference was relatively unimportant.[20]

Nature and the Judeo-Christian Worldview

Late twentieth century scholarship regarding the origins of Christianity and the Christian worldview has emphasized the continuity of Christianity with the Judaic tradition as well as with the late classical worldview. From this perspective Christianity is regarded as a Jewish heresy which grew out of a period of frustration under Roman domination, resulting in the proliferation of radical and apocalyptical Jewish sects which promised a savior God who would lead Zion to vanquish the Romans. Repudiated by the Jews as an inauthentic Savior, Jesus and the stories told about him in the Gospels became the nucleus of a growing body of interpretations about their spiritual meaning which created the Judeo-Christian worldview and its leading institution, the Roman Catholic Church.

In modern times, simultaneously with religious reinterpretation, the last few decades of the twentieth century have witnessed the rise of an entirely new kind of reinterpretation of Christian and Judaic scripture: the quest for the environmental significance and consequences of the Old and New Testaments as the core texts of Western civilization, interpreted by many environmental historians as the most ecologically destructive of all the world's civilizations overall, and indisputably the most destructive in recent centuries, as Spengler observed in Chapter One.

In 1967, Lynn White Jr., a medieval historian at UCLA, initiated an ongoing controversy regarding the influence of the Christian worldview upon Western European humanity's treatment of nature.[21] White argued that the Judeo-Christian belief in a divine creation in which humanity, created in God's image, was given dominion over all the creatures of the earth for his use, combined with the Christian rejection of pagan animism, with its belief in spirit or spirits in both animate and inanimate nature, set the stage for a peculiarly Western disregard for the intrinsic value of nature,[22] thereby opening the door to mistreatment of the environment by medieval and later Western European mankind. Our modern abuse of nature is particularly destructive because of Western humanity's unique success in developing both the manipulative technology and the scientific knowledge necessary to a thorough exploitation of the natural world. This idiosyncratic Western penchant for controlling and dominating nature has its roots in medieval science, in which nature was originally studied (in the early Middle Ages) in order to better understand God's messages to mankind contained in the structure of nature (i.e., the creation), which were considered to be complementary to revelation as contained in the Old and New Testaments. By the thirteenth century, however, Western science shifted from being an interpretation of symbols in nature to understanding God's mind through comprehending how His creation functions, ultimately leading to the Scientific Revolution and the discovery of natural laws. Thus, White understands even modern, contemporary science and technology to be "so tinctured with orthodox Christian arrogance toward nature that no solution for our ecologic crisis can be expected from them alone. Since the roots of our trouble are so largely religious, the remedy must also be essentially religious."[23] Clearly, this implies that that we must re-enspirit nature.

White's attack upon the belief system of Christianity from the vantage point of ecology had two major effects. For one, Christian churches began a process of self-scrutiny in an attempt to right the wrongs of environmentally destructive behavior. This "greening" of Christianity continues to the present day in the greatly expanded awareness of ecological problems on the part of many enlightened clergy, and their awareness has opened the minds of many of their parishioners. Secondly, White's critique caused an academic counterattack in defense of Christian values, particularly the value of stewardship of the land. In the study of medieval Christendom, for example, the Cistercian monks have been singled out for praise due to their practice of reclaiming malarial swamps and other lowland areas to develop their rich soils on a sustainable basis. Environmental historians and philosophers have also developed sophisticated arguments in defense of a Christian stewardship ethic. Nevertheless, as we shall see in the following chapter, the destruction of the Western European wilderness was so thorough by 1300 A.D. that scarcity of wood, famines, malnutrition, and plague dominated the fourteenth century.

If Lynn White, Jr. identified the roots of our modern ecological crisis in the Christian worldview, the environmental philosopher J. Baird Callicott located deeper roots in the pre-Christian classical beliefs that reached all the way back to early Greek philosophy. According to Callicott, the essential problem arose out of Greek speculation concerning the ultimate character or meaning of nature and the simultaneous quest for understanding regarding the nature of humanity. What is the essence of that out of which all things arise and into which all things are resolved? What is man? What is the body and soul of man? The fountainhead of answers to these questions came from the sixth century B.C. Ionian Greek philosopher and mathematician, Pythagoras. On the one hand, Pythagoras discerned that quantitative mathematical description was the key to understanding the order underlying nature and natural phenomena, and on the other he conceived the dualistic model of man in which the soul is not only separate and distinct from the body, but is also alien to it. The Pythagorean goal in life was to earn the release of the soul from the body at death, thereby allowing it to join its divine companions. This led Plato, who was more than a little influenced by Pythagoras, to argue that the body is the tomb of the soul as well as its place of imprisonment.[24] Callicott concludes: "The Pythagorean-Platonic concept of the soul as immortal and otherworldly, essentially foreign to the hostile physical world, has profoundly influenced the European attitude toward nature."[25] He adds that this concept of the soul was popularized in Pauline Christianity, and much later by Descartes in the seventeenth century.

Continuing his analysis of early Greek science, Callicott describes how the atomic theory developed by Leucippus and Democritus in the fifth century B.C. was combined, in early modern science, with the Pythagorean precept that the order of nature is determined quantitatively. However, these ideas, especially their interaction, lay dormant throughout the Roman and medieval periods and were overshadowed by theories developed by Plato and Aristotle to explain the variety of biological organisms occurring in nature. Plato explained the existence of species or types of organisms and other entities through his theory of ideas, the notion that each individual organism of a species reflected an eternal "form" or ideal prototype which exists in a higher, transcendental realm beyond the secular world. For Plato, ultimate reality exists in the higher world of forms, while the pale copies exist as ephemeral representations of the substantive higher order. Callicott describes this theory as a kind of "conceptual atomism" in contrast to the material atomism of Democritas and Leucippus. Plato's brilliant pupil, Aristotle, rebelled against his teacher's transcendentalism, but in his own secular classification of organisms into a hierarchy with man at the top and organisms of decreasing complexity further down the ladder of being, provided medieval theologians with a rationale for human superiority which meshed neatly with the idea of God's creation of species in order for them to serve his highest creation, man.

The authoritativeness of Plato and Aristotle's ideas about nature helped to shut down further scientific inquiry for nearly two millennia, from the Hellenistic era until the High Middle Ages, when practitioners of "natural theology," the study of nature for the better understanding of God, began to probe into how nature works, thereby laying foundations for the sixteenth and seventeenth century Scientific Revolution. Unfortunately, the "ecological" biology which Theophrastus, a student of Aristotle's, pursued, and which might have led to further breakthroughs in classical biological science, was not translated into Latin and was lost to the Middle Ages. J. Donald Hughes thinks that Theophrastus deserves the title, "Father of Ecology" because of the Greek scientist's empirical observations of the relationships between plants and their surroundings.[26]

Environmental philosopher Max Oelschlaeger agrees with Callicott that "the Greek conception of the soul suffused Judeo-Christianity."[27] He also supports Lynn White, Jr. in his observation that " the medieval mind conceived of nature as an earthly abode over which humankind had been given dominion by a beneficent God. Salvation of the soul was the ultimate goal, and thus life on earth was merely transitory. While ensconced in this vale of tears the human lot was to toil in order to bring forth the fruits of the earth."[28] In this reading of the Old and New Testaments the larger, ecological significance of Judeo-Christian belief was that it culminated rationalization of the transition from hunting and gathering to a sedentary life dependent upon agriculture. This new agrarian equilibrium required the argument, for its justification, that nature, though divine in origin, was not itself divine. Thus, during the Middle Ages Christians cut down the sacred groves worshiped by pagans and drove out the supposed witches and demons.

Environmental historian J. Donald Hughes supports Callicott and Oelschlaeger's correction of White's thesis that the roots of the modern environmental crisis may be traced to medieval Christianity. Hughes agrees with these philosophers that " both medieval and modern attitudes have ancient roots. Greece and Rome, as well as Judaism and Christianity, helped to form our habitual ways of thinking about nature."[29] In his later book, *Pan's Travail: Environmental Problems of the Ancient Greeks and Romans*, Hughes stressed the role of animism, which saw the natural world as full of gods and spirits, in constraining destructive attitudes and practices regarding nature, but explained how the power of animism weakened over time, particularly under the rule of the pragmatic Romans. In *Ecology in Ancient Civilizations* he described the further displacement of animism by the transcendent monotheism of Israel, writing:

> Instead of being divine in itself, nature was seen as a lower order
> of creation, given as a trust to mankind with accountability to God.
> But in the later history of that idea, people tended to take the
> command to have dominion over the earth as blanket permission to
> do what they wished to the environment, conveniently forgetting
> the part about accountability to God, or else interpreting most
> human activities as improvements in nature and therefore pleasing
> to God.[30]

Taken together, the ideas of the foregoing environmental philosophers and historians explain the complex multiple causation involved in changing human attitudes and behavior toward nature over long periods of historic time. The medieval worldview, it turns out, owes as much or more to Greek philosophy as it does to the Judeo-Christian scriptures. The influence of Plato upon early medieval Christianity, and of Aristotle in the High and Late Middle Ages, was profound. Before turning to an account of the ecological history of the Middle Ages, however, it would be worth our attention to examine in further detail the two categories of "nature" and "history" in the medieval worldview.

Only in the modern, post-Darwinian worldview do we understand nature and history to be inextricably connected. The gradual accumulation of empirical scientific evidence during the early modern centuries culminated in the discovery of geologic time and its application to the history of organic life explained in terms of evolutionary theory. Modern scientists cannot think about biological organisms or the physical landforms of the earth, or the origin of the cosmos and its celestial bodies, outside of the framework of the geologic time scale, which some environmental historians refer to as "deep time" in contrast to the brief six thousand year span of historic time, understood as the written record of human events since the invention of cities and writing. As we have seen, in the ancient classical civilization Plato and other philosophers had some grasp of periods of time far greater (although not geologic in scope) than the brief span of human history. Plato's account of the creation in the *Timaeus* begins with the Deity setting the cosmos in motion. Although it was perfect, it was not immortal and would decay over time until it dissolved into chaos at the end of 72,000 years (a number possibly based upon Pythagorean numerology). The first 36,000 years of Plato's cycle are maintained as a Golden Age of stability, when men were not subjected to toil, warfare, and disease (what we refer to as pre-civilized hunting and gathering societies today). During the second 36,000 years gradual decay and degeneration occurred contemporaneously with the rise of civilization and the arts, leading to the ultimate chaos that completes the world-cycle. The cycle then repeats itself indefinitely into the future.[31]

It is of enormous importance to know that the ancients were capable of thinking in terms of these considerable time spans because once the Christianized Roman Empire began to suppress ideas that were incompatible with the Judeo-Christian worldview (and its purposeful contraction of historical time to fit the chronology of the Old Testament), any larger, more sophisticated comprehension of time was not only virtually impossible, it was also heretical and punishable. Closely related to the ancient cyclical conception of time was the idea that the cosmos had no beginning and no end, that it always had existed and had not been created at some starting point in eternity. Aristotle adhered to this idea, and once his works were translated into Latin during the High Middle Ages, this heresy was quickly shunned by Scholastic theologians. Thus, the narrow view of time embraced by Christians to rationalize the "absolute" truths of the Old Testament created a powerful barrier to any conception of extended time until the eighteenth century. The Christian conception of linear time, extending from the Creation to the Millennium, is generally referred to as the doctrine or idea of providence. The alternative ideas of cycles and progress typically have been associated with the worldviews of Greco-Roman antiquity and the modern world since the seventeenth century, respectively. The ideas of cycles and providence co-existed in late classical civilization, and at least elements of the idea of progress have been recognized in the thought of classical civilization as well. In contemporary modern Western civilization ideas of cycles, providence, and progress exist side by side, although the cyclical interpretation of the pattern of history is limited to a minority of intellectuals.

The relationship of ideas of time to nature, i.e., of the prevailing philosophy of history (cyles, providence, progress) to nature is a crucial factor in determining the possibility of scientific progress (increases in our knowledge of nature). The possibility of scientific progress was closed down by two major factors in the ancient world. One was the culmination of Greek science in a model of nature as static and unchanging over time, largely due to the fact that Greek and Roman science was essentially rationalistic rather than empirical in its approach to the study of nature. Aristotle's biological investigations led him to the conclusion that species were eternal and unchanging, and he organized them into a ladder or chain of organic being after reflecting that it was " as though every single kind of living creature that could come into existence had in fact done so."[32] (In the seventeenth century this idea was related to the concept of plenitude, meaning that God had created every possible kind of living organism based upon the ultimate divine mathematical principles.)

The Aristotelian model of a static world made up of all possible organisms laid a respectable foundation for the Judeo-Christian worldview and its belief in a created world of nature. According to Toulmin and Goodfield, " by about A.D. 300 the historical framework of the Christian world-drama had been fixed in the form which it was to retain for some fifteen hundred years."[33] Beginning with the Council of Nicaea in A.D. 325, the sacred historian Eusebius made his influence felt through books and chronological charts rationalizing all of human history as factually summarized in the Old Testament. An earlier, simplified version of cosmic history by Julius Africanus took the idea of the "millennium," the thousand-year Kingdom of the Messiah, which would end the world according to Jewish prophecy, and used it as the key to interpreting the chronology of the Old Testament. Africanus saw all of history as equivalent to a cosmic week of 7,000 years, and interpreted the Old Testament accordingly. Following his time scheme, the Second Coming of Christ was expected around 500 A.D. Despite the failed prophecy, Africanus' millennial chronology survived, with modifications, into the Reformation, and was taken for granted by Martin Luther and other devout Protestants.[34]

The problem for science in the Middle Ages, and during the sixteenth and seventeenth century Scientific Revolution as well, was a total constraint upon the possibility of thinking of nature as changing dynamically over time. Medieval "science," particularly during the early Middle Ages, is an oxymoronic concept, as we witnessed in Lynn White Jr.'s account. Even those medieval scientists who dared to understand how nature works during the half-millennium that preceded Isaac Newton did so with the firm conviction that they were doing God's work. Newton, likewise, believed he was working to reveal the secrets of God's Creation, and adhered to the same 6,000 year history of the earth and its inhabitants as did his medieval forebears. Placed in this context (of Greek efforts to understand nature culminating in a model of a static cosmos, and its reinforcement by the belief in the fixity of nature since the Creation which Christianity brought to the late Roman Empire), the absence of scientific knowledge from medieval civilization is easily understood. This blindness to natural time underpinned the intellectual and environmental history of Western civilization during the Middle Ages, from the fall of the western Roman Empire to the discovery of the New World, a millennium of historical time during which the Western European civilization would rise slowly from the ashes of the ancient world and overrun the carrying capacity of the subcontinent.

NOTES

[1]. See Franklin Baumer, *Modern European Thought* (New York: Macmillan Publishing Co., Inc., 1977).

[2]. J.B. Bury, *The Idea of Progress* (New York: Dover Publications, Inc., 1955 (1932), p.351-352).

[3]. *The Random House Dictionary of the English Language*, Second Edition, Unabridged (New York: Random House, 1987).

[4]. William Ophuls, *Ecology and the Politics of Scarcity* (San Francisco: W. H. Freeman and Company, 1977), and *Ecology and the Politics of Scarcity Revisited: The Unraveling of the American Dream, with A. Stephen Boyan, Jr. (New York: W.H. Freeman and Company, 1992), pp.281-307.*

[5]. Alfred North Whitehead quoted in Henri Frankfort, *The Birth of Civilization in the Near East* (Garden City, New York: Doubleday and Company, Inc., 1956), p.VI. John C. Greene, *Science, Ideology, and World View* (Berkeley: University of California Press, 1981), p.3. Warren W. Wagar, *Worldviews: A Study in Comparative History* (Hinsdale Illinois: The Dryden Press, 1977), p.4.

[6]. Henri Frankfort, *The Birth of Civilization in the Near East* (Garden City, New York: Doubleday & Company, Inc., 1956 (1951)), pp.4-10.

[7]. Ibid., p.3.

[8]. Warren Johnson, *Muddling Toward Frugality: A Blueprint for Survival in the 1980s* (Boulder, Colorado: Shambhala, 1979), p.46.

[9]. Gilbert Murray, *Five Stages of Greek Religion* (Garden City, New York: Doubleday and Company, Inc., 1955, 1925).

[10]. Ibid., p.77.

[11]. Luther H. Martin, *Hellenistic Religions: An Introduction* (New York: Oxford University Press, 1987), p.136.

[12]. Ibid., p.158.

[13]. Ibid., pp.158-161.

[14]. Ibid., p.161.

[15]. See Charles Freeman, *The Closing of the Western Mind: The Rise of Faith and the Fall of Reason* (New York: Alfred A. Knopf, 2003), pp.163-201 for an extensive discussion of Arianism and the struggle to implement the Nicene Creed in the fourth century A.D. Freeman demonstrates the tenuous and highly politicized origins of the rigid doctrine that ruled the late Roman Empire and Middle Ages, and which persists in modern Catholicism.

[16]. C. Warren Hollister, *Roots of the Western Tradition: A Short History of the Ancient World* Fifth Edition (New York: McGraw-Hill, Inc., 1991), pp.196-206.

[17]. See, for example, Keith Hopkins, *A World Full of Gods: The Strange Triumph of Christianity* (New York: The Free Press, 2000).

[18]. C.Warren Hollister, *Roots of the Western Tradition*, p.200.

[19]. Gilbert Murray, *Five Stages of Greek Religion*, p.119.

[20]. Schiavone, Aldo, *The End of the Past: Ancient Rome and the Modern West* (Cambridge, Massachusetts: Harvard University Press, 2000), pp.162-164.

[21]. Lynn White, Jr., "The Historical Roots of Our Ecologic Crisis" in *Science*, Vol. 155, No. 3767, 10 March 1967, pp.1203-1207.

[22]. Ibid.

[23]. Ibid., p.1207.

[24]. J. Baird Callicott, *In Defense of the Land Ethic: Essays in Environmental Philosophy* (Albany: State University of New York Press, 1989), pp.181-182. Another author suggests the possibility of Pythagoras' influence upon the early Christian perception of the character of Jesus Christ. See Margaret Wertheim, *Pythagoras' Trousers* (New York: Random House, 1995), pp.19-20. On p.19 Wertheim writes:

> In many respects the mythico-religious dimension of Pythagoras bears an uncanny resemblance to the life of Christ depicted in the New Testament. Both men are said to have been the offspring of a god and a virgin woman. In both cases their fathers received messages that a special child was to be born to their wives--Joseph was told by an angel in a dream; Pythagoras' father, Mnesarchus, received the glad tidings from the Delphic Oracle. Both spent a period of contemplation in isolation on a holy mountain, and both were said to have ascended bodily into the heavens upon their deaths. Furthermore, both spread their teachings in the form of parables, called *akousmata* by the Pythagoreans, and a number of parables from the New Testament are known to be versions of earlier Pythagorean *akousmata*.

25. Ibid., p.182.

[26]. J. Donald Hughes, "Theophrastus as Ecologist," *Environmental Review*, vol. 9, No. 4, Winter, 1985.

[27]. Max Oelschlaeger, *The Idea of Wilderness* (New Haven; Yale University Press, 1991), p.73.

[28]. Ibid., p.70.

[29]. J. Donald Hughes, *Ecology in Ancient Civilizations* (Albuquerque: University of New Mexico Press, 1975), p.148.

[30]. Ibid.

[31]. J.B. Bury, *The Idea of Progress* , pp.8-10.

[32]. Stephen Toulmin and June Goodfield, *The Discovery of Time* (New York: Harper & Row, Publishers, Incorporated, 1965), p.51.

[33]. Ibid., p.60.

[34]. Ibid., p.61.

CHAPTER FOUR: THE ECOLOGICAL TRANSFORMATION OF WESTERN EUROPE

Perhaps the majority of first-time travelers to Western Europe do not experience the kind of cultural shock that I did when traveling through France, Switzerland, and Italy with a Trapper Nelson pack on my back containing Frommer's *Europe on Five Dollars a Day* in 1959. Having been educated as a scientist with degrees in geology, I was not mentally prepared for the aesthetic power of European architecture and pastoral landscapes. I was absolutely stunned in the most positive way by the richness of European culture in contrast to my youthful experience of America. Having earned a moderate amount of money in a short period of time as a mining geologist in Honduras, I followed the advice of a wise older friend and spent nearly every last cent on a four-month exploration of Europe's cities and landscapes. "Why should I go?" I had asked my friend. "Just take the money and go, and then you'll know why," he answered. When the trip was over and I set foot in New York City from the S.S. Ryndam with sixty-five cents in my pocket on Christmas Eve, 1959, I knew, with irrepressible joy, exactly what my friend had meant. I had contracted something tantamount to a kind of mental illness. I prefer to call it "Europaphilia." Gradually, I dissipated the cloud of ignorance which prevented me from understanding the fact that Americans, from Franklin, Jefferson and other of the Founding Fathers, to the great landscape painters Thomas Cole and Frederick Church in the nineteenth century, to the Hemingways and Henry Millers of the twentieth, had all made the transatlantic voyage by boat, and all, indeed, had contracted "Europaphilia."

Unfortunately, when one returns home from the European voyage and begins to read about the history of Europe, the romantic images of castles, chateaus, and Gothic spires are transformed into symbols of power, class struggle, endless warfare, religious faction and martyrdom, and a seemingly inexorable drive towards dominion over both other cultures and the natural environment. This almost "demonic" drive and passion to expand and dominate led Oswald Spengler to characterize Western European civilization as "Faustian" civilization, which he saw as having coalesced into a clearly recognizable agrarian-based culture around 1000 A.D. Modern scholarship, in both history and art history, generally substantiates Spengler's choice of ca. 1000 A.D. as the beginning point of Western civilization, although there has been disagreement as to precise timing.

British world historian and philosopher of history Arnold J. Toynbee, in his multivolume *A Study of History*, wrote that by the year 775, when the Roman Christian west had shrunk to the area of Charlemagne's dominions and the English successor states to the Roman Empire in Britain, a society recognizable as

"Western Christendom" had evolved.[1] Similarly, medieval historian C. Warren Hollister writes:

> In the course of the eighth century, Western Christendom began to emerge as a coherent civilization. It did so under the Carolingian Empire-a vast constellation of territories welded together by the Frankish king Charlemagne and his talented predecessors. Here for the first time the various cultural ingredients-Classical, Christian, and Germanic-that went into the making of European civilization achieved a degree of synthesis.[2]

In contrast, art historian Whitney S. Stoddard, in the course of explaining the development of "Romanesque art, the first truly European style,"[3] observes the decline of monasticism resulting from the barbarian invasions of the ninth and tenth centuries, and their corruption under the Franks, concluding that "The early tenth century marked the low ebb of medieval learning-a truly Dark Age."[4] Stoddard further observes that the rise of the Capetian line of kings, and the revival of trade and the growth of towns following the end of Viking and Magyar attacks, coupled with the excitement of the Crusades and the pilgrimage to the grave of Saint James, all contributed to the formation of the Romanesque style, the first truly European style in art to spread throughout Western European civilization.[5] Art historian Raymond Oursel agrees " that Romanesque art starts with the year 1000", culminating the slow social, economic, and political evolution of the ninth and tenth centuries.[6] In keeping with Henri Frankfort's idea of the "form" of a civilization, the Romanesque was the first ubiquitous expression of Western European form.

Medieval historian Norman F. Cantor supports the idea of the eighth and ninth centuries as a *formative* period of the "barbarian west" during which the great expectations of Carolingian kings and churchmen were not fulfilled but " formed a considerable portion of the more successful social order of the tenth and eleventh centuries."[7] In other words, the great efforts made to establish a Christian society during the eighth, ninth, and tenth centuries were finally realized by the late tenth and early eleventh centuries. The millennial date of 1000 A.D., coincident with the explosive spread of Romanesque monasteries and churches, symbolically represents the beginning of a fully crystallized Western European (Faustian) civilizational "form". This interpretation is fully consistent with Henri Frankfort's explanation of the "form" of the ancient Mesopotamian and Egyptian civilizations.

Spengler's depiction of Western civilization as "Faustian" in its dynamism, in its relentless striving for the infinite expressed from the soaring Gothic cathedral to the modern skyscraper and space exploration, is explained pragmatically by world historian William H. McNeill in *The Rise of the West* (1963). McNeill asserts that Western civilization " incorporated into its structure a wider variety of incompatible elements than did any other civilization of the world; and the prolonged and restless growth of the West, repeatedly rejecting its own "classical" formulations, may have been related to the contrarieties built so deeply into its structure."[8] The contrarieties or contradictions which gave rise to Spengler's Faustian behavior are summarized by McNeill as follows: the thoroughly warlike barbarian inheritance versus Christian pacifism; the tension between the territorial state and the Church's claim to govern human souls; tensions between faith and reason; and other tensions between naturalism and metaphysical idealism in art, between violence and law, vernacular and Latin, nation and Christendom. McNeill claims that this peculiarly Western propensity for cultural contradiction was geographically rooted in the advantageous location of Western Europe on the margins of Byzantine and Muslim civilization, and on the ruins of the Roman Empire. The new Christian civilization " inherited a society in which nearly all the overburden of civilized institutions had crumbled into agrarian simplicity. Europeans could therefore build anew, utilizing elements from the variegated cultural inheritances of their neighbors according to choice."[9] Europeans had no fear of such borrowing because they felt secure and superior in light of the universal claims of the Roman Catholic Church and Frankish military strength. In addition, other geographic and environmental factors contributed to Western civilization's rapid rise to power, including a wealth of fertile land developed under the Carolingian technology of the moldboard plow, an indented coastline, abundant navigable rivers, abundant iron and other metals, and abundant timber.

History and the Ecology of Europe

The ecology of Western Europe is similar to that of the temperate zone of eastern North America. Approximately 15,000 years ago, glacial ice began to retreat from the northern parts of both continents, and reforestation followed the retreating ice and tundra northwards. Since the beginning of agriculture ca. 9000 years ago, and perhaps earlier in some places, humans have been cutting down trees and changing the ecology of the Mediterranean Basin and Europe. By Roman times many of the plains and river valleys of Western Europe were farmed, with most wilderness pushed back to the hilly, mountainous, and swampy regions. The chaos, warfare, disease and depopulation which accompanied the decline and fall of the Roman Empire returned great tracts of farmland to the wild, as forests and their inhabitants, including bears and wolves, returned. The social, political, and economic instability of the period following Rome's collapse created a fortress Europe of abandoned or shrunken cities menaced by wandering barbarian tribes (Franks, Goths, Huns, Lombards, Visigoths) seeking lands to settle and integrating residual indigenous populations into warring fiefdoms. The hierarchical, feudal, Christianized kingdoms of the "Dark Ages" (500 A.D. - 1000 A.D.) grew up around walled fortresses which formed the cores of later medieval cities. Fortified monasteries also formed the centers of later towns and cities.

During the Dark Ages the return of the wilderness to abandoned Roman farmlands created a European frontier which would gradually be reconverted to farmland, especially between the 9th and 13th centuries. The Benedictine monks, from at least the time of Charlemagne (crowned Holy Roman Emperor, 800 A.D.), served as Christian missionaries who were sent out with Carolingian armies to the wild frontiers of what is now France and the low countries to convert pagan barbarian Germanic tribes to Christianity and the agricultural practices of which the Benedictines were major innovators. The monks usually followed Charlemagne's armies to establish order in the wake of conquest, much as the Dominicans, Jesuits, and Franciscans followed the Spanish imperial armies to Mexico and California from the time of Cortez in North America.

Traditionally, before and even after the revolution in thought introduced by the development of geologic science and evolutionary biology and their ecological implications, historians have written about the activities of mankind as though they were played out on planet earth as an essentially inert stage. In the light of our knowledge of geologic time, however, the stage of history has been anything but stable, migrating slowly over time by plate tectonic movement to create the impressive variety of landscapes and ecosystems to which humanoids have adapted over millions of years. The European subcontinent, separated from North America beginning roughly 200 million years ago, and intermittently glaciated and periodically reshaped in outline by rising and falling sea levels controlled by glaciation during the Pleistocene ice ages of the last 2.5 million years, has provided *Homo sapiens* and his ancestors with a peninsula-like land mass particularly favorable to historic mankind. During the 11,000 years and more since the retreat of the last (Wisconsin or Würm) ice sheets, the tundra and steppe which covered much of Northern Europe have receded to the far north as post-glacial forests have spread northwards. In historic time Western Europe has been part of a floral region known as the Boreal Kingdom, a temperate forest zone dominated by oak along with beech, elm, and other species. During glacial periods the genetic diversity of these temperate forests has been preserved in refuge areas of southern Europe in Spain, Italy, and the Balkans.[10]

Megafauna associated with the last major glaciation, such as the woolly mammoth, mastodon, and the woolly rhinoceros, appear to have been driven to extinction during the past 10,000-15,000 years by a combination of climate change and human hunting both in Europe and in North America. The grazing habits of these megaherbivores probably produced forests with more openings and fewer saplings than those inhabited by smaller herbivores during historic time. The residual fauna inhabiting these Neolithic and later European forests consisted of reindeer, red deer, forest bison, wild horse, aurochs (wild ox), elk, boar, bear, wolves and foxes, lynx, beaver, and many smaller species.[11] The general similarity of the majority of these moderate to small-sized mammals to those of North America is a reminder of the Siberian-Alaskan corridor across the Bering Strait which has allowed mammals to pass between Europe and North America during periods of glaciation (such as the last Wisconsin or Würm glaciation which ended 15,000 years ago) when sea level was about 350 feet lower than it is today.

The remarkable likeness of European and North American mammalian fauna is also preserved in remnant populations inhabiting the Bialowieza Forest on the Polish-Russian border. The Bialowieza is nothing less than the last small piece of primitive old-growth forest to survive in Europe. Largely as a result of preservation for hunting the forest bison, elk, and other mammals by the Polish and Russian nobility, the Bialowieza has almost miraculously survived into the twenty-first century. Although the forest has been damaged considerably by warfare, including World Wars I and II, a core area of fifteen square miles of essentially untouched primeval forest is surrounded by a utilized forest two hundred square miles in area. The major forest trees are oak, linden, hornbeam, and pine, along with ash, aspen, maple, willow, birch, elm, spindle trees and fir. The fauna inhabiting the primeval forest and its enveloping managed forest includes European moose, roe deer, red deer, wild boar, wolves, beaver, and stoat, as well as the recently reinhabited European forest bison and European wild horse or tarpan.[12]   Whereas the forest bison and tarpan have been re-introduced to the Bialowieza by breeding and back-breeding zoo animals, the aurochs or European wild ox became extinct in 1627. Botanist Edward Klekowski, who has studied the genetic composition of the ancient oaks of Bialowieza, describes " the steady-state biosystem of Bialowieza, and the 'fantastically diverse' habitats it engenders ..., a maze of windfalls, younger trees growing out of the rotting wood of old, fallen trees, a messy, dank forest which ... fits descriptions by the Romans of crossing the Alps into this impenetrable gloom that was all of northern Europe".[13]

Today, traversing the pastoral landscape of western and central Europe, the traveler rarely thinks of the ubiquitous primeval forest displaced by waves of human settlement, from the Celts and ancient Romans to the barbarian Germanic tribes and the Christianized feudal kingdoms which they became. The forests which we see today are thoroughly managed and emptied of most of their larger mammalian fauna, existing as utilitarian communal woodlots or state-managed watersheds. The change from primeval forest inhabited by dangerous beasts to the pastoral mix of farming, grazing, and managed forest is generally perceived as a positive one, culminating in a European countryside described by René Dubos as having been " shaped by more than a hundred generations of peasants out of the forests and marshes that covered most of western Europe before human occupation."[14]   Dubos admires these landscapes for the "charm and elegance" resulting from human management, and he considers them superior to the wilderness they have displaced. Writing of his native Île de France, he opines: "Humanizing the Ile de France has admittedly resulted in the loss of many values associated with the wilderness. From the human point of view, however, at least according to my taste and that of many other people, the region is now visually more diversified and emotionally richer than it was in its original forested state."

Thus, the vast majority of Europe has been transformed into a much-admired humanized landscape which is generally regarded as superior to the "wild nature" which preceded it. In other words, we think of the process of transforming natural or relatively more natural ecosystems into humanized systems as progressive or developmental, or at least we have done so in Western civilization until the wilderness and its lesser values have all but disappeared. What were the changing values, then, of Western Europeans at various stages in the process of turning their boreal forest into agrarian landscapes?

Changing Medieval Attitudes Toward Nature

        In the history of Western Europe and its surroundings, the Middle Ages cover the period of time extending from the fall of the western Roman Empire to the time of Columbus's discovery of the New World, approximately from 500-1500 A.D. Typically, the period from 500-ca.1000 A.D. has been referred to by medieval historians as the Dark Ages, and more recently the Early Middle Ages. The High Middle Ages extend from ca.1000-1050 to 1300, and the last two centuries from 1300-1500 are termed the Late Middle Ages. Given the thousand year duration of the Middle Ages, one could not expect a single outlook upon nature to persist over such a long period of time. On the other hand, the dominance of the Old and New Testaments as the ultimate source of truth and knowledge did continue over the entire period, and observations, descriptions, and explanations of natural phenomena could not contradict the divine truths recorded in the scriptures. These restrictions upon what we today refer to as "scientific inquiry" caused Copernicus to withhold his proof of a heliocentric, as opposed to the Ptolemaic geocentric, cosmology until he was safely on his deathbed. Copernicus's magnificent achievement under such conditions attests to how far Western Europeans had come in understanding nature since the Early Middle Ages. Contemporaneously with the scientific enterprise of Copernicus and his late medieval predecessors, Renaissance artists had come to appreciate nature for her order and beauty in ways that would have confounded and angered early medieval intellectuals, most of whom were bishops and monks. Thus, the historian must exercise caution in making generalizations about medieval attitudes toward nature.
        Despite the pitfalls involved in assessing medieval attitudes toward nature, in 1967 two brave souls in academia, medievalist Lynn White Jr. and American historian Roderick Nash, dared to make some powerful generalizations regarding medieval evaluations of nature. We have seen in the previous chapter that Lynn White identified an arrogance and destructiveness toward nature which received its impetus from Genesis to extend human dominion over the earth. White accused Christianity of being more arrogant towards nature than any other of the world's religions. Complementing White's attack upon the ecological destructiveness of the Western worldview, Nash pointed out the consistently negative references to nature or wilderness in the Old and New Testaments, and also emphasized the Old Testament perception of wilderness as the antithesis of

the Garden of Eden, a cursed land of "thorns and thistles" into which Adam and Eve had been banished for their sin against the Lord.[15] Nash also generalizes that:

> In early and medieval Christianity, wilderness kept its significance as the earthly realm of the powers of evil that the Church had to overcome. This was literally the case in the missionary efforts to the tribes of northern Europe. Christians judged their work to be successful when they cleared away the wild forests and cut down the sacred groves where the pagans held their rites. In a more figurative sense, wilderness represented the Christian conception of the situation man faced on earth. It was a compound of his natural inclination to sin, the temptation of the material world, and the forces of evil themselves. In this worldly chaos he wandered lost and forlorn, grasping at Christianity in hope of delivery to the promised land that now was located in heaven.[16]

Both White and Nash (under White's influence) concluded that the only medieval figure of importance who appreciated the value of nature or wilderness was St. Francis of Assisi. The heretical beliefs of St. Francis, including the idea of the animal soul, were stamped out by the Church, however, in favor of the long-accepted idea of human dominion over nature granted at the time of creation. Lynn White went a step further in nominating St. Francis as the patron saint for ecologists. It would not be long before the uniqueness of St. Francis as the lone medieval Christian capable of appreciating value in nature, value deriving from nature being part of God's creation, would be challenged. Susan Power Bratton published essays demonstrating positive attitudes towards nature on the part of the anchoritic desert monks of the late Roman Empire and the Irish Celtic monks of the Early Middle Ages, focusing particularly on the good relationships maintained between these early Christians and their animal neighbors in the wilderness.[17] Nevertheless, medieval Christian attitudes toward nature could still be categorized as generally negative and destructive, particularly during the Early and High Middle Ages, as shown by David J. Herlihy's diachronic or historical approach to medieval values in an essay titled "Attitudes Toward the Environment in Medieval Society."[18]

Herlihy divided medieval attitudes toward nature into four overlapping categories which he called eschatological, adversarial, collaborative, and recreational, and which appeared more or less successively over the course of medieval history. The first, the eschatological category, actually developed during the third, fourth, and fifth centuries of the Christianized Roman Empire. Essentially, this is the foundational period of Christian theology rooted in the millennial apocalypticism which it assimilated from radical Judaism. Within this perspective, as we saw in chapter two, nature is perceived as old and worn out in preparation for the second coming of Christ, the Last Judgment, and the end of the world (the end of nature) preceding the Millennium. Not only did St. Augustine and his theological predecessors perceive nature as worn out, they also saw it as little more than an ephemeral prop or stage upon which the great Judeo-Christian drama is played. Nature in itself had no value except to support the Christian God's project of sorting out good and evil human souls, separating the City of God from the City of Man. This is the ultimate foundation of the extreme anthropocentrism of Christian, Western European civilization and it was passed on even to modern historians, who have ignored nature and human-nature relationships to a remarkable extent. Later in this chapter we shall see how this medieval eschatology was greatly intensified during the sixteenth and seventeenth centuries of our "modern" Western world, supposedly modern in the sense of being post-medieval but proving otherwise in that medieval supernaturalism has dominated our values well into the nineteenth and twentieth centuries.

The adversarial attitude, Herlihy's second category, " implied a fear and awe of nature as the abode of mysterious monsters, spirits, and powers inimical to men."[19] As Herlihy suggests, the Early Middle Ages or Dark Ages of the sixth to the tenth centuries owed their fear of nature, embodied in the dark and gloomy forest of Europe north of the Alps, to both Judeo-Christian fear of the wilderness and residual pagan traditions which recognized enchanted or sacred aspects of the forest. In combination, this mixture of religious myths and fears stemming from Christian, Greco-Roman, and Germanic traditions was further exacerbated by the real hazards of barbarous men and dangerous beasts in the forest. One of the most celebrated cultural monuments of the early Middle Ages, the Anglo-Saxon epic poem *Beowulf*, perhaps best captures the primordial and culturally enhanced fear which kept peasants and serfs from fleeing their villages to freedom in the forest.

Herlihy's third attitudinal category, which he terms "collaborative", dominates the High Middle Ages of ca. 1000-1300. This view of nature reflects the optimism associated with a sustained period of forest clearing for agriculture and grazing which was simultaneous with the revival of trade, commerce and urban life. Romanesque and later Gothic cathedrals sprang up everywhere as testimony to the flourishing activity which was understood as the work of man doing God's work. As Herlihy writes: "The universe, in sum, was the work of artisans working in tandem: God, to whom we owe the original elements; nature, which forms and renews the cosmos; and man, who is conscious of his own needs but who makes, in conformity with nature, cheese, shoes, and the other products of his economy."[20] This collaborative attitude rejected the Augustinian belief in the senility of nature for " a kind of ecological triumphalism, a belief that man the maker can shape the world according to his needs and multiply his own numbers with impunity."[21] Herlihy does not locate the root of this optimism toward the material world in citations from Genesis but rather in the experiences of a young and growing frontier society. However, the explosive development of the High Middle Ages led towards a massive ecological crisis in the fourteenth century and the return of eschatological beliefs " that the world had grown old and that the end of time was approaching."[22]

The fourth set of attitudes toward nature described by Herlihy, the "recreational," consists of those in which "the natural world, real or imagined, promises psychological or spiritual renewal, a release from the tensions and melancholy of everyday life."[23] Recreational perceptions of nature could not arise in the Middle Ages until at least part of the foreboding wilderness had been tamed. By the twelfth century, enough people were living in cities and towns in isolation from any wilderness frontier to be able to appreciate the "tamed nature" of the pastoral landscape of fields, pastures, and woodlots which surrounded them. Recreational attitudes arose almost simultaneously with the collaborative attitudes associated with the great forest clearings of the High Middle Ages and continued throughout the late Middle Ages despite the eschatological fears that accompanied ecological collapse, plague, and chronic warfare. Celebration of an idyllic pastoral landscape and of the beauties of spring were dominant themes in the recreational experience of humanized nature, particularly by urbanites. Similarly, the great Roman poets Catullus, Vergil, Horace, Ovid, and Juvenal were worshiping the pastoral landscape when they "worshiped nature." Typically, Virgil (70-19 B.C.), greatest of the Roman poets, wrote of the happiness of farmers:

> ...still they live without care and live without deceit, rich with various plenty, peaceful in broad expanses, in grottoes, lakes of living water, cool dark glens, with the brute music of cattle, soft sleep at noon beneath the trees: they have forests, the lairs of wild game; they have sturdy sons, hard-working, content with little, the sanctity of God, and reverence for the old.[24]

Nature was so tamed in Virgil's Italy of the first century before Christ that he even dared express such sentiments, in the same poem, as " let me love the country, the rivers running through valleys, the streams and woodlands-happy, though unknown. Give me broad fields and sweeping rivers, lofty mountain ranges in distant lands, cold precipitous valleys, where I may lie beneath the enormous darkness of the branches!"[25] Many centuries of Roman conquest and domestication had passed before Virgil could express his unbounded love of nature. Much the same had transpired in medieval Europe before the sophisticated bard could make comparable pronouncements of nature's beauty (i.e., humanized nature's beauty).

Against this background of the character of European, Faustian civilization, the natural history of Western Europe and its original wild landscapes and ecosystems, and the range of European attitudes toward nature, shifting over time under civilizing influences and other factors, we shall consider highlights of the environmental history of Western Europe from the early Middle Ages to the sixteenth and seventeenth centuries.

Environmental Change in Early Medieval Europe

Perhaps "The Dark Ages" was not a misnomer for the early Middle Ages after all. The half a millennium between the fall of the Latin Roman Empire and the end of the second wave of invasions (Viking, Magyar, and Muslim) by around 1000, was a period of relatively low population which continued the population decline of the late Roman Empire for at least several centuries. Decline of agrarian populations throughout the provinces of the former empire continued as a result of: (1) chronic warfare as the Germanic tribes defined their territories; (2) the flourishing of plague and other diseases under conditions of malnutrition exacerbated by the decline of agriculture and trade; and (3) depopulated cities languishing behind their new walls and surrounded by limited agricultural land. Under these conditions the forest returned to abandoned fields and pastures and the native fauna followed its return. The one great lamp of culture which was lit by Christianity during this period was the establishment of the Benedictine monastic order by St. Benedict of Nursia (c.480-c.540) early in the sixth century. The original Benedictines and the reformed orders to which they gave rise would play an important role in clearing the forests of western and central Europe during the Early and High Middle Ages. Regarding new evidence from aerial photographs in 1990 indicating "that the Roman fields had reverted to wilderness," C. Warren Hollister wrote that " the silent catastrophe they record may well have resulted from the violence and depopulation following Rome's collapse in the West, aggravated by a great plague cycle that commenced in the 540s."[26]

Thus, there has been a general belief among medieval historians through the later twentieth century that Western Europe in the early Middle Ages consisted of a sprinkling of depopulated towns surrounded by a vast, unpopulated forest wilderness. Writing in the mid-twentieth century, H.C. Darby observed that during the Germanic barbarian invasions " the woods crept back over upon the many neglected fields. With the advent of more settled times, the attack upon the woodland was begun in earnest and was to gather force with the centuries."[27] Darby further observes that by around 800, the time when Charlemagne became Carolingian emperor, a considerable retreat of the forests had already occurred in parts of the empire. However, the barbarian assaults which followed over the next two centuries would have set back the process of deforestation. Writing in 1991, environmental historian Clive Ponting corroborated Darby: "Early medieval Europe was still a vast wilderness with a scattering of small, largely self-sufficient villages which had only very limited outside contacts."[28] Because the overall population within Western Europe remained low during and after the Carolingian Empire, forest clearance was limited until about the end of the tenth century. After that, forest clearance was completed in the Île-de-France about 1080 and in southeastern England shortly thereafter.

The two regions of Western Europe for which relatively good records of environmental change, particularly deforestation, exist are England and France.

In England for example, according to Peter Coates, accumulating evidence from aerial photography, non-excavatory archeology, pollen analysis, archeobiology, the derivation of place names, and pre-enclosure maps indicate the shift from forest to heath beginning with the fires of Mesolithic hunters more than 9,000 years ago. Half of England probably had lost its old growth forest by 500 B.C., and large mammals such as the lion, leopard, bear, and elk had vanished by that time. Some of the land reclaimed from marshes underwent afforestation after ca. 410 when the Romans left, but based upon the Domesday survey of land holdings under William the conqueror, only 15 percent of England was forested by 1086.[29] Roland Bechmann, in *Trees and Man: The Forest in the Middle Ages* (1990), supports Coates in estimating that " England at the time of the Norman Conquest included a percentage of forested land that ... was no more than 15 percent."[30] Therefore, in comparison to continental Europe, one can conclude that the ecology of England was humanized rather early in its history, and the conversion of fen and forest to farmland and pasture nearly complete at the beginning of the High Middle Ages.

Bechmann's account of forest clearing in France begins with his observations on afforestation during the late Roman Empire. Depopulation and the return of the forest tempted the people of Germania to occupy a land of rich soils. Following the invasions and abandonment of estates in the fifth century, aerial photographs show that many Gallo-Roman buildings were destroyed and their sites planted during the sixth century. Subsequently, forests were substantially reduced by clearings during the Merovingian and early Carolingian periods. Population appears to have stabilized in the ninth and tenth centuries, temporarily stopping the clearings, but around 950, after the Viking, Magyar, and Saracen invasions had ended, increased population led to further agricultural clearings. By this time, medieval records suggest that the Benedictine and offshoot Cluniac monasteries played a leading role in the clearings, but this is probably much exaggerated because history does not mention the modest but countless peasant clearings created both by enlarging areas at the edge of the forest and by creating new clearings deep in the woods.[31] Despite this caveat, Bechmann concludes that " the monasteries certainly had an important impact on the forests in France."[32] Their importance is also better documented during the High Middle Ages.

While considerable forest clearing occurred during the Early Middle Ages in much of what is now England and France, central and Eastern Europe remained heavily forested. The lands abandoned by the Germanic tribes during the late Roman Empire were occupied by Slavs migrating westward. Like the Germans who preceded them, they practiced swidden agriculture, which allowed the forest to return over farmed areas. This is known from the observations of the Roman historian Tacitus (ca. 55-120 A.D.), who, in the *Germania*, explains that the people of Germany " change their plough-lands yearly, and still there is ground to spare."[33] According to Darby, the Germans began to reoccupy these lands with permanent settlements in the tenth century. This began the development of an eastward-migrating "frontier" which continued throughout the High Middle Ages.

The European Conquest of Nature in the High Middle Ages

Following the second wave of barbarian invasions, the late tenth and early eleventh centuries witnessed the beginning of relentless and continuous deforestation within an emerging Western European civilization and on the eastern frontier in Germany. The established Christian medieval civilization gathered momentum from improvements in agricultural technology, population increase, the growth of towns, and the resumption of trade, opening up contacts with Islam and Byzantium which would lead to the diffusion of new knowledge and new perspectives into a culture which had existed for half a millennium under conditions of extreme insularity of thought and behavior. The absolute dominance of Christian belief over secular concerns during the Early Middle Ages gave to Western civilization a powerful sense of purpose and of superiority, but at a great price: "The demands of the next world occupied the attention of devout Christians, and so deterred any compelling interest in nature, science, history, literature, or philosophy for their own sake."[34] Despite Europe's isolation, however, new breakthroughs in European agricultural technology such as the redesigned horse collar, the tandem harness, and the heavy plow, as well as the watermill, produced food in greater abundance and variety, thereby allowing population to double from about thirty-five or forty million in the eleventh century to eighty million by 1300. Simultaneously, growing towns experiencing a commercial revolution had become cities of 100,000 in northern Italy by 1300, boosted by international trade in commodities such as cloth, grain, salt, slaves, and wine.[35]

Against this background of a surging Western European civilization which established frontiers in Iberia, the Levant, and in Germany during the High Middle Ages while building Romanesque and Gothic cathedrals in its burgeoning towns and cities, we return to the fate of the boreal forests and wetlands of England, France, and Eastern Europe. Darby has established that much forest clearing had already been accomplished on the European continent by 800 in parts of the great forests of Charlemagne's empire, such as the Ardenne. Also, shortly thereafter, the valleys of the Odenwald near the Rhine border of France and Germany were cleared and "forest villages" established. These towns on the German frontier differed from the nuclear village surrounded by radiating fields and pastures, such as were found in Gaul, in that " houses were laid out in a single or double row, usually along a valley bottom, and behind each house stretched its land in a long, narrow belt reaching back into the wood."[36] This "linear village" would be the prototype for agrarian development of central and Eastern Europe.

The high-medieval history of the British environment begins with the results of the Domesday survey, which showed only 15 percent of England in forest and wood-pasture by 1086, with a further 5 percent decline in forest cover by 1350. This period was characterized by temperatures several degrees centigrade warmer than they would be in succeeding centuries, and the population of England soared "from 1.5 to 4 million between 1086 and 1300, during which period vast tracts of increasingly marginal fenland, forest and hillside were cultivated."[37] Warmer temperatures also allowed cultivation at higher elevations than today. By the second half of the thirteenth century, however, farming on marginal lands was becoming unsustainable, and after 1300 colder and wetter conditions prevailed.

Describing the clearing of the woodland in France, Darby summarizes the detailed accounts of "l'âge des grands défrichements," the great heroic period of reclamation from c. 1050-1250. The monasteries were given credit for organizing this effort, with the older Benedictine monasteries having pioneered in the mountains of the Dauphiné, on the plains of the Île de France, and elsewhere in the eleventh century. In the twelfth century the newly created Cistercian order, with its stress upon manual labor in the field, "sought the wilderness and became the great farmers of the Middle Ages, and they were to find that solitude was but a poor defender of poverty."[38] Cistercian houses numbered five hundred by the end of the twelfth century. In addition to the work of the monasteries, peasant villages extended the arable land around existing villages and also founded subsidiary hamlets. Darby writes that these kinds of clearings "had come to have a definite place in manorial economy,"[39] as indicated by monastic and other documents. A third type of clearing activity involved the establishment of entirely new settlements, village colonies which were organized to cultivate not-too-distant forest or wasteland.

Bechmann provides a far more detailed account of deforestation in high-medieval France. Acknowledging that the monasteries played an important role in clearing the forest since the Early Middle Ages, Bechmann attempts to assess their relative importance. He points out that the monks were first portrayed by historians "as destroyers of the forest in the name of civilization and agriculture ," but that subsequently the importance of their role in the forest clearings was diminished by historians because they were the only ones who had left a written record of their acts, thereby overshadowing the greater, unrecorded role of the small peasant. Conceding that the monks appreciated the value of the forest in the medieval economy, and also that they wanted the isolation provided by the screen of forests, Bechmann still cautions against the extreme view that their role in clearings was negligible. Arguing for the importance of the monasteries, he first points out that nobles typically transferred lands that were forested to monastic orders rather than lands already cultivated. Secondly, it is clearly documented that the monasteries affected timberland by a combination of thinning, clearing, and grazing. Thirdly, villages which grew up around the monasteries, many of which grew to become important towns, resulted in a ring of cultivation taken from the original forests. "Almost everywhere one can notice the disappearance of forests near religious establishments, even if in many cases they prevented their total destruction."[40]

In addition to the role of the monasteries as promoters of forest clearances during the High Middle Ages, other important factors included widespread development of vineyards, which used large quantities of wood once areas were cleared, the promotion of large enterprises by the nobility (often acting in conjunction with the monastic orders, including the Templars), and the development of new towns or "villeneuves" and "villefranche." The demand for increasing vineyards resulted both from population growth and the degradation of local water supplies by pasturing of domesticated animals. The big pioneering operations were promulgated by the large landowners who possessed sizeable areas of "wasteland," i.e., fen, forest, and even coastal marshlands which could be reclaimed from the sea. Such operations often required the cash which only the monasteries possessed, and involved transfers of populations and equipment over long distances. The larger of the pioneering operations typically gave rise to the "villeneuves." In these new towns peasants or serfs were guaranteed a measure of autonomy which sparked incentives to produce surpluses which benefited their sponsors. These conditions also enticed inhabitants of ancient villages in the region to settle in peasant communities as nobles offered competing privileges in order to retain their farm laborers.[41]

In France, agricultural production kept up with the demands of a growing population to the extent that famines almost disappeared in the twelfth century. Simultaneously, people began to realize that the forests were almost gone, and rules aimed at conserving woodlands for hunting and other purposes began to appear in the early thirteenth century. Meanwhile, in central Europe, the "German frontier" continued to press eastwards into the great "primitive forest" or wilderness which, at least in remote and sparsely inhabited areas, probably resembled the Bialowieza old-growth remnant. The fact that the German tribes had migrated from this region in late Roman times under pressure from the Huns, only to be displaced by Slavs, suggests that this so-called wilderness, eighty percent forested at the time of Rome's fall, was not so wild as once thought. Simon Schama's *Landscape and Memory* and the writings of William Cronon and others in environmental history have made the general argument that humans have been inextricably entangled with nature far back in time before recorded history. A useful case history of the argument as applied to Europe is to be found in Stephen J. Pyne's *Vestal Fire: An Environmental History, Told Through Fire, of Europe and Europe's Encounter with the World* (1997).[42] The upshot of Pyne's thesis is that human interaction with post-glacial European ecology has so severely transformed indigenous flora and fauna since Mesolithic and Neolithic time that even the "primitive forest" or "wilderness" of Europe should be understood as a human artifact. Therefore, before completing our survey of the high-medieval conquest of nature on the German frontier, we should pay heed to some particulars of Pyne's argument.

Pyne believes that he has proved that Europe's " climate, biota, soils, and humans had all coevolved, woven into a fantastic tapestry of trees, river, livestock, humus, grasses, shrubs, and insects strung across the frame of this landscape loom."[43] Largely through agriculture and the use of fire Europeans had modeled their landscape, especially its plant cover, since the Pleistocene Ice Ages, and "landnam, the shock of first contact, was Europe's true creation story."[44] In Mesolithic times, the fires of hunters and foragers were constrained by a wet climate and a closed and shaded plant community, but the Neolithic revolution brought agricultural fire to central Europe some 4,000 years ago. Taking the pollen record (preserved in bogs) of Denmark as his model for central Europe, Pyne documents the transition from Late-Glacial tundra through steppe and pre-Boreal forests of birch and pine in which elk (European moose) and auroch replace reindeer, after which the Boreal hazel-pine forest dominates for millennia while providing a prolific food supply for Mesolithic humans. Subsequently, elm, oak, ash, and linden joined the hazel, to mature into the primeval European forest. Pyne then explains the spread of the linden or European lime tree to dominate the forest and turn it into a floral and faunal "desert." This equilibrium was broken after 3000 B.C., however, by a combination of climate cooling and the onslaught of Neolithic humans who "promoted shrubs, herbs, and forbs over trees, added new flora and fauna, served as a vector for disease and pest, and even altered the soil. They disassembled the linden forest as readily as they stripped away its bark and burned the residue."[45]

Pyne continues his Neolithic scenario by explaining how a combination of browsing by cattle, swine, sheep and goats, foraging, hunting, and under-burning " reworked a shade-tolerant high forest into a sun-flushed low forest, full of shrubs, grasses, forbs, and woody coppice."[46] Piecemeal restructuring of the forest continued into historic time along with minor climatic variations which helped to replace the linden with beech while fire-assisted agriculture altered both the structure and the composition of the forest. The heaths or wastelands of evergreen shrubs so common in northern and central Europe were also initiated at this time as anthropogenic creations. Thus was the original deciduous forest of western and central Europe transformed by the shock wave of landnam into grasslands, heaths, peat bogs, fields, and gardens, all of which the Roman Caesars dismissed as a "dismal wilderness."[47] Following the Germanic invasions of the late Roman Empire, Pyne describes the "Great Reclamation" when a more sedentary agriculture radiated out from centers in northern Gaul. Around 1050, the Great Reclamation, known to other historians as the "Age of Great Clearings," accelerated to produce a second wave of landnam in the primeval Hercynian forest created by the first wave of Neolithic agriculture. On this final frontier "monasteries supplied the agrarian model, villagers the greatest labor, and east-migrating Germans the most dramatic saga."[48]

Readers of Alfred Crosby's *Ecological Imperialism* will immediately recognize the similarities between Pyne's first and second waves of forest transformation and destruction by Germanic peoples and Crosby's first wave of Amerindian hunter-gatherers gradually practicing swidden as well as settled agriculture, and his second wave of post-Columbian Europeans. In both cases the forest has been burned and restructured by the first wave, and then largely removed by the second. In both cases the second wave has perceived the forest as a "primitive wilderness" even though it had actually been greatly modified by the first wave of human inhabitants. Consequently, there were true ecosystems or biomes only prior to the initial arrival of *Homo sapiens* on the scene. Once mankind arrives, "nature" is transformed into a human artifact, and over time his fellow, competing large mammals are annihilated, and the second wave of inhabitants has only smaller and less dangerous species to compete with. If this contemporary interpretation of environmental history is valid, then where is the much-worshiped purity of wild nature in these "wildernesses" that turn out, after all, to be anthropogenic creations?

Returning to the German frontier of central Europe and the last remnant of the primeval wilderness of Poland's Bialowieza, described by botanist Edward Klekowski as "the forest our ice age ancestors saw,"[49] one wonders to what extent humans, whose presence Klekowski acknowledges, had already changed the Bialowieza by the time the region was generally converted to sedentary agriculture. Is the Bialowieza raw, wild nature, i.e., a natural terrestrial ecosystem or biome, or have its flora and fauna been selected to some extent by humans? Moreover, does hominid interaction with a biome in pre-historical times disqualify it as wilderness? In the new, deconstructive environmental histories it would seem that few regions on the earth, including the oceans, have not been tampered with by *Homo sapiens* or his hominid ancestors, leaving little that we can call wild nature in the sense of not having been manipulated by humans. It seems, however, that if any ecosystem is disturbed by humans but nevertheless remains an essentially alien environment to *civilized* humans, then it still deserves the appellation of "wilderness," just as the Romans perceived the northern European forest. On the other hand, given our recent knowledge of humanity's ubiquitous presence in nature, is wilderness better understood as a matter of degree of interference, measured perhaps in terms of the persistence of floral and faunal species despite some limited human impact? These issues will be further considered in a later chapter.

Even Darby, writing half a century ago of the age of great clearings, had some awareness of forest clearing by those who preceded the high-medieval German frontier when he writes that "we must not forget the work of the Slavonic peoples themselves ... Slavonic lords and peasants may have been responsible for much clearing," also, "in Poland, Bohemia, and elsewhere, they founded new villages and reclaimed the wilderness."[50] The point is that despite his reference to "wilderness," Darby may have been aware of shifting populations in the primitive forest further back in time. Acknowledging the continuing importance of the monasteries during the Great Reclamation in Germany, Darby reminds us of the analogy which "has been drawn between this advance and the expansion of the American people westward from the Atlantic seaboard."[51] Not only were the forests cut down and burned, but marshes were drained, streams channelized, and mines developed in mineral-rich mountainous regions in Bohemia and Austria. By 1300, however, the eastward march of the German frontier, like the interior frontiers of western Europe, had exhausted itself, and the vicissitudes of the Late Middle Ages were about to begin.

Ecological Crisis in the Late Middle Ages

Medieval values during the High Middle Ages were dominated by "collaborative" as well as "recreational" attitudes, the former expressing the sense of doing God's work on the forest frontiers of Europe, and the latter representing a newfound enjoyment of the tamed nature resulting from the transformation of woodland into fields and pastures. In the fourteenth century these attitudes would persist, but the older, "eschatological" attitude associated with the desert saints and Saint Augustine would return with a vengeance. Commenting on the values of the High Middle Ages, Clarence Glacken opines that the ambition of the nobility and churchmen alike "called for activity and change as a part of economic expansion and of conversion." Moreover, Glacken recognizes a yearning for control over nature beginning with the age of great clearings. "In the later Middle Ages the interest in technology, in knowledge for its own sake either to improve thinking or to better the human condition, in clearing, and in drainage and the like betrayed an eagerness to control nature."[52] The illusion of controlling nature under the auspices of the Christian God created a climate of certainty and optimism which, combined with the increase of wealth in burgeoning cities, produced the magnificent High Gothic cathedrals. During the fourteenth and fifteenth centuries, however, the Age of Faith would be challenged materially by climatic, epidemiological, and environmental crises, and spiritually by the deterioration of the papacy and the rise of Renaissance humanism in Italy.

Hollister observes, for example, that the Late Middle Ages "were marked by a gradual ebbing of confidence in the values on which high-medieval civilization had rested. General prosperity gave way to sporadic depression, optimism to disillusionment, and the thirteenth-century dream of fusing the world of matter and spirit faded."[53] On the other hand, Norman Cantor calls our attention to the creativity of the Late Middle Ages despite "devastation from pandemics, war, climatic deterioration, and economic depression .... No era in western civilization left a heritage of more masterpieces in literature and painting or seminal works of philosophy and theology."[54] Cantor also reminds us that the period 1270-1500 gave rise to two distinct and contradictory cultures, the Renaissance in Italy versus the northern European extrapolation of high medieval values, whose incompatibility "produced a monumental intellectual and moral crisis around 1500 that brought on the Protestant Reformation."[55] At the earthy, substructural level, the agricultural crisis of the fourteenth century and attendant economic depression underlies these dynamic shifts in worldview. The key factors contributing to this crisis included the end of the medieval "warm" period and the onset of the "Little Ice Age," tripling of the European population during the High Middle Ages, soil exhaustion on overworked farmland and marginal lands, the gradual breakdown of feudalism into a market economy, and, both cause and effect, increasing famines.[56] Thus, the agricultural crisis of the fourteenth century preceded the Black Plague and interacted with it during the century that followed, thereby compounding the demographic collapse of the Late Middle Ages. Let us, then, consider the contributing environmental factors, beginning with the Little Ice Age, as part of what some historians have called "the ecological crisis" of the fourteenth century.

The Little Ice Age is a well known natural phenomenon, spectacularly recorded in the events of the Late Middle Ages. Barbara Tuchman, in her highly acclaimed *A Distant Mirror: The Calamitous Fourteenth Century* (1978) devoted less than half a page to it, describing it as having been caused "by an advance of polar and alpine glaciers and lasting to about 1700."[57] Neither Hollister nor Cantor even mention it in their medieval histories. Enter the ubiquitous Brian Fagan and a history of the period 1300-1850 organized around climate change: *The Little Ice Age: How Climate Made History, 1300-1850.*[58] In *The Little Ice Age* Fagan argues that a complex state of flux has always existed between short-term climatic fluctuations and human relationships to the natural environment, reminding us that we tend to forget how little historical time has passed since Europeans were at the mercy of harvest failures resulting from combinations of climatic shifts, human misuse of the land, and misguided political or economic policies. Although he rejects environmental determinism as simplistic, Fagan argues that breakthroughs in paleoclimatology during the last decades of the twentieth century have rescued climate change from being " the ignored player on the historical stage.... Now we know that short-term climatic anomalies stressed northern European society during the Little Ice Age, and we can begin to correlate specific climatic shifts with economic, social, and political changes ...."[59]

Fagan explains the effects of irregular shifts in atmospheric pressure over the north Atlantic, known as the North Atlantic Oscillation or NAO, upon the climate, especially the rainfall, of much of Europe. The cause of the NAO is not understood, but variables such as sea surface temperatures, the rate of sinking of northern surface waters, and the strength of the Gulf Stream are thought to be major controlling factors in its functioning. The NAO produced unpredictable alterations of warm-wet and cold-dry weather patterns in cycles lasting up to seven years or more, or even decades, or they can be much shorter. These cycles were superimposed upon the long-term cooling that lasted approximately from 1300-1850, although increasing cold had already begun in the thirteenth century, creating difficulties for Norse ships voyaging between Iceland and Greenland. By 1312 a "high" NAO, which shifts the Atlantic storm track southward to produce mild winters and above-average rainfall, began a warm-wet cycle which by 1315 started a spring deluge across northern Europe. The rains were almost incessant throughout May, July, and August, causing floods that swept away villages in central Europe and eroded deep gullies in hillside fields that had been cleared during the High Middle Ages. As Fagan notes, "Villages throughout northern Europe paid the price for two centuries of extensive land clearance."[60] The following year heavy spring rains prevented the sowing of most crops, crops failed, herds were decimated, and famines spread.

By the end of 1316 the peasantry of northern Europe was impoverished and malnourished, even reduced to eating cats, dogs, and worse. A bitterly cold winter in 1317-1318 gave rise to famine and malnutrition amongst farm animals, followed by outbreaks of disease which decimated the herds of northern Europe, but somehow leaving pigs relatively unaffected. Fagan states that up to 10 percent of the urban population of Flanders died in the Great Famine. The year 1317 witnessed yet another wet summer, and European society began to believe that divine retribution for ongoing warfare and other sins of humanity was responsible for the terrible weather and resulting famines. This climatic cycle lasted from 1315 to 1322, ending with a severe winter and followed by unpredictable weather as the Little Ice Age took hold. "There were arctic winters, blazing summers, serious droughts, torrential rain years, often bountiful harvests, and long periods of mild winters and warm summers."[61]

Human demography was influenced not only by climatic change and famine, but also by other causes of population decline in the fourteenth and fifteenth centuries, including the intermittent but savage Hundred Years War (1337-1453), and the Black Plague of 1347-1351, followed by succeeding outbreaks which continued into the seventeenth century. Taken together, these events, combined with preceding environmental degradation, have produced what medieval historian Charles R. Bowlus has designated an "ecological crisis in fourteenth century Europe."[62] Bowlus compares the effects of the Black Plague or "Black Death" to those which might result from a nuclear holocaust in our own day. Although only one of a series of crises which undermined the high medieval foundations of European civilization, the plague appeared to many contemporaries to represent God's wrath over the sins of Latin Christendom. Bowlus, however, noting that the medieval preoccupation with theodicy has been long abandoned by modern historians, believes that the fourteenth century crisis, including the Black Death, "were caused at least partially by the sins that Europeans had committed against their natural environment during the twelfth and thirteenth centuries ... that an ecological crisis was the root cause of the disasters that plagued Europe during the 1300s."[63] He agrees with the view of other medieval historians that chronic undernourishment of the late medieval peasantry contributed to high mortality rates during the plague. This view is challenged by David Herlihy in *The Black Death and the Transformation of the West,* in which he denies that there are "direct linkages between famine and plague, malnutrition and disease."[64] He argues that under certain conditions malnutrition can prevent infection by not providing the nutrients that germs require in order to multiply.[65] Despite Herlihy's dissenting opinion, the consensus of medieval historians has been that severe malnutrition resulting from famine renders most individuals more prone to disease.[66]

Viewed from an economic perspective, the boom of the twelfth and thirteenth centuries was followed by a depression in the fourteenth century and a slow recovery in the fifteenth. Some of the wealth which funded Renaissance artists would have been invested in commerce and forest clearings during the economic boom of the High Middle Ages. Bowlus argues that an over-expanded economy, relative to available resources and existing technology, created an environmental crisis by the end of the thirteenth century. The expansion of agriculture into marginal lands such as the Welsh highlands, the Black Forest, and the upper watersheds of Alpine regions gave rise to slope erosion, siltation of stream and river channels, and flooding which destroyed crops and soils along the floodplains of large rivers in Europe. A wood shortage also developed, along with rapid population growth supported by a more efficient system of agriculture resulting from such innovations as the heavy, wheeled, mould-board plow, increased use of horse power aided by an improved horse-collar, and the three-field system of crop rotation in northern Europe.[67]

Urbanization accompanied sustained population growth and by 1300 some European cities, such as Florence, were forced to import more than half of their grain from distant sources. This situation also resulted from regional specialization in commodities such as wine and wool. Bowlus thinks that the loss of European forests and ensuing shortages of wood were central to the fourteenth century economic and ecological crises. Timber shortages were common in mining districts, and wine often could not be shipped for lack of oak casks. However, the most spectacular example of ecological limits having been overreached in the fourteenth century was, according to Bowlus, general and recurring famine:

> There had been famines in the twelfth and thirteenth centuries, but those had largely occurred in regions where local crops had failed and cereals could not be imported from elsewhere because of lack of port facilities or financial strength in the region affected. There are no recorded examples of general famines in Europe between 1100 and 1300. In the fourteenth century general famine became common, however, the most serious coming between the years 1315 and 1317. The summer of 1314 had been an unusually wet year and yields were low all over Europe.[68]

Considering the important role that climate played in the famines of the early fourteenth century and later (as Fagan has demonstrated), Bowlus admits that they may not have occurred if climate had not worsened in Europe during the fourteenth century. Nevertheless, he infers from abundant evidence "that many marginal lands had been brought into cereal production during the age of agricultural expansion and that these lands were losing their fertility before the fourteenth century had dawned. Even some prime grain lands experienced declining yields prior to the end of the thirteenth century."[69] In addition, limits to animal husbandry appear to have been reached. Certainly pig production was constrained by the loss of forest cover, and increased numbers of horses (for plowing) took oats away from the peasant diet. Transhumance also kept manure far from grain-producing lowlands, and removed vegetative cover, thereby increasing erosion in hilly and mountainous regions. During the thirteenth century the Alpine districts of central Switzerland were colonized, and "communities reached agreements with one another after 1300 to prevent further clearing and overgrazing because of the danger of avalanches which descended without warning into deep, narrow valleys."[70] Thus, the evidence leads Bowlus to conclude that Europe's agricultural economy was approaching natural limits at the close of the thirteenth century.

The economic stagnation that accompanied these constraints did not cause, but contributed to the growing chaos of the fourteenth century. Declining affluence forced the Avignon papacy to husband its resources more efficiently, contributing to its unpopularity. Also, economic competition between commercial centers intensified, and the credit structure of medieval Europe collapsed in 1341. Almost simultaneously, diabolism, the inquisition, and the Black Death exploded upon the European scene, collectively testing and undermining the already weakened authority of the medieval church. Europe would not recover from the demographic impact of the plague until the second half of the fifteenth century, and the psychological impact of plague, famine, and war gave rise to spiritual disillusionment.[71]

Early Modern Sequel: Demographic Explosion and Apocalyptical Christianity

The "form" or fundamental structure and worldview of Western European civilization was severely strained by the overwhelming concentration of natural and historical events of the fourteenth century. However, none of these events was more destructive of both human life and of the morale of European civilization than the Black Death, although recurring famine, continuing into the seventeenth century in England and the eighteenth century in France, was comparable to the plague in its demoralizing effects. Historians have long debated the question as to what degree the Black Death was essentially a natural catastrophe or a largely anthropogenic event. William McNeill, in *Plagues and People*, has argued that most contagious diseases, with the exception of the Black Death, represent "microparasitism," the infection of peripheral cultures by diseases carried by invading cultures which have grown relatively immune to the disease over time, with diseases such as measles and smallpox being far more effective than bubonic plague in killing conquered peoples while passing over immunized conquerors. The Black Death differs in requiring a combination of rats and fleas as vectors of the disease to human populations. Macroparasitism, the control of conquered peoples to the advantage of the conquerors, initially has microparasitism on its side during the time of conquest. However, the Black Death was lethal to both conqueror and conquered through the vectors of black rat and flea.[72] Furthermore, in order for the combination of plague virus, black rat, and flea to develop, the black rat first had to escape from its original habitat, most likely in India, which it probably managed to do as a passenger on ships trading between Egypt and India. The black rat appears not to have reached northern Europe in Justinian's time, when bubonic plague first struck the Mediterranean basin, thus temporarily confining the plague to coastal regions of the Mediterranean.[73]

Bowlus is not the only historian to describe the fourteenth century as one of ecological crisis. David Arnold, in *The Problem of Nature: Environment, Culture and European Expansion*, writes: "It has been argued that the greatest single environmental crisis to hit Europe ... in the past two thousand years was the Black Death" which killed twenty million people in Europe alone between 1346 and 1351. He also cites Robert S. Gottfried as calling the Black Death Europe's "greatest ecological upheaval" and a major turning point in Western Civilization.[74] In 1361 and 1369 bubonic plague returned and killed another 10-20 percent of Europe's population, thereafter returning every ten to twelve years for some time until its final convulsions in the Great Plague of London in 1665 and a serious epidemic in Marseille in 1720.[75] Regarding individual susceptibility to the plague, Arnold considers the thesis of Herlihy and others that one's health or state of nutrition did not determine individual vulnerability, but prefers patterns of social organization as of greater importance than individual health, with the concentration of populations in cities, especially if increased by influxes of panicked and infected peasants, as the key factor in determining susceptibility.[76]

In addition to vacating enormous areas of agricultural and grazing lands and allowing the re-growth of forests during the fourteenth and early fifteenth centuries, undermining the feudal system as peasant wages increased, and shifting the medieval worldview towards apocalyptical extremism, the demographic impact of the plague must be appraised against the long-term population trend of Europe, which was steadily upwards despite setbacks due to chronic warfare, famine, and plague. Also, the steady expansion of Europeans into the Mediterranean basin and down the west coast of Africa as new techniques of navigation were developed and acquired, the rise of Renaissance humanism in Italy, and many other positive developments coincided with the intermittent plagues and famines of the fifteenth century. Nevertheless, a deep sense of pessimism permeated the late medieval world view before Columbus's discovery of America, and Columbus himself manifested this spiritual pessimism. Reflecting upon the general disillusionment with the papacy and the condition of Europe at this time, John Opie writes:

> Personal faith was confused. The visible world was an endless succession of violence, scarcity, pestilence, injustice, and human cupidity. The horrors of the papacy gave ordinary people a sense of frenzied desperation and anguished loss. The people of Western Europe felt themselves deceived at the highest levels; their spirituality was submerged in visions of death and future suffering,

as in the demonic scenes of horror and punishment in the paintings of Grünewald and Bosch.[77]

To men of more penetrating insight, the late medieval church displayed " a vast bureaucratic superstructure, unrelated to the beatific vision."[78] Adding to the uncertainty and cynicism that the failure of the church produced, by mid-sixteenth century Copernicus would rearrange the heavens themselves, ultimately another enormous embarrassment to the church. The inability of the church to prevent the horrors of plague and famine also led to the rise of a more magical Christianity, i.e., the cult of the saints and their purported protective and healing powers. This development led to religious dispute and ultimately contributed to the Reformation.[79]

The environmental history of the sixteenth and seventeenth centuries is summarized by Carolyn Merchant in *The Death of Nature: Women, Ecology and the Scientific Revolution*. Merchant's fundamental thesis is that the late medieval view of nature as an organism (acquired from the philosophers of antiquity, particularly Plato's *Timaeus*) was transformed during the sixteenth and seventeenth centuries into the modern, mechanistic model of nature which has dominated Western European attitudes toward and treatment of nature down to the present day. In the pre-modern paradigm, nature was also understood as being feminine rather than masculine in character. As Merchant writes: "Between the sixteenth and seventeenth centuries the image of an organic cosmos with a living female earth at its center gave way to a mechanistic world view in which nature was reconstructed as dead and passive, to be dominated and controlled by humans."[80] Summarizing the major ecological changes which occurred during this historical period, Merchant emphasizes the importance of the shift from medieval subsistence agriculture to agriculture increasingly under a growing capitalist system. Growing crops for profit led towards new technologies which increased efficiency of production and increased specialization in types of crops grown, and extended farming and grazing into fens or marshlands, with Holland and England leading the way.

The demographic collapse caused by the Black Death and famine reduced European population to its lowest between 1400 and 1450, thereby allowing soils to recover their fertility and the forests to return over abandoned marginal lands. As peasants gained greater freedoms in the following century, the sixteenth and seventeenth centuries witnessed numerous peasant revolts and the gradual destruction of the feudal system except in eastern Europe. Simultaneously, the spread of capitalistic economic practices from the city-states of Renaissance Italy to northern Europe expanded the market economy, but with destructive ecological consequences. In England, enclosures of the common lands for sheep pasture forced subsistence farmers off arable lands, followed by overgrazing when control of grazing animals was lax. However, the destruction of European marshlands through reclamation was the most significant ecological impact of the early modern centuries, with widespread disappearance of wildlife, especially water birds and related species. The recently expanded forests of Western Europe fared just as badly under the increasing demands of mercantile capitalism. The greatest demands for wood came from shipbuilding and the mining and smelting of iron, copper, and other metals. Although the Venetians had practiced conservation as early as 1470, demands for timber and metals for the endless warfare between early modern nation-states, typically over religious differences and struggles for dominance in the colonies abroad, overrode attempts to manage and conserve forest resources. In England, deforestation led to the gradual substitution of coal for wood as fuel, producing smoke and soot in London that vexed Queen Elizabeth to such a degree in 1578 that she asked the brewers and other industries of London to shift back to using wood, which they could not and did not do.[81]

Merchant's 1980 account of early modern European environmental history makes no mention of climate changes, which does not appear to be a major factor in turning Europe's ecosystems into a network of farms, grazing lands, and wood lots. Two decades later, Fagan's *The Little Ice Age* makes climate the major cause of environmental and social change in western Europe, as indicated by its sub-title, *How Climate Made History: 1300-1850*. His account of the sixteenth and seventeenth centuries tells of the transformation of sixteenth century subsistence agriculture, little different from that of the Middle Ages, though a gradual "agricultural revolution" involving "intensive commercial farming and the growing of animal fodder on previously fallowed land." The revolution began in Flanders and the Netherlands in the fifteenth and sixteenth centuries and spread to England during a period of erratic climate and intense cold. On the continent, France lagged behind despite the climatic vicissitudes and poor harvests. "Millions of poor farmers and city dwellers lived near the edge of starvation, as much at the mercy of the Little Ice Age as their medieval predecessors."[82] Although climatic variation indirectly influenced Europe's ecology by habitat destruction as agriculture became both more intensive and extensive, Fagan is interested in the human historical consequences of the Little Ice Age rather than its ecological effects. The end result of Europe's successful response to the Little Ice Age has been a policy of "total farming" which has maximized agricultural production while destroying most of Europe's ecosystems.

The sixteenth and seventeenth centuries, while witnessing the rise of absolutist nation-states and the drama of Renaissance and Reformation, likewise were an era of renewed demographic explosion. Approximately between 1500 and 1650 Europe's population doubled after its decline in the fourteenth and fifteenth centuries in the shadow of the Black Death. Not that the Black Death disappeared in early modern times. Indeed, it persisted in epidemic form across the European continent from 1494-1648 about once every nine years, and was at least present in some part of Europe during every year of that period.[83] Simultaneously, such diseases as smallpox, typhus, and syphilis took their toll on the population. Andrew Cunningham and Ole Peter Grell recount the history of this period in *The Four Horsemen of the Apocalypse: Religion, War, Famine and Death in Reformation Europe*, in which they demonstrate the powerful impact of chronic warfare, famine and disease upon the worldview of Western European civilization, creating an age of apocalyptical religious fanaticism which saw the apocalypse, last judgment, and millennium at every turn. The recently invented printing press helped to fan the flames of popular belief and the witch hunts and atrocities which ensued. Even by around 1550 the perception that Europe was full of people, that even the densest forests or highest hills were cultivated, had become commonplace. "Indeed, this demographic change, this population pressure, underlay all the crises of the age: it can be said to have created the crisis mentality which made the Four Horsemen of the apocalypse the popular image of the age."[84] Once again the perception that the earth was worn out with age and use was used as evidence that renewal through Christ's return was imminent. Describing the early modern obsession with omens and portents as representing "the eschatological decay of nature," Cunningham and Grell write that "signs and portents of approaching apocalyptic disasters were not seen as restricted to the sky. Nature itself in its accelerating decay and old age was providing similar signs of the impending End."[85]

Thus we hear in sixteenth and seventeenth century Europe the same familiar complaints that we heard from Tertullian and Augustine during the imperial centuries of Rome. The truth in both the early modern and late Roman observations of decay, of course, is that the landscapes and ecosystems in question had indeed been used to the point of severe environmental degradation. Within the apocalyptical Christian worldview, however, in the absence of the slightest degree of ecological knowledge, the interpretation of the worn-out earth as being ready for renewal, under the new order of the millennium upon Jesus' return, made perfect sense. At the least, any anomalies or dissonances in nature, typically in the form of extended bad weather, floods, famines, or plagues, had to be understood as punishments meted out by the Christian God for human misbehavior. Along with recurrent plague and other disease outbreaks, periodic famine represented God's wrath in the sixteenth and early seventeenth centuries.

Cunningham and Grell quickly dispatch the view that famines are acts of nature which bring population back into balance with the food supply, as Reverend Thomas Malthus thought, Nature being for him a manifestation of the divine order of the cosmos. They reject this Malthusian view for the interpretation that " famines were a side-effect of the fundamental social and economic structure of the society, in the context of a continued population increase."[86] The way out of the famine trap, of course, was the development of a commercial society in which food is grown for sale to markets rather than for subsistence, which was first accomplished by England in the mid-seventeenth century. Today's global marketplace is the end result, or perhaps the *reductio ad absurdem* of this transformation. At any rate, the English were the first to break the famine cycle, with the side effect of stabilizing population as well, because subsistence economies inevitably produce growing populations.

By the middle of the seventeenth century the "apocalyptical age" of Western European civilization, c.1500-1650, was past, and the obsessive preoccupation with recognizing disease outbreaks, periodic famine, chronic warfare, and astronomical and other natural anomalies as portents of the apocalypse began to subside on the European continent while lingering a few decades longer in England. Medieval Catholicism and its Augustinian vision of heaven and hell persisted in the popular culture, along with its Protestant variants, and Christianity, the organizing principle of Western civilization, maintained its dominance even over the leading minds of the era. Commercialization and secularization of daily behavior in the burgeoning towns had not yet undermined the Christian worldview, and dissenters would not be tolerated. As for nature, little of it remained in Europe, but European explorers and colonists were constantly bringing back reports of new animals and plants as well as newly discovered societies in the Americas and elsewhere. At the end of the fifteenth century Christopher Columbus thought that he had been doing God's work in converting the remaining inhabitants of the earth before Christ's return. Shortly thereafter, Martin Luther saw his own challenge to corrupt Rome as preparation for the Second Coming in a weary world worn out with age. Gradually, however, in the century and a half that followed, the achievements of a small coterie of intellectuals focusing their attention upon the natural world, epitomized by Copernicus's earth-shaking discovery that Earth was not the center of the universe, gave rise to skeptical ideas about the nature of things, including both the natural world and Christian society itself.

Within the context of the sixteenth and seventeenth century Scientific Revolution, largely a revolution in astronomy and physics, we shall consider the development of early modern attitudes toward nature, the transformation of an organic to a mechanistic model of nature described by Carolyn Merchant and others, and the modern roots of the still prevalent belief in human superiority and dominion over the natural world. We have already seen that the idea of human dominion over nature has deep roots in classical antiquity and in the Judeo-Christian tradition. These entrenched attitudes would receive powerful reinforcement from the early modern intellectuals who laid the foundations of modern science, particularly because they built these foundations upon the presuppositions of the Christian worldview which assumed a static, unchanging cosmos since the time of the Biblical "Creation."

NOTES

[1]. Toynbee, Arnold J., *A Study of History* (New York & London: Oxford University Press, 1947, Abridgement of Volumes I-VI by D.C. Somervell), p.8.
[2]. Hollister, C. Warren, *Medieval Europe: A Short History* (New York: McGraw-Hill Publishing Company, 1990), p.77.
[3]. Stoddard, Whitney S., *Monastery and Cathedral in France* (Middletown, Connecticut: Wesleyan University Press, 1966), p.11.

[4]. Ibid., p.7.

[5]. Ibid., p.11.

[6]. Oursel, Raymond, *Living Architecture: Romanesque* (New York: Grosset and Dunlap, 1967), pp.14, 16.

[7]. Cantor, Norman E., *Civilization of the Middle Ages* (New York: Harper Collins Publishers, 1963), p.173.

[8]. McNeill, William H., *The Rise of the West*. Chicago: The University of Chicago Press, 1963, p.539.

[9]. Ibid., p.538.

[10]. Bengtsson, Jan, Nilsson, Sven G., Franc, Alain, Menozzi, Paolo, "Biodiversity, disturbances, ecosystem function and management of European forests," *Forest Ecology and Management* 132 (2000) 39-50, pp.40-41.

[11]. Simmons, I.B., *Changing the Face of the Earth: Culture, Environment, History* (Oxford: Basil Blackwell, 1989), p.52; Fagan, Brian, *The Journey from Eden: The Peopling of Our world* (London: Thames and Hudson, 1990), p.151.

[12]. Wright, Patricia, "Forest Primeval: Life and Death among the Ancient Trees of Poland's Bialowieza Forest," *Massachusetts* (Massachusetts Alumni Quarterly), Fall 1991, pp.6-8, and Schama, Simon, *Landscape and Memory* (New York: Alfred A. Knopf, 1995), pp.37-53.

[13]. Ibid., p.9.

[14]. Dubos, René, *The Wooing of Earth: New Perspectives on Man's Use of Nature* (New York: Charles Scribner's Sons, 1980), pp.49-50.

[15]. Roderick Nash, *Wilderness and the American Mind* (New Haven: Yale University Press, 1967, 1982), pp.13-15.

[16]. Ibid., pp.17-18.

[17]. Susan Power Bratton, *Christianity, Wilderness, and Wildlife: The Original Desert Solitaire* (Scranton: University of Scranton Press ; London and Toronto: Associated University Presses, 1993), pp. 157-216.

[18]. David J. Herlihy, "Attitudes Toward the Environment in Medieval Society," in *Historical Ecology: Essays on Environment and Social Change*, ed. Lester J. Bilsky (Port Washington, New York: National University Publications, Kennikat Press, 1980).

[19]. Ibid., p.101.

[20]. Ibid., p.112.

[21]. Ibid.

[22]. Ibid., p.113.

[23]. Ibid.

[24]. Gilbert Highet, *Poets in a Landscape* (Harmondsworth, Middlesex: Penguin Books Ltd, 1959), Virgil quoted from the *Georgics*, p.66.

[25]. Ibid.

[26]. Hollister, *Medieval Europe* (1990), p.52.

[27]. H.C. Darby, "The Clearing of the Woodland in Europe," in *Man's Role in Changing the Face of the Earth* (Chicago: University of Chicago Press, 1958), p. 190.

[28]. Clive Ponting, *A Green History of the World* (New York: Penguin Books, 1991), p.120.

[29]. Peter Coates, *Nature: Western Attitudes since Ancient Times* (Berkeley: University of California Press, 1998), pp.43-46.

[30]. Roland Bechmann, *Trees and Man: The Forest in the Middle Ages* (New York: Paragon House, 1990), 159.

[31]. Ibid., pp.76-78.

[32]. Ibid., p.81.

[33]. Tacitus, *On Britain and Germany* (Harmondsworth: Penguin Books Ltd., 1948), p.122.

[34]. Richard Tarnas, *The Passion of the Western Mind: Understanding the Ideas That Have Shaped Our World View* (New York: Ballantine Books, 1991), p.172.

[35]. Hollister, *Medieval Europe*, pp. 142-145.

[36]. Darby, "The Clearing of the Woodland in Europe," p.193.

[37]. Coates, *Nature*, p.45.

[38]. Darby, "The Clearing of the Woodland in Europe," p.194.

[39]. Ibid.

[40]. Bechmann, *Trees and Man*, p.83.

[41]. Ibid., pp.88-103.

[42]. Stephen J. Pyne, *Vestal Fire: An Environmental History, Told Through Fire, of Europe and Europe's encounter with the World* (Seattle: University of Washington Press, 1997).

[43]. Ibid., p.148.

[44]. Ibid.

[45]. Ibid., p.152.

[46]. Ibid., p.153.

[47]. Ibid., p.156.

[48]. Ibid., p.160.

[49]. Patricia Wright, "Forest Primeval," p.6.

[50]. H. C. Darby, "The Clearing of the Woodland in Europe," p.197.

[51]. Ibid., p.196.

[52]. Clarence Glacken, *Traces on the Rhodian Shore* (Berkeley: University of California Press, 1967).

[53]. Hollister, *Medieval Europe*, p.304.

[54]. Cantor, *The Civilization of the Middle Ages*, p.529.

[55]. Ibid., p.530.

[56]. Robert Worth Frank, Jr., "The 'Hungry Gap,' Crop Failure, and Famine: The Fourteenth Century Agricultural Crisis and *Piers Plowman*," *Agriculture in the Middle Ages: Technology, Practice, and Representation* (Philadelphia: University of Pennsylvania Press, 1995), p.228.

[57]. Barbara W. Tuchman, *A Distant Mirror: The Calamitous Fourteenth Century* (New York: Alfred A. Knopf, Inc., 1978), p.23.

[58]. Brian Fagan, *The Little Ice Age: How Climate Made History, 1300-1850* (New York: Basic Books, 2000).

[59]. Ibid., Preface, p.xv.

[60]. Ibid., p.38.

[61]. Ibid., p.48.

[62]. Charles R. Bowlus, "Ecological Crisis in Fourteenth Century Europe" in *Historical Ecology: Essays on Environment and Social Change*, ed. Lester T. Bilsky (Port Washington, NY: Kennikat Press, 1980), pp.86-99.

[63]. Ibid., pp.86-87.

[64]. David Herlihy, *The Black Death and the Transformation of the West* (Cambridge, Massachusetts: Harvard University Press, 1997), p.33.

[65]. Ibid., p.34.

[66]. Hollister, *Medieval History*, p.329, writes that the plague spread swiftly "among a population whose resistance may have been weakened by malnutrition." Tuchman, *A Distant Mirror*, p.84, notes that the plague in Florence and Siena was preceded by the famine of the same year, but makes no judgment regarding causation. Fagan, *The Little Ice Age*, p.82, remarks that the high mortality rate from bubonic plague in France was "among people who had suffered from malnutrition during the great famine a generation earlier."

[67]. Bowlus, "Ecological Crisis," pp.87-89.

[68]. Ibid., pp.95-96.

[69]. Ibid., p.96.

[70]. Ibid., p.97.

[71]. Ibid., pp.97-99.

[72]. William McNeill, *Plagues and People* (Garden City: Anchor books, 1976), pp.53-56.

[73]. Ibid., pp.124-125.

[74]. David Arnold, *The Problem of Nature: Environment, Culture and European Expansion* (Oxford: Blackwell Publishers, 1996), p.62.

[75]. Ibid., pp.63-64.

[76]. Ibid., pp.72-73.

[77]. John Opie, "Renaissance Origins of the Environmental Crisis," *Environmental Review* 11 (Spring 1987), p.8.

[78]. Ibid., p.9.

[79]. David Herlihy, *The Black Death and the Transformation of the West*, p.81.

[80]. Carolyn Merchant, *The Death of nature: Women, Ecology, and the Scientific Revolution* (San Francisco: Harper and Row, Publishers, 1980, 1989) Preface: 1990, p.xvi.

[81]. Carolyn Merchant, *The Death of Nature*, pp. 42-68.

[82]. Brian Fagan, *The Little Ice Age*, preface, xvii; p.103.

[83]. Andrew Cunningham and Ole Peter Grell, *The Four Horsemen of the Apocalypse: Religion, War, Famine and Death in Reformation Europe* (Cambridge: Cambridge University Press, 2000), pp.15, 274.

[84]. Ibid., p.15.

[85]. Ibid., p.79. Also, pp. 82, 87, 89.

[86]. Ibid., p.205.

CHAPTER FIVE: THE SCIENTIFIC REVOLUTION AND THE DOMINATION OF NATURE

Ancient Science

While the vicissitudes of famine, disease, warfare, and religious disputation wore on through the sixteenth and seventeenth centuries, and Europeans generally succumbed to a mood of apocalyptical paranoia, a small coterie of brilliant men rekindled the ancient quest to understand the nature of the material world. The Scientific Revolution of the sixteenth and seventeenth centuries had deep roots in both the attempts of Greek philosophers to understand nature and the scientific undertakings of the High and Late Middle Ages which built upon the knowledge of Aristotle and other Greek authors, whose work reached Western Europe through contacts with Islamic civilization between the eleventh and fifteenth centuries. Scientific inquiry into how nature works had expired with classical civilization, but was revived as an aspect of the Renaissance, building upon the intellectual achievements of Greco-Roman civilization.

From the earliest Western probings into the meaning of nature by the Ionian Greek philosophers, science and religion had been intertwined, and medieval and early modern science was likewise practiced in a religious context. Not only the medieval scientists, but Copernicus, Kepler, Galileo, Bacon, Descartes, and Newton were all deeply committed to belief in Christianity. This is why Lynn White Jr., in "Historical Roots of Our Ecologic Crisis," found modern Western science tinctured with Christian arrogance toward nature, and traced our abuse of nature to biblical attitudes of human dominion over the natural world. Subsequent controversy over White's thesis has focused on the question of whether the idea of human domination of nature was continuously present and effective in Western civilization or, rather, took hold with the early modern Scientific Revolution, as Carolyn Merchant argues. Most likely, the Scientific Revolution reinforced the Christian idea of human dominion over nature. In essence, what was the Scientific Revolution, and what continuity with medieval science does it appear to manifest?

The origins of the early modern Scientific Revolution are far more complex than most textbook accounts would lead us to believe, largely because attempts to understand nature have been interwoven with the Western philosophical quest into the meaning of God, soul, and spirit since the time of the Ionian Greeks of the sixth century B.C. Propounders of the Orphic mysteries, such as Pythagoras and Empedocles, envisioned the organic unity of the world, the Pythagoreans believing "that the universe is spherical, animate, ensouled, and intelligent."[1] Plato followed the Pythagoreans in his conviction that the cosmos is "a living creature, one and visible, containing within itself all living creatures which are by nature akin to itself."[2] In these beliefs, early and later Greek philosophers thought of the cosmos as the Earth and its surrounding realm of sun, planets, and stars, with no grasp of the immensity of the physical universe. Earth-centeredness was the natural consequence of the lack of any sense of the size of Earth relative to other celestial bodies. As to the motions of these bodies, in the fifth century Anaxagoras held that they were the result of the operation of a mind or soul, which he extrapolated from Thales' belief that a lodestone possesses a soul because it can move a piece of iron.[3]

Greek scientific inquiry lost momentum with Socrates and Plato, who emphasized ethical, political and spiritual concerns over understanding nature. Plato was much influenced by the Pythagorean idea of the immortal soul, although he also made astute observations of changes in the Greek landscape, as described in Chapter Two, as well as having speculated on the origins of the cosmos. However, his successor, Aristotle, pursued more systematic inquiry into astronomy and biology, the former through a speculative approach, and the latter through extensive empirical observation. In astronomy, Aristotle contributed the idea that the heavenly bodies are carried on actual physical spheres rather than simple geometrical paths, as Eudoxas had proposed. Aristotle also postulated that the outermost sphere of fixed stars was moved by an "Unmoved Mover," as were the inner spheres, the "movers" being "spiritual in character, the relation of a mover to its sphere being akin to that of a soul to a body."[4] This mix of naturalistic and spiritualistic explanation probably reflects Pythagoras' influence upon Plato, and although Aristotle did not agree with Plato's idea of the creation of the cosmos, arguing rather that the cosmos always had existed and would continue to exist for eternity, he had nevertheless imbibed the Pythagorean idea of the immortal soul, although he would modify it and reject its immortality.

Aristotle's "scientific" model of the cosmos proposed an absolute difference in kind between the composition of heavenly bodies and terrestrial matter, which was made up of earth, water, air, and fire, the four "elements" of the lower, sublunar realm. The heavens, in contrast, were composed of the "quintessence," a fifth and purer element. Furthermore, the heavenly bodies were incorruptible and eternal, as were their circular and uniform motions. Like Plato, Aristotle believed "that intellectual designs and purposes were the formative and guiding principles of all natural processes."[5] To his four terrestrial elements Aristotle added four main types of cause. Material causes came from the primary matter out of which things were made. Formal causes consisted of the patterns and forms which were impressed upon primary matter. Thirdly, efficient causes provided the mechanisms which created the designs imposed upon matter, and final causes, the fourth category, "were the purposes for which objects were designed."[6] Related to his rigid theory of causation, in the field of physics Aristotle explained bodily motion as the result of "direct contact with a continuously operating mover."[7]

Aristotelian biology was discussed in Chapter Three in relation to developments in ancient Greek thought that contributed to an unscientific temper of mind during the Middle Ages, both before and after Aristotle's influence upon medieval natural philosophers. It should be made clear at this point that there is no intention here to malign Aristotle as an early scientist. He has been under attack by modern scientists since the seventeenth century, and environmental historians and philosophers berate him today. But these perspectives all look backwards over two millennia of historical change in the West. Viewed from the perspectives of his predecessors, as historian of science David C. Lindberg points out, Aristotle's philosophy of nature, encompassing physics, astronomy, biology and human anatomy and linked to humanistic perspectives in philosophy, was the astonishing achievement of an unbounded genius.[8] In biology alone, Aristotle's accomplishment was extraordinary if one thinks back over the speculations made about nature in the several centuries that preceded him. His scientific method included direct observation, both in the field and in the laboratory, and an amassing of data by himself and his assistants as essential to his empirical evaluation of facts. He was also relatively skeptical, if not skeptical enough at times, and although he classified types of "souls" in living things (under Plato's influence), he also maintained that the soul is the "form" of the organism and that, as such, was not immortal, including the souls of human beings, since the organism disintegrates at death, its form evaporating into nonbeing.[9]

During the Middle Ages Aristotle's un-Christian observations regarding the human soul were, of course, expunged from his philosophy of nature, which otherwise lent itself neatly to the Christian worldview as a geocentric model of the cosmos in which God's rationally designed creatures were neatly ordered in a chain of being which reached from God and the angels on downward to progressively simpler forms of life and, finally, inert matter. The medieval church's qualified approval of Aristotle as the official philosopher of nature, after cleansing by Scholastic theologians, thus transformed him from groundbreaking avant garde scientist of antiquity to stale, orthodox pillar of medieval Christianity, and authoritarian obstacle to scientific inquiry during the Scientific Revolution of the sixteenth and seventeenth centuries. Even then his genius was highly regarded by many who disputed the particulars of his natural philosophy. Francis Bacon did not hold Aristotle in high regard, however, writing in the *Novum Organum* (1620) that Aristotle was a striking example of a philosopher " who corrupted natural philosophy with his dialectic, when he constructed the world out of categories...."[10]

The Death of Nature

The authority of Ptolemy in astronomy, and Aristotle in biology as well as in astronomy and physics, were powerful forces of resistance to original thinking in early modern scientific speculation about how nature functions. There were also animistic Renaissance philosophies of nature which competed to explain the natural world, both within the framework of the established Ptolemaic and Aristotelian systems, and also in relation to Greek beliefs rooted in Platonism and Stoicism. These cosmic models of the world as an organism have been described by Carolyn Merchant in *The Death of Nature*, in which she concludes that the Renaissance organicist philosophies of the Florentine Neoplatonists, and also of Tomasso Campanella, Giordano Bruno, Paracelsus, and others provided useful ideas to early modern mechanistic scientists and philosophers (e.g., Paracelsus' preference for empirical methodology over the books and authority of the ancient Greeks) such as Francis Bacon, who rejected the idea of an animistic nature as radical and atheistical. Thus, Christian orthodoxy held sway simultaneously with the new mechanistic philosophy of nature, leading Merchant to conclude: "The rejection and removal of organic and animistic features and the substitution of mechanically describable components would become the most significant and far-reaching effect of the Scientific Revolution."[11] Thus was nature in the modern world, once and for all, so it seemed, divested of any indwelling intrinsic value and transformed, in the view of modern science, into a gigantic, spiritless machine, the clockwork mechanism of Descartes, Newton, and most of the scientists who have followed them down to the present day.

Historian of science David C. Lindberg essentially supports Merchant's conclusion regarding the significance of the new mechanistic view of nature:

> In exchange for the purposeful, organized, and (in many ways) organic world of Aristotelian natural philosophy, the new metaphysics offered a mechanical world of lifeless matter, incessant local motion, and random collision. It stripped away the sensible qualities so central to Aristotelian natural philosophy .... And for Aristotelian teleology, which discovered purpose *within* nature, it substituted the purposes of a creator God, imposed on nature from outside.[12]

The organicist belief in an ensouled nature could not possibly be allowed as part of the new mechanistic paradigm. Lindberg's statement is best understood in the context of the "continuity debate," the ongoing argument over whether or not medieval science contributed significantly to the early modern Scientific Revolution. Was medieval science continuous with or discontinuous with early modern science? Early modern scientists themselves generally acknowledged the Greek scientific achievement but judged the Middle Ages as a period of philosophical stagnation at best. Francis Bacon wrote that: "The sciences which we have came down to us mostly from the Greeks; for the additions made by Roman or Arab writers, *or those of more recent times* [Italics mine], are few and of little importance; and such as there are have been built on the foundation of Greek discoveries."[13] In the eighteenth century Voltaire continued to denigrate the Middle Ages as barren of scientific knowledge, and the debate has continued to the present. Lindberg, after thorough analysis, suggests a middle ground, weighing the medieval achievement of having given "serious, critical attention to the details of the Aristotelian methodology"[14] against their inability to reject the Aristotelian system as a whole. However, he maintains that a stronger case for discontinuity can be made on the basis of a shift in worldview in which the leading "new" scientists of the seventeenth century (Galileo, Descartes, Gassendi, Boyle, Newton, and others) rejected Aristotle's metaphysics and revived and reformulated "the corpuscular philosophy of the ancient atomists. This produced a radical conceptual shift, which destroyed the foundations of natural philosophy as practiced for nearly two thousand years."[15] Thus, in the terms of philosopher of science Thomas Kuhn (*The Structure of Scientific Revolutions*) a major paradigm shift had occurred by the seventeenth century, with the ancient-medieval model of "science" (the word was not in use before the nineteenth century) yielding to a new paradigm of scientific practice which carried with it profound implications for how we perceive reality in general. In other words, an entire new worldview would unfold from the new scientific perspective, ultimately demolishing the medieval worldview to the satisfaction of secularized scientists and intellectuals in general.

Kuhn himself characterized scientific revolutions as changes of worldview.[16] Lindberg points out that Kuhn saw the Scientific Revolution of the sixteenth and seventeenth centuries not as a single revolution in thought but as a group of independent revolutions within several specific disciplines. Kuhn thought that in order for a revolution to occur, well-developed theoretical antecedents must exist, and that therefore the only truly revolutionary changes in the early modern period were in the "classical" sciences of astronomy and physics (specifically, mechanics and optics).[17] Taken together, these two revolutions were to have enormous consequences. The revolution in astronomy initiated cautiously by Copernicus, and linked to the synthesis of celestial and terrestrial mechanics in Newton's laws by the scientific achievements of Tycho Brahe, Johannes Kepler, Galileo Galelei and others, gradually fractured and deconstructed the medieval worldview and its Ptolemaic-Aristotelian model of the cosmos. Christian belief has waged a defensive battle against accumulating scientific knowledge ever since, and a secular worldview has emerged in competition with the Christian belief system, causing a permanent schism in the thought of modern Western civilization.

Building upon the Copernican revolution, the Newtonian synthesis and the physical laws of nature (based upon the mechanistic or mechanical philosophy of nature) have reduced our conception of nature to that of bodies possessing the properties of weight, size, and shape in space, and dependent upon the arrangement and motion of small particles (corpuscles or atoms). Rejecting the sensory, "secondary" properties of color, odor, taste, etc. as subjective, the primary physical properties can be measured and mathematically quantified according to equations representing the physical laws of nature. According to Carolyn Merchant, these developments "transformed the body of the world and its female soul, source of activity in the organic cosmos, into a mechanism of inert matter in motion, translated the world spirit into a corpuscular ether, purged individual spirits from nature," thereby resulting in a dead, mechanical system which had been set in motion by the Creator according to the laws of inertia and motion.[18]

> The removal of animistic, organic assumptions about the cosmos constituted the death of nature-the most far-reaching effect of the Scientific Revolution. Because nature was now viewed as a system of dead, inert particles moved by external, rather than inherent forces, the mechanical framework itself could legitimate the manipulation of nature. Moreover, as a conceptual framework, the mechanical order had associated with it a framework of values based on power, fully compatible with the directions taken by commercial capitalism.[19]

Merchant's account of the rise of the mechanical order in the thought of the French mechanists Marin Mersenne (1588-1648), Pierre Gassendi (1592-1655), and René Descartes (1596-1650) relates scientific physical theory to the rationalizing tendencies of early modern European governments, particularly in France and England, under which " nature came to be viewed as a resource to be subjected to control with human beings as her earthly managers."[20]   In England, Thomas Hobbes developed this idea into a political system in which the original state of chaos of humans living in the "state of nature" (a chaotic free-for-all which is the opposite of what is known today of the behavior of hunter-gatherers) is brought under control by a benign authoritarian monarchy in order to allow naturally aggressive, violent, and competitive humans to pursue and develop the resources of the state in an orderly manner.[21]

Underlying this activity of transforming nature-as-resources into economic wealth in the orderly Hobbesian state was the Baconian doctrine of dominion over nature, Hobbes having been secretary and friend to Francis Bacon (1561-1626) during the last five years of Bacon's life.[22] Merchant's entire chapter on dominion over nature in *The Death of Nature* is focused on Bacon's thought, embellishing the idea that "Bacon fashioned a new ethic sanctioning the exploitation of nature."[23]   A major theme in this account relates Bacon's literal Christian belief in the Fall of Adam and Eve from God's grace in the Garden of Eden to the Scientific Revolution as a means of regaining the dominion over nature which was granted to Adam and Eve at the Creation and lost during the Fall. The reclamation of power over nature through the new method of empirical science pioneered by Bacon and the seventeenth century community of scientists would regain the conditions of the Christian paradise in time for the Second Coming of Christ, Last Judgment, and Millennium. Thus, the Scientific Revolution engenders human "progress" within the Christian framework of limited linear time leading to the end of the imperfect postlapsarian Earth before Christ's reign. From Bacon to Isaac Newton and later, this literal, essentially "fundamentalist" belief in Biblical truth created the larger cosmological and historical framework within which science, technology, and progress, interacting with bourgeois capitalism practiced by the pious and the saved, presumably led to the end of history and the triumph of the City of God.

Merchant's thesis regarding the Christian origins of both dominion over nature and the modern idea of progress are developed fully in her later work, to be discussed shortly. Her two major themes of: (1) the despiritualization of nature during the Scientific Revolution (thereby making nature vulnerable to the early modern combination of Christian eschatology, mechanistic science and technology, and capitalism) and; (2) the Christian perception of modern science and technology as the means of regaining dominion over nature since the Fall, together present a powerful combination of elements in the early modern worldview, lethal enough to unleash a process of severe environmental and sociocultural degradation upon Western Europe and her recently acquired colonies. However, as described in detail in Chapter Four, in Western Europe the death of nature, in the sense of the actual destruction of nature, had already occurred to a great extent when these beliefs and practices took hold.

Modern Science and Christian Myth

Carolyn Merchant's explanation of the destruction of the organic paradigm of nature in the sixteenth and seventeenth centuries, thereby clearing the way for the mechanical or mechanistic paradigm, appears to be valid and has not been challenged by serious criticism. However, the significance of her thesis is perhaps diminished within the following perspective. The debunking or demythologizing of the organic worldview was at least the third major assault upon Greek mythological explanations of nature over the course of more than two millennia. During the Hellenistic period of Greco-Roman civilization, philosophers following the materialistic and atomistic tradition established by Democritus and Leucippus attacked the nature mythologies of Pythagorean and other philosophies which recognized spirit or soul in nature. Subsequently, Christian theologians during and after the Roman Empire eschewed the idea of an enspirited nature in order to locate spirituality entirely within the transcendental realm of the Judeo-Christian God, as Lynn White, Jr. and others have shown. During the sixteenth and seventeenth centuries, as we saw in the preceding chapter, the Christian despiritualization of nature was thorough enough in the belief systems of the masses of peasants, burghers, and ecclesiastics that we can conclude that it dominated the early modern world view. The major exceptions to this perception consisted of the minority of intellectuals and renegade priests, monks, and theologians who dabbled in the ancient Greek ideas of organicism which had been rekindled by the Renaissance re-reading of the classics. Intellectuals who, like Giordano Bruno, openly confessed their belief in these resuscitated ancient concepts were dealt with severely by the ubiquitous Christian authorities.

The main point of this discussion of Merchant's thesis in *The Death of Nature* is that the most powerful obstacle to the mechanistic paradigm of modern science was and continues to be Christian religious myth (as well as the mythological worldviews of all traditional, pre-modern religions), even though the modern mechanistic paradigm was forged in the context of the Christian belief system. Chapter Four made quite clear the thorough destruction of nature, i.e., of ecosystems, in Western Europe even by the end of the thirteenth century, when few or no organicist explanations of enspirited nature had yet been reclaimed by the Renaissance turning to the wisdom of the ancients. Of course, pagan myths of both the ancient and Germanic cultures persisted in areas not thoroughly Christianized, but overall the inculcation of Christian dogma was well established in Western Europe by the High Middle Ages, the period during which the destruction of nature on a colossal scale took place. No widely held belief in the existence of spirit in nature existed at this time to prevent the mindless destruction of European ecology. Rather, in the early modern period, concurrently with the development of mechanistic science, desperate actions to preserve or conserve what remained of the European forest were undertaken, if only to retain hunting grounds for royalty and nobility.

Even as the Greek and other pagan nature myths dissolved before the onslaught of mechanistic science, all effectuated by true believers in the Christian worldview, from Copernicus to Newton, the grand myth of Judeo-Christianity, upon which Western civilization was founded, managed to survive by giving up its monopoly on the meaning of nature gradually and retreating to the safe enclosure of its sacred texts, where verbal ambiguity and factual uncertainty could not be assaulted by the new scientific methodology. Geocentrism, Aristotelian teleology, and eventually, even the mythological history of the Earth itself, confined in the temporal straight-jacket of Biblical chronology, all had to be abandoned like so many outer fortifications surrounding a citadel under siege.

The gradual defeat of the Judeo-Christian worldview, at least on the intellectual battlefield of natural philosophy, was a more significant event in the history of modernity than the desacralization of nature through the destruction of pre-modern organicist beliefs, as explained above. Nature continues to be destroyed globally on a massive scale because the process of demythification simply has not proceeded far enough. While demythologized, mechanistic science has spread around the globe, traditional cultural myths, like Judeo-Christianity, Mohammedanism, Hinduism, etc. have persisted to a great extent. Depending upon where in the world one was raised, modern secular culture, which has carried the implications of the findings of modern, mechanistic science on into individual, personal belief systems, has become a criterion of judgment and "truth" which is potentially threatening to traditional belief systems. This cultural schizophrenia, characteristic of modern, open societies, is a critical factor in the contemporary ecological crisis. The scientific method has acted as a demythologizing tool which has proven its reliability in the course of the triumph of modern science, although, as Thomas Kuhn and Carolyn Merchant have shown, scientific paradigms contain myths of their own, such as the mechanistic model of nature based upon early modern celestial and terrestrial mechanics, which was an inadequate paradigm for explaining terrestrial life and organic processes. Only the extension of modern scientific method to the study of Earth itself, and of the organisms which inhabit it, could provide a more objective, less metaphorical or mythological basis for understanding that at least the organic part of nature is enspirited in the sense that organisms, to varying degrees, possess sentience, the ability to feel pleasure, pain, and the experience of life, and also carry the genetically encoded information of the evolutionary process. Modern, mechanistic science ultimately produced the disciplines of geologic and evolutionary biological science which tell us far more about the intrinsic value of organic nature than the simple metaphors of the ancient Greeks ever could. Ironically, the mechanistic scientific paradigm designed to study the inorganic processes at work in the realms of astronomy and physics provided students of Earth and its life with an objective methodology which has solved the mysteries of the origin and history of Earth and of the evolutionary processes which have produced the species and ecosystems which we now *value* through this relatively recent knowledge of the nineteenth and twentieth centuries.

The "death of nature," in the tangible sense of destruction of species and ecosystems, had already been largely accomplished in Western Europe by 1300 A.D. Shortly thereafter, nature died off in the islands of the "Mediterranean Atlantic" (the Azores, Madeiras, and Canaries) as Europeans sought new worlds to conquer before discovery of the Americas, a sad story told so well in Alfred Crosby's *Ecological Imperialism*. All of this ecological destructiveness happened long before the desacralization of nature associated with the modern Scientific Revolution, which leads me to conclude that a factor other than ideas of an ensouled nature invented by Greek philosophers was and continues to be a more powerful cause of Western European destructiveness towards nature, just as Lynn White Jr. intuited. This factor is, of course, the complex mythology of Judeo-Christianity which underlies the cultural schizophrenia of the Western world today. Carolyn Merchant deconstructs this mythology in terms of Western dominion over nature expressed through the uniquely Western idea of progress in an essay titled "Reinventing Eden: Western Culture as a Recovery Narrative."[24] As she writes, "beginning in the seventeenth century and proceeding to the present, New World colonists have undertaken a massive effort to reinvent the whole earth in the image of the Garden of Eden."[25] According to Merchant, this process is built around the three subplots of Christian religion, modern science, and capitalism, representing, respectively, the beginning, middle, and end of the history of Western civilization. In this story, Genesis I advocates recovery through human domination of nature, whereas Genesis II recommends stewardship, or human management of the garden carved out of wild and desolate nature. Merchant's analysis of capitalism and the idea of progress in this context is discussed in the following chapter.

Complementing Merchant's interpretation that the desacralization of nature associated with the Scientific Revolution led to massive exploitation of the natural world under the endorsement of the mechanistic scientific paradigm, which definitely applies to the last three centuries, the position taken here is that Christianity had already desacralized nature long before the Scientific Revolution. As we read in Chapter Two, Saint Augustine saw the material world of nature as base and worthless, essentially a mere prop underlying the cosmic drama of salvation. Philosopher Max Oelschlaeger agrees that: "A constant theme of medieval theology is the insistence that nature, though proof of God's existence (the argument from design), was not divine. The sacred groves worshiped by pagans, and reputedly the denizen of witches, shamans, and Lucifer himself, had to be eliminated from the face of the earth."[26]

Thus, recognizing that it was heresy to revere wild nature, and that destruction of the wilderness fulfilled God's plan by exercising human dominion over nature while ridding the world of pagan beliefs, Oelschaeger supports Lynn White Jr.'s thesis that Judeo-Christian religion is the root cause of Western destructiveness toward nature. Oelschlaeger, however, digs far more deeply down into the origins of Western religious tradition to show that it contained, from the very beginning, the seeds of dominion over nature. Let us consider Oelschlaeger's analysis of the Judaic origins of human domination over the environment before relating it to his interpretation of the Scientific Revolution.

According to Oelschlaeger, reading the Old Testament as a rationalization of agriculture, the stories in *Genesis* justify the beginnings of mastery over nature resulting from the Neolithic Revolution, the shift from hunting-gathering to cultivating the soil in the Biblical "sweat of one's brow." This is only *one* cause of Western domination of nature, however, for "only in the context of the agricultural revolution, and the later Protestant Reformation, and industrial, scientific, and democratic revolutions, can any sense be made of the claim that environmental malaise is rooted in Judeo-Christianity."[27] Nevertheless: "The Hebrews desacralized nature and viewed it as the creation of a transcendent God who had given them an exclusive claim to the land."[28] Thus, to a great extent Hebrew identity was forged by rejecting the nature religions of their ancient rivals, Babylonian and Egyptian civilization. In quoting the famous passage from Genesis I.26-31, in which man is given dominion over the fish, birds, cattle, and all the wild animals on earth, Oelschlaeger identifies Adam as the cultivator of the soil, the paradigmatic representation of the agriculturalist point of view. The language of Genesis I.26-31 (known as the P source in the texts from different historical periods which were amalgamated to form the *Old Testament*) reveals "the roots of an intense anthropocentrism. *Man*, of all the animals, is alone made in Yahweh's image, and the remainder of creation-the wild animals-is given by God to his *son* to rule over. Furthermore, man is 'to subdue' the earth, literally to be free from nature's tyranny *and* idolatry of things in nature, and to be fruitful and multiply ...."[29]

In his analysis of the Scientific Revolution, Oelschlaeger corroborates Merchant's explanation of the significant link between modern science and Christianity represented by Francis Bacon's perception of the scientific enterprise as the means of regaining human dominion over nature long after Adam's fall from grace in the Garden of Eden. Following the centuries of struggle of postlapsarian Christians against the destructive forces of nature (as witnessed in the brief history of medieval and early modern Christendom in Chapter Four), fallen mankind, through faith and perseverance, had at last gained the means to re-create the Garden of Eden on Earth through the scientific methodology acquired in the sixteenth and seventeenth centuries. Oelschlaeger asserts that "Bacon's ideal was no less than a complete mastery of nature," and "that everything in the world could be fashioned to human purpose through science."[30] "Civilized humans-in-the-modern-age would employ the power of science to remake the wilderness, the world with which humans-in-the-archaic-age had empathetically identified themselves. The *modern* project Bacon envisioned was to convert wild nature as rapidly as possible into the New Atlantis."[31] For Bacon this was *no analogy*. The *New Atlantis* produced by modern science would be the *actual New Jerusalem*, literally the Kingdom of Heaven on Earth with humans fully recovered from their fallen condition.

Thus, given God's grace, it was inevitable that humanity would ultimately redeem itself, following a predetermined historical course which required mankind to dominate nature in order once again to be free in the reclaimed Garden of Eden. However, Oelschlaeger also reminds us that once the new scientific methodology was released, Christianity, seemingly the very crucible of its origins, would be subjected, inadvertently, to its scrutiny, thereby contributing to the rise of secular humanism in its place. As Oelschlaeger writes, "natural reason--that peculiar combination of logic and observation--was to run roughshod over mystical faith in things illogical and unseen: that is, the entire mythology of Judeo-Christianity."[32] As discussed in Chapter Three, this demythologization occurred quite gradually, ultimately culminating in the geological and evolutionary biological scientific paradigms of the late eighteenth and nineteenth centuries. The Scientific Revolution which had taken strength from Christian dogma gradually generated a body of knowledge of the natural world understood in the context of *natural time*, rather than *human chronology*, this shift in chronology delivering a crushing blow to the tottering Christian worldview. The demolition continues to the present day, of course, given the reluctance of any great civilization to discard the mythological foundations upon which it was constructed. Belief in the Greek Olympian gods and their Roman derivatives also persisted for centuries after their deconstruction by the Ionian philosophers.

An even less sanguine view of Christianity and its contribution to attitudes of human dominion over nature is presented by Frederick Turner in *Beyond Geography: The Western Spirit Against the Wilderness*. Describing Christianity as a "crisis cult" which arose in response to the gloom and despair of an oppressive Roman Empire, and which in the long run was betrayed in its

fundamental beliefs by the fathers of the Church (who led believers away from
belief in the divine spark within themselves and in all creation towards a historical
myth focused on the life of Jesus). "Thus the crisis cult that had arisen to deliver
its believers from the fate of the Roman Empire turned them over to the greater
and more inescapable, 'terror of history.' "[33] Turner means by the "terror of
history" Christianity's turn away from the timeless cycle of myth and nature to
"the hope of recovering in an apocalyptic future what it had once had in the past.
The historical interpretation of Christian mythology thus became the very engine
of history."[34] The implications of the ideas concerning time and Christian
eschatology presented by Merchant, Oelschlaeger, and Turner for the modern idea
of progress will be discussed in the following chapter. For now, let us turn to
Turner's view of the relationship between Christianity and the Scientific
Revolution.

The major theme of Turner's book is the peculiar destructiveness of the
Christian, Western European civilization which was unleashed upon the New
World during the age of exploration. This destructiveness was of both nature and
indigenous culture, as we have learned in the present age from the perspective of
multiculturalism. Turner observes that during the period of exploration and
colonization "Christians of all nationalities and persuasions were united in a
conception of the earth as a divinely created *thing*, there for the enjoyment,
instruction, and profit of man."[35] Correctly attributing this perspective to Saint
Augustine, Turner points out that its ultimate derivation is "Old Testament
scripture as rendered through Christian exegetes."[36] This view is thoroughly in
agreement with Oelschlaeger's interpretation of the Christian origins of the idea
of human dominion over nature. Given that these attitudes were already well
established at the time of the great explorations in the New World, this supports
my contention that nature, as conceived of by Western Europeans, was already
dead long before the Scientific Revolution. Thus, Merchant can only claim that
the Scientific Revolution and its conception of a mechanical, spiritless, clockwork
nature gave further corroboration and impetus to the belief in a "dead nature" and
greater impetus to its further destruction. It would seem that literally "the death
of nature" had occurred already over much of the Mediterranean world by the end
of classical antiquity, and in Western Europe by ca.1300.

Turner cites Max Weber in support of his own thesis, noting that Weber
recognized a "gradual, inexorable elimination of the magical or numinous from
the world" resulting from the influence of the Old Testament.[37] This Christian
view of the world as spiritless resulted in its being understood as an open field for
human activity acceptable to a God removed from it. Thus, when the West did
expand after Columbus, it was into a New World perceived as non-sacramental
and devoid of all spirit life. Consequently, little or no resistance to the
investigation of nature was offered during the Scientific Revolution. However,
the unexpected entailment was the erosion of the authority of the Christian
religion, once the scientific wisdom of the ancients had been undermined. Turner
doubts that the courageous scientists of the early modern period, probing nature

under the shadow of the Inquisition, could have guessed that their work "was preparing the way for a mechanistic conception of nature and man that would have the practical effect of denying the operation of God in a world He putatively had created."[38] The mechanistic paradigm was far enough in the future in terms of widespread acceptance that it would be anachronistic to project it back upon the seventeenth century practitioners of science. These men were still likely to believe in the diabolical spirits which Christianity had identified as its enemies, a state of mind in the early modern world to which the previous chapter rigorously attests. Turner describes this decadent early modern Christian worldview with restrained disdain:

> ... the forms and observances of a great religion die slowly, and long after its positive and life-enhancing aspects have fallen into practical disuse, its hag-ridden residues are keenly felt. So, in the middle of the seventeenth century, one can find Christians who seemed utterly consumed by the life of the spirit and as convinced of the operations of the devil as they were of divine providence. And yet these same Christians were denying life to much of the world and acting upon that world as if it were a passive configuration of matter devoid of its own interior life, laws, and spirit and existing only to be "civilized" for gain.[39]

Turner has made a powerful argument that Western civilization was imbued with attitudes of arrogance and dominance toward nature, not to mention towards other cultures, at the time of the great discoveries, prior to the Scientific Revolution. This view is fully supported by evidence presented in Chapter Four, which shows the general destructiveness of nature by Western Europeans throughout the Middle Ages, under the strong influence of the Augustinian version of the Christian worldview. Attitudes of dominion rather than stewardship over nature tended to prevail in general, although good agricultural stewardship practices, as exemplified by those of the Cistercian monks, were not entirely uncommon. The Scientific Revolution further contributed to the death of nature by providing the knowledge and techniques necessary to more intensive manipulation of the natural world by European nation states and their colonies in the New World. There, the "four great ills of civilization" described by Ophuls in Chapter Two (overutilization of nature, violence against outsiders, despotic rule over insiders, and slavery) were fully realized by Western European civilization. Augustinian Christian attitudes of dominion, whether in Catholic or Protestant variants, provided a rationalization for brutal manipulation of nature and indigenous peoples, subliminally and overtly reassuring colonists of the rightness of their actions. Much of the environmental and human degradation wrought in the colonies was accomplished well in advance of the spread of the mechanistic paradigm and the technological progeny resulting from the Scientific Revolution.

Complementing Oelschlaeger's emphasis upon the Judaic origins of human dominion over nature, Environmental philosopher J. Baird Callicott has traced the origins of both modern science and the attitudes of dominion over nature found in Genesis back to ancient Greek philosophy, the most deeply rooted source of Western attitudes toward nature. Beginning with the Ionian Greeks of the sixth and fifth centuries B.C. and culminating in the atomic theory of matter proposed by Leucippus and Democritus, Greek inquiry into the meaning of nature and natural processes produced two crucial insights which were passed on to early modern natural philosophers or scientists. First, the atomic theory provided the idea of "corpuscles," the equivalent of atoms, and secondly, Pythagoras, in the sixth century B.C., proposed that " the *order* of nature can be successfully disclosed only by means of a quantitative description, a rational account in the most literal sense of the word...."[40] Early modern science, according to Callicott, can be represented as essentially consisting of " a merger of the Pythagorean intuition that the structure of the world order is determined according to ratio, to quantitative proportions, and the Democritean ontology of void space (so very amenable to geometrical analysis) and material particles."[41]   Here we have represented the very heart of the mechanical philosophy of nature which was coalesced by Newton.

The same Pythagoras who ultimately inspired the mathematical foundations of modern science was also a proselytizer, if not the possible inventor of, the Western concept of the soul "as not only separate and distinct from the body, but as essentially alien to it."[42]   For Pythagoras and the religious sect that followed him, the goal in life was to earn the release of the soul to another, transcendental world at death. Plato was strongly influenced by this idea of an immortal soul uncomfortably encased in an earthly, organic body antipathetic to its otherworldly nature. Callicott concludes that this Pythagorean-Platonic concept of the soul "has profoundly influenced the European attitude toward nature." In this model of man and nature the human being consists of an organic body in which is entombed an immortal soul striving to escape the body and the inferior material world surrounding it. Add to this the themes of human dominion over nature found in Genesis, "and we have a very volatile mixture of ingredients set to explode in an all-out war on nature, a war which in the twentieth century has very nearly been won. To victors, of course, belong the spoils!"[43]

The modern legacy, to which the thought of Pythagoras and other Greek philosophers has contributed, includes a faith in modern mechanistic science *and* belief in an immortal human soul (much strengthened both by Christianity and the dualism of René Descartes). This combination of beliefs has produced a condition of intellectual and social schizophrenia, a schizophrenic culture with one mental foot planted in the mythological, super-naturalistic realm, and the other in the naturalistic faith in mechanistic science and technology. Thus, we live in what I prefer to call "a schizoid culture," a condition which I will describe more fully in a later chapter, since this condition did not manifest itself with full force until the Enlightenment and the following centuries. The schism between modern science and religious belief was already recognizable during the Renaissance, [44] but it subsequently intensified to produce a science-based secular humanist counterculture within traditional Christian culture in the West. Although secular humanism has become the dominant social paradigm in Western Europe, it still remains a minority subculture in the United States and central Europe, where traditional Christian religion holds sway with the majority. Thus, in the twenty-first century, traditional Christian belief acts in consort with the machines and theories produced by mechanistic science to bend nature totally to human needs and wants. I will further explain the ecologically destructive role of Christianity in chapter Thirteen.

Having demonstrated the powerful influence of Greek natural philosophy upon the modern Scientific Revolution, Callicott next turned to the Holy Bible to identify similarities between Genesis-P and the nature philosophy of the pre-Socratic Greeks. Biblical scholarship during recent centuries has identified three narratives of different authorship and historical age known as the J or Yahwist, P or Priestly, and E or Elohist sources. The E narrative is not important for Callicott's argument, the J or Yahwist narrative dates from c. 1000 B.C., and Genesis-P dates from c. 500 B.C., roughly contemporary with the era of the Ionian nature philosophers. Also, Genesis-P precedes the text of Genesis-J in the Old Testament. Callicott bases his argument on the classical scholarship of F. M. Cornford of Cambridge, who recognized the similarity of Ionian creation myths to the account of creation found in Genesis-P. Both are abstract, quasi-scientific accounts of the evolution of the cosmos. There is no need to recapitulate the likenesses of the creation accounts here. The main point is that, like Greek humanism which saw man as "the measure of all things," Genesis-P manifests an extreme anthropocentrism resulting in the attitudes of dominion which have been passed on to Judeo-Christian, Western civilization. It is Genesis-P which Oelschlaeger and Turner have described so disparagingly in the preceding sections of this chapter.[45]

Whether or not the Cornford-Callicott hypothesis is correct, namely, that early Greek nature philosophy influenced Genesis-P and intensified its arrogance toward nature, the P-version of Genesis has had a major influence upon Western attitudes toward nature over an extended period of time. As Oelschlaeger has suggested, if *Old Testament* attitudes of dominion over nature are viewed within the context of later events contributing to Western arrogance toward nature, then the foregoing analyses of Genesis-P tend to support Lynn White Jr.'s contention that the Judeo-Christian tradition deserves considerable blame for the modern ecological crisis, but, à la Callicott, with the assistance of classical Greek *hubris*.

There would appear to be little doubt that embedded cultural attitudes toward nature have consequences regarding our treatment of nature, but the question is, to what extent? Obviously, the authors whose ideas I have summarized generally believe that these attitudes exert a strong influence over our behavior. My own opinion is that humans naturally manipulate and transform nature in destructive ways, but that cultural attitudes may reinforce destructive behavior or constrain it. This viewpoint is probably implicit to some degree in each of the perspectives presented in this chapter. Nevertheless, it should prove useful to expand upon the idea in order to further clarify the actual historical *practice* of human domination of nature, from the pre-agricultural millennia of hunting and gathering to the modern Scientific Revolution. In a sense, we will be adding the effects of cultural practices and beliefs in agrarian civilization onto pre-existing Paleolithic behavior and attitudes which were already well-established at the time that the earliest civilizations, Mesopotamia and Egypt, first came into existence.

The earliest pre-historical record of the interaction of *Homo sapiens* with nature, i.e., with ecosystems, suggests that we had a great deal to do with the massive extinction of large mammals and birds in virtually all parts of the world *and at different times,* with extinctions occurring within centuries or millennia after the arrival of humankind. This phenomenon is now well-documented, whether for mammoths and mastodons in Europe and North America, or moas in New Zealand.[46] The geologically recent period of human destruction of giant "game" species (i.e., large sources of meat for primitive and later hunters) ranges from the late Pleistocene through the Holocene to as recently as a millennium ago, when the first humans arrived in New Zealand to begin the slaughter of its giant birds. This massive slaughter of animal species was complemented by the pre-historical practice of burning so well described by Stephen Pyne in Chapter Four. Thus, direct hunting of large animals was complemented by the transformation of habitat through burning, and later on, in many regions, pastoralism and transhumance. Some of this ecological change benefited herbivores, at least for a while, as in the case of increased grasslands created by Amerindian cultures in the Central Lowlands and Great Plains. However, in most regions the net effect over time was the extinction of large animals, which was well under way in many regions before agriculture became established. Despite the Paleolithic assault of humankind upon mammals and other animals, Oelschlaeger and other authors have argued that there was no sense of dominion involved, but rather a Paleolithic identification with nature, the Magna Mater or Great Mother. Hunter-gatherer societies utilized myth and ritual to reconcile the reality of hunting with the myth of living in harmony with nature, which itself was considered sacred.[47] Only with the Neolithic Revolution, which culminated the gradual shift from hunting and gathering through horticulture to agriculture and the extensive burning and clearing that it required, did attitudes of human dominion over nature arise, and even then only when the transitional nature gods of Egyptian and Mesopotamian civilization were displaced by Yahwism, which, despite " lingering reverberations of the Paleolithic mind, especially as revealed in the symbolic significance of shepherd and wilderness...sounds the death knell of mythic consciousness and the rise of historic consciousness as the dominating sensibility in the West."[48]

This new consciousness, which saw God as located transcendentally above a nature devoid of spirit, and which saw mankind as God's privileged son, also introduced the cosmic drama of Adam's fall from the Garden of Eden, on the one hand condemned to a struggle with nature, and on the other given the free will to reclaim his place in the Garden after millennia of pain and suffering in a "wilderness" of thorns and thistles (i.e., the agrarian environment). This process, the reclamation of the Judeo-Christian Eden, is interpreted by some environmental historians as having become the driving force of Western history, virtually requiring humankind to dominate nature according to God's plan. As Merchant has shown, only with Bacon and the Scientific Revolution was the basis for the goal of total dominion over nature finally achieved.

Although the supernaturalistic raison d' être of Judeo-Christian, Western European civilization may explain, or appear to explain, Western domination of nature, in Chapter Two we saw that Greco-Roman and other civilizations dominated nature to the point of massive environmental degradation under totally different worldviews. Does this not suggest that the tendency to control and manipulate nature in all civilizations, given their dependence upon agriculture and pastoralism, virtually guarantees dominion over nature *in practice* no matter what the worldview of a given civilization? If it is inherent in the very nature of civilization to dominate and destroy ecosystems, then what is all the fuss about attitudes and worldviews? The answer to this question is, as noted earlier, that the attitudes toward nature implicit in all worldviews can have either a restraining or an enhancing influence upon the tendency of human civilizations to despoil their natural surroundings. The Greeks were more reverent of nature than the Romans, as J. Donald Hughes has demonstrated. To what extent did this constrain and slow down their destructiveness toward nature? The Western world in the second half of the twentieth century was far more sensitive to environmental degradation than in the nineteenth and early twentieth. How great a difference has this made in our behavior towards the natural world? I would answer, more or less, that industrial capitalism and communism, both stemming from the Industrial and Scientific Revolutions, as well as unconstrained population growth, *combined with* the attitudinal factors intrinsic to the Western worldview, from Judeo-Christian arrogance toward nature to the mechanistic paradigm of nature produced by the Scientific Revolution, resulted in the massive environmental devastation created by the modern world. The extraordinary destructiveness of *modern* Western civilization would not have been possible without these ideas, innovations, and beliefs. Environmental degradation in communist countries was intensified by a top-down industrial system determined to outdo the West, thereby resulting in even more severe destructiveness, and the prevention of protest and reform by totalitarian governments. However, since the 1960s, especially in the West, an impressive rear-guard action has been fought in defense of nature, only to be overwhelmed by explosive economic growth accompanying expansion of the "global market place."

Unfortunately, despite environmentalist critiques of entrenched ideas and attitudes of dominion over nature residing in the worldview of modern Western civilization, this perspective continues to rule the thinking not only of the West, but also of the many Westernized nations who have followed our lead into industrialization, modernization, and the global marketplace. Should a change in worldview or metanoia occur in this century, the worldwide dissemination of the entrenched modern worldview will make it much more difficult to displace. Ideas do have consequences, and the idea of domination which grew up with the Scientific Revolution has greatly intensified the natural human tendency to dominate and destroy the environment.

Western Dominion Over Nature

A final objective of this chapter is to explain more thoroughly the relationship between the idea of dominion over nature and human nature itself. A major contribution to our understanding of this relationship is to be found in William Leiss's classic *The Domination of Nature* (1972),[49] which has been an important source for most of the scholarly works discussed in this chapter. Although Leiss's book professes the reformist Marxist attitudes of the period in which it was written, it is nevertheless a careful and thoroughgoing analysis of the idea of the domination of nature, particularly as related to science, human nature, utopian thought, and the idea of progress, the latter topic to be discussed in the next chapter.

As the most successful mammalian species on the planet, *Homo sapiens* has been highly adaptable to the great variety of ecosystems found on Earth. When the combination of post-Pleistocene climate change and an increasing human population contributed to the Neolithic Revolution, humankind began to manipulate nature more intensively through the practices of agriculture and pastoralism. Gradually, as civilizations took hold in different environments, attitudes toward nature also changed. Underlying this profound transformation was the genetic make-up of *Homo sapiens*, a set of endowed traits and abilities so protean as to create the potential to adapt to virtually any ecological circumstance to be found on the planet. And yet this genetic endowment is not nearly as rigid as one might imagine. As Paul Ehrlich has written, "human nature is not the same from society to society or from individual to individual, nor is it a permanent attribute of *Homo sapiens*. Human natures are the behaviors, beliefs, and attitudes of *Homo sapiens* and the changing physical structures that govern, support, and participate in our unique mental functioning."[50] Here Ehrlich is emphasizing the role of cultural evolution in generating the great diversity of human nature, as well as the point that there is a considerable range in human genomes, and therefore, for this combination of reasons, in human nature itself. For Ehrlich, the idea of a fixed, single human nature, still prevalent in common parlance, is egregiously wrong and thoroughly outmoded. Nevertheless, one commonality which all human beings seem to share, and which is found ubiquitously in all cultures, is the need for some kind of myth or religion. Religion may be considered as a kind of myth, although typically not entertained as such because it requires faith or true belief regarding sacred texts or pronouncements. However, considering the definition of myth (A traditional or legendary story, usually concerning some being or hero or event, with or without a determinable basis of fact or a natural explanation, especially one that is concerned with deities or demigods and which explains some practice, rite, or phenomenon of nature.),[51] religion does qualify as a variety of myth typically associated with an historic figure of great spiritual insight.

Religion, as a subset of myth, usually attempts to respond to the theodicy problem, the question of why God allows evil to exist in the world. Certainly theodicy has been a central issue in Judeo-Christian religion, and has remained so down to the present day. When a landslide or earthquake suddenly kills and maims thousands of people, the question as to why an allegedly omnipotent, omniscient, and omni-beneficent Christian God would allow such suffering, such physical evil to happen is guaranteed to be explored the following Sunday in church. (A non-believing secular humanist, on the other hand, would simply acknowledge the unpleasant side effects of natural processes at work.) The great maverick American intellectual Lewis Mumford[52] recognized, in the responses of humanity to the perennial problem of physical evil or suffering in the world, the basis for two kinds of utopian vision which gave rise to what he called utopias of escape and utopias of reconstruction, a utopia being an account of an ideal world as in the original *Utopia* of Thomas More published in 1515. More coined the neologism "utopia" from the Greek, meaning u-topia or no place, or eu-topia, the good (ideal) place. A century later Francis Bacon published *The New Atlantis*, an incomplete account of an ideal Christian society in which experimental science plays a major role.

Both More's *Utopia* and Bacon's *New Atlantis* are classified by Mumford as *utopias of reconstruction*, meaning that their response to the existence of evil in the world is to reconstruct the world itself, from social institutions and class structure to improvements in our physical surroundings to protect us from natural disasters. In contrast, *utopias of escape* such as Saint Augustine's *City of God* seek to abandon the secular world and its evil for life in a transcendental world of perfection. During the century following More's *Utopia*, a number of Christian utopian works were published as utopias of reconstruction depicting ideal Christian societies in this world. One of them, *Christianopolis* (1624), included a scientific college based on experimental science. Francis Bacon was familiar with these utopias when he published *The New Atlantis* in 1627.[53] In all of these utopias the fundamental characteristic of human nature was, of course, the presence of an immortal soul in every individual. Although the ultimate purpose of human life was the salvation of the soul, sixteenth and seventeenth century Christian utopias nevertheless depicted the best possible life for Christians in this world until the release of the soul at death. As the Scientific Revolution unfolded, utopian authors increasingly turned to science and technology as the means of achieving the ideal society on Earth.

New Atlantis is an imaginary Christian kingdom located on the island of Bensalem, somewhere in the Atlantic Ocean. The kingdom is described as having been informed of the truth of Christianity by a divine miracle twenty years after the ascension of Christ. Prior to Christianization, King Salomana, the lawgiver of Bensalem, established Saloman's House, which was dedicated " to the study of the works and creatures of God...."[54] Sometimes referred to as the "college of the Six Days works" (referring to God's creation), the Society of Saloman's House is considered "the very eye" of the kingdom of Bensalem. The larger society of the ordinary citizens of Bensalem is depicted as rigorously pious even though its members possess private property and money, and the members of Saloman's House, living together in a monastery-like community, are exemplars of moral perfection. Bacon clearly assumes that Christianity is responsible for this general perfectibility of the citizens of New Atlantis. Members of Saloman's House also recite daily prayers regularly, as in monastic orders. The utilitarian value of science, practiced by these holy men, for the kingdom of New Atlantis is unequivocally clear:

> Lastly, we have circuits or visits, of divers principal cities of the kingdom; where as it cometh to pass we do publish such new profitable inventions as we think good. And we do also declare natural divinations of diseases, plagues, swarms of hurtful creatures, scarcity, tempest, earthquakes, great inundations, comets, temperature of the year, and divers other things; and we give counsel thereupon, what the people shall do for the prevention and remedy of them.[55]

What is most striking about this statement is that it no longer views natural catastrophes as simply "acts of God," but rather as events which can be rationally understood and therefore avoided or managed by humans. Bacon's insight into the utilitarian value of scientific knowledge stands in remarkable contrast to the superstitious views of natural processes which dominated both the medieval and early modern periods of European history, and which led to religious fanaticism, persecutions of "witches," and other behavior considered aberrant by modern standards. Thus, with no recognizable conflict between science and religion, the sincerity and strength of Bacon's belief in traditional Christianity coexists with the scientific and rationalistic elements of the modern mind in his scientific philosophy.

Prima facie, all would seem to be well in Bacon's *New Atlantis*, but William Leiss claims to find roots of our destructive modern technological society in this seventeenth century utopia. After the concept of mastery over nature, Bacon's most influential idea was that of organized scientific research, presented most strikingly in the example of Saloman's house. For Bacon, mastery of nature would be the goal of modern society, and scientific research would be the means of achieving it. As Leiss points out, the idea of human dominion over nature would take on an entirely new significance in the context of "the economic, social, political, scientific, and technological changes which capitalism fused together into a system of expanding productivity...."[56] Because Christian faith remained strong despite the social changes which tended to undermine it, Bacon's formula for mastery became widely accepted and then was gradually secularized as the influence of religion diminished. Building upon his belief that religion and science together were redeeming humanity since the Fall from Eden, Bacon could maintain that the mastery of nature through science was in conformance with God's plan. It is here that Leiss recognizes a fundamental flaw in the Baconian system. Bacon believed that the arts and sciences were the innocent or value-free means of generating social progress, i.e., he believed in the separation between scientific knowledge and moral knowledge which we refer to today as the fact-value dichotomy. Regarding this view, Leiss asks why the recovery from the Fall was " not the result of moral progress, rather than scientific progress."[57] In the *New Atlantis* the idea of mastering nature through scientific progress is presented in terms of its relationship to social progress. This leads Leiss to answer his own question by means of a comparison between More's *Utopia* and Bacon's *New Atlantis*.

Leiss's argument against Bacon's vision of social progress resulting from scientific and technological progress goes as follows. Thomas More envisioned an improved society based upon a more rational social division of labor. His *Utopia* demonstrated the primary need for a high level of material satisfaction in order to achieve the social objective of individual self-development for all citizens through a combination of continuing education and productive labor. In the *New Atlantis*, however, science and society as a whole are divided into the two major classes of priestly, highly moral scientists on the one hand, and ordinary, less pious and moral citizens on the other. Some modern scholars have argued that the high moral expectations of the priest-scientist rest " on the assumption that the qualities essential for scientific research--impartiality, disinterestedness, analytic rigor, and so forth--would be operative also in the scientist's social role."[58] Leiss finds this interpretation anachronistic if applied to *modern* societies, and sees instead, as Bacon appeared to, religion as the ethically restraining force controlling the uses to which scientific knowledge might be put, but even religion has its limits in controlling pride, self-interest, and ambition. Obviously, organized scientific research in our contemporary world has grotesquely exceeded the limits imagined by Bacon in the early seventeenth century. Leiss concludes:

> In *Utopia* individual self-development pursued during the time not
> devoted to necessary labor-a time continually made more ample by
> the applications of scientific knowledge-is the primary objective:
> the "moral progress" of individuals was to be the mediating link
> between scientific progress and social progress, the third term in
> which is manifested the rationality of the whole. And it is just this
> element of moral progress that is so conspicuously absent in *New
> Atlantis*.[59]

Thus, Leiss's scholarly analysis of the idea of human domination of nature leads ineluctably toward the modern philosophy of history known as the idea of progress, an idea which became a veritable secular faith in the late eighteenth and nineteenth centuries, and which haunts us to the present day despite all the contrary evidence of twentieth century history. Leiss's major concern in contrasting the ideal societies of More and Bacon was for the fate of both the individual and society in modern nation-states driven to rapid and inexorable change by the powerful impetus of scientific and technological development. If Thomas More offered us a glimpse of a future society which allowed the fullest possible development of individual human potential, Francis Bacon inadvertently provided us with an ominous caveat against a society distorted and barbarized by science and technology run amok and overrunning the potential moral development of a citizenry distanced from its esoteric techniques. Leiss has made it clear, then, that both the natural environment and human nature itself are at stake given relentless scientific and technological progress unconstrained by a morally responsible polity. The questions he has raised lead us directly towards the examination of the idea of progress itself in the following chapter.

NOTES

[1]. J. Donald Hughes, *Pan's Travail: Environmental Problems in Ancient Greece and Rome* (Baltimore: Johns Hopkins University Press, 1944), pp.54-55.
[2]. Ibid., p.55.
[3]. Stephen F. Mason, *A History of the Sciences* (New York: Collier Books, 1962), p.35.
[4]. Ibid., p.42.
[5]. Ibid., p.43.
[6]. Ibid., p.44.
[7]. Ibid., p.43.
[8]. David C. Lindberg, *The Beginnings of Western Science* (Chicago: The University of Chicago Press, 1992), p.67.
[9]. Ibid., pp.64-68.
[10]. Francis Bacon, *Novum Organum*, translated and edited by Peter Urbach and John Gibson (Chicago and Lasalle, Illinois: Open Court, 1994).
[11]. Carolyn Merchant, *The Death of Nature: Women, Ecology and the Scientific Revolution* (San Francisco: Harper and Row, Publishers, 1980), p.125.
[12]. David C. Lindberg, *The Beginnings of Western Science*, pp.361-362.
[13]. Francis Bacon, *Novum Organum*, p.80.
[14]. David C. Lindberg, *The Beginnings of Western Science*, p.361.
[15]. Ibid.
[16]. Thomas S. Kuhn, *The Structure of Scientific Revolutions* (Chicago: The University of Chicago Press, 1970), pp.111-135.
[17]. David C. Lindberg, *The Beginnings of Western Science*, p.359.
[18]. Carolyn Merchant, *The Death of Nature*, p.195.
[19]. Ibid., p.193.
[20]. Ibid., p.205.
[21]. Ibid., pp.206-215.
[22]. Ibid., p.206.
[23]. Ibid., p.164.

[24]. Carolyn Merchant, "Reinventing Eden: Western Culture as a Recovery Narrative" in *Uncommon Ground: Toward Reinventing Nature*, William Cronon, ed. (New York: W.W. Norton and Company, 1995), pp.132-159.

[25]. Ibid., p.143.

[26]. Max Oelschlaeger, *The Idea of Wilderness* (New Haven: Yale University Press, 1991), p.72.

[27]. Ibid., p.44.

[28]. Ibid., p.45.

[29]. Ibid., p.52.

[30]. Ibid., p.81.

[31]. Ibid., pp.81-82.

[32]. Ibid., p.76.

[33]. Frederick Turner, *Beyond Geography: The Western Spirit Against the Wilderness* (New Brunswick, New Jersey: Rutgers University Press, 1983), p.63.

[34]. Ibid., p.65.

[35]. Ibid., p.174.

[36]. Ibid.

[37]. Ibid., pp.174-175.

[38]. Ibid., p.176.

[39]. Ibid.

[40]. J. Baird Callicott, *In Defense of the Land Ethic: Essays in Environmental Philosophy* (Albany: State University of New York Press, 1989), p.181.

[41]. Ibid.

[42]. Ibid., p.182.

[43]. Ibid., p.183.

[44]. John Opie, "Renaissance Origins of the Environmental Crisis," *Environmental Review*. Vol. 11. No. 1, pp.3-17, Spring, 1987.

[45]. J. Baird Callicott, *Beyond the Land Ethic: More Essays in Environmental Philosophy* (Albany: State University of New York Press, 1999), pp.201-206.

[46]. For example, see review by John Terborgh, "The Eternal Frontier: An Ecological History of North America and its Peoples" (by Tim Flannery), *The New York Review of Books*, Vol. 48, No. 14, Sept. 20, 2001, pp.44-46.

[47]. Max Oelschlaeger, *The Idea of Wilderness*, pp.16-21.

[48]. Ibid., p.42.

[49]. William Leiss, *The Domination of Nature* (New York: George Braziller, 1972).

[50]. Paul Ehrlich, *Human Natures: Genes, Cultures, and the Human Prospect* (Washington, D.C.: Island Press, 2000), p.12.

[51]. *The Random House Dictionary of the English Language*. Second Edition (New York: Random House, 1987).

[52]. Lewis Mumford, *The Story of Utopias* (New York: The Viking Press, 1962, 1922).

[53]. Louise Marie Berneri, *Journey through Utopia* (London: Routledge and Kegan Paul Ltd., 1950), pp.126-128.

[54]. Francis Bacon, *The New Atlantis* in *Ideal Commonwealths* ed. by H. Morley (New York: The Colonial Press, 1901), p.118.

[55]. Ibid., p.137.

[56]. William Leiss, *The Domination of Nature*, p.49.

[57]. Ibid., p.53.

[58]. Ibid., p.66.

[59]. Ibid., pp.69-70.

CHAPTER SIX: ENLIGHTENMENT I: THE IDEA OF PROGRESS

Introduction

The transformation of the Christian, medieval worldview into the secular, modern worldview involved several centuries after 1500 A.D., and the schism it produced in Western thought remains markedly strong in the United States today in comparison to Western Europe. By the Age of Enlightenment (1715-1789) the challenge to the medieval worldview was well established, and the modern philosophy of history known as the idea of progress began to take hold among European and American *philosophes* (popularizers of the new philosophy) such as Voltaire, David Hume, Thomas Jefferson and Benjamin Franklin. The factors contributing to this shift in the Western worldview were many, some of which have been considered in the previous chapter. Certainly, the popularization of science was at the heart of the Enlightenment, along with a deep reverence for the art and thought of classical antiquity, and together they produced the scientific humanism described so well by historian Peter Gay.[1]

Contemporary with the Scientific Revolution, of course, was the great age of discovery and geographic exploration which allowed Europe and its system of beliefs to be understood anew within the comparative perspective of other cultural mores, in the long run leading to the corrosive effects of cultural relativism upon the medieval worldview. Eventually, the presuppositions of modern science laid down by Francis Bacon (1561-1626) and René Descartes (1596-1650), i.e., of empiricism and inductive method on the one hand, and systematic doubt on the other, contributed to the rise of the idea of progress. Descartes postulated the supremacy of reason over revelation as a means of attaining truth about how nature works, insisted on the invariability of the laws of nature, and required strict, logical proofs in the search for truth in his *Discourse on Method* (1637). These ideas contributed to the "Quarrel of the Ancients and Moderns" in which Charles Perrault (1628-1703), in *The Age of Louis the Great* (1687), maintained that all historical ages are capable of producing men of equal natural talent, from Euripides, Plato and Aristotle in antiquity to Shakespeare, Moliere and Galileo in Perrault's own time. Building on Perrault's principle that genius occurs in all ages, Bernard de Fontenelle (1657-1757) in *Digression on the Ancients and Moderns* (1688), argued that the modern seventeenth century mind was composed of all the minds or ideas of preceding ages. Fontenelle thus refuted the idea that the human race was in its "old age," a concept traceable to Saint Augustine's analogy between the human individual and the human race, which was also accepted by Francis Bacon. In place of the Augustinian analogy, Fontenelle substituted the idea of humanity's continuing youthful vigor in ages to come. Taken together, the insights of Bacon, Descartes, Perrault, and Fontenelle laid the foundations for belief that progress in knowledge is assured for the indefinite future because human nature cannot retrogress. Therefore, the progress of knowledge could be looked upon as a necessary and inevitable consequence of the structure of the human mind, as the Marquis de Condorcet (1743-1794) was to observe at the end of the eighteenth century.

A further consequence of the impact of the new scientific method upon historical thought in the seventeenth century was that natural law came to be understood as God's creation. As Alexander Pope put it: "Nature and Nature's laws lay hid in night; God said, let Newton be, and all was light." Thus, deism or natural religion, and religious tolerance, began to flourish, at least among literate Europeans and colonials, during the eighteenth century. The intellectual illumination supplied by science and cultural relativism shone unprecedented light upon the entrenched religious and sociopolitical beliefs of Europeans, and the Enlightenment spread like wildfire amongst the literate middle and upper classes until the French Revolution. During the eighteenth century, Locke's empirical psychology, Voltaire's critiques of the *ancien régime*, and Rousseau's reflections on the natural perfectibility of humankind through education all fanned the flames of dissent leading to cultural crisis in Europe. Within this historical context the idea of progress arose, flourished, and even mixed with the old idea of providence during the time of conservative reaction after the Revolution and Napoleon.[2]

Progress and Providence

What an obvious idea it seems at first, that mankind has progressed or developed over the course of history from pre-agrarian simplicity to the builders of computers and space stations. And to the present day the modern, liberal capitalist enterprise appears to continue the trajectory of progress in its spread around the globe. Unique to Western European civilization (until the gospel of progress spread over much of the globe to transform other traditional societies) and developing as an intellectual movement during the European age of Enlightenment, the idea of progress was understood during the first half of the twentieth century as being the consequence of the triumph of the accumulation of knowledge, boosted by scientific rationalism in the seventeenth century, when the "Quarrel of the Ancients and Moderns" took place. The classicist J. B. Bury explicated this idea in his scholarly *The Idea of Progress* in 1920. This explanation of the idea of progress as a child of the Scientific Revolution, following upon the triumph of the seventeenth century moderns over the defenders of the intellectual superiority of the ancient Greeks and Romans, was generally accepted by scholars until the second half of the twentieth century. The one indisputable key to the superiority of moderns over ancients was modern science, the methodical, empirical, and inductive study of nature which had begun to challenge the foundations of Western European society during the sixteenth and seventeenth centuries.

During the mid-twentieth century, however, philosophers of history focusing on biblical studies, such as Karl Löwith, continued the project of reclaiming the idea of progress for Christianity, leading some authors to redefine the belief in progress as the bastard progeny of Christian eschatology and Enlightenment optimism regarding human nature. This view of progress as a modern variation on the Christian idea of providence is known as the "secularization hypothesis" or "secularization thesis." This idea resonates with what we learned about Francis Bacon in the last chapter, namely that in the *New Atlantis* and elsewhere he viewed the new scientific method and its power to penetrate the secrets of nature as a means to regain the human dominion over nature that was lost with the fall from the Garden of Eden. Thus, supposedly, we could trace the Christian roots of the idea of progress back to Joachim de Fiore in the twelfth century, and following Robert Nisbet's *History of the Idea of Progress* (1980), even back to Saint Augustine and others in the ancient world. Environmental historian Carolyn Merchant has continued this tendency to Christianize the idea of progress by explaining modern progressivism as a "recovery narrative" in which modern humanity, empowered by the quasi-magical methodology of science, has regained the equivalent of the lost Garden of Eden in the secular world, thereby fulfilling the prophetic vision of Bacon's *New Atlantis*. America in particular has been the site and focus of this providential progressivism. Some years ago, when I encouraged a number of progress scholars to lobby a publisher to bring back J. B. Bury's classic *The Idea of Progress*, which had gone out of print, I was shocked to see the cover photograph of the new printing which showed Christian saints sculpted at the entrance portal of Rheims Cathedral. Was this a gaff or had the modern idea of progress been completely Christianized, at least in the mind of the designer of the new printing of Bury's classic. Poor Bury would roll over in his grave at the thought of this bizarre turn of events in historical interpretation displayed on the cover of his own book by an illustration representing the Christian idea of providence.

Having studied Bury and his critics carefully over the years, I have arrived at the conclusion that the reclamation of the idea of progress to its alleged Christian origins is at least partially based upon a simple fallacy which I shall attempt to explain. If we follow the generally accepted periodization of Western European civilization, c. 1000-2000 A.D., it is obvious that the philosophy of history which has dominated our Faustian culture is the Christian idea of providence which we have considered in earlier chapters. To summarize briefly, the idea of providence was perhaps best formulated by Saint Augustine in *The City of God*, and explains human history as having begun with the Creation, having incurred a major disaster with the fall of Adam and Eve, and having as its end or grand finale the Apocalypse, Last Judgment, and Millennium. Thus, breaking with Greek philosophies of historical flux and cycles while incorporating the late classical idea of a guiding, divine providence in the cosmos, the Christian idea of providence is resoundingly teleological in nature, i.e., sees all of history, from beginning to end, as manifesting the Christian God's divine purpose imposed upon an otherwise indecipherable world history. Certainly, the idea of providence initiated the Western European tradition of organizing history according to a linear rather than a cyclical pattern. In the secular, modern idea of progress the beginnings of civilization and technics would substitute for the Creation, the fall of ancient civilization might represent the fall from grace, and the end of history would be a moving point in time during which mankind becomes more knowledgeable and happier on indefinitely into the future.

Beyond this link (through creating an analogy between the idea of providence and the idea of progress) the claim for Christian roots underlying the idea of progress becomes problematical. Christian visions of the triumphant millennium at the end of history show us the outcome of history as envisioned and planned in the mind of God. The modern idea of progress, as the dominant philosophy of history in the modern Western worldview, envisions meliorative change in human affairs as a result of the strictly secular actions of mankind, unaided by any supernatural power. Francis Bacon serves ideally as a transitional figure between belief in the idea of providence and belief in the idea of progress. Bacon recognized modern science as the mechanism whereby fallen mankind would regain his rightful place in a recovered Garden of Eden made possible through human dominion over nature with the aid of the instruments of science and technology. However, although humanity would be comfortable and happy through utilization of the new science, it would be as part of God's plan for a mankind created in his image. Bacon believed in the idea of providence and simultaneously had a grasp of meliorative change over historical time. Thus, he contributed to the idea of the progress of knowledge which was fully developed by the Abbé de Fontenelle by the end of the seventeenth century. Did Bacon adhere to a "Christian idea of progress"? Here lies the fallacy. As a literal believer in the Christian worldview and its idea of providence, Bacon had no belief in human progress as a consequence of human effort unaided by God, which is a concept at the very core of what was originally intended by the late seventeenth idea of progress. "Progress" towards a future Christian millennium does not represent the modern *idea of progress*, nor is it a causal antecedent to it. The idea of providence and the idea of progress are two distinctively different conceptions of the nature of history, each situated in a particular historical context within which it developed. This does not mean that individuals today cannot believe simultaneously in both the idea of providence and the idea of progress, since the contradiction involved is generally overlooked in the popular mind.

Compounding the fallacy which conflates the idea of providence and the idea of progress by ignoring their respective historical contexts, once the idea of progress became established in the late seventeenth and eighteenth centuries, philosophers of history, beginning with A.R.G. Turgot and the Marquis de Condorcet, began to introduce the secular equivalent of teleological speculation into the idea of progress, as reflected in the title of American historian Carl Becker's *The Heavenly City of the Eighteenth Century Philosophers* (1932). Thus, a hybrid idea of progress, which incorporated teleology in terms of historical destiny or the "hidden hand," actually a compound of the idea of progress and the idea of providence, came into vogue during the late eighteenth and much of the nineteenth century, thereby laying the foundations for the "secularization thesis." I distinguish this "millennarian" progressivism from the late seventeenth and eighteenth century idea of secular, reason and science-based progress, or "scientific" progressivism, the latter perspective held by the majority of historians today.

Christopher Lasch, in *The True and Only Heaven: Progress and Its Critics,* supports my view and also agrees with German historian Hans Blumenberg when he writes that the modern conception of history can be distinguished from the Christian conception by " the assertion that the principle of historical change comes from within history and not from on high and that man can achieve a better life 'by the exertion of his own powers' instead of counting on divine grace."[3] Lasch argues that the "secularization thesis" has obscured essential differences between the idea of providence and the idea of progress, particularly in Nisbet's *History of the Idea of Progress,* which claimed to find highly developed theories of progress in the thought of Seneca, Lucretius, and the Christian fathers:

> Nisbet assumes that Roman and Christian philosophers shared our high opinion of material comforts. But although they admired the ingenuity that produced those comforts, they believed that moral wisdom lay in the limitation rather than in the multiplication of needs and desires. The modern conception of progress depends on a positive assessment of the proliferation of wants. Ancient authors, however, saw no moral or social value in the transformation of luxuries into necessities.[4]

Lasch continues his attack on the "secularization thesis" with the observation that even the idea of the Christian millennium as a source of the idea of progress was never a dominant view among Christians. Rather, as in Saint Augustine's view, it was the wretched, exhausted state of the world that portended the end of the world and the return and reign of Christ. Lasch's rejection of the "secularization thesis" leads him to conclude that the real meaning of the idea of progress is " not the promise of a secular utopia that would bring history to a happy ending but the promise of steady improvement with no foreseeable ending at all."[5] Lasch goes on to re-state what is in essence the Bury thesis, perceiving progress as largely the consequence of accumulated knowledge furthered by modern science, and its model of relentless inquiry leading to continual change, and thereby a sense of inevitability, without ever mentioning Bury's classic, *The Idea of Progress.* In any case, why should it matter for the environmental implications of the idea of progress whether it is uniquely modern and largely science-based or, rather, a modern variation on the Christian idea of providence and the teleological view of history which it represents? It does matter because contemporary, twenty-first century humanity is at a critical juncture in terms of our ability to make a moral decision regarding our planetary ecosystems, and our collective perception of whether or not history is predetermined or in our own hands is crucial to making the right decision. Our understanding of the nature of progressive change normally affects most of us subliminally in a barrage of commercial and political propaganda. The nature of "progressive" change itself is rarely considered or examined by the general public or our political leaders, which results in the general acceptance of the perception of "progress" as endlessly accumulating scientific and technological knowledge creating a society of unlimited material abundance. As Lasch would agree, this conception of the nature of human progress is one of the root causes of today's ecological crisis, but it is now called into question by the ecological crisis itself. However, the inertia of outmoded belief prevents the spread of an idea of progress reconstructed in conformity with natural limits.[6]

Historian Warren Wagar's caveat that there is no single, monolithic, gold-plated idea of progress but, rather, a variety of ideas of progress[7] in different times and climates of opinion, is excellent advice for the intellectual historian. The history of the idea of progress since its inception in the late seventeenth century uncovers a belief in the certainty of the progress of knowledge adhered to by Fontenelle in the seventeenth century, and a consequent belief in the *possibility* of moral or general progress adhered to by many of the eighteenth century philosophes. Subsequently, and anticipated by Turgot in the middle of the eighteenth century, this *secular* idea of progress resulting from human thought and action *without the aid of divine providence in its various guises* (continuing to the present day in the form of Adam Smith's "hidden hand") began to be mixed or conflated with the idea of providence, providence being converted to posterity in Condorcet's *Esquisse*, but becoming increasingly attached to progress throughout the nineteenth century. The Christian providential view of history, which continued to flourish and resist the philosophical implications of Charles Lyell and Charles Darwin's scientific achievements, assimilated the original secular idea of progress (of Fontenelle and the philosophes) into its teleological worldview. It was this integration of the secular, scientific or Fontenellian idea of progress with Christian providentialism that created what J. B. Bury has called "the complete idea of progress" because of its belief in the necessity of progress, i.e., that it was certain and foreordained as a reflection of the order of the cosmos.

Bury made the mistake of defending the uniqueness of his "complete" idea of progress during the early decades of the twentieth century, long before Wagar's insightful realization that no single idea of progress could be defended above all others once the history of the idea had been explored more fully. By Wagar's time, beginning in the nineteen sixties and seventies, a variety of ideas of progress began to be defended by various scholars. Nisbet carried this trend to the extreme in claiming to have discovered ideas of progress in classical antiquity and the Middle Ages as well as in the modern period. These developments in progress scholarship led me to construct a classification of ideas of progress during the mid-seventies which identified three major types of the idea of progress. These included: (1) The *scientific* (or Fontenellian) idea of certain cumulative progress in knowledge, particularly since the Scientific Revolution, which entailed *possible* general progress; (2) the *millenarian* idea of progress which required belief in its *necessity*, which was generally provided by some teleological driving force such as divine providence; and (3) *utopian* progressivism, which finds the course of historical change ambivalent, as in Rousseau's *Discourses*, and then attempts to design appropriate principles to re-track history onto a truly progressive course.

Viewed from the perspective of environmental history, any idea of progress involving a teleological component must be rejected as unsound on the grounds of scientific naturalism. Thus all variants of millenarian progressivism and any utopian progressivism appealing to supernaturalistic intervention, however well camouflaged as a "hidden hand," must be rejected as inconsistent with science-based knowledge of both natural processes and historical change, as explained in Chapter One of this book. The same objective treatment is applied to human history as is applied to geological and biological processes. Given this context, the only ideas of progress acceptable to objective, science-based analysis are Fontenellian scientific progressivism and utopian progressivism which eschews teleological arguments.

Theodore Olson, in *Millennialism, Utopianism, and Progress*,[8] at first might appear to have developed a tripartite classification of ideas of progress similar to my own. However, Olson's entire argument is constructed to demonstrate the contribution of millennialism and teleology-based utopianism to the rise of a teleological idea of progress similar to Bury's "complete" idea of progress with its emphasis on the immanence of progressive change within history: "The doctrine of progress proposes that there is a blind force in history, a force uncontaminated by historical contingency, yet dedicated to the continual improvement of man."[9] On the other hand, Olson is aware that a spectrum of types of ideas of progress exists when he writes that "God (and his will) are variously smuggled in and out of the progress doctrine; and it is crucial for the effectiveness of the result that we forget having made these moves."[10]

One of the most recent sophisticated analyses of the nature of the idea of progress has been presented by historian Bruce Mazlish in *Progress: Fact or Illusion* (1996).[11] Mazlish understands the idea of progress as having its most significant origins in the seventeenth century, when several streams of thought merged to create a modern explanation or philosophy of history. These streams included first and foremost the Scientific Revolution, secondly, the secular replacement of religious sectarianism and salvationism by the nation-state and the hopes for freedom and plenty which a peaceful Europe would engender, and thirdly, the millenarians and the religious inspiration for the Scientific Revolution. Mazlish writes that "Providence now took on the face of Progress." He also remarks that the sense of the inevitability of progress found in Francis Bacon's writings was based upon belief in Christian prophecy. Here again we find the idea of providence conflated with the idea of progress, as in Theodore Olson's

Millennialism, Utopianism, and Progress. (Mazlish, however, found a way out of this trap). Finally, there was the powerful stream of the Quarrel of the Ancients and Moderns, in which modernity triumphs first on literary grounds, and then on the basis of science.[12] In addition to these four major factors, the idea of progress was further propelled by: the application of scientific certainty concerning natural phenomena to the reconsideration of human phenomena as subject to natural laws, the view that there was a natural progression in human belief from myth to religion to science, seemingly verified by the improvement of humanity's material condition resulting from the Industrial Revolution, c. 1760-1860.[13] Mazlish adds that because of increasing disputation between religion and science, the "millennial" source of progress was called into question. Thus, the idea of progress which he wishes to defend as worth preserving, now and in the future, is detached from the idea of providence. Reason, science, and truth are the criteria established by Mazlish as "three major parts comprising the idea of progress."[14] On these grounds, he writes: "This faith in reason, which is ineluctably connected to truth and science, as I have defined them, is the bedrock on which the idea of progress can, and must, rest. Progress in knowledge is possible, and real. The history and practice of science demonstrate this fact beyond reasonable doubt."[15]

The idea of progress which Mazlish has chosen to defend is obviously the type which I have classified as "scientific" or "Fontenellian" progressivism in which there is belief in the certainty of the progress of knowledge, particularly scientific and technological knowledge, but uncertainty regarding the possibility of progress in morality and society. There is no millenarian or teleological component to this idea of progress, which is fully detached from the idea of providence. In fact, in the long run, scientific progressivism is closely associated with the secular revolt against the idea of providence which has produced modernity, for good and for ill. This point was strongly argued by W. Warren Wagar in *Good Tidings: the Belief in Progress from Darwin to Marcuse* (1972). Wagar took J. B. Bury's side in rejecting the "secularization thesis" and argued that the modern idea of progress is a belief peculiar to modern, secular society, although Christianity and other religions subsequently adapted progressivism to their own belief systems. As Wagar writes: "The rational and liberal humanism of the modern era has always been, at its roots, a religion of man, a faith in man and human possibility, which ultimately evolved into a faith in history, or, what amounts to the same thing, a faith in progress."[16] It is this secular, scientific or Fontenellian idea of progress described by J. B. Bury and defended as our most authentic conception of progress by Christopher Lasch, Leonard Marsak, Wagar, Raymond Duncan Gastil[17] and others, that Mazlish explicates as a worthy foundation for belief in at least limited progress today. As Mazlish writes, in terms of the *possibility* of moral progress (which is presumably enhanced by reason, science, and truth):

> That possibility is that cultures, in terms of their institutions, may move in a progressive direction, even though the individuals in

them are themselves not necessarily morally better. This complicated argument moves in the opposite direction from Rousseau's assertion that man is naturally good, but is made evil by social institutions. Instead, I am arguing that people are necessarily both good and evil in an Antigone-like way, but that their institutions may move in a progressive moral direction. In short, culture may become more moral (allowing for the fact that there is no absolute morality), although necessarily in terms of its own values.[18]

Anticipating Mazlish, Marsak pointed out Fontenelle's conviction that individual morals are unchanging, leading him to pessimism regarding the possibility of social progress. Despite Fontenelle's tragic view of humanity, "science was made to serve as the basis for his idea of progress."[19] Thus, the moral behavior of scientists would serve as a model of selflessness and dedication to knowledge for the self-interested individuals who make up the majority of humankind. Unfortunately for historical optimists, this is similar to Francis Bacon's utopia, *The New Atlantis*, considered at the end of the last chapter. What is most important in Marsak's treatment of Fontenelle as a defender of science is the recognition that science is the essential foundation of the modern, secular idea of progress, a point which was implicit in Bury's explanation of Fontenelle's theory of the progress of knowledge. Not surprisingly, the dominant idea of progress adhered to today is essentially Fontenelle's scientific progressivism. Modern history and anthropology have not been kind to our view of human nature, particularly in the twentieth century, and millennialism and the idea of providence for the most part have been detached from the idea of progress and relegated to church sermons and the intellectual limbo of religious fundamentalism. The roots of the contemporary progressivism defended by Wagar, Mazlish, and Marsak are clearly traceable to Fontenelle.

Progress and the Environment

We have seen how much the history of ideas has to do with ecological problems and environmentalism. The idea of human domination over nature is obviously important in understanding attitudes toward and treatment of natural environments. This idea has received careful treatment by Carolyn Merchant, Max Oelschlaeger, William Leiss and others discussed in the preceding chapter. Overall, however, the literature on this subject is not extensive. The idea of progress, which has been the dominant Western philosophy of history in modern times, has also received relatively little systematic treatment in relation to environmental problems. Leiss [20] analyzed the relationship between scientific and technological progress and the idea of human mastery of nature. His analysis illuminated the disparity between social progress and scientific-technological progress, which is closely related to a dichotomy between environmentalist and anti-ecological conceptions of progress. For example, he contrasted the moral purposes of Thomas More's *Utopia* with the scientific preoccupation of Francis Bacon's *The New Atlantis* to suggest two alternative ends of progressive change, one emphasizing the improvement of individuals and society within a relatively stable-state environment, the other stressing dynamic pursuit of scientific knowledge and mastery over nature to produce material abundance, with the expectation of concomitant improvement in society and morals. In the analysis and defense of the idea of progress summarized above, Bruce Mazlish did not think that any reasonable and useful alternative grand narrative could be substituted for the modern idea of progress, yet he does note parenthetically that "one might try to speculate on a stable-state society."[21] This will be the major theme of the rest of this chapter.

While historians of ideas have been developing a more sophisticated appreciation of the history and origins of the modern idea of progress, the basis for an alternative vision of human progress has been germinating in the writings of a group of environmental philosophers referred to in recent years as "deep ecologists." The term "deep ecology" was proposed by the philosopher Arne Naess in 1973 to distinguish the holistic, biocentric, and religious understanding of nature from "shallow ecology" or environmentalism which attempts to solve environmental problems from the perspective of traditional attitudes, including beliefs in human superiority over nature, mastery of nature through science and technology, and the probability or inevitability of progress based upon unconstrained technological development and economic growth. The deep ecology movement is attempting to challenge these strongly held anthropocentric attitudes with a vision of progress which emphasizes individual self-awareness and respect for the intrinsic, non-utilitarian value of nature, supported by a low-energy, labor intensive technology and a fixed population. To deep ecologists, most of what is called progress today is a spurious progress which can only lead to ecological destruction:

> The ultimate value judgment upon which technological society rests--*progress* conceived as the further development and expansion of the artificial environment necessarily at the expense of the natural world--must be looked upon from the ecological perspective as unequivocal *regress*.[22]

If we are seriously to evaluate and to understand this recent challenge to contemporary attitudes toward progress, we must begin with a critical analysis of scholarly work on the idea of progress itself. Twentieth century studies, from J. B. Bury's *The Idea of Progress* (1920) to Robert Nisbet's *History of the Idea of Progress* (1980), have generally viewed the idea of progress as a speculative philosophy of history involving an act of faith or belief in the certainty of human progress in science, technology, society, and morals.[23] Bury defined it as "the idea that mankind has moved, is moving, and will move in a desirable direction."

Exceptions to this premise, that the *historical necessity* of progress is intrinsic to the very definition of the idea of progress, have increased as the idea of progress has fallen into disrepute among intellectuals and lost its power to compel belief. Its presuppositions, particularly of the beneficent effects of social development and rapid increases in scientific and technological knowledge, have been challenged by a congeries of twentieth century nightmares---fascism, totalitarianism, social alienation, overpopulation, warfare, resource struggles, environmental degradation, terrorism, and the threat of nuclear holocaust. Nevertheless, belief in progress, as part of an implicitly accepted Western worldview, remains a powerful folk myth, strongly affecting our social and economic behavior. Most frequently, the idea of progress has been conflated with the process of economic development. This problem will be explored in a later chapter.

The conception of a monolithic idea of progress requiring faith in its immanence in the world-historical process surely is insupportable today. The loss of faith of intellectuals in the idea of progress explicated by Bury and Nisbet, however, should compel us to re-examine the questions of whether the idea of progress is best understood: (1) as Bury argues, a seventeenth century science-based optimism which was transformed into a teleological philosophy of history by Turgot and Condorcet in the eighteenth century; (2) as Nisbet argues, as the teleological philosophy of history of the Western tradition since antiquity; (3) as Lasch would have it, the expectation of indefinite, open-ended improvement, through human effort, but with no foreseeable ending; or, (4) following Wagar, as a spectrum of philosophies of history, varying in degrees of optimism, which reflect responses to the actualities of rapid technological, scientific, and social change in Western Europe and its colonies since the seventeenth century. In other words, can we properly refer to *ideas* of progress which acknowledge the value of past historical development but which perceive future progress as only *probable* or *possible* rather than certain and necessary? To take the earliest historical example, is the Abbé de Fontenelle's seventeenth century theory, based upon the continuing improvement of scientific knowledge as a basis for social and moral progress, acceptable as a philosophy of history comparable in validity to the idea of progress propounded by Turgot and Condorcet? The above discussion of this problem suggests that the Fontenellian idea of progress is more authentic because of its detachment from the idea of providence. To cite an even less orthodox alternative idea of progress, is Jean-Jacques Rousseau's interpretation of cultural development (as a two-edged sword capable of rendering more harm than good to industrialized and sophisticated societies *unless* men redesign social objectives and institutions in a revolutionary way based upon new knowledge of nature and human nature) as acceptable a philosophy of progress as that of Turgot or Condorcet? During the past few decades, studies of the history of ideas in seventeenth and eighteenth century Europe and analyses of the idea of progress itself strongly suggest that the Bury-Nisbet ideas of progress disallow other Enlightenment ideas of progress by defining the idea too narrowly. Alternative definitions and interpretations of the historical meaning of human progress could assist in opening new avenues of thought beneficial in planning for environmentally acceptable progress in a world now suffering from rapid historical change, i.e., development, rationalized as progress.

Bury enunciated the historical origins of the modern idea of progress in 1920, before ecological and resource constraints were identified and recognized as seriously affecting future economic and social progress. Nevertheless, he viewed the idea of progress from a perspective of historical relativism. In discussing the "illusion of finality" (the sense that the ultimate truth is known, certain, and unchallengeable) which characterizes belief in all philosophies of history, Bury[24] concluded that "the process of change, for which Progress is the optimistic name," must also "fall from the commanding position in which it is now, with apparent security, enthroned." Central to the "complete" idea of progress described by Bury was belief in its historical necessity. Such a belief promoted the illusion of finality which Bury recognized as the Achilles heel of all philosophies of history. Thus, it came as something of a surprise to some scholars when Robert Nisbet argued passionately, in his *History of the Idea of Progress*, for the need to restore faith or belief in the traditional idea of progress in order to rejuvenate a declining Western civilization. He concludes his book with the statement that the historical record shows it is only "in the context of true culture in which the core is a deep and wide sense of the *sacred* are we likely to regain the vital conditions of progress itself and of faith in progress-past, present and future...."[25] Thus, Nisbet's idea of progress requires its amalgamation with the idea of providence.

It is Nisbet's appeal to belief or faith which represents what Bury termed "the illusion of finality." Such an appeal to faith is particularly disconcerting when Nisbet alloys it with a veneration of laissez-faire capitalism (perhaps an illusion of finality in itself) and a sometimes thinly veiled but usually unequivocal rejection of the significance of ecological and resource constraints to economic growth as a consideration in assessing human progress. It suggests that Nisbet's definition of progress, like those of Turgot and Condorcet, is designed to meet the requirements of human psychological needs characterized by Eric Hoffer's "true believer" mentality[26]

In contrast to the teleological element which pervaded nineteenth century ideas of progress, the compulsion to believe in progress as a certainty was distinctly *not* characteristic of the *philosophes* as a group, and twentieth century scholarship on the Enlightenment has solidly established the existence of strong currents of pessimism in eighteenth-century thought, resulting in a general wariness of any *faith* in progress on the part of Voltaire, Diderot, Rousseau, and Fontenelle [27] Can we not then view the idea of progress which assumes that it is an inevitable pattern of history, the idea so eloquently expressed by Turgot and Condorcet as a minority view of Enlightenment thought, as but one of several ideas of progress developed by the *philosophes*? The rapid development of science, technology, industrialization, and economic growth in the succeeding post- Enlightenment years buoyantly carried Turgot's and Condorcet's optimistic interpretation through the nineteenth century and into the twentieth. The less optimistic philosophies of progress of Fontenelle and Rousseau lay in the limbo of unused ideas until the unpleasant facts of twentieth century historical change called upon them as organizing principles of history perhaps more durable than those elaborated by Turgot, Condorcet and their nineteenth and twentieth century disciples.

A Classification of Ideas of Progress

As explained at the beginning of this chapter, the most notable challenge to Bury's thesis that the idea of progress is a distinctly modern philosophy of history has been the attempt, particularly of post World War II Christian theologians, to depict the doctrine of progress as a Christian heresy; Emil Brunner[28] called it "the bastard offspring of an optimistic anthropology and Christian eschatology." As W. Warren Wagar has pointed out, Bury's defense of the modernity of the idea of progress became the "old orthodoxy" of progress scholarship until the 1950's. Subsequently, Christian and lay scholars of the "new orthodoxy," from Karl Löwith and Reinhold Niebuhr to Robert Nisbet, have identified the idea of progress defined by Bury as emerging from an earlier spiritualistic stage of development during the Middle Ages.[29] An intricate defense of the old orthodoxy has been written by Hans Blumenberg (1983), and Christopher Lasch has reinforced this position (1991).

Perhaps the endless debate over when the idea of progress first was articulated in Western civilization involves a logical fallacy: the assumption of a single standard or definition of the idea of progress when in fact many different ideas have existed. Scholarly debate over when the idea of progress emerged may be shifting towards questions of what *elements* comprise the modern idea or ideas of progress in contrast to those held in other historical periods. It may be that we are beginning to heed Wagar's caveat against attaching our studies of progressive thought to the conception of a single, exclusive idea of progress:

> There is not one true, monolithic, gold plated Idea of Progress, which emerges at a particular point in time, and against which all other so-called ideas of progress must be measured. Some of the difficulty in the historiographical debate over the origins of the idea of progress is certainly rooted in the too abstract and narrow definitions of that idea which scholars have demanded. There are, rather, many ideas of progress and where the lines are to be drawn and how the various types are to be discriminated historically is exceedingly difficult to determine.[30]

In my own research on Rousseau's thought on progress, the necessity of placing his ideas within a spectrum of Enlightenment attitudes toward human progress directed me towards a tentative classification of late seventeenth and eighteenth century ideas of progress. [31] As early as 1925, Frederick J. Teggart stressed the importance of distinguishing *belief in progress* from *belief in the possibility of progress:*

> To believe in progress is to adopt a supine attitude toward existence; is to cultivate an enthusiasm for whatever chance may bring; is to assume that perfection and happiness lie ahead, whatever may be the course of human action in the present. To restrict belief to the possibility of progress implies recognition of the fact that change may result in destruction as readily as in advancement; implies consciousness of the precariousness of human achievement. [32]

Fontenelle and Rousseau were clearly believers in the *possibility* of progress as Teggart described it, but are these *ideas of progress*? Several Enlightenment scholars have attempted to demonstrate that they are. Norman Suckling, for example, distinguished an early Enlightenment belief in the possibility of progress from a belief in the necessity of progress which arose in the second half of the eighteenth century. [33]   Leonard M. Marsak maintained that the Enlightenment conception of progress was distinct from the teleological idea of progress which arose in the nineteenth century (adumbrated by Turgot and Cordorcet), and J. H. Brumfitt (1972) argued that the majority of the *philosophes* adhered to a philosophy *for* progress while a small minority held a philosophy *of* progress. [34] Thus, with the exception of Turgot, the philosophes, prior to the advent of the French Revolution, generally subscribed to a belief in the *possibility* of social and political progress. In this belief, Rousseau was in agreement with Fontenelle, Voltaire, and the Encyclopedists. The crux of Rousseau's disagreement with his contemporaries, however, was his insistence that progress in knowledge and science is more likely to lead to regress than to progress in morals and society, and that only the establishment of ethical principles for improving social behavior would allow increased knowledge and control of the physical world to improve the quality of human existence. This is similar to Mazlish's argument earlier in this chapter. For Rousseau, the combination of progress and regress resulting from historical development required a utopian vision to increase the possibility of progress in the future.

From Fontenelle to Turgot, Condorcet, Saint-Simon, Comte, and Spencer, from belief in the possibility of progress to belief in the necessity of progress, there is the common thread of developmentalism, of recognizing in the history of humanity's development the unequivocal evidence that progress in knowledge or general progress has occurred in the past and may or must take place in the future. This emphasis upon the *terminus a quo*, i.e., the end from which or beginning point, of a linear conception of time allowed Fontenelle to establish a scientific basis for possible general progress in the future, with the progress of knowledge assured by the achievements of the Scientific Revolution.[35] In portraying the accomplishments of the entire sweep of civilization as a progressive development, Turgot and Condorcet passed beyond the idea of progress as essentially beginning *de novo* in the seventeenth century to the historicist conception of progress as a law, immanent in the historical process. I have termed the earlier, predominant Enlightenment idea of progress *scientific progressivism* and the later historicist idea of Turgot and Condorcet, with its belief in the historical necessity of progress, *millenarian progressivism*. Both are *developmental* ideas of progress, projecting progress into the future on the basis of perceived human successes in the past, whether short-term or long-term. Further, millenarian progressivism is equivalent to Bury's "complete" idea of progress, while scientific progressivism was interpreted by him as a stage en route to the appearance of millenarian progressivism in the work of Turgot and Condorcet. The idea of intellectual and social development is central to all ideas of progress, but different interpretations of the meaning of development may give rise to alternative ideas of progress. This point has been well made by Charles Van Doren[36] in his distinction between *progress* as "irreversible meliorative change" and *process* as "irreversible cumulative change."

It was Rousseau's perception, in the *Discourse on the Arts and Sciences* and the *Discourse on the Origins of Inequality*, that advances in knowledge, technology, and economic development had diminished rather than increased the moral condition and happiness of a large proportion of the populations of Western European and other "highly civilized" cultures, leading him to construct an intricate utopian proposal for moral, social, and political reform in the *Discourse on Political Economy, Emile,* and *The Social Contract.*[37] Recognizing the gains and losses of a contingent process of historical development, Rousseau wished to redirect humanity towards a more genuine progress in the future on the basis of striving towards utopian goals which would gradually assure men of greater empathy, compassion, fraternity, and above all, justice and equality in their social, political, and economic relations and institutions.

I have termed this understanding of progress *classical utopian progressivism* or simply *utopian progressivism* because it challenges the ambivalent results of past development with a utopian vision of future change based upon rational principles which are designed to counteract the undesirable effects of historical contingency. Its orientation towards the future and the *terminus ad quem,* i.e., end to which or end point of linear time, distinguish it from developmental ideas of progress. It also differs from developmental progressivism in projecting alternative or countercultural goals into the future rather than a trajectory of established practices and institutions. This is one reason why, in the face of new obstacles, the traditional Western scientific and millenarian ideas of progress are little more than rationalizations of today's status quo: goals for future development are nothing more than an extension of applications of scientific and technological knowledge implemented by the institutions of democratic and socialistic capitalism which benefited from their explosive development in recent centuries.

The American cultural historian Leo Marx tends to corroborate my utopian-developmental dichotomy of ideas of progress. Marx distinguished between the "goal-oriented," essentially political progressivism of the American founding fathers and the later, "technocratic" concept of progress which has dominated American attitudes toward progress in the nineteenth and twentieth centuries.[38] The latter concept emphasizes "the most rapid possible rate of technological innovation as the essential criterion of social progress."[39] The goal-oriented (utopian) idea of progress manifested in Thomas Jefferson's vision of an uncorrupted agrarian polity is traced by Marx through the thought of Emerson, Thoreau, Hawthorne, and Melville as anti-technocratic and adversarial.

Implying the presence of both millenarian (historicist) and utopian elements in the modern idea of progress, the philosopher Hans Jonas wrote that for modern Western society:

> progress is a law as well as an ideal. This combination of a *perceived* vector property in the movement of history as progress-bound with *adopting* this vector into the conscious will, which thereby makes itself its further agent, is perhaps the most uniquely Western trait. In the way of self-fulfilling prophecies, it makes to some extent its view of history come true by being a historical force itself.[40]

In this statement Jonas revealed the ultimate conservatism which underlies rigid developmental ideas of progress, whether expressed in capitalist or Marxist terms. The basis for past progress, once institutionalized, crystallizes into an ideology which is expected to control the future. In this context, utopian visions of different and better societies become radical challenges to the certainty of the ordained trajectory of progress. Only an emphasis on re-examining and redefining the goals of progress can counteract the environmental destructiveness of an outmoded "progress," in which traditional societies and their bases in nature are destroyed by the world market and its technological infrastructure.

It is for this reason that blueprints for ecological and nuclear survival have proliferated during the past three decades, providing a utopian counterweight to subliminally entrenched historicist convictions about the direction and goals of progress. Late twentieth century progress scholarship, such as Theodore Olson's *Millennialism, Utopianism, and Progress* (1982), and essays by Bruce Mazlish, Leo Marx and others in *Progress: Fact or Illusion* (1996) may help to illuminate this schism in ideas of progress. Although, like Bury and Nisbet, Olson adhered to the concept of a single monolithic idea of progress, his importance lies in having demonstrated the role of the utopian tradition in modern progressive thought (just as Tuveson[41] elaborated upon the importance of millennialism.)

Olson raised the question of how the two disparate traditions of utopianism and millennialism have contributed to the emergence of the idea of progress in eighteenth-century Europe. My own typology of Enlightenment ideas of progress clearly reflects the influence of these traditions in creating a spectrum of ideas of progress ranging from moderately optimistic scientific progressivism to the wildly optimistic and teleological millenarian progressivism which Bury and Nisbet adopted as a norm, to the utopian progressivism which rejects modern development as regressive and offers steady-state alternatives.

As suggested earlier in this chapter the Scientific Revolution must be considered as the most important influence on the development of progressive thought during the Enlightenment. Together, and sequentially, the scientific and millennial traditions provide a line of development from late seventeenth and eighteenth century belief in the *possibility* of general human progress to the late eighteenth and nineteenth century faith in the *certainty* of progress. This is the genealogy of the *complete* idea of progress described by Bury and traced through the thought of Perrault, Fontenelle, St. Pierre, Turgot, Condorcet, and on to Saint-Simon and Comte. However, Bury allowed no place in his scheme for the role of the utopian tradition subsequently shown to be an important element in modern progressive thought by Olson and other scholars. Bury could not reconcile utopianism with the idea of progress as he understood it because he viewed progress essentially in terms of recognition of past cultural development and the projection of such development indefinitely into the future.[42] Nisbet has likewise adhered to developmentalism, with its focus on the *terminus a quo*, the beginnings of a culture, rather than on the *terminus ad quem* and the future possibilities of culture, which is the major concern of writers of utopias. Yet, down to the present day, the perception of the idea of progress as a single, monolithic conception combining science- and technology-based optimism with Christian millenarianism persists, as in Richard B. Norgaard's *Development Betrayed: The End of Progress and a Coevolutionary Revisioning of the Future* (1994).[43] Nevertheless, Norgaard does appreciate the application of utopian thought to the idea of progress in his project of revisioning the future from an ecological perspective.

If we learn to reject the interpretation of a single, amalgamated idea of progress, the idea of progress in the eighteenth century can then be expressed as three theories of progress manifesting three psychological types or predilections. The *scientific progressivism* represented in the thought of Fontenelle reflects the objectivity, skepticism, and qualified optimism of the experimental, rationalistic scientist or follower of science. The *Millenarian progressivism* of Turgot and Condorcet acknowledges the value of science and the power of its method, but within the context of a recognized pattern of historical development that includes science as one of several mechanisms of progress, all functioning as manifestations of a progressive tendency immanent in the process of historical change. It is the "true believer's" faith in the certainty of historicist progress which characterizes the millenarian mentality and its quest for historical truths.

Millenarian progressivism *followed* scientific progressivism and introduced the teleological component associated with the Christian idea of providence during the nineteenth century. Millenarian progressivism fits the amalgamated idea of progress accepted by Bury, Nisbet, Norgaard, and others as the *only* idea of progress. Finally, the *utopian progressivism* of Rousseau, Jefferson, and twentieth century deep ecologists is the product of a desire to take things as they have developed to the present time for good and for ill and to "set things right." Thus, Norgaard fits my classification as a utopian progressivist. The process of development has produced a mix of positive and negative conditions which should not be projected into the future but which must be altered if the human condition is to improve rather than deteriorate. This "utopian propensity" is generally concerned with controlling or modifying undesirable human traits, beliefs and institutions, and with planning for the future. The utopia is thus a model or blueprint for the unachievable perfect society which can nevertheless be approximated despite the sometimes inimical forces of human nature and historical contingency. To the degree that these forces appear reasonably controllable to the utopist, his hopes for progress are sustained by striving towards his ideal society.

Environmentalist Expressions of Utopian Progressivism and the Redefinition of Progress

Much of the disagreement among scholars over the definition and nature of progress in relation to environmental problems stems from the various ways in which authors link the idea of progress to the concept of utopia. From Lewis Mumford to Frank Manuel, utopia has been variously defined. Mumford saw the relationship of progress and utopia as depending upon whether one emphasizes the *terminus a quo* (end from which) or *terminus ad quem* (end to which) of progress, i.e., the process of *development* from a starting point as opposed to the future *goal* or *utopia*, and this provided a workable path through the labyrinth of contradictory evaluations of the relationship.[44] This approach also helps to explain the strongly opposed evaluations of utopia as a threat to progress or, alternatively, as creating the possibility of progress in the future. Conservative ideologues who argue in support of developmental progressivism (essentially as a projection of the *status quo* into the future) naturally see utopias and utopian thought as subversive and dangerous because to them such writings represent the unattainable ideal of a perfected society, potentially at the cost of individual human freedom or the restructuring of social classes.[45] Whether or not their fears are justified, the conservative denigration of utopia stands in stark contrast to Ernest Callenbach's purportedly benign use of a fictional utopia to establish new principles of progress or William Ophuls's enumeration of the elements of a potential but undescribed environmental utopia. Discussion of these two examples, the first a descriptive utopia and the second an exercise in utopian thought, will be followed by a concluding analysis of the significance of utopian thought for an ecologically-based idea of progress.

Ecological writing expounding utopian principles dates from at least 1950,[46] but the classic ecological utopia is Ernest Callenbach's *Ecotopia* (1975). *Ecotopia* describes an ecologically sound polity established in the seceded Northwest at the end of this century as a society in which deep ecology values, appropriate technologies, feminism, and autarky distinguish the steady-state alternative from growth-oriented technological societies. The fundamental goal of Ecotopia is stable-state life systems in which the costs of human use are not "ignored, or passed on through subterfuge to posterity or the general public."[47]

When questioned by the book's protagonist, a skeptical reporter from the United States, as to whether they have given up "any notions of progress," an Ecotopian citizen replies:

> It may sound that way, but in practice there's no stable point. We're always striving to approximate it, but we never get there. And you know how much we disagree on exactly what is to be done-we only agree on the root essentials, everything else is in dispute.[48]

Thus, Ecotopian agreement on the first principle, or goal, of ecological stability and on the difficulty of maintaining it expresses the principle of utopian progressivism. The impossibility of perfection is clearly acknowledged, and progress is understood as the continuing attempt to realize the ideal, albeit imperfectly, in the face of natural and human contingencies. Following Oscar Wilde's aphorism, "Progress is the realization of utopia," Ecotopians would indefinitely attempt to realize the implementation of a peace-loving culture in maximum harmony with the planet's ecosystems. Another Ecotopian citizen is described by the reporter as "strongly optimistic about the future. Believes that the nature of political power is changing, that the technology and social structure can be put at the service of mankind, instead of the other way round."[49] This observation distinguishes between the dubious contemporary perception of scientific and technological change as progress (indeed, "inevitable progress") and the idea of true progress as requiring general human amelioration, including moral progress, even if it means the diminution or scaling down of scientific and technological development to technologies in harmony with nature. Conversion to the steady-state "would mean sacrifice of present consumption, but it would ensure future survival- - which became an almost religious objective, perhaps akin to earlier doctrines of 'salvation.' People were to be happy not to the extent they dominated their fellow creatures on earth, but to the extent they lived in balance with them."[50]

Finally, Callenbach makes it clear that the great cultural transformation necessary for the survival and progress of humanity may mean economic disaster for plutocrats but not necessarily survival disaster for persons because "the new nation could be organized to devote its real resources of energy, knowledge, skills, and materials to the basic necessities of survival."[51] The enormous obstacles which stood in the way of the political and social transformation of modern China and other traditional societies suggest that the relatively peaceful restructuring of smaller, less densely populated states would be feasible.

William Ophuls's *Ecology and the Politics of Scarcity* (1977, 1992) stands an excellent chance of being looked upon as a classic in the twenty-first century, criticism of his political viewpoint notwithstanding.[52] Its explication of ecological complexity and of the failure of existing political and economic paradigms to prevent wholesale degradation of natural systems is a powerful critique of developmental progressivism. More importantly, Ophuls's guidelines for the building of a steady-state society clearly illustrate the method and importance of utopian progressivism. In the context of the geologic time scale and the evolution of homeostatic ecosystems, the scientific and millenarian ideas of progress of the eighteenth and nineteenth centuries appear to be naively anthropocentric and anachronistic. However, "to challenge endless scientific and technological progress amounts to a kind of secular heresy."[53] Ophuls does not call for an abandonment of technology but for its humanization as a historical force and a recognition of its limits:

> This judgment certainly does not mean that all technological solutions are anathema. Indeed, to counter single-minded technological optimism with an equally single-minded neo-Luddite hostility to technology in all its forms is absurd, for a non-technological existence is impossible. The question at issue is what kind of technology is to be adopted, and to what social ends it is to be applied. The whole subject of technology needs to be demythologized, so that we have a realistic view of what technology can and cannot do and of what its costs are.[54]

Ophuls supports environmentalist demands for an alternative technology "diametrically opposed to *autonomous* technological growth of the kind that has produced an ecological crisis." He recognizes the prevalence of developmental progressivism in our belief system when he observes "that during the last 300 years society has adapted to technology rather than vice versa."[55] He observes that many have repudiated feasible alternative technology because "its adoption will require a revolutionary break with the values of the industrial era."[56] There is no doubt in his mind as to the significance of these technological alternatives and the urgent need to shift to them immediately: "The epoch we have already entered is a turning point in the ecological history of the human race comparable to the Neolothic Revolution; it will inevitably involve racking political turmoil and an extraordinary reconstruction of the reigning political paradigm throughout most of the modern world."[57]

In considering probable sociopolitical responses to increasing ecological scarcity, Ophuls implicitly makes a distinction between developmental progressivism gone awry and utopian progressivism: "There is no escape from politics. As a consequence of ecological scarcity, major ethical, political, economic, and social changes are inevitable whatever we do. The choice is between change that happens to us as a "side effect" of ever more stringent technological imperatives and change that is deliberately selected to accord with our values."[58]

The need to define ecologically preferable values and goals may offer humanity "a grand opportunity to build a more humane post-industrial society."[59] Such an undertaking would require reconsideration of existing bourgeois social contract theory in order to create an "ecological contract" theory which would promote "harmony not just between men, but also between man and nature."[60]

It is clear that Ophuls prefers a planned economy to present corporate laissez-faire practice as the basis of an ecological society. He states that capitalism is "founded on hidden social costs, in which development...would not have occurred if all the costs had been counted in advance.... In brief, honesty and progress may not be compatible."[61] In this context, "progress" means process or irreversible cumulative change. Recognizing economic growth as the "secular religion of American society," Ophuls notes the desire of individuals, businesses, and even government to maintain growth at any cost. Thus, political questions concerning society's goals and "general will" (decision-making in the service of the common good rather than individual or factional interests) are evaded by perceiving the "will of all" (self interest) to be the "general will" and construing political processing of economic development to be the end of politics. The result is a "muddling through" which ignores long-term goals. On the other hand, socialism has not provided a solution "because the ideology of growth and belief in the power of technology are even more strongly entrenched in the U.S.S.R. than in the West".[62]

What basis, then, does Ophuls offer us for retracking a misdirected "progress"? First, he is adamant that " 'metanoia,' a fundamental transformation of worldview, must precede concrete action".[63] Second, rejection of current political reality is essential because "ultimately *politics is about the definition of reality itself,*" and, "in its truest sense, *politics is the art of creating new possibilities for human progress.*" Third, we can expect the transformation to take at least several decades to produce a result which "will be a mosaic of many elements, some designed by man, and others fashioned by the accidents of history." Fourth, "the hour is very late," and finally, "the process of tearing down the old reality and constructing the new has already begun."[64]

Well aware of the potential dangers of premature planning for, or prognostication of, the form of the ecological culture of the future, Ophuls offers only general principles as guidelines for the building of a steady-state society. These include: *communalism* as opposed to individualism; *authority* versus liberty (to destroy nature); *aristocratic principles* versus egalitarian democracy; *politics* versus economics; *stewardship* versus exploitation; *modesty* versus "progress" or economic development; *diversity* and *autarky* versus centralization; *holism* versus scientific reductionism; *Socratic morality* versus instrumental ethics; and *post-modernity* as *pre-modern values* versus the rule of modern technology and economics.[65]

Ophuls thinks that it is at least possible to meet the challenge of ecological scarcity "by creating a genuinely post-modern civilization that combines the best of ancient and modern."[66] His skepticism regarding the past several centuries of "progress" does not preclude his belief in the possibility of spiritual and intellectual progress combined with material frugality. He is convinced "that nothing of real value would be lost if development were to cease."[67] Comparing the cultural achievements of ancient Greece and pre-modern Japan with our own, he concludes that "development appears to be virtually irrelevant to cultural richness and progress; social arrangements, not wealth in itself, seem to determine the level of social amenity."[68]

Writing two decades later in *Requiem for Modern Politics: The Tragedy of the Enlightenment and the Challenge of the New Millennium* (1997), Ophuls still had not given up hope that the modern liberal polity, child of the Enlightenment, was still capable of reforming itself through a renewed, "Thoreauvian" connection with nature: "Thus, the problematique of modern civilization has a genuine 'solution.' It is by becoming Thoreau's more experienced and wiser savages that we can fulfill our true destiny as civilized beings, guarantee the continuity of civilization itself, and preserve the real political achievements of the Enlightenment from the forces of endarkenment now threatening to engulf them."[69]

Ophuls is not so certain that his goal will be achieved, however, given that the vast majority of modern humanity "are credulous believers in a secular monotheism that deifies Science, the mighty god that will lead us to the promised land of progress."[70] Given this reality, it is hardly surprising that utopian blueprints for an ecologically benign future peaked in the 1970s and 1980s and declined in an era of technologically-induced wealth generating a new wave of suburban sprawl, large houses, and SUVs.

Conclusions

The function of utopian thought has always been critical as well as reconstructive. Callenbach and Ophuls tell us as much about the blindness towards nature and intransigence of contemporary ideals and institutions as they do about the possibility of a future society established upon principles of ecological cognition. Karl Mannheim emphasizes this aspect of utopia in contrasting it with ideology. He defines utopia as an "orientation which transcends reality and which at the same time breaks the bonds of the existing order.... The utopian state of mind is incongruous with the existing order."[71] Ideology, on the other hand, he characterizes as rationalization of an existing order, as knowledge which "fails to take account of the new realities applying to a situation" and "attempts to conceal them by thinking of them in categories which are inappropriate."[72] In this sense, the modern idea of progress is our own unanalyzed ideology, the most powerful element in the worldview of the West dangerously simplified to equate development of interacting science, technology, and industrial capitalism with a grand human effort, or even a historical design, for human betterment. The philosopher Frederick A. Olafson lends strong support to the importance of Mannheim's distinction between ideology and utopia as applied to the idea of progress.[73] Asserting the ideological character of the established idea of progress, Olafson argues that "its principal disadvantage is that it also tends to 'close down' for ethical purposes the very situations it has opened up by declaring that change in the general conditions of life is possible. It closes them down by offering assurances based on the scientific/technological character of the source of change to dissuade us from insisting upon a genuinely independent, ethical evaluation of what is, in fact, coming to pass."[74]

Olafson goes on to conclude that two centuries of American life under the ideology of progress have failed to develop the humanistic education necessary to the healthy functioning of a democratic society, in recent decades creating a crisis of basic literacy. In this context, he perceives the idea of progress to have been "an obstacle in the way of a more deeply conceived and adequate view of human betterment."[75]

Having unmasked the ideology of progress, i.e., developmental ideas of progress, as actually regressive in both ecological and social terms, and having sketched the necessary ecological basis of future progress, can the environmentalist prophets of post-modernity effect action? The past two decades of what we might call "post-environmentalism" seem to suggest that they cannot.[76] The deep ecology arguments for redefining and reorienting human progress are less likely to influence the existing technological society than appeals to prudence in our treatment of nature in order to assure long-term human survival. Given the imperatives of the burgeoning world market and its exponentially increasing demands for scarce resources, we are losing wilderness and other habitats at a terrible rate. Even worse, the conversion of diversified, labor intensive farming to machine and energy intensive monocultures will be highly destructive of both habitat and traditional social orders. The transformation of rural, agrarian man into the nature-alienated individual of megalopolis will continue the building of the technological society, possibly in the form of the anti-progressive dystopias of Huxley and Orwell.

The attempts of Callenbach, Ophuls, and other environmentalists to educate us in a radical, utopian way have not yielded impressive results thus far, yet the teaching of values, particularly the values of a new ecological humanism, is our best hope of avoiding a world technological society almost completely alienated from nature. At present, these values appear to be losing ground to those of economic efficiency and the short-term quest for wealth at the expense of the needs of posterity. Indeed, an environmentalist idea of progress may not gain widespread recognition until well into the twenty-first century, but the foundations are being created now, and there are several things which academics and intellectuals can do to accelerate a transformation of values.

First of all, we must all struggle constantly against the intellectually stupefying effects of academic specialization, which counts the introductory Western Civilization course and the discipline of intellectual history among its victims. Humanists even more than scientists are often psychologically alienated from nature by the heavily anthropocentric and multicultural orientation of their disciplines. In addition to personal effort, raising ecological consciousness will require restructuring both survey and advanced classes in the humanities and social sciences to include the study of man's place in nature as well as in human experience. Ultimately, these perspectives also should be assimilated into primary and secondary education.

Secondly, within this broad framework, our understanding of the history of ideas must be rejuvenated. An awareness of the need to control and to buffer the effects of science and technology, rather than a faith in their infallibility must become an element in any post-modern idea of progress. Rousseau, perhaps the earliest serious critic of developmental progressivism and its relation to human domination of nature (and of his own species through the use of power over nature), has provided us with the earliest model of utopian progressivism as an alternative vision of human progress. This will be elaborated in the following chapter.

Finally, a cultural perspective which recognizes geologic (natural) as well as historic (cultural) time is sorely needed. Our explosive "progress" of the past three centuries may be regarded as a frightening natural anomaly in the history of ecosystems, for it has released the energy and mineral resources concentrated on our planet for hundreds of millions of years within an iota of natural time. Understood in this perspective, the idea of progress is based upon an anthropocentric conception of time almost as naive as the Judeo-Christian idea of providence. What is required is a "new Enlightenment," a general awareness of the cultural meaning of the scientific perception of time. The history of the development of modern, economically oriented cultures must be seen against the chronology of nature, with its ecosystems possessing a genealogy almost infinitely grander than that of human history. Only by recognizing the moral significance of the disparity between our cognition of cultural and natural time can we find the humility to redefine our idea of progress so that the conditions of humanity are ameliorated within a healthy and sustaining natural order. If we ignore the disparity, existing belief systems may lead to wholesale destruction of the planet's ecosystems. As the twenty-first century grapples with the meaning and implications of definitions of "progress," the thinkers of the Enlightenment will continue to provide the historical basis of the dialogue on how, and whether, the ecological systems of the planet will survive.

NOTES

[1]. Peter Gay, *The Enlightenment: An Interpretation: The Rise of Modern Paganism* (New York: Alfred A. Knopf, 1967).

[2]. Much of this introduction to the idea of progress is based upon J. B. Bury, *The Idea of Progress* (New York: Dover Publications, 1932), Chapters 3, 4, 5.

[3]. Christopher Lasch, *The True and Only Heaven: Progress and Its Critics* (New York: W. W. Norton & company, 1991), p.45.

[4]. Ibid.

[5]. Ibid. , p.47.

[6]. Ibid.

[7]. W. Warren Wagar, ed., *The Idea of Progress since the Renaissance* (New York: John Wiley and sons, 1964), p.36.

[8]. Theodore Olson, *Millennialism, Utopianism, and Progress* (Toronto: University of Toronto Press, 1982).

[9]. Ibid., p.9.

[10]. Ibid., p.10.

[11]. Leo Marx and Bruce Mazlish, eds*., Progress: Fact or Illusion?* (Ann Arbor: The University of Michigan Press, 1996), pp.27-44.

[12]. Ibid., pp.29-30.

[13]. Ibid., pp.30-31.

[14]. Ibid., p.35.

[15]. Ibid., p.37.

[16]. W. Warren Wagar, *Good Tidings: The Belief in Progress from Darwin to Marcuse* (Bloomington: Indiana University Press, 1972), p.14.

[17]. Raymond Duncan Gastil, *Progress: Critical Thinking about Historical Change* (Westport, Connecticut: Praeger, 1993).

[18]. Leo Marx and Bruce Mazlish, *Progress: Fact or Illusion?,* p.39.

[19]. Leonard M. Marsak, "Bernard de Fontenelle: In Defense of Science," in *The Rise of Science in Relation to Society*, Leonard M. Marsak, ed., (London: Collier-Macmillan Limited, 1964), p.77.

[20]. William Leiss, *The Domination of Nature* (New York: George Braziller, 1972).

[21]. Leo Marx and Bruce Mazlish, eds., *Progress: Fact or Illusion?,* p.35.

[22]. Bill Devall and George Sessions, *Deep Ecology* (Salt lake City: Gibbs M. Smith, 1985), p.48.

[23]. J.B. Bury, *The Idea of Progress*, p.2.

[24]. Ibid., pp.351-352.

[25]. Robert Nisbet, *History of the Idea of Progress* (New York: Basic Books, 1980), p.357.

[26]. Eric Hoffer, *The True Believer* (New York: Time, Inc., 1963). Although Hoffer focuses on fanaticism in mass movements, his analysis is also useful for understanding less zealous adherence to vaguely stated elements in modern worldviews, such as the idea of progress.

[27]. Henry Vyverberg, *Historical Pessimism in the French Enlightenment* (Cambridge: Harvard University Press, 1958).

[28]. W. Warren Wagar, ed., *The Idea of Progress since the Renaissance*, p.31.

[29]. Ibid., pp.29-35, and Robert Nisbet, History of the Idea of Progress, pp.2-3, 5.

[30]. W. Warren Wagar, *The Idea of Progress since the Renaissance*, p.36.

[31]. Gilbert F. LaFreniere, *Jean-Jacques Rousseau and the Idea of Progress* (Ann Arbor: University Microfilm, 1977). Gilbert F. LaFreniere, "Rousseau's First Discourse and the Idea of Progress," *Willamette Journal of the Liberal Arts*, Vol. I. No. 1 (Fall) 1983, pp. 2-26. Gilbert F. LaFreniere, "World Views and Environmental Ethics," *Environmental Review*, Vol. 9, No. 4 (Winter) 1985, pp. 307-322.Gilbert F. LaFreniere, The Redefinition of Progress," *The Willamette Journal of the Liberal Arts*, Vol.4, No. 2, (Summer) 1989, pp. 73-93. Gilbert F. LaFreniere, "Rousseau and the European Roots of Environmentalism," *Environmental History Review*, Vol. 14, No. 4 (Winter) 1990, pp. 41-72.

[32]. Frederic J. Teggart, *Theory and Processes of History* (Berkeley: University of California Press, 1962, p.220.
[33]. Norman Suckling, "The Enlightenment and the Idea of Progress." *Studies in Voltaire and the Eighteenth Century*, No. 58, 1967: 1461-1480.
33. J.H. Brumfitt, *The French Enlightenment* (London: MacMillan, 1972). Leonard Marsak, *The Nature of Historical Inquiry* (New York: Holt, Rinehart and Winston, 1970).
[35]. Leonard Marsak, "Bernard de Fontenelle: The Idea of Science in the French Enlightenment," *Transactions of the American Philosophical Society* 49, Part 7 (December 1959): 1-64, p.46.
[36]. Charles Van Doren, *The Idea of Progress* (New York: Frederick A. Praeger, 1967), pp. 4-8.
[37]. See Jean Jacques Rousseau, *The Social Contract and Discourses*, Translated by G.D.H. Cole (New York: E.P. Dutton and Company, Inc., 1950), and *Emile*, translated by Barbara Foxley (London: Dent, 1969).
[38]. Leo Marx, "Does Improved Technology Mean Progress?" *Technology Review* (January 1987) , pp.33-41, 71.
[39]. Ibid. , p.39.
[40]. Hans Jonas, "Reflections on Technology, Progress, and Utopia," *Social Research* 48 (Fall, 1981): 411-455.
[41]. Ernest Lee Tuveson, *Millennium and Utopia: A Study in the Background of the Idea of Progress* (New York: Harper & Row, 1964).
[42]. J.B. Bury, *The Idea of Progress*, pp. 78-312.
[43]. Richard B. Norgaard, *Development Betrayed: The End of Progress and a Coevolutionary Revisioning of the Future* (London: Routledge, 1994), pp.49-52.
[44]. Lewis Mumford, *The Condition of Man* (New York: Harcourt, Brace, 1944), p.265.
[45]. Following publication of conservative Robert Nisbet's *History of the Idea of Progress* in 1980, a spate of analyses of conservative attitudes toward the idea of progress appeared: Frank E. Manuel, "Men of Ideas," *New York Times*, March 16, 1980, VII, 1, 24-25; J.M. Cameron, *New York Review of Books*, April 17, 1980, 36-38; Gertrude Himmelfarb, "In Defense of Progress," *Commentary*, Vol. 69, 53-60, June, 1980; J.A. Guegen, *The American Political Science Review*, Vol. 75, 176-177, March 1981; W. Warren Wagar, *American Historical Review*, 568, June 1981; Jerome Himmelstein, "The Two Nisbets:The Ambivalence of Contemporary Conservatism," *Social Forces*, vol. 60, 231-36, September, 1981; Gorman Beauchamp, "The Politics of Progress," *Michigan Quarterly Review*, Vol. 21, 658-73, Fall 1982. These analyses from the political left and right appear to have culminated the political debate over progress which led to conservative domination of progressive thought in practical terms, even linking progressive change (once again) to the idea of providence and the continuing era of religious revivalism in the United States.
[46]. Baker Brownell's *The Human Community* (New York: Harper & Row, 1950) is recognized as the first ecotopian vision of the post-World War II era by Devall and Sessions, *Deep Ecology*, 166-69.
[47]. Ernest Callenbach, *Ecotopia* (New York: Bantam Books, 1975), p.23.
[48]. Ibid., p.39.
[49]. Ibid., p 42.
[50]. Ibid., pp.55-56.
[51]. Ibid., p.56.
[52]. See Robert W. Hoffert, "The Scarcity of Politics: Ophuls and Western Political Thought", *Environmental Ethics*, Vol. 8, No. 1 (Spring 1986): 5-32. Hoffert's critique charges Ophuls with placing too much of the blame for contemporary resource scarcity on the Western liberal tradition and with seeking a potentially dangerous political solution to the problems of ecological scarcity by applying the political thought of Hobbes and Rousseau. Although Hoffert's own grasp of the idea of progress within the liberal tradition is unsure (see especially p.14), he is aware of the influence of the utopian tradition, through Plato and Rousseau, upon Ophuls's attempt to seek a basis for social and political reconstruction (redirected progress or what I hve been calling utopian progressism) in a world of scarcity (pp.8, 21-3, 25-30). Hoffert recognizes two possible responses

to environmental limits to growth, one grounded in a moral revolution and the other involving unpalatable Hobbesian coercion, which Ophuls recommended only as a last resort. Both alternatives were stated by Ophuls, but Hoffert seems to misunderstand and also misrepresents the choices which Ophuls presented: "Metanoia" (a change in worldview entailing a transformation of moral and religious values) or leviathan.

[53]. William Ophuls, *Ecology and the Politics of Scarcity: Prologue to a Political Theory of the Steady State* (San Francisco: W.H. Freeman and Company, 1977), p.116. Revised as *Ecology and the Politics of Scarcity Revisited: The Unraveling of the American Dream*, with A. Stephen Boyan, Jr. (New York: W.H. Freeman and Company, 1992).

[54]. Ibid., p.126.

[55]. Ibid., p.128.

[56]. Ibid., p.129.

[57]. Ibid., p.137.

[58]. Ibid., pp.161-162.

[59]. Ibid., p.164.

[60]. Ibid., p.165.

[61]. Ibid., p.176.

[62]. Ibid., p.204.

[63]. Ibid., p.223.

[64]. Ibid., pp.223-225.

[65]. Ibid., p.225-232.

[66]. Ibid., p.232.

[67]. Ibid., p.239.

[68]. Ibid., p.240.

[69]. William Ophuls, *Requiem for Modern Politics: The Tragedy of the Enlightenment and the Challenge of the New* Millennium (Boulder, Colorado: Westview Press, 1997), p.280.

[70]. Ibid., p.207.

[71]. Karl Mannheim, *Ideology and Utopia* (New York: Harcourt, Brace & World, Inc., 1936), p.192.

[72]. Ibid., p.96.

[73]. Frederick A. Olafson, "The Idea of Progress: An Ethical Appraisal." In *Progress and its Discontents*, ed. Gabriel A. Almond, Marvin Chodurow, and Roy Harvey Peace (Berkeley: University of California Press, 1982).

[74]. Ibid., p.527.

[75]. Ibid., p.545.

[76]. See, for example, Benjamin Kline, *First Along the River; A Brief History of the U.S. Environmental Movement*, Second Edition, (San Francisco: Acada Books, 2000), and chapter nine ("The Early 1990s: Government Retrenchment and Public Apathy"). Ambivalent as he was about the future of environmentalism, by 2005 Kline would be aghast at the domination of national politics by Republican anti-environmentalists, the struggle to open the Arctic National Wildlife Reserve to feed the thirst of American gas guzzlers, and even the collapse of the cynosure of statewide land-use planning in Oregon after passage of a measure (Measure 37, which requires that any potential financial loss in the development potential of land resulting from land-use planning must either be compensated to the landowner by the local government or the landowner is allowed to go ahead with his or her project, despite destructive environmental and social consequences. In short, the "taking issue" has been used to undermine land-use planning. Developers in other states have advanced similar initiatives since the defeat of sound land-use planning in Oregon. This is an enormous defeat for environmentalists nationally) in November, 2004 which essentially crippled the Land Conservation and Development commission, thereby opening the door to the "Californization" of Oregon, and setting an example for the rest of the country: Americans desire land development and economic growth at any cost. Periodical and website articles pronouncing the "Death of Environmentalism" appeared concurrently.

CHAPTER SEVEN: ENLIGHTENMENT II: NATURE AND THE IDEA OF NATURE IN EIGHTEENTH CENTURY EUROPE

Nature in the Age of Enlightenment

What is man? In the eighteenth century this question continued the Renaissance search for the essence of humanity which had ended with antiquity. During the Middle Ages it appeared that this question had been answered once and for all in western Europe, namely that Man was God's creation at the top of a chain of organic and inorganic matter created as a home for him. Christian myth brilliantly rationalized Man's position at the apex of the natural world, and this myth held sway for more than a thousand years after the fall of Rome. Then, as the foundations of modernity developed out of the Renaissance, Reformation, and Scientific Revolution, a serious alternative conception of Man took shape in the minds of eighteenth century intellectuals disenchanted with the explanatory power of Christian myth. Man, after all, may not have sprung from divine origins, but may be nothing more than another animal, more intelligent than other animals, to be sure, but nevertheless an organic being, unsouled and mortal like all others, as Aristotle believed. Thus, the same irreverent ideas about mankind which Greek philosophers had entertained two millennia earlier were on the loose in Europe, and secular humanism gradually established itself amongst Europeans in the modern world, leading historian Peter Gay to dub the Enlightenment "the rise of modern paganism." A battle for nothing less than the soul of Man was underway, and his place in the natural world would shift from that of "Lord Man," as John Muir mocked the Christian myth of creation, to that of highly successful and intelligent primate produced by evolutionary change over millions of years. Although it would require the efforts of Darwin and Wallace during the nineteenth century to make the case for man's biological ancestry, the Enlightenment did produce forerunners of evolutionary theory known as transformists, men such as the Comte de Buffon, Denis Diderot, and Jean-Jacques Rousseau, who envisioned biological changes or transformation in the human species over time. However, thinkers of the eighteenth century functioned within a chronological limbo because the science of geology did not establish a viable scientific paradigm until the end of the century.

As a manifestation of the scientific foundations of secular humanism, contemporary environmental history views *Homo sapiens* as nothing more than another species of mammal. From this perspective, *Homo sapiens,* disenthroned from its special place in the creation myth of Christian religion, is perceived by many biologists and environmentalists as the most perniciously destructive animal in the course of mammalian evolution, the mammal which has either eliminated or enslaved the majority of other large, competing mammals. It is now commonplace knowledge for scientists to have recognized the extinction of large species of mammals, birds, and reptiles on islands and continents throughout the world shortly after the arrival of humanity. In earlier chapters we have seen that even the habitat of many species has been thoroughly transformed in the Mediterranean basin and in western Europe through *landnam*, the shock of first contact between man and the rest of the natural world. The further back we look into prehistoric time, the more impressed we are with the devastating ecological impact of human intrusion. In many cases, as in western Europe, wave upon wave of human invaders have added their impacts of hunting, fire, and eventually agriculture and civilization to any given region. Knowledge of this deep past is crucial to understanding the experience and perception of nature by western Europeans during the Age of Enlightenment. We have seen that by 1300 A.D. most of western Europe had been transformed into a pastoral environment of agricultural fields, grazing land, and diminishing woodlots and hunting grounds surrounding the growing medieval towns and villages. Wildlife, especially large mammals, had been driven to the limited remaining areas of high mountains, rugged, forested hills and plateaus, and marshes, or, in short, unfarmable and inhospitable lands. During the several centuries that followed until the dawn of the Enlightenment, some mammals became extinct or nearly so, and others were greatly reduced in numbers. This meant that the "nature" which Europeans encountered in the late seventeenth and eighteenth centuries was an almost thoroughly humanized environment, free of the dangers of wolves, bears, and large ungulates such as the aurochs or the forest bison (wisent).

Before considering early modern attitudes toward nature (in its diminished condition), it would be a useful indicator of the degree of its domestication to survey the displacement and extinction of the large mammals of western Europe, beginning with the British Isles and proceeding eastwards in the tracks of the "Great Frontier" of the High and Late Middle Ages. Like all islands, large and small, the British Isles were more vulnerable to species extinctions due to the lack of contiguous continental regions from which locally extirpated species might return. Following demographic decreases in human population due to plagues and other causes, the forest returned to large areas of abandoned farmland and grazing land on the European continent, and large ungulates, wolves, and bears returned with the re-establishment of their habitat. On islands, on the other hand, this replenishment is not possible. On most islands humans also have often established new populations of exotic species, frequently causing harm to what remained of already depleted ecosystems. This has been the case in the British Isles. The history of today's mammalian fauna in Britain began when Ice Age (Pleistocene) mammals were isolated from the European continent by rising sea levels ca. 7,500 B.C. In this discussion we will concern ourselves chiefly with the larger mammals, with passing reference to the smaller mammals. The Mesolithic fauna (ca. 8000-2700 B.C., northwest Europe) of lowland Britain greatly resembled the fauna of present-day Bialowieza National Park in eastern Poland, including brown bear, red deer, roe deer, wild boar, aurochs ("wild cattle"), elk (equivalent of the

American moose), lynx, wolves, otters, weasels, polecats, and stoats. The forest bison did not return to Britain in Post-Glacial times. With the exception of the aurochs, this fauna also resembles that of northeastern North America. During the Mesolithic, the lynx and elk appear to have become extinct, possibly from human hunting along with ecological factors.[1] During the 3000 years from Neolithic to Roman times the British countryside changed from woodland with small clearings to farmland with small woodlands, and domestic sheep and cattle began to dominate the countryside. Simultaneously, the aurochs and brown bear became extinct, although perhaps a bit later in Scotland.[2] The invading Anglo-Saxons came upon a farmed landscape dominated by cattle and sheep, and further cleared the woodland which was still home to deer and wolves. The Norman invaders who succeeded them also continued farming but expressed a strong interest in hunting game species, which led to the establishment of protected forests and enclosed parks. The Normans also introduced fallow deer and rabbit, and hunted the wolf as vermin.[3] The beaver survived into the medieval period but disappeared from Wales by 1200 and from Scotland by 1600. The wild boar appears to have become extinct in Scotland by the early seventeenth century and in England by the end of the century, although they were kept for hunting since about 1300. Wolves also had become extinct by the end of the seventeenth century.[4] The several British deer species declined steadily during the Early Modern period, the largest, the red deer, becoming extinct except in deer parks by the end of the eighteenth century.[5] Wild mammals suffered similar fates in Ireland and the smaller islands.

Derek Yalden, author of *The History of British Mammals* (1999), from which the above account was taken, also documents the rather extensive colonization of the British Isles by exotic mammals such as the black rat, rabbit, fallow deer, and many others, concluding that at least some balance has been restored to the variety of the wild mammalian population. Another author concludes that " big wild herbivores and carnivores have suffered heavily, while adaptable rodents and the rabbit have thrived and multiplied-some well enough to kill trees, undermine river banks and generally change the local vegetation."[6] Regarding the Age of Enlightenment in the British Isles, one ecological biologist writes:

> Ours is not the first century to have witnessed the rapid wholesale destruction of many wildlife habitats in the interests of agricultural gain. The losses during the seventeenth and eighteenth centuries were unprecedented. As the demand for food increased with the growing population, so it became viable to farm heath, moor and fen-the 'wastes' as these wilder parts of the countryside were called. The miles of new hedgerows created by the awards of the Enclosure Acts clearly made the large fields they divided more hospitable to the local woodland wildlife but their contribution was almost irrelevant to the overall balance.[7]

So it was that extensive habitat destruction in the early modern period gave the death blow to most remaining large wild mammals in the British Isles. Human biologist Garrett Hardin once reflected, "Is it better for an animal to be small and hideable or large and powerful? Nature commensurates the incommensurables."[8]   This evolutionary principle may well have been generally operative prior to the evolution of hominids and *Homo sapiens*, but, as the British Isles have shown, since our arrival on the scene it was far preferable for wild mammals to be small and hideable than to be large and powerful, and all the world has followed this new evolutionary path down to the present day. Large mammals, as signature species representing particular ecosystems or complexes of ecosystems, at various times in history have captured our imagination because of their beauty and power. It is indeed unfortunate that our admiration for these creatures, at least since Early Modern times, has led us towards preserving them at such a late hour in history. Even utilitarian motives for preservation, such as hunting, have preserved only one last piece of European old growth forest with its large mammals in the Bialowieza National Park of Poland. Let us now cross the Straits of Dover to the European mainland to assess the fate of large mammals there.

Although we tend to associate bison with the Great Plains of North America, the European bison, also known as the forest bison or wisent, once ranged from the British Isles and western Europe to Siberia and the Pacific coast of Asia. This impressive ungulate, somewhat taller and leaner than its North American cousin, was present in southern Sweden around the time of Christ, and populations existed in the Ardennes and Vosges Mountains of Belgium and France until the seventh and fourteenth centuries respectively. In what is now Germany, they survived until the fourteenth century in Pomerania, the fifteenth in Brandenburg, and were not eradicated from eastern Prussia until the eighteenth century. Hunted out of Hungary and central Poland in the sixteenth century, European Bison inhabited Transylvania until the eighteenth century. The last wild forest bison was shot in Poland early in the twentieth century. Subsequently, breeding of zoo populations allowed their reinhabitation into the Bialowieza old growth forest on the Polish-Russian border. In sum, European bison had been hunted to extinction in western Europe before the Age of Enlightenment, when small numbers remained in parts of central Europe.[9]

The Aurochs was a large ungulate even more formidable to humans than the forest bison, bulls weighing about 2200 pounds and with a four foot spread of horns. There were various subspecies of aurochs, which ranged from Ireland to the East China Sea, and numerous breeds of cattle are descended from them. The European bison is related to the yak of Asia rather than the aurochs. The aurochs was domesticated as early as 6500 B.C. or before. In part because they competed with domestic cattle in forests and on grazing lands, and also because of their dangerous size, they were actively hunted, leading to their demise in France in the thirteenth century, in Bavaria in the late fifteenth century, and finally their extinction in Poland in the seventeenth, the last animal having been killed in 1627. The large ungulate next in size in Europe was the elk, the equivalent of the American moose. However, their range was eastwards and northwards of western Europe, from Scandinavia to Poland and eastwards through Siberia to the Pacific. The next in size was the European red deer, the approximate equivalent of the American elk or wapiti.[10] By the eighteenth century the red deer was for the most part the largest wild mammal in western Europe, and although not usually dangerous, they have occasionally killed hunters. Traveling through the Vosges Mountains in 1977, I was surprised to find a memorial to someone killed by a red deer, which is also called "le grand cerf" (the great stag) in France. Thus, by the Age of Enlightenment the wildest parts of France still supported red deer and other species of deer, wild boar, wolves and even the brown bear in the Alps and Pyrenees. The brown bear was common in central Europe until the eleventh century frontier movement.

Not only were the large mammals extirpated from most of western Europe, but the smaller ones were gradually depleted as well, beginning in the Middle Ages, due to the demand for furs both for warmth and as status symbols. As Clive Ponting points out, in medieval Europe the skins of small mammals such as various species of foxes, sable, martin, ermine and squirrels were much in demand for making cloaks, robes, and other garments. Several hundred squirrel or sable furs might be used to make a single piece of expensive clothing.[11]

During the Late Middle Ages English merchants looked to Scotland and Ireland for furs of marten, beaver, rabbit, and hare. Furs also came from southern Italy, Spain, Burgundy and Germany. As early as the ninth century Viking traders in Russia had developed an extensive fur-trapping network, first shipping furs of sable, black fox, ermine, beaver and squirrel to the Byzantine Empire, but also to western Europe by the twelfth century. In the Middle Ages alone, hundreds of millions of animals were killed at unsustainable rates in Russia and western Europe. Even by the 1460s London merchants complained about inadequate supplies. By 1600 the beaver was almost extinct in southern Europe and trappers in Russia began to exploit the unexplored vastness of Siberia to meet the demands of western Europe, thereby providing the Russian state with a third of its income. Even such a vast area as Siberia was exhausted of fur-bearing animals by the end of the eighteenth century.[12] As Ponting observes, "The virtual extermination of fur-bearing animals in western Europe and the western parts of Russia by the early sixteenth century meant that from the start of settlement in, and trade with, North America the search for furs was one of the driving forces behind the European expansion across the continent."[13] In 1743, in the middle of the Age of Enlightenment, the French port of La Rochelle imported from Canada 127,000 beaver skins, 30,000 martens, 12,000 otters, 110,000 racoons and 16,000 bear skins.[14] Garret Harden's "is it better to be small and hideable or large and powerful" clearly had become irrelevant to the mammalian populations of western Europe and the regions which traded with them for furs. All mammals, great and small, had been driven to extinction or greatly reduced in population. With the removal of most of the boreal forest and the majority of its animal inhabitants, western Europe had almost entirely replaced its original post-glacial ecosystems with an intensively managed pastoral landscape. Thus, when people reflected upon "nature" in eighteenth century western Europe, they were actually responding to an artifact of their own making. Nevertheless, new attitudes towards this truncated "nature" were developing during the seventeenth and eighteenth centuries.

The Idea of Nature in the Eighteenth Century

It is a suitable irony that about the time that many members of Faustian civilization began to develop an affection for nature, most indigenous mammals had been annihilated from all but a few corners of western Europe. As we read in Chapter Four, Western European civilization had developed a sequence of overlapping attitudes toward nature during the Middle Ages, all of them expectably reflecting pervasive Christian beliefs. The *eschatological* attitude rationalized the deteriorating environment of the Late Roman Empire as a sign of the coming Apocalypse. With the fall of Rome and the return of wildness to abandoned lands, the wilderness, especially the forest, was perceived as the abode of witches, demons, and the devil himself, making nature a formidable *adversary*. Once forest clearing had begun in earnest, medieval ecclesiastics, nobles, and peasants all had the sense of working *collaboratively* with God to make the world a better place for (Christian) humanity. During the Late Middle Ages and Renaissance, famine, war, and plague contributed, along with obvious environmental degradation, to revival of the eschatological view of nature, but simultaneously there were occasional signs of an increasing awareness of the *recreational* value of nature, as in Petrarch's enjoyment of his climb up Mt. Ventoux in southern France in 1336. This was almost a subversive act at the time, as Petrarch himself observed upon reading a passage in Saint Augustine's *Confessions* at the summit, and he quickly turned from enjoying the things of this world, in this case the mountain scenery, to contemplation of God and the fate of his own soul.[15]

The Scientific Revolution of the sixteenth and seventeenth centuries occurred within the context of a rejuvenated Christian faith fueled by the controversies of the Reformation and its aftermath. Early modern scientists were, as we have seen, devout Christians seeking to understand how God's creation functions, all to the greater glory of God. Once fully articulated by Newton, the mechanistic paradigm of nature which arose from their arduous pursuits was understood as the design of creation formulated by the deity and stamped upon the natural world as a fixed, unchanging, static home for God's highest creation, mankind. The Scientific Revolution gave rise to new attitudes toward nature which included, on the one hand, the Baconian-Newtonian model of Nature-as-mechanism (useful for further transforming wild nature into a new Eden for god's chosen species), and on the other the perception of nature as the locus of aesthetic beauty reflecting the Creator's intelligence and beneficence. Working scientists often embraced both of these attitudes, pursuing further knowledge of mechanistic Nature while admiring the mathematical and aesthetic beauty of God's creation.

At the beginning of the eighteenth century this static model of Nature as God's unchanging universe designed for man's benefit was the basis for many works on "physico-theology." Books such as the Abbé Pluche's *The Spectacle of Nature* (1732) were widely distributed throughout Europe in many languages, spreading the word of how the universe was created and designed for man by the Creator, from the simplest substances and organisms to the orbits of the planets in the solar system. Under the providential guidance of a wise deity, a plenitude of creatures and their surroundings had been designed for the benefit of humans in this best of all possible worlds.[16] Physico-theology or the "argument from design" began to appear in print in the 1670s, gradually becoming the " widely acceptable basis for reconciling faith in a divine Creator, and Providence, with the advances in science." As Jonathan Israel observes, such arguments " proved the strongest single intellectual pillar buttressing the mainstream Enlightenment." The "argument from design" was also a major weapon in the arsenal of revealed religion in their cultural war against philosophical irreligion associated with Spinoza and the freethinking atheists and deists of what Israel calls the "Radical Enlightenment." In the eighteenth century this confrontation replaced the centuries-old struggle between Catholicism and Protestantism. Physico-theology dovetailed nicely with mainstream, moderate versions of the Early Enlightenment, such as Newtonianism, Malebranchism, Neo-Cartesianism, and Leibnizianism.[17]

The defenders of revealed religion in the Early Enlightenment could not have imagined a more frightening specter than the challenge to their stable, created Christian universe which would emerge from scientific inquiry and speculation in the Middle to Late Enlightenment. By the middle of the eighteenth century thinkers such as Buffon, Diderot, Maupertuis, and even Rousseau had begun to think of Nature in terms that temporalized the static, Aristotelian Great Chain of Being. The sixteenth and seventeenth century scientific revolutions in astronomy and physics lent themselves perfectly to the "argument from design" in that they dealt with structures and forces, e.g., the solar system and gravity, which appeared to have remained in static harmony since the time of creation in 4004 B.C. However, by the middle of the eighteenth century questions were being raised about organic nature, fossils, and extinct species which suggested the possibility of change or development over time. With the rise of geological science during the eighteenth century, culminating in James Hutton's uniformitarian paradigm, explained in his *Theory of the Earth* (1795), all was not for the best for the physico-theologists by the end of the century. Even the history and pre-history of man himself had been tainted by theorizing about transformations in man's physical form over time in Rousseau's *Second Discourse, On the Origin of Inequality among Men* (1755).

Eighteenth century speculations about changes in organisms over time are often referred to as "transformism" in order to distinguish them from later, more sophisticated evolutionary theory. Of particular importance to this Enlightenment temporalization of nature was the contribution of the Comte de Buffon (1707-1788), whose *Epochs of Nature*, published near the end of his life, combined cosmological, geological, and biological insights to explain the history of the earth and its inhabitants. This groundbreaking history of nature began with an explanation of how a molten earth was formed by a collision of a comet with the sun. Dividing earth's history into six epochs as naturalistic counterparts to the six days of creation, Buffon gave an account of the cooling and consolidation of the first rocks, the inundation of the planet with water, the formation of the continents (anticipating the theory of continental drift), and the origin of life. As superintendent of the Jardin du Roi, Buffon dared to expound transformist concepts in his encyclopedic *Natural History* (1749 and later) until he was warned by the ecclesiastical authorities of the Sorbonne that he was treading on dangerous ground. Subsequently, he was careful to couch his ideas within a teleological framework compatible with physico-theology.[18]

Establishment of the cosmological and geological temporal framework for biological evolution by Buffon, Hutton, and the early proto-geologists encouraged such eighteenth century scientists as Erasmus Darwin (1731-1802) and Jean Baptiste Lamarck (1744-1829) to speculate on the causes and mechanisms of evolutionary change. In *Zoonomia* (1794), Darwin described the age of the earth in millions of years and considered life to have "originated from a primordial protoplasmic mass."[19] Lamarck, the most important of the eighteenth century evolutionists or transformists, studied botany throughout Europe, and published his *Dictionary of Botany* and *Flora of France* prior to turning to invertebrate zoology. His best known work was *Philosophical Zoology* (1809). In Lamarck's view of evolution, the animal kingdom was a transitional series that graded from simple to complex forms. Building upon the ancient Greek theory of inheritance of acquired characteristics, further developed by Buffon and Erasmus Darwin, he argued: that plants and animals are modified by changes in surrounding environments; that old organs were modified and new ones developed in response to new needs resulting from environmental changes; and that use and disuse modify organic development and that the resulting modifications are inherited.[20]

Half a century earlier, several of the *philosophes* reflected the growing transformist intellectual climate of opinion, particularly under the influence of Buffon, of whom the best known are Diderot and Rousseau. Diderot made the transition from Christian (educated by Jesuits) to philosophical materialist and quasi-scientific evolutionary theorist, and through his close friendship and exchange of ideas with Rousseau, passed transformist ideas on to him. Rousseau was also, like Diderot, directly influenced by reading Buffon. Unlike Diderot, however, Rousseau never gave up his deism, and adhered to a belief in physico-theology, the "argument from design," to the end of his life despite his tranformist speculations regarding the origins of humanity in the *Second Discourse*.[21] Diderot's transformist ideas, on the other hand, were scattered throughout his writings. In the *Letter on the Blind, for the Use of Those Who See* (1749), for example, he wrote: "What is this world? A complex whole subject to revolutions of beings which follow each other, thrust forward and disappear...."[22] In *D'Alembert's Dream* he wrote, more specifically: "Let the present race of existing creatures pass away; leave the great inert sediment to work for a few million centuries. It may be that the renewal of species takes ten times longer than their allotted span of life."[23]

Rousseau appears to differ from Diderot and Buffon in applying transformist ideas to the process of human development in the *Second Discourse* while not applying evolutionary concepts to other animals. Perhaps, being under the strong influence of Diderot and Buffon's transformist ideas at the time that he published the *Discourse on the Origin of Inequality Among Men* in 1755, after his break with Diderot Rousseau withdrew from those radical ideas which conflicted with his deism. In any case, it is instructive to examine Rousseau's transformism in the *Second Discourse* as an index of the degree to which radical protoevolutionary ideas had penetrated the circle of moderate *philosophes* by the middle of the eighteenth century.

Most of Rousseau's transformist ideas are to be found in the preface to the *Second Discourse*, as when he proposes to understand mankind by distinguishing what is fundamental in his nature as opposed to "the changes and additions which his circumstances and the advances he has made have introduced to modify his primitive condition...."[24] Further noting "the changes happening to the constitution of the body," Rousseau observes "that it is in these successive changes in the constitution of man that we must look for the origin of those differences which now distinguish men...."[25] Several pages after the preface Rousseau speaks rhetorically to mankind: "The times of which I am going to speak are very remote: how much are you changed from what you once were!" On the same page, beginning the first part of the formal discourse, Rousseau makes an even stronger transformist statement:

> Important as it may be, in order to judge rightly of the natural state of man, to consider him from his origin, and to examine him, as it were, in the embryo of his species, I shall not follow his organization through its successive developments, nor shall I stay to inquire what his animal system must have been at the beginning, in order to become at length what it actually is. I shall not ask whether his long nails were, at first, as Aristotle supposes, only crooked talons; whether his whole body, like that of a bear, was covered with hair; or whether the fact that he walked on all fours, with his looks directed toward the earth, confined to the horizon of a few paces, did not at once point out the nature and limits of his ideas. On this subject I could form none but vague and almost imaginary conjectures.[26]

Rousseau goes on to remark that the science of comparative anatomy and the observations of naturalists in his day were too uncertain to provide him with a picture of human development over time, thus forcing him to speak of early, primitive, pre-civilized man as though his physical appearance was the same as that of mankind in Rousseau's own time, well knowing that this was almost certainly not the case in historical reality. Perhaps the most thorough analysis of Rousseau's transformism in the late twentieth century is Asher Horowitz's *Rousseau, Nature, and History* (1987). Attesting to Rousseau's importance as a transitional thinker about the idea of nature, Horowitz states that "Rousseau's theory of human nature cannot be understood in isolation from the first, tentative steps of contemporary natural science towards becoming historical and dynamic."[27] Rousseau's evolutionary conception of human history was a step towards a scientific anthropology even though his deism appears to contradict the idea of development in nature. For example, Horowitz points out Rousseau's 'Lettre a Philopolis' in response to Christian naturalist Charles Bonnet's critique of the *Second Discourse*, in which Rousseau contended "both that providence is an unacceptable principle of explanation and that belief in providence leads to intractable difficulties in dealing with the problem of evil...."[28] Observing that the materialism of Diderot and d'Holbach more likely led to a chaotic picture of dynamic life rather than to a developing or evolutionary system, Horowitz concludes that in contrast:

> Rousseau's deism allowed of a large amount of transformation in the system of living beings. Although as a theological device he may not have dispensed with the argument from design, he took sufficient pains to exclude it from the realm of natural history, where no question can be decided on *a priori* grounds and where empirical methods and discoveries must remain unchecked and unencumbered by theological considerations.[29]

In my own study of Rousseau and the European roots of modern environmentalism, I came to a similar conclusion, namely that: "Rousseau's 'Nature' was an intellectual construct stressed between the static paradigm of the Great Chain of Being and the emerging, dynamic evolutionary paradigm of modern biology."[30] Thus, Rousseau's apparent contradictions turn out to be a manifestation of the eighteenth century struggle between proponents of physico-theology and the first modern stirrings of evolutionary biology or transformism. In the subsequent modern centuries the "argument from design" would persist, but in the shadow of the new evolutionary biology.

In addition to witnessing the temporalization of the idea of nature, the Age of Enlightenment is also credited with the beginnings of modern ecology, as explained, for example, in Donald Worster's *Nature's Economy* (1977). Worster applied a dichotomy of two ecological traditions, the "imperial" and the "arcadian," to assist in explaining the roots of modern ecology. The "imperial" tradition, represented in the works of Carl Linneaus (1707-1778), pursued the goal of human dominion over nature, thereby following the dictates of Christian theology and Francis Bacon's project of reclaiming a "New Eden." The "arcadian" tradition, on the other hand, "advocated a simple, humble life for man with the aim of restoring him to a peaceful coexistence with other organisms."[31] Worster chooses Gilbert White, the parson-naturalist of Selborne in southwestern England, as the epitome of arcadianism. In relation to the preceding dichotomy between the static-mechanistic worldview which embraced physico-theology and the dynamic-transformist perception of nature which led to evolutionary biology, both Linnaeus and White were associated with the former, as were Newton and Voltaire. Thus the scientific classification developed by Linnaeus, and the observations of local natural history made by White were accomplished within the static paradigm associated with the "argument from design" and the idea of "plenitude," the latter meaning that a beneficent deity has filled the natural world with the greatest possible variety of organisms, creating maximum variety in God's creation with the minimum of natural laws.

Reflecting back to the beginning section of this chapter, one begins to understand why nature appeared to early modern Europeans to manifest such a harmonious set of interrelationships, particularly in relation to mankind, when one considers the thorough decimation of nearly the entirety of all large mammals, including herbivores, carnivores, and smaller fur-bearing mammals, by early modern times in western Europe, and first and foremost in the British Isles. Thus, in Gilbert White's *The Natural History of Selborne* the author describes the rich variety of birds and small mammals such as foxes, hedgehogs, rabbits, weasels, stoats, ferrets and polecats, as well as managed red deer, with no reference to the diminution of mammalian species in the Selborne region, although possibly he had been aware of the extinction of large mammals in the recent past. Nevertheless, God's plenitude is lauded: "It is curious to observe with what different degrees of architectonic skill Providence has endowed birds of the same genus, and so nearly correspondent in their general mode of life!"[32] White also probably saw God's concern for mankind in the existence of forests and wastes which 'are of considerable service to neighborhoods that verge upon them, by furnishing them with peat and turf for their firing; with fuel for the burning their lime; and with ashes for their grasses; and by maintaining their geese and their stock of young cattle at little or no expense."[33] In sum, White appears to have had no grasp of the enormous changes in his surrounding environment during the centuries and millennia preceding him, as though a pastoral landscape populated with birds and small mammals and humans and their farm animals and pests had existed in Selborne almost from the time of creation. How fitting that a biota shaped by humanity and its captive creatures for millennia (since the Neolithic Revolution and even before) should appear to be so well suited to human needs! There is no reason to expect White to have had a diachronic perspective of nature when it was generally assumed that the creation had occurred a mere few millennia ago. In this context the "argument from design" takes on an amusing, almost comic aspect viewed from the intensely historical perspective of today. Thus, when Worster writes of "the abiding symbiosis that nature and man enjoy in Selborne,"[34] it is a vastly diminished nature having absolutely nothing in common with the natural ecosystems it has displaced that is "symbiotic" with a species that has transformed nature almost totally to its own needs. Small wonder then, that "nature," i.e., pastoral nature, appeared to have been designed by God for his favorite species! As Worster writes of White: "The productions of nature, he was sure, exist partly, if not chiefly, to provide a benign and profitable environment for mankind."[35] However naive White's reflections upon final causes in nature might appear to us, he nevertheless made an important contribution to a burgeoning ecological consciousness through his painstaking observations of climate and the daily lives of the organisms around him in Selborne, leading him to an awareness of intricate interrelationships in nature. Donald Worster remarks that *The Natural History of Selborne* "was a book that laid the foundations for the natural history essay in England and America. It was also one point of origin, representative if not seminal, for the modern study of ecology."[36]

The idea of nature in the Enlightenment is often singled out as the dominant unifying concept of the age, despite the vexing range of meanings associated with the word 'nature'. Asher Horowitz believes that the leading thinkers of eighteenth century Europe were "united in opposing the rationality and immanent order of nature to the authority of the Church, tradition, and the absolutist state." The *philosophes* were optimistic that the observable order and regularity of nature would gradually become "the ultimate and only source of truth and authority."[37] However, as we have seen, both the inorganic, mechanistic nature of the Newtonian universe and the organic nature of Gilbert White's Selborne were embedded within the theological context of the physico-theologists throughout the eighteenth century. Nonetheless, "Nature" did gain other important meanings during the Enlightenment through extension and secularization of the ideas which linked "natural philosophy" to physico-theology. As one contemporary scholar writes:

> The 'natural' was seen as the 'good,' meaning original, authentic, simple, uncorrupted, and, by extension, in the works of Rousseau and others, as a state opposed to 'civilization' with all its artificiality and corruption. Thus, 'nature' became a description of a moral ideal as well as of a scientifically discernable order, and was thus seen as something which could reside in the hearts of men, as much as being an external order visible and tangible and measurable to natural philosophers.[38]

In contemporary usage, the moral concept of 'nature' inherited from the Enlightenment has become the less value-laden concept of 'human nature,' whereas the concept of external nature has diverged into several fundamental concepts, including: nature as the concept of the non-human, through which humanity perceives its difference; the structures, processes, and causal powers operative within the physical world; and the natural observable features of the world as opposed to urban or industrial environment.[39] Since the advent of modernity during the Enlightenment, these several ideas of nature have emerged while detaching themselves from "the argument from design." We should keep these differences in mind as we further examine the eighteenth century idea of nature. In addition to describing the natural world and moral goodness, nature gave rise to natural law: "the entire universe was basically part and parcel of one all-encompassing, fundamentally benevolent and rational whole: the realm of nature, including the physical, biological, social, and moral worlds, which were to be grasped in one unified system of natural law."[40] Thus did the physiocrats seek to identify a natural order in the economic realm, concluding to their great satisfaction that a combination of unconstrained trade and individual property rights would be guided by natural law to produce the maximum social well-being. Thus, the moral shackles were removed from the pursuit of self-interest, while the harmony of competing interests was maintained by nature. Adam Smith's *The Wealth of Nations* (1776) culminated this rationalization of modern economic activity towards the end of the Enlightenment.

Rousseau and the Roots of Environmentalism

Against the background of the transformation of western Europe into a pastoral landscape and the rise to prominence of the idea of nature in the eighteenth century, Jean-Jacques Rousseau emerges as the outstanding paradigmatic figure of the era who wrote at length in defense of nature. What raises Rousseau above his Enlightenment peers as a pre-environmentalist or proto-green figure is the comprehensiveness of his critique of all civilization, and particularly Western, Faustian civilization, as intrinsically destructive of nature, as responsible for the alienation of humanity from nature, and consequently as responsible for alienating human societies and individuals from one another. If the Enlightenment was the period in Western history when modernity laid its foundations, Rousseau was the first great critic of its presuppositions of progress and the domination of nature through the application of science and reason. Explicating the perceived ambiguities of Rousseau's system of ideas, in *Jean-Jacques Rousseau: Transparency and Obstruction* (1971), Jean Starobinski summarized Rousseau's argument that the disparity between men's words and their actions, between appearance and reality, was rooted in the tyrannization of nature by the social order of civilization, thereby giving rise to the vices of men and the ills of society. This theme lies at the heart of the argument in Rousseau's first two *Discourses*. Furthermore, Starobinski writes that Rousseau's critique of civilized society stems from the fact that society is contrary to nature:

> Society, which *negates* nature (or the natural order), has not eradicated nature. Society and nature remain, rather, in permanent

conflict, and it is this conflict that gives rise to all man's ills and vices. Rousseau's critique thus begins a "negation of the negation." He *accuses* civilization, which is characterized by its *negativity* with respect to nature. The established culture denies nature: This is the dramatic claim put forward in the two *Discourses* and *Émile.*[41]

The perspective of environmental history suggested by Starobinski's bold assertion, builds upon considerable evidence in support of Rousseau's prominence as the pre-eminent proto-Green thinker of the Enlightenment. Within twelve years, from 1750-1762, Rousseau constructed a system of ideas ranging from a critique of the beliefs and practices of Western civilization to a set of utopian models for reform in the realms of politics, society, education, and religion, all within the context of a deep respect for nature. His romantic novel, *Julie, ou La Nouvelle Héloïse*, would change Western aesthetic attitudes toward nature irrevocably.

Alfred Biese, in *The Development of the Feeling for Nature in the Middle Ages and Modern Times*, wrote:

> The most epoch-making event in European feeling for Nature was the appearance of *La Nouvelle Héloïse* (1761). ...Rousseau's influence upon feeling in general, and feeling for Nature in particular, was an extraordinary one, widening and deepening at once. By his strong personal impulse he impelled it into more natural paths, and at the same time he discovered the power of the mountains.... His feelings woke the liveliest echo, and it was not France alone who profited by the lessons he taught. He was no mountaineer himself, but he pointed out the way, and others soon followed it.[42]

Although Haller had written poetically of the beauty of the Alps in 1750 and de Saussure had begun climbing in 1760,[43] Biese nevertheless affirms the overwhelming importance of Rousseau:

> Rousseau was the real exponent of rapture for the high Alps and romantic scenery in general. Isolated voices had expressed some feeling before him, but it was he who deliberately proclaimed it, and gave romantic scenery the first place among the beauties of Nature. He did not...discover our modern feeling for Nature; the great men of the Renaissance, even the Hellenic poets, fore-ran him; but he directed it, with feeling itself in general, into new channels.[44]

The French intellectual historian Daniel Mornet agrees that "it was chiefly the *Modern Heloise* that suddenly made Switzerland and the Swiss mountains fashionable."[45] In *Le Sentiment De La Nature En France* Mornet focused on the psychological aspects of the new aesthetic response to nature initiated by Rousseau in *Julie*. Mornet argued that, while there has been a desire to escape to the tranquility and simplicity of the country or pastoral nature since the beginnings of urban life, a second and entirely modern mode of relating to nature was introduced by Rousseau. This modern sensibility would bring to the spectacle of nature a refinement of feeling and a moral purity of the soul or heart. Thus, for Jean-Jacques and kindred spirits "nature is a condition of the soul. She reflects everything that fills the hearts of her lovers."[46] The love of nature as a manifestation of the psychological or spiritual condition of the individual would, as such, limit the number of her worshippers in this more profound, second type of perception to individuals who were solitary by nature. Mornet writes that even if Rousseau's contemporaries have not admitted owing their new ways to him, that "if they have gone beyond the rustic idyll, if they have brought to nature a soul finally open to solitude, to reverie, and to all the resulting emotions, it is because Rousseau has given them this new soul."[47]

The passage in *Julie* which best illustrates Mornet's argument is as follows (Saint-Preux has just led Julie, now Wolmar's faithful wife, to an isolated spot where Saint-Preux adored her years before):

> This solitary place formed a retreat, wild and deserted but full of those kinds of beauties which please only sensitive souls and appear horrible to others. A torrent caused by the thawing snows rolled in a muddy stream twenty feet from us and noisily carried along dirt, sand, and stones. Behind us a chain of inaccessible crags separated the flat place where we were from that part of the Alps called the glaciers, because of the enormous peaks of ice which, incessantly increasing, have covered them since the beginning of the world. To the right, forests of black fir afforded us a gloomy shade. A large wood of oak was to the left beyond the torrent, below us that immense body of water that the lake forms in the midst of the Alps separated us from the rich shores of the Vaud region, and the peak of the majestic Jura crowned the landscape.[48]

It is not difficult to imagine how solitary and sensitive souls, among others, were attracted to Rousseau's vivid descriptions of Alpine scenery. Judith H. McDowell, translator of the above passage, agrees with Mornet that a second, moral sense of "nature" was introduced by Rousseau, who believed "that though most men have been perverted by society, there remain a few, the 'sensitive souls,' in whom 'nature' still persists."[49] This observation should lead us back to where we began, to Rousseau's condemnation in the first two *Discourses* of the corrupting, nature-alienating attitudes and practices of eighteenth century

European civilization and its false conception of progress. Rousseau and fellow possessors of a deeper sensibility and a "higher" morality could become critics of the established, commercial society around them by virtue of having avoided denaturization and by maintaining a continuing close relationship with nature, the ultimate source of moral virtue beneath the facades and masks of a dehumanizing culture.

D.G. Charlton, in *New Images of the Natural in France* (1984), carefully differentiates Rousseau's influence upon the appreciation of pastoral landscapes from his influence upon the appreciation of the "wild sublimity" of mountain landscapes. Charlton also delineates the role of *Julie* in initiating a trend towards gardens that imitate wild nature to take the place of the rigid, formal gardens that reflected the anthropocentrism of classicism. Concerning the new pastoralism of the eighteenth century, Charlton finds Rousseau "the undeniably outstanding figure, certainly in France and arguably in Europe...."[50] Remarking on Rousseau's sense of unity in nature as of greater importance than scientific detail, which Rousseau also knew as a scientific botanizer, Charlton notes the importance of Rousseau's sense of psychological, aesthetic, moral, religious, and even economic harmony with nature.

> More than any other French writer of the century Rousseau discerned and expressed a sense of the potential harmony between man and nature. In the course of the eighteenth century ... the notion of nature as a mechanism, as a divinely-designed order, perfect in its essentially passive regularity, gradually yielded, in some minds at least, to a notion of a no less divinely-intended 'harmony' between two creative, developing 'organisms' - man and the natural world. No-one in France prior to the Romantic Movement illustrates that change more clearly than Rousseau.[51]

Charlton emphasizes the point that the nature which Rousseau loved and admired was predominantly pastoral even though he was "one of the first French writers to describe wild mountain scenery."[52] Acknowledging the importance of *Julie* for the aesthetic appreciation of mountains, Charlton characterizes Rousseau's interest as "focused far less on mountain peaks than on mountain slopes...an undoubted and sincere affection for mountain *scenery*, a sense of affinity, or psychological harmony, between a rather generalized wild nature and some of man's more elevated or 'sublime' inner experiences...."[53]

In sum, my own published research and Charlton's analysis of Rousseau's view of man and nature essentially corroborates the claims made by Biese and Mornet earlier in the twentieth century. Charlton's interpretation of Rousseau's strongly moral and spiritual experience of nature is also suggestive of Rousseau's similarity to Wordsworth and Emerson, both of whom were strongly influenced by Rousseau's thought.

Rousseau's contribution to the development of an aesthetic appreciation of nature owed much, of course, to the earlier contributions of English authors during the late seventeenth and early eighteenth centuries. Christopher Thacker recognizes Shaftesbury's *The Moralists* (1709) as the earliest work leading towards the Romantic love of nature,[54] observing that "in the mid-1750s several books appeared which 'codify' the developing approval of wildness. The principal writers are Burke, Young, Diderot and Rousseau. They are all indebted to Shaftesbury."[55]

Having established the basis for Rousseau's prominence as the leading defender of nature and pre-environmentalist of the Enlightenment, I begin with the *Discourses* to assess the depth of his critique of the environmental destructiveness and social malaise associated with the rise of civilizations in general and Western civilization in particular. For the subject of environmental history the *Discourse on the Origins of Inequality* or *Second Discourse* (1755) is Rousseau's most significant work, but it should be considered in the context of the *First Discourse* or *Discourse on the Arts and Sciences* (1751). In the *First Discourse* Rousseau attacked the general assumption held in the eighteenth century (as well as today) that increased knowledge and technology generally contribute to an increase in human well-being. Professing to argue in defense of virtue, he maintained that the opposite is true, that as civilizations become more complex and ornate men become more corrupt and abusive of one another, having lost the transparency or openness of communication that existed in a more primitive state. "As the conveniences of life increase, as the arts are brought to perfection, and luxury spreads, true courage flags, the virtues disappear; and all this is the effect of the sciences and of those acts which are exercised in the privacy of men's dwellings."[56] Citing the growing decadence of ancient Egypt, Greece, and Rome over time, Rousseau condemned the sophistication and luxury produced by the arts and sciences, and claimed to have unmasked the so-called "progress" of civilization as a regressive process in relation to human morals and happiness. In the last chapter we identified this false idea of progress, at least from Rousseau's perspective, as "developmental progressivism." Rousseau's condemnation of developmental progressivism was surely influenced by his assimilation of Plato's interpretation of the "rise" of civilization as actually a fall from a Golden Age of simplicity and virtue. Numerous references to Plato are made in the text and footnotes of the *First Discourse*.

Four years passed before Rousseau would turn to nature and a natural history of civilization for an explanation of the apparent contradiction that the increase of trade, commerce, urbanization, and the arts and sciences could only lead to human decadence and inequality. The *Discourse on the Origins of Inequality* is a remarkable work for its time, described by Jonathan Bate as "modernity's founding myth of the 'state of nature' and our severance from that state."[57] Following the dedication of the *Second Discourse* to Geneva as an exemplary republic, the preface presents the transformist perspective which Rousseau shared with Buffon and Diderot. At the beginning of the discourse itself he proposes to discover in the course of history "the moment at which right took the place of violence and nature became subject to law" and to further explain how the strong came to serve the weak and "the people to purchase imaginary repose at the expense of real felicity."[58] Envisioning man in the state of nature when he could satisfy his hunger on fruits and nuts, drink from the nearest brook, and sleep at the foot of a tree, Rousseau presents an image of mankind's pre-civilized habitat:

> While the earth was left to its natural fertility and covered with immense forests, whose trees were never mutilated by the axe, it would present on every side both sustenance and shelter for every species of animal. Men, dispersed up and down among the rest, would observe and imitate their industry...."[59]

Thus, by learning to live on a great variety of foods, as an omnivore man would find his sustinence more easily than other animals. Blessed with a robust constitution and capable of defending himself against most animals or fleeing to the safety of a tree, primitive man was fearful of novelty but was generally comfortable in his surroundings because in a state of nature "the face of the earth is not subject to those sudden and continual changes which arise from the passions and caprices of bodies of men living together."[60] Although challenged by the natural infirmities of infancy and old age, primitive man was generally free from illness, which is mainly found in a state of society. Differentiating man from other mammals on the basis of his free will, Rousseau explained that the human quality of free agency gives rise to *perfectibilité*, "the faculty of self-improvement, which, by the help of circumstances, gradually develops all the rest of our faculties, and is inherent in the species as in the individual...."[61] "It is this perfectibility which, successively producing in different ages his discoveries and his errors, his vices and his virtues, makes him at length *a tyrant both over himself and over nature*."[62] (Italics mine.) At this point a footnote refers to an appendix wherein one finds Rousseau's most powerful attack on the human abuse of nature, or, as we would say today, the environment. Here he looks ahead to the abuses stemming from civilization.

> It has indeed cost us not a little trouble to make ourselves as wretched as we are. When we consider, on the one hand, the immense labors of mankind, the many sciences brought to perfection, the arts invented, the powers employed, the deeps filled up, the mountains levelled, the rocks shattered, the rivers made navigable, the tracts of land cleared, the lakes emptied, the marshes drained, the enormous structures erected on land, and the teeming vessels that cover the sea; and, on the other hand, estimate with ever so little thought, the real advantages that have accrued from

all these works to mankind, we cannot help being amazed at the vast disproportion there is between these things....”[63]

All this, Rousseau continues, i.e., all these assaults upon the environment, have been carried out largely at the whim of human pride and self-admiration. However, despite the fact that men are actually wicked in their endeavors and relationships, Rousseau claims that they are naturally good. “What then can have depraved him to such an extent, except the changes that have happened in his constitution, the advances he has made, and the knowledge he has acquired?”[64] He then goes on to further document the ills of civilization, from mental illness, violent passions, and the drudgery of labor which burdens the poor, to the horrors of urbanism: epidemics bred by the contaminated air of crowded cities; the adulteration of food and drink by unhealthy substances; the destruction of cities by fire and earthquake; crime, warfare, and shipwrecks and other disasters related to endless trade. To this he adds the unhealthiness of industrial trades, where miners are exposed to lead, copper, mercury, cobalt, and arsenic, and innumerable tilers, masons, and carpenters die in accidents.[65] This litany of urban suffering stems largely from the proliferation of luxury among the wealthier classes, the root evil of all politics, “for, in order to maintain all the servants and vagabonds it creates, it brings oppression and ruin on the citizen and the laborer....”[66] On top of this, luxury and the industries which support it undermine the practice of agriculture by depleting the countryside of its farmers. The appendix ends with a plea for better governmental practices to mitigate urban problems.

Returning in the text to the condition of savage man, Rousseau reflects on the question of what might have induced mankind to leave the state of nature (as hunter-gatherers) for a life of continual labor as farmers. In modern terms, he is attempting to understand the causes of the Neolithic Revolution, and he begins by surmising upon the possibility that humans may have reproduced to the point of exceeding their food supply. “In a word, how could such a situation induce men to cultivate the earth till it was regularly parcelled out among them; that is to say, till the state of nature had been abolished?”[67] Many problems needed to be solved en route to an answer to this question: man’s overcoming his natural antipathy towards continual labor; the origin of language, before or after the origin of society; the dilution of natural human empathy or compassion as socialization and intergroup warfare arose. Having responded to these issues to the best of his ability, Rousseau proposed to prove that the inequality of mankind is the result of the progress of human knowledge, consistent with his thesis in the *First Discourse* that societies grow more corrupt as they increase in sophistication as a result of improvements in the arts and sciences.

The first man who, having enclosed a piece of ground, bethought himself of saying “This is mine,” and found people simple enough to believe him, was the real founder of civil society. From how many crimes, wars, and murders, from how many horrors and misfortunes might not anyone have saved mankind, by pulling up the stakes, or filling up the ditch, and crying to his fellows: "Beware of listening to this imposter; you are undone if you once forget that the fruits of the earth belong to us all, and the earth itself to nobody."[68]

It is the "fall" into property ownership that completes the severance of mankind from nature. Jonathan Bate's noteworthy exegesis of the *Second Discourse* explains Rousseau's narrative as analogous to the classical myth of the Four Ages, with the falls into deforestation, meat-eating, and language preceding the fourth fall into property.[69] On the way towards the invention of agriculture, which precipitated the fall into property, Rousseau describes how humans frequented the shorelines of the oceans and rivers and invented the hook and line, and how in the forests they invented bows and arrows to become huntsman. The techniques of snaring and trapping other animals soon followed, and huts were made of branches, mud, and clay. These and many other changes preceded mankind's fall into agriculture and property, which together with metallurgy produced the first "civilized" men.

> But from the moment one man began to stand in need of the help of another; from the moment it appeared advantageous to any one man to have enough provisions for two, equality disappeared, property was introduced, work became indispensable, and vast forests became smiling fields, which man had to water with the sweat of his brow, and where slavery and misery were soon seen to germinate and grow up with the crops.[70]

The new condition of agriculturally-based civilization caused the natural empathy or compassion of individuals to weaken, and ambition or *amour-propre*, the desire to surpass others, roughly what Plato had called *thymos*, asserted itself under the new social circumstances of ruler and ruled, master and slave. As Rousseau put it: "Usurpations by the rich, robbery by the poor, and unbridled passions of both, suppressed the cries of natural compassion and the still feeble voice of justice, and filled men with avarice, ambition, and vice."[71] Rousseau appears to have been equally aware of the devastating environmental impact of the new ecological regime of agriculture. In a footnote in which he drew upon Buffon's *Natural History*, he stated that humans make great demands on forests for firewood and other uses of wood, and that over periods of human use the soil cover underlying the forest becomes depleted and barren like the soils of Arabia and other desiccated areas of the Orient (the Middle East).[72] He also states clearly in this footnote that the destruction of soil should accelerate in proportion to how long the soil is cultivated. Furthermore, not only does civilized life ultimately destroy nature, it also creates the conditions of human suffering, inequality, and alienation, for, unlike the savage, civilized man "is always moving, sweating, toiling, and racking his brains to find still more laborious occupations: he goes on in drudgery to his last moment, and even seeks death to put himself in a position to live."[73] In sum, fully consistent with the *Discourse on the Arts and Sciences*, Rousseau concludes that inequality "owes its strength and growth to the development of our faculties and the advance of the human mind, and becomes at last permanent and legitimate by the establishment of property and laws."[74] As Jonathan Bate puts it, "the paradox of the human condition is that our very freedom to transform and transcend the state of nature is the source of our enslavement."[75]

Throughout Rousseau's subsequent writing, in the *Émile, The Social Contract, The New Eloise* or *Julie, The Confessions,* and *The Reveries of the Solitary Walker* and other works, his primary objective of returning modern society to a life in harmony with and closer to nature is clear. At times his awareness of the historical and environmental context of his program for reforming society and the idea of progress in accordance with this goal is reiterated in resonance with the *Second Discourse*. In the *Émile*, for example, he explains why differences between European nations and their people, based upon soil and climate, were less noticeable in his own time because "the fickleness of Europe leaves no time for natural causes to work, when the forests are cut down and the marshes drained, when the earth is more generally, though less thoroughly, tilled...."[76] His awareness of the thoroughness of the humanization of nature in Enlightenment Europe, even in the mountains of Switzerland, is made clear in *The Reveries of the Solitary Walker* when he tells of discovering the wildest spot he has ever seen in a mountain canyon covered with fir and gigantic beech trees and bounded by sheer rock faces above him. "From the mountain gorges came the cry of the horned owl, the little owl and the eagle, while from time to time some more familiar birds lightened the horror of this solitary place."[77] Shortly thereafter, while imagining himself to be in a place never seen by other men, he hears a repetitive clicking noise and follows it to a stocking mill little more than twenty yards from where he had been enjoying his "wilderness" experience!

From the *Second Discourse* to his final work, the unfinished *Reveries*, Rousseau never lost sight of the importance of nature and the idea of the human "state of nature" as norms for judging how far mankind had fallen away from a relatively happy condition of existence. The *Second Discourse* expands the equivalent of what William Ophuls called "the four great ills of civilization" (See Chapter Two). First of all, Rousseau leaves no doubt whatsoever concerning human overexploitation of nature, to the point that wilderness or wild nature was practically non-existent in eighteenth century Europe. Secondly, he was particularly sensitive to European enslavement and abuse of the people of other cultures through trade, colonization, and empire building by the European nation-states in his own time, and recognized violence towards outsiders as a mark of civilization in general. Thirdly, the *Second Discourse* explained the rise of political and religious despotism as a consequence of the shift from the savage, hunter-gatherer condition to civilization based upon agriculture, metallurgy, and sociopolitical organization and control. Finally, perhaps more than any other of the four great ills of civilization, Rousseau was outraged by socioeconomic inequality among men, not only in the egregious form of slavery which existed from antiquity to his own time, but also in the class structure of his own era inherited from the Middle Ages.

Rousseau's denunciation of civilized life at the apogee of the Age of Enlightenment came as a psychological shock to his fellow philosophes, and in the same year that the *Second Discourse* was published, the physical shock of the great 1755 Lisbon earthquake set European intellectuals of all persuasions to thinking about the cause of that devastating temblor. The Jesuits argued that God was punishing Europeans for their worldliness and wicked ways, and the destruction of Lisbon was discussed most frequently within the context of the "argument from design," which, for some deists as well as Christians, still allowed for divine intervention into the natural processes of the cosmos. Rousseau responded that if men lived simpler lives in a more natural pastoral setting, people would not be crushed to death by collapsing buildings and cathedrals, all of which attested to the pride and ambition of civilized mankind.

Only a few years after the Lisbon earthquake controversy, another philosophe active in the Scottish Enlightenment, Adam Smith, published a book titled *Theory of Moral Sentiments*, in which he gave the impression of applying the "argument from design" to economic theory. Describing the selfishness and rapacity of rich proprietors pursuing only "their own vain and insatiable desire," Smith observed that these same men did society the service of employing thousands of poor workers in their factories to produce luxury goods for the wealthy, leading him to conclude that: "They are led by an invisible hand to ....without intending it, without knowing it, advance the interest of the society."[78] A similar reference to the "invisible hand" is made by Smith in *The Wealth of Nations*, seventeen years later, when he writes that a merchant is "led by an invisible hand to promote an end which was no part of his intention."[79] Down to the present day there has been an ongoing controversy as to whether Smith intended that the invisible hand represented the intrusion of divine providence into the secular world of economic activity, or whether he was ironically metaphorical in his use of the term. The fact that Smith was amused by philosophers who believed in systems of divine order strongly suggests that he intended the hidden hand as a metaphor for the mechanism of the market system which drives capitalist enterprise.[80] In any case, one can clearly see that Rousseau and Smith drew opposite conclusions regarding the production of luxury items for the wealthy by the laboring poor, and that the development of commerce and industry lauded by Smith as a manifestation of human progress was for Rousseau an indication of decadence and inequality.

Christopher Lasch, some of whose views on the idea of progress we encountered in Chapter Six, thought that the influence of theorists of progress such as Fontenelle and Condorcet upon the spread of the belief in progress since the Enlightenment was far less than the influence of the idea that human wants are insatiable and that rising acceptance of that idea along with rising standards of comfort and luxury would power the indefinite expansion of the productive forces needed to satisfy the revolution of rising expectations. In other words, Lasch understood the modern idea of progress as the product of economic change rather than of science and reason, the more idealized aspects of the Enlightenment. Lasch argues that the idea of progress, i.e., the progressive ideology, was neither the product of the secularization of the Kingdom of God nor of the new views on the processes of historical development, but that:

> Its original appeal and its continuing plausibility derived from the more specific assumption that insatiable appetites, formerly condemned as a source of social instability and personal unhappiness, could drive the economic machine--just as man's insatiable curiosity drove the scientific forces. The moral rehabilitation of desire, even more than a change in the perception of time as such, generated a new sense of possibility, which announced itself most characteristically not in the vague utopianism of the French Enlightenment but in the hard headed new science of political economy.[81]

Moreover, it was Adam Smith who made the important point that in the eighteenth century it was not the luxury items of the rich that fueled the modern productive machine, but the modest expenditures of more numerous ordinary consumers. The desire and effort of the common man to better his condition was for Smith the "principle from which public and national, as well as private opulence is originally derived."[82] He also laid the foundations for belief in modern society as capable of indefinite expansion, freed from the traditional cultural cycle of growth, maturity, senescence, and death by the newly discovered understanding of the social benefits of the unlimited pursuit of wealth under the guidance of the "invisible hand." Reading *The Wealth of Nations,* one searches in vain for any sense of spatial or ecological limits to the course of the unlimited economic development so heartily recommended by Adam Smith, not to mention the consequent alienation of humanity from nature as a result of that development. Smith took Rousseau's moral critique of humanity in the *Second Discourse*, along with similar critiques made of the ancient Romans in classical times as well as of early moderns in the Renaissance, and stood them all on their heads. Pre-modern vices were henceforth transformed into the economic virtues that would turn the wheels of modern societies.

Today, the democratization of European nation-states and their colonies having been largely achieved since the disorder promulgated by the French revolutionary banner of "liberté, égalité, fraternité," it would seem that most of Rousseau's utopian (in the eighteenth century) sociopolitical ideals have been fulfilled. However, thanks to the overwhelmingly thorough realization of the ideas of Adam Smith and the liberal economic theorists who followed him, social and political reform have been achieved at enormous social and ecological costs. William Ophuls explains why this relationship is more complex than it might first appear to be in terms of the "four great ills of civilization." Enlightenment thinkers, Thomas Hobbes and Adam Smith included, set out to remedy the ills of political tyranny and socioeconomic oppression, but they would accomplish this by worsening the two other great ills, the exploitation of nature and violence against outsiders through conquest and colonization. Ophuls credits, or rather, blames Hobbes for having laid down the theoretical foundations for the modern consumer society, a polity capable of transforming political conflict into economic competition: "instead of fighting over irreconcilable matters of principle such as religion, men and women contend in the marketplace for carnal satisfaction under the aegis of a powerful but dispassionate and supposedly benign sovereign that facilitates their pursuit of earthly happiness."[83] Unfortunately, the side effects of the modern, economically driven world order are devastating both to the natural systems of the planet and to the traditional organic communities of people subjected to relentless economic development. As Ophuls observes, the civilizational ill of human slavery has subsequently given way to the enslavement of nature, perceived by the modern mind as dead matter conveniently providing mankind with an assemblage of exploitable resources. Western and westernized people inhabit the modern "megacity" or megalopolis devoid of any psychological connection to nature, so that "the monetary system of megalopolis has come to be very tenuously related to or even totally decoupled from the real world that it is supposed to track."[84] Here, Ophuls echoes Spengler's reflections on the destructive role of money in the late stage of a great culture, as described in Chapter One. In sum, Ophuls recognized the tragedy of the Enlightenment in the disastrous trade-off involved in escaping the ills of political and religious despotism and slavery at the cost of destroying nature and colonizing non-European countries for their trade and resources, the latter problems setting the stage for disaster in the twenty-first century (to be discussed in the next chapter).

Not surprisingly, one of the staunchest defenders of modernity and its exploitation of nature and the (relatively) undeveloped countries, Francis Fukuyama, in *The End of History and the Last Man* (1992), looks back to the Age of Enlightenment and the thought of Jean-Jacques Rousseau as the root source of modern environmentalism. He recognizes Rousseau as "the first modern philosopher to question the goodness of historical progress," and proceeds directly to the *Second Discourse* to document the importance of Rousseau's belief that historical change had made men profoundly unhappy because human needs are naturally few and easy to satisfy without science and technology. Aside from the basic needs of food, shelter, and sex, humans require little more to be happy. "All other human wants are not essential to happiness, but arise out of man's ability to compare himself to his neighbors and feel himself deprived if he does not have what they have. The wants created by modern consumerism arise, in other words, from man's vanity, or what Rousseau calls his *amour-propre*."[85] Furthermore, for every want satisfied, modern economies produce a new need, and the cycle continues indefinitely to power the technological-industrial society. Fukuyama also asserts that almost all anti-technological doctrines have a common ancestry in Rousseau's thought, from those of the Romantics of the early nineteenth century to those of the counterculture of the 1960s and the Environmental Movement, whose radical element has "attacked the entire modern project of mastering nature through science," and suggested that mankind would be happier if he returned to a pre-industrial condition.[86]

> Rousseau's attack upon civilized man raised the first and most fundamental question mark over the entire project of conquering nature, the perspective that sees trees and mountains as raw materials rather than as places of rest and contemplation. His criticism of the Economic Man envisioned by John Locke and Adam Smith remains the basis of most present-day attacks on unlimited economic growth, and is the (oftentimes unconscious) intellectual basis for most contemporary environmentalism.[87]

Nevertheless, Fukuyama argues, it is highly unlikely that any new variation on radical environmentalism inspired by Rousseau's critique of economic modernization will be able to reject technological civilization and its conquest of nature, first of all, because a widespread rejection of uncontrolled science and technology would most likely turn Europe, America, and Japan into impoverished Third World countries, secondly, because freezing technological development would create worse socioeconomic inequalities than presently exist, and finally, because contemporary defense of the environment appears to require developed, science and technology-based societies. It is in the Third World that the worst environmental offenses occur, from deforestation of tropical rain forests to disposal of toxic wastes. For these reasons it appears highly unlikely to Fukuyama "that our civilization will voluntarily choose the Rousseauian option and reject the role that modern natural science has come to play in our contemporary economic life."[88] Perhaps he is correct in assessing our inability to control the "progress" or, more properly, "process" of irreversible cumulative change resulting from modern science and technology, in effect a Pandora's Box of good and evil producing everything from antibiotics to nuclear war. Combined with the modern laissez-faire political system designed to facilitate economic development, we face, in Ophuls's words, a veritable "tragedy of the Enlightenment" leading us towards the thorough annihilation of Earth's ecosystems, and severe resource constraints and attendant hardships for future generations. On the other hand, without the knowledge of the natural world gained from the Scientific Revolution and the Enlightenment we would not even be able to grasp the full significance of the human impact upon nature in an overpopulated world. Is it possible that these foundations of modernity may provide us with the basis for avoiding ecological Armageddon? An attempt will be made to answer this question in the following chapters.

NOTES

[1]. Derek Yalden, *The History of British Mammals* (London: T & AD Poyser Ltd. 1999), pp.80, 72-75.

[2]. Ibid., pp.128-129.

[3]. Ibid., p.163.

[4]. Ibid., pp.165-167.

[5]. Ibid., pp.171-172.

[6]. David Lambert, *Wild Mammals of the Countryside* (Covent Garden, London: Concertina Publications Limited, 1979), p.78.

[7]. Peter Crawford, *The Living Isles: A Natural History of Britain and Ireland* (New York: Charles Scribner's Sons, 1985), p.183.

[8]. Garrett Hardin, "The Tragedy of the Commons," *Science* 162: 1243-1248, 1968.

[9]. Sybil P. Pakev, ed., *Grzimek's Encyclopedia of Mammals* (New York: McGraw-Hill Publishing company, 1990), Vol. 5, p.405.

[10]. Ibid., pp. 386, 397, 409-410, 582.

[11]. Clive Ponting, *A Green History of the World: The Environment and the Collapse of Great Civilizations* (New York: Penguin Books, 1991), p.178.

[12]. Ibid., pp.178-180.

[13]. Ibid., p.180.

[14]. Ibid., p.181.

[15]. Roderick Nash, *Wilderness and the American Mind*, 3rd Ed. (New Haven: Yale University Press, 1982), pp.19-20.

[16]. Franklin L. Baumer, *Modern European Thought: Continuity and Change in Ideas 1600-1950* (New York: Macmillan Publishing Co. Inc., 1979), p.204.

[17]. Jonathan I. Israel, *Radical Enlightenment: Philosophy and the Making of Modernity 1650-1750* (Oxford: Oxford University Press, 2001), pp.456-461.

[18]. Baumer, *Modern European Thought*, pp. 209-210.

[19]. Eldon J. Gardner, *History of Biology* (Minneapolis, Minnesota: Burgess Publishing company, 1972), p.294.

[20]. Ibid., pp.294-296.

[21]. Marc F. Plattner, *Rousseau's State of Nature* (DeKalb: Northern Illinois University Press, 1979), p.37.

[22]. Jonathan Kemp, ed., *Diderot, Interpreter of Nature: Selected Writings* (New York: International Publishers, 1963), p.29.

[23]. Ibid., p.75.

[24]. *The Social Contract and Discourses*, Jean Jacques Rousseau, Translated with an Introduction by G.D.H. Cole (New York: E.P.Dutton and Company, Inc., 1950), p.189.

[25]. Ibid., pp.189-190.

[26]. Ibid., p.199.

[27]. Asher Horowitz, *Rousseau, Nature, and History* (Toronto: University of Toronto Press, 1987), p.52.

[28]. Ibid., p.57.

[29]. Ibid., pp.57-58.

[30].Gilbert F. LaFreniere, "Rousseau and the European Roots of Environmentalism," *Environmental History Review*, V.14 (Winter, 1990), pp.41-72, p.66.

[31]. Donald Worster, *Nature's Economy: A History of Ecological Ideas* (Cambridge: Cambridge University Press, 1977, 1985, p.2.

[32]. Gilbert White, *The Natural History of Selborne* (London: Cassell and Company, 1908), p.174.

[33]. Ibid., p.27.

[34]. Donald Worster, *Nature's Economy*, p.4.

[35]. Ibid., p.8.

[36]. Ibid., p.5.

[37]. Asher Horowitz, *Rousseau, Nature, and History*, p.37.

[38]. Dorinda Outram, *The Enlightenment* (Cambridge: Cambridge University Press, 1995), p.50.

[39]. Jonathan Bate, *The Song of the Earth* (Cambridge University Press, 2000), pp. 33-34. The classification is taken from Kate Soper, *What is Nature? Culture, Politics and the Non-Human* (Oxford and Cambridge, Mass.: Blackwell, 1995).

[40]. Asher Horowitz, *Rousseau, Nature, and History*, p.39.

[41]. Jean Starobinski, *Jean-Jacques Rousseau: Transparence and Obstruction* (Chicago: The University of Chicago Press, 1988 (1971)), p.23.

[42]. Alfred Biese, *The Development of the feeling for Nature in the Middle Ages and Modern Times* (London: George Routledge and Sons, Ltd., 1905), pp.274-78.

[43]. Ibid, p.278 and Daniel Mornet, *French Thought in the Eighteenth Century* (New York: Prentice-Hall, Inc. 1929), p.246.

[44]. Biese, *The Development of the Feeling for nature in the Middle Ages and Modern Times*, p.266.

[45]. Daniel Mornet, *French Thought in the Eighteenth Century*, p.246.

[46]. Daniel Mornet, *Le Sentiment de La Nature en France, de JJ. Rousseau à Bernardin de Saint-Pierre* (Genève-Paris: Slatkine Reprints ([1907], 1980), p.187.

[47]. Ibid., pp.203-4.

[48]. Jean-Jacques Rousseau, *Julie, ou La Nouvelle Hèloïse*, ed. and transl. by Judith H. McDowell (University Park: The Pennsylvania State University Press, 1968), p.335.

[49]. Ibid., Intro., p.12.

[50]. D.G. Charlton, *New Images of the Natural in France* (Cambridge: Cambridge University Press, 1984), p.34.

[51]. Ibid., p.37.

[52]. Ibid., p.39.

[53]. Ibid., pp.47-48.

[54]. Christopher Thacker, *The Wildness Pleases: The Origins of Romanticism* (London: Croom Helm, 1983), p.12. Also see pp.12-18.

[55]. Ibid., p.77. Also see Marjorie Hope Nicolson, *Mountain Gloom and Mountain Glory: The Development of the Aesthetics of the Infinite* (New York: W.W. Norton & Company, Inc., 1963). Although Nicolson acknowledges Rousseau's appreciation of nature along with Thomas Gray, James Thomson, and William Collins (p.20), she ignores him to focus on the English tradition of an aesthetic of the Sublime.

[56]. Jean-Jacques Rousseau, *A Discourse: Has the Restoration of the Arts and Sciences Had a Purifying Effect Upon Morals?* In G.D.H. Cole, ed., *The Social Contract and Discourses* (New York: E.P. Dutton and Company, Inc., 1950), p.164.

[57]. Jonathan Bate, *the Song of the Earth*, p.42.

[58]. Jean-Jacques Rousseau, *A Discourse on the Origin of Inequality* in G.D.H. Cole, ed., *The Social Contract and Discourses*, p.197.

[59]. Ibid., p.200.

[60]. Ibid., p.202.

[61]. Ibid., pp.208-209.

[62]. Ibid., p.209.

[63]. Ibid., p.273.

[64]. Ibid., p.274.

[65]. Ibid., pp.276-278.

[66]. Ibid., p.279.

[67]. Ibid., p.212.

[68]. Ibid., pp.234-235.

[69]. Jonathan Bate, *The Song of the Earth*, p.42.

[70]. Jean-Jacques Rousseau, *A Discourse on the Origins of Inequality*, p.249.

[71]. Ibid., p.249.

[72]. Jean-Jacques Rousseau, *Oeuvres Complètes, Tome III, Du contrat Social, Écrits Politiques* (Paris: Éditions Gallimard, 1964), p.198.

[73]. Jean Jacques Rousseau, *A Discourse on the Origins of Inequality*, p.270.

[74]. Ibid., p.271.

[75]. Jonathan Bate, *The Song of the Earth*, p.45.

[76]. Jean-Jacques Rousseau, *Émile*, translated by Barbara Foxley (New York: Dutton, 1911), p.417.

[77]. Jean-Jacques Rousseau, *Reveries of the Solitary Walker* (Harmondsworth, Middlesex, England: Penguin Books, Ltd., 1979), pp.117-118.

[78]. Emma Rothschild, *Economic Sentiments: Adam Smith, Condorcet, and the Enlightenment* (Cambridge, Massachusetts: Harvard University Press, 2001), p.117.

[79]. Ibid.

[80]. Ibid., pp.116-118.

[81]. Christopher Lasch, *The True and Only Heaven: Progress and Its Critics* (New York: W.W. Norton and Co., 1991), p.52.

[82]. Ibid., p.53.

[83]. William Ophuls, *Requiem for Modern Politics: The Tragedy of the Enlightenment and the Challenge of the New Millennium* (Boulder, Colorado: Westview Press, 1997), p.98.

[84]. Ibid., p.100.

[85]. Francis Fukuyama, *The End of History and the Last Man* (New York: Avon Books, 1992), p.83.

[86]. Ibid.

[87]. Ibid., p.84.

[88]. Ibid., p.86.

CHAPTER EIGHT: EUROPEAN COLONIZATION AND THE EXPLOITATION OF RESOURCES ON A GLOBAL SCALE

The preceding three chapters have examined the sixteenth, seventeenth, and eighteenth centuries from the perspectives of the Scientific Revolution, the development of the uniquely Western idea of progress, and shifting attitudes toward nature against the background of European exploitation of nature during the Early Modern era. This period of ca.1500-1800 roughly coincides with the great Age of Exploration which embraces the discoveries of Columbus, Vasco de Gama, Jacques Cartier, De Soto, Magellan and other famous early explorers on through James Cook's Pacific explorations of the late eighteenth century. During these three centuries all but a few corners of the globe were opened up to European trade, exploitation, and development. Having exhausted most of their own accessible resources in many categories, such as timber and fur-bearing animals and scarce metals such as gold and silver, Europeans were highly motivated to trade with newly discovered regions in the Orient or to plunder those lands which offered less resistance to the combination of long-distance sailing ships, guns, and steel swords which their explorers and colonists brought with them. The major purpose of this chapter is not to summarize the great explorations and the subsequent development of colonies, but rather to consider pre-colonial ecological conditions and their transformation by Europeans into developed and intensively utilized and managed landscapes. Also, given limited space for this extensive topic, it will be necessary to focus on North America, with some references to Central and South America, rather than other parts of the colonized world.

An extensive literature concerning the relationship between ecology and colonization in North America has accumulated in recent decades. Roderick Nash's *Wilderness and the American Mind* (1967) explored American colonial attitudes toward nature and Joseph Petulla's *American Environmental History* (1977), emphasized the exploitation of resources while Alfred Crosby's *The Columbian Exchange* (1972) introduced the epidemiological factor related to interactions between explorers, colonials, and Native Americans. In his sequel to *The Columbian Exchange, Ecological Imperialism: The Biological Expansion of Europe, 900-1900* (1986), Crosby further explored the complex relationships existing between epidemiology, human demographics, and native ecosystems. Also during the 1980's Frederick Turner's *Beyond Geography: The Western Spirit against the Wilderness* (1983) detailed the connection between Western Christianity and the brutal exploitation and domination of the Americas during the period of exploration and early colonization. Kirkpatrick Sale continued the assault upon the destructiveness of Columbus and the other heroes of the Age of Discovery in *The Conquest of Paradise: Christopher Columbus and the*

Columbian Legacy (1990), and Clive Ponting's *A Green History of the World: The Environment and the Collapse of Great Civilizations* (1991) extended the critique of European expansion and its ecological consequences to encompass the entire globe. More recently, David Arnold built upon the epidemiological insights of Crosby to uncover the roots of third world poverty and population problems in *The Problem of nature: Environment, Culture and European Expansion* (1996), and J. Donald Hughes embedded the period of European discovery and conquest within his *An Environmental History of the World: Humankind's Changing Role in the Community of Life* (2001). Many more major books and a voluminous periodical literature complement the few titles mentioned here, which are at least indicative of the kinds of environmental histories and related works which have greatly influenced our perception of the Age of Discovery and the centuries of colonization from 1500-1800.

The majority of Americans are not aware of the new insights deriving from environmental history, and, like the theory of evolution, it will require a long period of time for this kind of revisionist history to be assimilated into the public consciousness. I recall the brain-numbing tediousness of my high school American history class, essentially taught as an exercise in dogmatic patriotism. At the time, I was hiking and collecting minerals in the heavily wooded hills of western Massachusetts, where colonial times are recalled by the dates of incorporation of towns in the 1600s, and restaurants, hotels, and shops popularizing colonial themes. In the mid-twentieth century there appeared to be no connection between the names and dates of human events and the landscape itself, but the old stone walls lost deep in the woods were artifacts that begged for explanation, particularly as they were everywhere to be found in a forest that covered most of the western two-thirds of the state. Obviously, given the abandoned farms and stone walls, the entire region had once been almost entirely deforested and farmed and subsequently had been largely abandoned during the nineteenth century. Although, given my interest in natural history, I wondered about changes in the New England landscape, I didn't think much at all about the Native Americans of the region, despite the Indian place names which abound in the Berkshires and the Connecticut Valley. The indigenes themselves had been so thoroughly eradicated that only names such as "Mohawk Trail," or the "Redmen" sports logo of the University of Massachusetts (politically corrected to the "Minutemen" decades ago) reminded one of their long tenure upon this land before the arrival of the Mayflower.

It was only when, in my early twenties, I was hired to evaluate the potential of a gold claim in central Honduras that I experienced first-hand the legacy of European colonization. We spent about ten days on the Caribbean coast at La Ceiba gathering supplies for that expedition. We then flew to Olanchito in the interior where we picked up a Land Rover for the seventy mile drive over an unmaintained dirt track which served as the road to a trading post in the interior of central Honduras. Everywhere in the country I was astonished by the poverty, from the Carib villages on the Caribbean Sea where the local black population survived largely on fish and bananas, to the Indian and mestizo populations of the interior who lived on beans and maize supplemented by the occasional slaughter of a chicken, cow, or pig. For the next month we would live with the Indians in one of their thatched huts in a small village on a minor stream. The effects of malnutrition on the native population and their animals were everywhere in evidence, and half a century after my stay, the pine forest (on south slopes) and jungle (on north slopes) have been burned or logged while the population of Honduras has increased by several times. Bandits roamed the back country when we were there, not much different in appearance from the ones portrayed in B. Traven's *The Treasure of the Sierra Madre*. There was evidence of repeated burns on many of the hills and mountains that we traversed in the course of prospecting for mineral deposits. Not only had chickens, cattle, and pigs been introduced to the Indians of central Honduras, along with manufactured clothing, but even the Asiatic plant called "Elephant's Ears" was widespread in some areas, its giant leaves being useful as umbrellas during heavy tropical downpours.

Having come close to death from a tropical water-borne disease as well as from the much-weighed decision of bandits not to kill us for our supplies, not to mention several close encounters with snakes such as the Fer-de-Lance, Coral Snake, and a pit viper called a "tomigarth," my Honduras experience influenced my perception of the third world greatly, if not profoundly. For one thing, reinforced by later trips to Mexico, it set me to studying Latin American history, including the histories of the ancient Maya and Aztec civilizations. For another, it caused me to see Latin America and its cultural, economic, demographic, and environmental problems in a fresh, more sympathetic light. Finally, it contributed to my curiosity concerning historical explanations of cultural phenomena in general, which in the long run led to my interests in the philosophy of history and environmental history.

North American Ecosystems and Early Human Impact

Any ardent reader of *National Geographic* magazine, as I was in my teens, inevitably must become aware of the great differences in the mammalian populations of different continents. The mammals of North America and South America alone are strikingly different when one thinks of moose, elk, caribou, and black and grizzly bears in contrast to tapirs, llamas, alpacas, and jaguars. These differences are somewhat blurred by the geologically recent land bridge that we call Central America. Even more strikingly different from American mammals are the elephants, giraffes, lions and zebras of Africa, and the kangaroos, koalas, and dingoes of Australia. Yet look back only 12,000-15,000 years and North America was home to mammoths, mastodons, giant bison, giant ground sloths, camels, horses, and saber-tooth tigers, all in addition to the mammalian species that inhabit the continent today. At the beginning of the preceding chapter and in Chapter Four we learned about the recent and earlier post-Pleistocene mammals of western Europe. European mammals, both the extinct giants and remaining moderate-sized ones, are so similar to those of North America because of the Pleistocene land bridge at the Bering Straits, which allowed a steady flow of mammals between Eurasia and the Americas.

The giant Pleistocene mammals or "megamammals" and the present large mammals share an ancestry extending back to the Mesozoic Era, or dinosaur age, when mammals were small and seemingly inferior creatures. The ancestral mammals evolved side-by-side with the later dinosaurs and both were subjected to the geographic effects of long-term shifting of the earth's rigid plates, initially referred to by meteorologist and geophysicist Alfred Wegener (1880-1930) as "continental drift." Plate tectonics, the theory which explains continental drift, envisions three types of plate boundaries: (1) oceanic ridges, where molten material rises towards the surface as plates are spread apart, presumably by convection cells in the asthenosphere, the plastic and locally molten material that underlies the rigid, lithospheric plates (literally the shell of solid rock that envelopes the planet); (2) transform faults, which are roughly perpendicular to the spreading ridges, and which shear the crust by horizontal offsets; and (3) subduction zones, where plates collide in combinations of dense oceanic crust and lighter continental crust, with oceanic crust diving beneath continental or oceanic crust to melt and become part of the asthenosphere, while rocks are compressed and uplifted to form mountain ranges. Several times in the course of Earth's 4.5 billion year history, continental crust, which has slowly formed from the differentiation of lower density granitic (continental) rocks by partial melting of diving oceanic plates at subduction zones, has been bumped together by convective movement into a single landmass or supercontinent. The latest of these supercontinents is called Pangaea, meaning literally "all lands." Several hundred million years ago, animals and plants were evolving on this supercontinent without any ocean barriers separating them. About 200 million years ago, early in the Mesozoic era, Pangaea began to be rifted apart. Tensional forces, probably due to convection in the asthenosphere which had shifted direction, first separated Laurasia (present day North America and Eurasia) from Gondwanaland (present day South America, Africa, India, Australia, and Antarctica) about 150 million years ago, and, as this separation increased, by 100 million years ago began to separate North and South America from Europe and Africa along a north-south spreading ridge (the Mid-Atlantic Ridge).

The process of continental drift away from the center of the Pangean supercontinent has continued to the present day. For millions of years the dominant dinosuars and, eventually, small early mammals evolved side by side until the occurrence of a catastrophic event which drastically changed the course of evolution on planet Earth. Sixty five million years ago a sizeable asteroid collided with Earth to create a massive dust cloud, tremendous initial heat, and gigantic tsunamis up to 1000 feet high. The impact created Chicxulub crater in the northern Yucatan region of Central America and left a thin layer of irridium-rich sediment which marks the Cretaceous-Tertiary boundary, noteworthy above all for recording the extinction of the dinosaurs. The extinction of those large, totally dominant animals opened the way for the gradual diversification of the class Mammalia to fill the new ecological niches of the post-Cretaceous continents. By 50 million years ago the continents were yet further apart and the Age of Mammals was underway, eventually giving rise to the magnificent array of orders that includes the rodents, hoofed mammals, carnivores, whales, fin-footed marine mammals, and many other orders. While these diverse organic forms sprang from the original primitive mammals (evolved from mammal-like reptiles during the Mesozoic era), the drifting continents continued to crumple and uplift mountain ranges along subduction zones, creating a spectrum of ecological conditions, and influencing climate as well as ocean currents. Several million years ago, at the beginning of the Pleistocene Epoch, a young mountain range connected North America and South America, resulting in animal migrations in both directions. Also during Pleistocene time, the Bering Strait land bridge began to allow animal and plant migrations between Eurasia and North America, in both cases gradually obscuring the uniqueness of each continent's separately evolved mammals by introducing species of exotic provenance.[1]

The preceding background is essential to understanding how complicated the interactions of humans and other mammals have been over long periods of time, not only in North America but on other continents as well. Despite the climatic vicissitudes of the Pleistocene Ice Ages, the mammals of North America had evolved to a level of diversity comparable to those of Africa prior to the arrival of humans. In fact, all of the continents once boasted a mammalian diversity comparable to pre-European Africa. However, over the past 50,000 years the large mammals or megamammals have suffered mass extinctions on all continents, resulting in major changes in plant communities and ecosystems. Paleontologist Peter Ward observes that the megamammals of North America went extinct at the end of the Pleistocene Epoch, about 15,000-12,000 years ago, when the continental glaciers were beginning to retreat. The large mammals which died out included mammoths, mastodons and gomphotheres (all elephant-like members of the order Proboscidia), seven genera of giant ground sloths and their armadillo relatives, at least ten species of horses, two species of tapirs, a camel and two llamas, three types of pronghorns, saber-toothed tigers, the giant short-faced bear, and many others.[2] These animal extinctions were simultaneous with both drastic changes in vegetation and the recent arrival of *Homo sapiens* across the Bering Strait approximately 15,000-20,000 years ago. Changes in forest types definitely decreased the carrying capacity for large mammals, but there is also evidence that human hunting of large mammals played a role in their extinction. A comparable mass extinction of megamammals took place in South America between 15,000 and 10,000 years ago. In Europe and Asia, on the other hand, there were fewer extinctions than in the Americas and Australia, perhaps because of the much longer history of hominids and their association with megamammals in the Old World. Ward concludes that we may not be able to discern whether the New World mass extinctions were largely the result of changes in vegetation or of human hunting and habitat manipulation by fire.

Whatever the cause, "the Ice Age mass extinction resulted in a major reorganization of terrestrial ecosystems on every continent save Africa."[3] What Peter Ward and other paleontologists have learned from the study of mass extinctions scattered throughout the geologic record is that ecosystems remain relatively stable for long periods of geologic time, literally millions or tens of millions of years, until the appearance of some anomalous change or combination of changes such as asteroid collisions, earth processes including continental drift, mountain building and volcanism, all of which affect climate, or the rapid development of new organisms such as the relatively recent hominid mammals like ourselves. Then, through one or some combination of these factors, sudden or relatively sudden ecosystemic changes occur along with species extinctions, and relatively rapid evolutionary changes occur in other species in adjustment to the changed environmental conditions. It was into these recently changed conditions in the ecology of North and South America that the early European explorers and colonists stumbled, misperceiving the American landscape as a "wilderness" largely unaffected by human activities despite the abundant evidence of large numbers of indigenes in many regions, and even moderately advanced civilizations in Mexico, Central America, and Peru. Yet from the point of first contact with Native American populations, "virgin soil" epidemics of European diseases (which had become endemic to Old World humans since the Neolithic Revolution) began to decimate indigenous populations, thereby reinvigorating the myth of untamed wilderness as explorers moved westward into depopulated regions.

Ecological Imperialism

Alfred Crosby's *Ecological Imperialism* begins by summarizing the influence of plate tectonics upon the distribution of Earth's flora and fauna. He also devotes a chapter to the Neolithic Revolution and its ecological significance, particularly as a source of endemic European diseases such as plague, typhus, typhoid, smallpox, cowpox, and chickenpox, which along with pneumonia, influenza, venereal disease (syphilis was probably acquired from the indigenes), and other chronic illnesses, would be spread to the natives of the New World in various combinations by European explorers, conquerors, and colonists. Crosby also recounts the unsuccessful pre-Columbian forays of Europeans to Iceland and Greenland to the north, and to the Holy Land of the Levant to the south before turning to the subject of how Columbus's discovery of America was to a great extent predetermined and inevitable.

The northward thrust of the Vikings and the southward drive of the Crusaders during the Middle Ages complemented the major expansion of Christian, Western civilization on the eastern frontier which drove through central Europe during the High Middle Ages and stabilized in the Late Middle Ages of the fourteenth and fifteenth centuries. Recalling Chapter Four, the psychological consequences of endemic warfare and the Black Plague of the fourteenth century, accompanied by the famine-inducing Little Ice Age which intensified late in the fourteenth century, included a morbid pre-occupation with death and salvation in addition to the apocalyptical foreboding that the end of the world was nigh. Despite the somber mood of the age, the sailors of Iberia and Italy continued to explore beyond the Pillars of Hercules into the waters described by Crosby as the "Mediterranean Atlantic," that body of water bounded by the Azores, Madeiras, Canaries, and Cape Verde islands on the west and the Portuguese and African coasts on the east. It was here in the island-bounded east Atlantic that European sailors learned to sail the open blue waters of the Atlantic, and also learned the wind patterns (trade winds and easterlies) necessary to crossing the Atlantic Ocean. Encounters with Muslims in the course of the Crusades and later trading had equipped European sailors with the lateen sail which allowed them to tack into the wind, as a result of which they discovered the way back to Europe by tacking into the trade winds until they reached the westerlies which carried them eastward to Lisbon and other ports. The lateen sail made the crabwise slide to the northwest, the *volta do mar*, possible and was crucial to the success of Columbus' crossing of the Atlantic from the Canary Islands, where he also had made the transition to blue-water sailor or *marinheiro*.

Not only did the Portuguese and Spanish learn their long-distance voyaging upon the Mediterranean Atlantic, but in the conquest of the Canary Islands and the colonization of the Azores and Madeiras they designed the template for the discovery, conquest, and colonization of the Americas. In the Azores and Madeiras the Portuguese burned the forests, "seeded" the islands with feral animals-goats, sheep, rabbits, and, inadvertently, rats-and generally destroyed the pre-existing ecology, after which they planted wheat and other crops, and above all, sugar, which was in great demand in Europe from the fifteenth century onwards. In the Canaries, in the course of struggling to subdue the intractable indigenes called the Guanches, the Portuguese and Spaniards finally found themselves succeeding as the Gaunches died off in great numbers from diseases imported by the Europeans. To the conquerors these epidemics appeared to be acts of God clearing the way for the triumph of Christianity, much the same as they would be interpreted later on in Mexico and Massachusetts. The brutal destruction of the indigenes and the ecosystems of the islands of the Mediterranean Atlantic were the result of what Crosby called the European "portmanteau biota," the suite of organisms including farm animals such as cows, pigs, and sheep, the seeds of weedy European plants which propagate rapidly on disturbed ground, such as ferns, thistles, nettles, and tough grasses, and the assortment of disease organisms associated with post-Neolithic Europeans and their domesticated animals. As Crosby writes:

> The Europeans crossed the waters to the Canaries, as to the Azores
> and Madeiras, with a scaled-down, simplified version of the biota
> of Western Europe, in this case of the Mediterranean littoral. This
> portmanteau biota was crucial to their successes in these island

groups and to their successes, and failures, later and elsewhere. Where it "worked," where enough of its members prospered and propagated to create versions of Europe, however incomplete and distorted, Europeans themselves prospered and propagated.[4]

Of all the weapons of the portmanteau[5] biota described by Crosby, certainly the most powerful was that of disease (diseases to which Europeans were, at least to some degree, immune, but which were deadly and virulent to non-European indigenous populations). However it was that the megamammals were destroyed in the Americas, there is no doubt that Native American humans were destroyed chiefly by European diseases. Native Americans appear to have flourished both as hunter-gatherers preying upon the megamammals and, after extinction of the giant mammals, through a combination of hunting and gathering in conjunction with various levels of slowly developed agricultural technology, typically cultivating corn, squash, and beans. The irony in the European conquest of North America lies in the European perception of the North American continent as thinly inhabited, in ignorance of the fact that from the first years of conquest, from the time of Columbus in the West Indies and Cortez in Mexico, European diseases spread outwards along trade routes and began to wipe out up to ninety percent of the indigenous population. Thus it was easy for John Locke to rationalize the taking of property from the Indians in the seventeenth century, and for Adam Smith to write in the eighteenth century that: "The colony of a civilized nation which takes possession, either of waste country, or of one so thinly inhabited, that the natives easily give place to the new settlers, advances more rapidly to wealth and greatness than any other human society."[6]

The native North American population at the beginning of European conquest is estimated by Crosby to have been well up in the millions, the population of Florida alone at the beginning of the sixteenth century amounting to perhaps 900,000 Indians.[7] Clive Ponting estimates that even during the period of the Roman Empire the population of the Americas had reached about five million, and Charles Mann gives a range of estimates for North America in 1491 from 1.15 million to 18 million.[8] Geographer J. M. Blaut has estimated the population of the Americas in 1492 at somewhere between 50 million and 200 million![9] Obviously, given the wild range of scholarly estimates of native American populations at the time of Columbus' discovery, all we can really say with any degree of confidence is that the population of North America alone, based upon documentary and archeological evidence, was somewhere in the millions and subsequently vastly reduced by pandemics of European diseases of which smallpox was perhaps the worst. In his account of Hernando de Soto's 1539-1542 expedition through the southwestern United States, based upon contemporary chronicles, Crosby writes of populous and "stratified societies ruled with an iron hand from the top, and of scores of temples resting on truncated pyramids, which though often stubby and made of earth rather than masonry, remind one of similar structures in Teotihuacán and Chichen Itzá."[10] And yet when the French explorers came, ca.1700, the region was sparsely populated and vacant relative to the numbers of Indians encountered by De Soto. Similar changed conditions can be documented throughout North America, and are the basis for the high end of the population estimates stated above.

If this scenario for massive die-offs of Native Americans in the sixteenth, seventeenth, and eighteenth centuries is correct, and it seems to be the consensus of recent scholarship, then the ecological consequences would have been enormous, and the later explorers and colonists would have encountered an "untamed wilderness" full of abundant wildlife and thinly populated by Indians, with the exception of areas where the population had begun to recover after a period of pandemic disease. This situation also would have greatly benefited the fur trappers who followed the explorers westward across the North American continent. Many other factors affected by the return of wildlife habitat and animal populations following decimation of Native Americans would be involved. Reforestation certainly would have been common in naturally forested areas, and associated animal populations would shift accordingly. Perhaps the bison or American plains buffalo's large population of 50-60 million was to some degree a consequence of the decreased Indian population, although the re-introduction of the horse to North America by the Spaniards must have played an important role in the shifting ecology of the Great Plains. The major point here is that so much ecological disturbance most likely resulted from the reduction of the Native American population at different times and places that early European encounters with "wilderness" were actually encounters with a spectrum of once-managed landscapes in various stages of restoration towards unmanaged ecosystems. The import of this condition of flux for European attitudes was profound. Perceiving the "wilderness" as timeless and "natural" even though their own diseases had been the chief cause of its humanly depopulated state (what became Hobbes' and Locke's "state of nature"), European colonizers could easily rationalize their right to the lands of a continent in such an untrammeled, pristine, virgin condition.

Criticisms of Crosby's Ecological Imperialism

Crosby's thesis, that after unsuccessful attempts at colonization in Greenland, the Levant, and the Asian and African tropics (essentially because disease worked against the Europeans in these regions), Europeans were able to colonize the temperate and subtropical regions of the Americas, Australia, and New Zealand and turn them into food-exporting "Neo-Europes" chiefly on the basis of the introduced portmanteau biota of European animals, plants, and diseases, has not been without strong criticism. Historian David Arnold, for example, argues that Crosby proselytizes a "colonizing" or biological environmental determinism, a monocausal explanation which ascribes "too much to biology and too little to conscious human agency," and also that "Crosby projects an air of biological inevitability about what was often the result of deliberate human intention."[11] Implicit in Crosby's biological determinism there is also to be found, according to Arnold, a powerful Darwinian tendency "that identifies European expansion with racial difference and the supposed inevitability of white racial supremacy."[12] Citing arguments by earlier authors that America was conquered largely by disease, Arnold avers that Crosby's comparable view of Amerindians, Maori, and Australian Aborigines as biologically defenseless is an assertion of "their failure to survive competition from a 'superior' race."[13] Arnold's judgment is perhaps a bit harsh, given that Crosby explained the factors of geographic isolation, differences in domesticated animals and stage of agricultural and pastoral development between the Old and New worlds, and the blood type of New World indigenes which made them vulnerable to exotic European diseases. In the light of Crosby's explanation, New World people may be seen as different, but not necessarily inferior, although this is a moot point. A third criticism of Crosby made by Arnold is that although India, Africa and other parts of the world did not become Neo-Europes, European explorers and colonists nevertheless were successful there in gaining political control and establishing plantations and mines despite the environmentally unfriendly conditions of these typically tropical regions. Military control and forced labor in various forms allowed the extraction of resources and trade even though the portmanteau biota had failed to annihilate the wild animals and native populations of tropical, heavily populated regions. Arnold concludes that Crosby's *Ecological Imperialism* "seems in the final analysis to be a strangely one-dimensional, Eurocentric reading of the process of European expansionism, a cautionary tale in the dangers of ascribing to biology what more properly belongs to human agency."[14]

Perhaps Arnold is largely correct to the extent that the impacts of the portmanteau biota account for many changes, particularly massive die-offs of indigenous populations under New World conditions of isolation, but fail to account for the at least equally powerful psychological and material factors propelling Europeans toward expansionism. These derived both from the idiosyncrasies of Western Christianity, burgeoning capitalism and mercantilism, and diminished European resources in the Late Middle Ages, as well as the desire to develop more extensive trade with the Orient that led Columbus to an unknown continent while seeking a new trade route. Once arrived in the Americas, the Spanish encountered a frontier quite different from what the Portuguese found in India and Indonesia, namely, cultures in relatively early stages of development in the Americas as compared to older, more sophisticated, densely populated civilizations in Asia. The American cultures lent themselves more readily to conquest, both because of their stages of development and because of their greater vulnerability to European disease.

Jared Diamond, in *Guns, Germs, and Steel* (1999), lends support to Arnold's critique of Crosby's environmental determinism when he balances the impact of germs on Native American populations and societies with the importance of such complementary causal factors as European technology, political organization, and writing. "Rivaling germs as proximate factors behind

Europe's conquest of the Americas were the differences in all aspects of technology. These differences stemmed ultimately from Eurasia's much longer history of densely populated, economically specialized, politically centralized, interacting and competing societies dependent on food production."[15] Diamond goes on to document the immensely superior metallic military arsenal and machines of the Europeans, including the wheel and relatively sophisticated sea transport in relation to military actions. He also emphasizes the greater social cohesiveness of the Europeans based upon official, politically approved religious beliefs, and their improved efficiency stemming from the organized military and economic bureaucracies which writing allowed in comparison with the logistics of the Aztecs and Incas. "Thus, Eurasian societies in the time of Columbus enjoyed big advantages over native American societies in food production, germs, technology (including weapons), political organization, and writing. These were the main factors tipping the outcome of the Post-Columbian collisions."[16] In *Nature: Western Attitudes Since Ancient Times*, Peter Coates supports Diamond in insisting on the importance of European prowess in battle, acquisitiveness, cruelty, and technological superiority as factors too much ignored by Crosby's ecological imperialism, observing that "the role of environmental factors and processes of environmental change are more dramatic and palpable in 'frontier' situations where technologically advanced capitalist cultures wrested control from more subsistence-oriented and less technologically potent ones."[17] Granting the refreshing provocativeness of Crosby's thesis, and the fact that feral animals such as pigs caused devastating damage to New World fauna and flora, Coates nevertheless hesitates to accept ecological imperialism as the major cause of European success in converting much of North and South America into Neo-Europes. "Crosby bombards his readers with seductive images of plants, animals and germs marching remorselessly across new worlds, threatening to overwhelm us through sheer force and his argument's compelling simplicity."[18] On the other hand, Coates does acknowledge the great importance of European diseases in facilitating the conquest of the Americas, and most historians would likely agree, for Crosby is strongly supported by other scholars on this point.

One of those scholars is geographer J.M. Blaut, who is a strong critic of diffusionism, the idea that Europe was more advanced than all other regions prior to 1492, and that with colonialism the innovative ideas and technologies of Europe flowed outward from Europe at the center to the relatively backward colonies at the periphery. Blout argues that Europeans distinctly did not possess this historical superiority over what we now call the Third World, and that in the process of colonization capitalism became concentrated in Europe partly as a result of the wealth, beginning with gold and silver, which poured into Europe and stimulated its economy to "kickstart" the mercantilist-capitalist developments of early modern Europe into the revolutionary economic changes that contributed to the origins of the Industrial Revolution. Regarding the factors leading to European conquest of the Americas, Blout strongly invokes "the massive depopulation caused by the pandemic of Eastern Hemisphere diseases that were introduced to America by Europeans."[19] He also considers the second factor of the European advantage in military technology, but admonishes the reader that military technology itself, after the initial battles, would not have assured the Europeans of victory against the large populations of civilized Aztecs and Incas. "The point is that history went in a different direction because of the incredibly severe and rapid impact of introduced diseases. Resistance collapsed because the Native Americans were dying in epidemics even before the battles were joined. Probably 90% of the population of central Mexico was wiped out during the sixteenth century...."[20] When Cortés arrived in Tenochtitlán, European diseases, especially smallpox, which already had been transmitted from Cuba to Mexico by Native American traders, were beginning to decimate the Aztecs.

The criticisms of Crosby's ecological imperialism cited above would seem to suggest that, except for diseases, ecological damage resulting from the portmanteau biota was not an important factor in the conquest of the Americas. This conclusion may be warranted regarding the early stages of European conquest, particularly when they happened swiftly, as in Mexico and Peru. However, once Europeans established a foothold and began importing their animals and plants, conversion of ecosystems upon which indigenous inhabitants depended into Europeanized environments facilitated the continuation of colonization and displacement or Europeanization of native cultures. In support of Crosby's ecological imperialism, J. Donald Hughes demonstrates in his *An Environmental History of the World* that most of the Mexican environment was transformed or destroyed by the portmanteau biota during the sixteenth and seventeenth centuries: "From the beginnings of European exploration, colonization, and trade in the fifteenth century, the organisms they carried with them had a worldwide impact. Mexico offers an example of the way in which their onslaught altered ecosystems and reduced the abundance of native species or made them extinct."[21]

Given the great diversity of habitats in Mexico, from tropical rainforests of the Atlantic coast to the high volcanoes and plateaus of the interior to the pine and oak forests of the western mountains, the biological diversity of Mexico was still rich in the time of the Aztec Empire. The empire itself was unstable, in part due to overpopulation of the plateau on which the capitol city of Tenochtitlán was built, and also due to the imposition of taxes, and much worse, demands for victims for human sacrifice from the conquered peoples dominated by the Aztecs. Teotihuacán, an earlier city, had collapsed six hundred years before the Aztec Empire due to overpopulation, deforestation, and shortages of water. The Spaniards under Cortez, aided by the spread of smallpox when Tenochtitlán was under siege, had stumbled upon a Mexican civilization in disequilibrium and vulnerable to conquest, which is not to take away from the heroic efforts of Cortez and his soldiers, but they were at the right place at the right time. Having dealt with the depredations of disease upon the human population, Hughes then documents the massive environmental impacts stemming from burgeoning herds of European pigs, cattle, horses, burros, and sheep.

Feral pigs devoured everything from vegetation and acorns to reptiles, small mammals, and ground-nesting birds, being omnivores like the primates. They reverted to a form approximating the ancestral European wild boar in their feral state. Domestic cattle had already exploded in numbers by 1540, causing prices of beef to drop. By 1565 cattle had increased to the degree that they destroyed extensive grasslands in northern Mexico, and after 1586 cattle died of starvation as ranges were overtaken by unpalatable brush, leaving large areas severely degraded. Escaped horses also formed large herds, and by 1580 their use by Indians was spreading northwards, adding to the impacts on grassland ecosystems. The burro, on the other hand, in the feral state was attracted to shrubs and other plants in the northern mountains. More destructive than cattle and horses were the herds of Merino sheep raised for wool. With them from Europe came even more destructive goats which browse on bushes and small trees. More resilient than sheep, they became feral and adapted to Mexico's mountainous terrain. Meanwhile, herds of domestic sheep overran areas of the intensive irrigation agriculture of Indians and downgraded these lands into a mesquite-covered desert. Even in the tropical lowlands cattle and sheep had caused severe ecological damage by the end of the sixteenth century. In addition to the rapid and extensive damage done to native ecosystems and agriculture by grazing ungulates, the rats carried inadvertently by Spanish ships killed off native rodent species and devoured Indian stores of maize, in addition to carrying diseases such as bubonic plague and typhus.[22]

Clearly Hughes' assessment of the ecological and agricultural impacts of the portmanteau biota favors Crosby's thesis of ecological imperialism far more than Crosby's critics. Another global environmental historian also supports Crosby's thesis. Clive Ponting, in *A Green History of the World* (1991), documents the massive ecological damage caused by pigs, cattle, horses, camels, rabbits, European starlings, as well as honeybees and European weeds, in many regions of South America, Australia, and New Zealand as well as in Mexico and the United States. Ponting draws the general conclusion that: "Europeans also made their mark on the ecosystems of the rest of the world (besides North America) by introducing new species as well as exterminating many of those they encountered and severely reducing the numbers of many more."[23] The worst of these ecological disasters in Australia resulted from the introduction of the rabbit as a game animal in 1859. With no natural predators the rabbit population grew so rapidly that crops in Victoria were devastated and the rabbits rapidly moved into South Australia where they affected sheep farming as well as crops. It took many unsuccessful campaigns of rabbit extermination before the disease myxomatosis was introduced from Brazil, finally causing a massive die-off. However, the rabbit population has recovered considerably during the second half of the twentieth century. Following numerous additional examples, Ponting writes:

> The effects of the expansion of Europe, its people, its plants and animals, were far-reaching and irreversible. The wildlife of the

world was never the same again. Many species were driven to extinction or so reduced in numbers that they could barely survive in a few limited areas. Many animals and plants of European origin were spread around the world, disrupting natural ecosystems and again causing the extinction or decline of many native species.[24]

Perhaps some of the controversy regarding the destructive and transformative efficacy of the portmanteau biota is caused by variations in impacts upon ecosystems and individual species which are affected by climate. As we have seen in Chapters Two and Four, arid, semi-arid, and tropical climatic regions tend to be more vulnerable to disturbance than temperate zones which are dominated by deciduous and coniferous forests. The extent of human impact has been highly variable from one geographic region to another during the age of exploration and colonization. Crosby's thesis is, at the very least, a useful intellectual probe which can help us to understand cultural and ecological change resulting from European colonialism. My own experiences in the jungles of Honduras and the mountains and deserts of Mexico seems to corroborate the extensiveness of permanent ecological damage to forests and streams and their animal populations in tropical regions, and even worse damage to more fragile plant communities in deserts and dry mountainous areas such as Mexico and the southwestern United States. More than half a century of observing the forests of New England has taught me how resilient those forest ecosystems are in comparison, and in a region far more densely populated than the jungles of central Honduras or the southwestern deserts. Humans and their animals and plants have left a wreck in their wake in the fragile tropics and deserts, but have not permanently destroyed the ecosystems of the northeastern United States because of their capacity for self-restoration once destructive human use ceased. Given enough time, tropical forests can also recover to a great extent, minus a given number of species, as in the lands of the Mayan civilization.

Explaining European Expansion: Three Levels of Causation

In Chapter One we were introduced to three levels of environmental history, including the substructural, superstructural, and ideological, i.e., (1) natural processes as related to human utilization of nature, (2) social institutions related to making a living from nature, and (3) the ideas which rationalize the role of humans in the world. All cultures are dependent upon nature, build a particular complex of institutions and technologies for exploiting nature, and rationalize their activities through myth, religion, philosophy, history and science. In attempting to explain why Western civilization in particular has spread out from western Europe to conquer nature and primitive indigenes, thereby spreading Christian culture around the globe, Alfred Crosby has emphasized the substructural category of environmental history. If we link all three of these categories of environmental history to levels or kinds of causal explanation, at the substructural level (the level of nature and natural processes), we find geological causes such as plate tectonics, biological causes such as mammalian evolution, the spread of diseases from other mammals to humans, and meteorological-geographic causes such as the distribution of wind patterns all at work over vast epochs of pre-historical time. At the superstructural level, from the Neolithic Revolution to the rise of Mediterranean civilizations, we can identify the cultural breakthroughs which have contributed to the rise (and fall) of ancient civilizations and to the origins of our own Western European or Faustian civilization as well. Finally, ideology, as myth, religion and philosophy, has had long-term effects upon successive cultures, exemplified by the powerful influence of Judeo-Christian religion and Greek philosophy upon the later Western European civilization. Our contemporary attitudes toward nature are to a great extent indebted to these venerable ideas. In Crosby's *Ecological Imperialism*, however, we see negligible attention paid to superstructural or ideological causes of European conquest and colonization. Crosby focuses almost entirely upon substructural causes, from plate tectonics to the prevalence of endemic diseases among the peoples of Eurasia. This emphasis on long-term natural causes is the basis for criticisms of Crosby's thesis for falling under the category of "environmental determinism," to the neglect of causes stemming from technological innovation, political organization, economic systems such as capitalism and mercantilism, and belief systems such as Western Christianity. We read above that Jared Diamond complemented Crosby's environmental determinism by emphasizing the importance of Western military superiority, political organization, and written documentation. Diamond thus stressed the importance of substructural causation during the Age of Discovery and subsequent colonization. How important, then, was the role of ideology, the third level of causation, during this period of history?

The most powerful and sophisticated argument that ideology was the leading cause of European expansion is presented by Frederick Turner's *Beyond Geography: The Western Spirit against the Wilderness*. Acknowledging Western superiority in ocean navigation and firearms as well as its mercantilist ambitions, Turner argues that there is something unique in the Western, Christian worldview that contributed more than anything else to the success of its imperialistic expansion during the Early Modern period. Thus, he writes: "With the spectacle before us of the invasion of a new world by a whole civilization, we are led as by a dowser's wand beneath politics, technics, and economics to the world of the spirit. And the spirit of the West is to be found in the history of that religion to which all Europe subscribed."[25] Embedded in Christian religion is the belief that all nature was created for human use (we have explored this theme in detail particularly in Chapters Three, Five, and Seven). This presupposition has led Western civilization to emphasize the capacity of rational thought in order to exploit or utilize nature as the Christian God wished, and giving rise to the idiosyncratic historical development of science and technology in tandem in Western Europe. During the Age of Exploration "Christians of all nationalities and persuasions were united in a conception of the earth as a divinely created *thing*, there for the enjoyment, instruction, and profit of man."[26] Although Saint Augustine was explicit about this belief late in the Roman Empire, its ultimate derivation is from Old Testament scripture as rendered by Christian exegetes. (We have encountered this connection in Chapters Four, Five and elsewhere, beginning with Lynn White, Jr.'s accusation that Christian religion is largely responsible for the modern ecological crisis, and in Carolyn Merchant's assessment of the Christian origins of early modern science and the pursuant idea of progress.) What is unique to Turner's view is that he understands Western exploration and conquest (along with all that these changes mean for the contemporary global marketplace) as an inevitable consequence of the Western, Christian idea of nature. Reflecting upon the courage of early Western scientists, Turner writes:

> It is doubtful that any of them could have guessed that his work was preparing the way for a mechanistic conception of nature and man that would have the practical effect of denying the operation of God in a world He putatively had created.... For the majority there were still spirits about, but these were mostly malignant ones, such as the fallen angels who had become the nature deities of the pagans. But the forms and observances of a great religion die slowly, and long after its positive and life-enhancing aspects have fallen into practical disuse, its hag-ridden residues may be keenly felt. So, in the middle of the seventeenth century, one can find Christians who seemed utterly consumed by the life of the spirit and as convinced of the operation of the devil as they were of divine providence. And yet these same Christians were denying life to much of the world and acting upon that world as if it were a passive configuration of matter devoid of its own interior life, laws, and spirit and existing only to be "civilized" for gain.

Is Turner correct? Can it be that the Western imperialistic drive during the post-medieval centuries was powered not so much by "ecological imperialism" or "guns, germs and steel" as by a European consciousness intent upon bending nature and all "primitive" peoples associated with it to the will of a Christendom convinced of its absolute right to the domination of nature? This question presupposes the operation of all three levels of causation upon Western conquest and colonization, the substructural, superstructural and ideological. Perhaps no definitive answer can be given to this question, but an approach is suggested by Kirkpatrick Sale's *The Conquest of Paradise* and recent environmental histories such as David Arnold's *The Problem of Nature*.

Sale supports Turner's view of the dark and destructive side of Western civilization in describing the grim, pessimistic apocalyptical Christianity of the late fifteenth century, which owed its somber worldview to the famine, warfare, and plague which overshadowed Western European civilization in the fourteenth and fifteenth centuries. Christopher Columbus was more than usually taken with the eschatological vision of a world approaching its end, much as Saint Augustine had been a millennium before. While artists like Albrecht Dürer depicted macabre images of the Dance of Death in northern Europe, to the south the Spaniards were ending a triumphant holy war against the infidel Muslims. Torture, executions and royal direction of the Inquisition since 1483 further hardened a people accustomed to "the savagery of brother against brother and cousin against cousin, complete with kidnapping, torture, mutilation, fratricide, patricide, assassination, and fomented rebellion."[27] Simultaneously, the Black Death, although considerably abated, continued during the Age of Discovery, along with leprosy, scurvy, smallpox, measles, diptheria, typhus, tuberculosis and influenza. The debilitation and death caused by ubiquitous disease was accompanied by frequent grain crises and famine through much of the fifteenth century, further contributing to the widespread sense of impending doom. A general response to these oppressive and bewildering circumstances was a turning towards materialism and its economic expression, capitalism. "The impulse to treasure the material here-and-now, the tangible, in a world of both corporeal uncertainty and spiritual vacuity would seem to be perfectly normal, as we see it today, and yet it appears to have been something quite new for Europe."[28] This turn to materialism is actually a phenomenon well known to historians of Europe, and is, in fact, closely associated with the Renaissance, the revival of art, literature, and learning which extended from the fourteenth to the seventeenth century. Environmental historian John Opie thinks that the Renaissance contributed to the modern ecological crises by providing a sizeable proportion of Western human beings with a sense of liberation from the Church at the same time that they became aware of their increasing power over nature. "The medieval concept that individual persons belonged to a larger community, both natural and supernatural, disappeared."[29]

The combination of apocalyptical Christianity with egocentric humanism produced larger than life heroes and tyrants. In the Spanish discovery and conquest of Mexico and Peru, Columbus is more representative of the apocalyptical, and Cortez and Pizarro more typical of the Renaissance versions of Christianity. Apocalyptical, fearful, arrogant and aggressive all at once, the Christianity of the explorers and conquerors was a volatile mix capable of driving more resilient individuals to almost incredible acts of boldness and courage as viewed from our time. Cupidity was also common among these men. Sale suggests that, although this was not the first era in history when men strove for wealth, it was "perhaps the first in which the possession of material goods began so markedly to replace other values at the center of ethical and religious pantheons." Noting the proliferation of the desire for material goods and wealth in late fifteenth century Florence, Sale writes:

> This straightforward materialism, developed over long decades
> with sophisticated humanism and rationalism as its companions,
> created the essential conditions for the success of that economic

system we have come to call capitalism. ... Other contributory
elements, to be sure, had also come into being by the late fifteenth
century, credit lines, currency transfers, bills of exchange,
maritime insurance, international banking, and the accumulation of
metals and moneys themselves, but it was materialism's pattern of
mind, its order of values, its reinterpretation of the world, that
really permitted all these other instruments to develop in Europe
and mesh and flourish.[30]

Finally, at the superstructural level, the rise of European nationalism, still
in its infancy in the time of Columbus and Cortez, contributed the financial
resources, centralized organization, and competitive impetus necessary to
guarantee the long-term success of the colonies representing the more powerful
nation-states, which, since the decline of the papacy in the early fourteenth
century, began to claim much of the wealth which formerly flowed from the
feudal aristocracy and growing bourgeoisie to the papal coffers and the
monasteries.[31]

It should be rather obvious at this point that the success of European
conquest and colonization was the result of multiple causes, substructural,
superstructural, and ideological, operating in tandem and so interwoven that it is
virtually impossible to isolate any particular level of causation as dominant or as
the major cause, whether ecological imperialism, superior Western institutions
and technology, or the "Western Spirit" or worldview of the West. David
Arnold's assessment of theoretical arguments or explanations of European
conquest and colonization supports my approach based upon Donald Worster's
three explanatory levels of substructure, superstructure, and ideology. Arnold
compares the biologically driven model proposed by William McNeill and Alfred
Crosby (the substructural level of causation) with the driving forces of capitalism,
science, and technology which make the environmental history of the Americas
an extension of European economic history (the superstructural level), and with
the interpretation "that the expansion of the ecological frontier was a product of
cultural determinism, of values derived directly from Europe which, through their
transference to the New World, had a transforming effect on the American
frontier."[32] (the ideological level). Arnold displays a preference for ideological
and superstructural causation when he observes that the Aztec, Incan, and other
sedentary Amerindian societies were not destroyed by the cattle, pigs, and sheep
themselves, but rather by the very different cultural system which they
represented (especially pastoralism), including the Judeo-Christian view of human
dominion over nature and antipathy toward wilderness.

The Great Frontier and the Spread of European Civilization

British Environmental historian David Arnold, a strong critic of Alfred Crosby's *Ecological Imperialism* in the preceding section of this chapter, is also critical of Frederick Jackson Turner's 1893 "frontier thesis," which was written, at least in part, to challenge historians who traced the origins of American democracy to Germanic folk institutions. To Turner's mind American democracy had emerged as a product of the frontier, " the meeting point between savagery and civilization." Indian lands were viewed by Turner, like the seventeenth century English political philosopher John Locke, as "free" land available for the taking. Defining the term "frontier" so loosely that later historians have struggled to understand it, Turner wrote: "The frontier is the line of most rapid and effective Americanization. The wilderness masters the colonist. It finds him a European in dress, industries, tools, modes of travel, and thought. It takes him from the railroad car and puts him in the birch canoe."[33] Given that the railroad came into common use in the United States in the 1830s, Turner is writing about the far western American frontier. He had little use for the "savages" who populated that region or any other part of what became the United States as the line of the frontier moved westward from the Atlantic coast. The positive value he saw in the Indians and Indian Wars was as a stimulus to unification of the frontiersmen who fought against them. They were a part of nature to be overcome in order to create the values and institutions of white America. In his racist vision of progress, Turner saw America evolving from unworthy primitivism to superior agrarian-based communities of Americanized Europeans.

If Frederick Jackson Turner was blind to the fact that the American frontier was an extension of the European frontier described in Chapter Four, American historian Walter Prescott Webb's *The Great Frontier* (1952) was an effective antidote to Turner's provincialism: "If I could export one thing American to European scholars ... it would be an understanding of the frontier, not the American frontier but their own, and its significance in their history and in their present lives."[34] Webb wished to apply the American frontier concept to all of modern Western civilization. Taking the late medieval frontier for granted, Webb was referring to a European frontier extending from 1500-1950, especially recognizing the importance of the Age of Exploration (1500-1800) for the "Boom Era" which boosted European resources and power to hitherto unimaginable levels. Webb might appear to prefigure late twentieth century environmentalist critics of Faustian modernity and its wasteful and destructive misuse of nature when he wrote:

> ... the character of the modern age is due in large measure to the fact that it had a frontier setting, that it grew up in an economic boom induced by the appropriation and use of frontier resources, and that its institutions were designed and modified to meet the demands of a booming society. If this be true, then the whole

modern age would appear to be an abnormal one. By its very nature a boom is a temporal thing, something out of the ordinary run of life, an abnormal state of affairs which is destined to end when the forces that caused it cease to operate. Though this truth is obvious, people do not bear it in mind, even in the briefest periods of unusual prosperity. They act as if the "good times" are permanent, as if the abnormal activity is the normal thing.[35]

However, Webb was anything but a "pre-environmentalist" thinker, even when he wrote that the Age of the Frontier was "a strange historical detour in which men developed all sorts of quaint ideas about property for all, freedom for all, and continuous progress."[36] He was totally development-oriented, completely lacking in ecological awareness, and in favor of such projects as using the flood waters of the Mississippi River to convert the Great Salt Lake into an irrigation reservoir, or turning the entire Amazon basin into a gigantic farm, "developing the remaining frontier areas so that they could be used ...."[37] Despite his approval of promethean projects that would drastically affect the planet's ecosystems, Webb did provide us with the insight that "modernity," the last five hundred years of European domination of the planet, was coming to a close, that severe resource shortages were on the near horizon, and that humanity would have to adapt its attitudes and institutions to a coming era of scarcity. This message was given contemporary relevance by political philosopher William Ophuls in his classic *Ecology and the Politics of Scarcity* (1977), revised in 1992. Ophuls found Webb's *The Great Frontier* a superlative statement on the theme of scarcity and abundance of resources in modern history, "but scarcely balanced, for Webb means to disabuse us of the notion that progress, plenty, and democracy are our eternal and immutable birthright."[38] Viewing the Great Frontier in an ecological context, Ophuls criticizes the assumption of ecological abundance in the political theories of John Locke, Adam Smith, and Karl Marx, which has led political theorists of all persuasions in our time to take abundance for granted and assume the necessity of further growth.[39] We will further explore these implications of the Great Frontier in Chapter Ten. For the present, the salient point is to recognize the pervasiveness of the idea of development (rationalized and alloyed with the theoretical idea of progress since the Enlightenment) in Western civilization. During the medieval period Western European civilization pressed on its eastern frontier to establish its limits where it abutted the margins of other civilizations. Beginning with Columbus and the great discoveries, however, Western civilization began to carry the idea and practice of development from a continental to a global level of activity, thereby establishing the infrastructure of the modern world, composed of "developed" countries such as those of Europe and the United States, and "undeveloped" or "underdeveloped" countries in Latin America, Africa, and parts of Asia. With such venerable roots, it is hardly surprising that contemporary nation-states which are the heirs of the Great

Frontier have envisioned a global economic marketplace (of haves and have nots) extending indefinitely into the future. "Plus ça change, et rien ne change pas."

NOTES

[1]. See Edward J. Tarbuck and Frederick K. Lutgens, *Earth: An Introduction to Physical Geology* (Upper Saddle River, New Jersey: Prentice Hall, 2002), pp.531-536, and 236. Also, National Geographic Society, *Wild Animals of North America* (Washington, D.C.: National Geographic Society, 1960), pp.36-51.

[2]. Peter Ward, *Future Evolution* (New York: Henry Holt and Company, 2001), p.41.

[3]. Ibid., p.45.

[4]. Alfred Crosby, *Ecological Imperialism: The Biological Expansion of Europe, 900-1900* (Cambridge: Cambridge University Press, 1986), p.89.

[5]. *Portmanteau* is the French word for a carrying case for clothing.

[6]. Quoted in Alfred Crosby, *Ecological Imperialism*, p.195.

[7]. Ibid., p.212.

[8]. Clive Ponting, *A Green History of the World* (New York: Penguin Books, 1991), p.92. Charles C. Mann, "1491," The Atlantic Online, http://www.theatlantic.com pp.4, 7. (*The Atlantic*, March 2002). Mann's book, *1491: New Revelations of the Americas before Columbus* (New York: Alfred A. Knopf, 2005) tends to lean towards the higher end of population estimates, but he fairly represents the limitations of such estimates, given the relatively crude methodologies from which they are derived. See pp. 108-109, 101-104, and 143-145. The larger the estimates of pre-Columbian Native American populations, the more Mann can argue that the Americas were predominantly a human artifact rather than wilderness. Although Alfred Crosby and others had already demolished the "pristine myth" of the Americas as a wilderness containing small populations of humans, Mann presents an excellent summary of the evidence that populations of Native Americans throughout the Americas were far greater than most historians have assumed.

[9]. J.M. Blaut, *The Colonizer's Model of the World: Geographical Diffusionism and Eurocentric History* (New York: The Guilford Press, 1993), p.184.

[10]. Alfred Crosby, *Ecological Imperialism*, p.211.

[11]. David Arnold, *The Problem of Nature: Environment, Culture and European Expansion* (Cambridge, Massachusetts: Blackwell Publishers, Inc., 1996), p.89.

[12]. Ibid., p.89.

[13]. Ibid., p.90.

[14]. Ibid., p.91.

[15]. Jared Diamond, *Guns, Germs, And Steel: The Fates of Human Societies* (New York: W.W. Norton and Company, 1999), p.358.

[16]. Ibid., p.360.

[17]. Peter Coates, *Nature: Western Attitudes since Ancient Times* (Berkeley: University of California Press, 1998), p.100.

[18]. Ibid., p.101.

[19]. J.M. Blaut, *The Colonizer's Model of the World*, p.184.

[20]. Ibid.

[21]. J. Donald Hughes, *An Environmental History of the World: Humankind's Changing Role in the Community of Life* (London: Routledge, 2001), p.118.

[22]. Ibid., pp.113-119.

[23]. Clive Ponting, *A Green History of the World*, p.170.

[24]. Ibid., p.174.

[25]. Frederick Turner, *Beyond Geography: The Western Spirit against the Wilderness* (New Brunswick, New Jersey: Rutgers University Press, 1983), p.174.

[26]. Ibid.

[27]. Kirkpatrick Sale, *The Conquest of Paradise: Christopher Columbus and the Columbian Legacy* (New York: Alfred A. Knopf, 1990), p.33.

[28]. Ibid., p.42.

[29]. John Opie, "Renaissance Origins of the Environmental Crisis," *Environmental Review*, Vol. 11, No. 1 (Spring) 1987, p.10.

[30]. Ibid.

[31]. Ibid., pp.44-45.

[32]. David Arnold, *The Problem of Nature*, p.130.

[33]. Frederick Turner quoted in David Arnold, *The Problem of Nature*, p.101.

[34]. Walter Prescott Webb, *The Great Frontier* (Austin: University of Texas Press, 1952), p.7.

[35]. Ibid., p.140.

[36]. Ibid., p.415.

[37]. Ibid., pp. 416-417.

[38]. William Ophuls, *Ecology and the Politics of Scarcity* (San Francisco: W.H. Freeman and Company, 1977), p.164.

[39]. Ibid., p.145.

CHAPTER NINE: MODERNITY AND ITS DISCONTENTS

The Modern Worldview

The question to be asked in attempting to determine when an essentially "modern" worldview came into existence in Europe and its colonies is: When do the voices recorded from the historical past begin to sound familiarly contemporary, reflecting presuppositions of thought quite similar to our own? My reading of the historical sources regarding this question suggests that not until late in the Enlightenment, after ca. 1760, do we begin to recognize ideas expressed by members of the intelligentsias of Europe and the European colonies which reflect our own modern values (as broad as the spectrum of values may be at the beginning of the twenty-first century).

Marshall Berman, in his thorough analysis of the experience of modernity in *All That is Solid Melts into Air*, characterizes contemporary modernity and its antecedents as follows:

> The maelstrom of modern life has been fed from many sources: great discoveries in the physical sciences, changing our images of the universe and our place in it; the industrialization of production, which transforms scientific knowledge into technology, creates new human environments and destroys old ones, speeds up the whole tempo of life, generates new forms of corporate power and class struggle; immense demographic upheavals, severing millions of people from their ancestral habitats, hurtling them halfway across the world into new lives; rapid and often cataclysmic urban growth; systems of mass communication, dynamic in their development, enveloping and binding together the most diverse people and societies; increasingly powerful national states, bureaucratically structured and operated, constantly striving to expand their powers; mass social movements of people, and peoples, challenging their political and economic rulers, striving to gain some control over their lives; finally, bearing and driving all these people and institutions along, an ever-expanding, drastically fluctuating capitalist world market.[1]

In describing modernity, Berman does not explicitly mention *secularization*, the shift away from religious other-worldliness towards a more this-worldly, materialistic view of reality, presumably because he finds it implicit in the characteristics of modernity which he describes. However, many contemporary historians emphasize secularization as the key to understanding modernity, and most recognize its importance as comparable to and derivative from such antecedents of modernity as the great geographical discoveries, the scientific revolutions, commercialization, industrialization, urbanization, the rise of individualism, and the spread of democratic political institutions. Secularization will be further explored later in this chapter.

Above all, Berman does stress the frenetic pace of scientific and technological development and associated historical change as unique to modernity and its worldview, modernism. He divides the history of modernity into three periods, from 1500-ca. 1800, 1790-1900, and the twentieth century. His first period includes the era of discovery and colonization, the Renaissance, the Reformation, the Scientific Revolution, and the Enlightenment. During this period the foundations of modernism are laid down to the point that the new worldview is articulated by some voices during the second half of the eighteenth century: "If there is one archetypal modern voice in the early phase of modernity, before the American and French Revolutions, it is the voice of Jean-Jacques Rousseau. Rousseau is the first to use the word *moderniste* in the ways in which the nineteenth and twentieth centuries will use it; and he is the source of some of our most vital modern traditions, from nostalgic reverie to psychoanalytic self-scrutiny to participatory democracy."[2]

Berman goes on to describe Rousseau's account, in *Julie* (1762), of frenetic and explosive pre-Revolutionary Paris as an urban whirlwind, *le tourbillon social*. The shocked response of Rousseau's hero, Saint-Preux, to the dizzying pace and variety of Parisian life is for Berman an archetype for the experience of the millions of people in the nineteenth and twentieth centuries destined to be displaced from the rural countryside into rapidly growing cities. Early modern Europeans would be subjected to even greater vicissitudes of whirlwind change during Berman's second period of modernity, the nineteenth century. A virtual blitzkrieg of industrial growth fueled by science-based technology would transform European and colonial landscapes and society relentlessly, subjecting them to "an ever-expanding world market embracing all, capable of the most spectacular growth, capable of appalling waste and devastation, capable of everything except solidity and stability."[3]

Berman finds the nineteenth century to be the heart of modernity, and *All That is Solid Melts Into Air* focuses on the complexity of nineteenth century modernism, particularly as expressed in the thought of Goethe, Marx, and Nietzsche, the father of postmodern rebellion against the modern age. The basically contradictory nature of modern life and its constantly self-revolutionizing tendency was, as Berman observes, most vigorously articulated by Marx: "At the same pace that mankind masters nature, man seems to become enslaved to other men or to his own infamy."[4] For both Marx and Nietzsche the long-term solution to the chaos of modern life and its destructiveness of traditional value systems would be the creation of a new kind of man, proletarian man or superman, capable of creating new values responsive to the conditions of life in the modern age.

Berman himself is also optimistic that a reformed humanity will adjust to the *tourbillon* of the twentieth and twenty-first centuries: "To be a modernist is to make oneself somehow at home in the maelstrom, to make its rhythms one's own, to move within its currents in search of the forms of reality, beauty, of freedom, of justice, that its fervid and perilous flow allows."[5] He sees the twentieth century, his third period of modernity, as having fulfilled the hopes of nineteenth century modernism beyond its wildest expectations, making it what "well may be the most brilliantly creative in the history of the world, not least because its creative energies have burst out in every part of the world."[6] However, recognizing the twentieth century also as an "Age of Extremes," as one historian has dubbed it,[7] Berman cautions contemporary modern man to study more carefully the contradictions of his age, lest uncritical acceptance of change as desirable for its own sake lead to continued barbarism and anomie in the twenty-first century and beyond. Berman does not attempt an anatomy of the modern Western worldview directly; rather, he presents us with the virtually dumbfounding complexity of modernity through examining its contradictions as expressed by the great, ambivalent geniuses of the nineteenth century, of whom Baudelaire was one of the greatest, along with Marx and Nietzsche.

In sharp contrast to Berman's literary approach to penetrating to the essence of modernity, several more focused attempts to discern the worldview of modernism are made by historians Crane Brinton, Richard Tarnas, and W. Warren Wagar.

Whereas Berman's chronology of modernism distinguished three periods, from 1500 to ca. 1800, the nineteenth century, and the twentieth century, Crane Brinton thought that the modern worldview was established by the beginning of the Enlightenment, although modified over the past two centuries. Brinton also recognized the period 1450-1700 as transitional, "essentially the years of preparation for the Enlightenment. In this transition humanism, Protestantism, and rationalism (and natural science) do their work of undermining the medieval, and preparing the modern, cosmology."[8] Although humanism and Protestantism were highly corrosive of medieval authority during the earlier years of 1450-1700, rationalism and natural science are recognized by Brinton as having been more powerful forces in the long run in undermining the medieval worldview and constructing a new cosmology.[9]

Above all, rationalism, "the belief that the universe works the way a man's mind works when he thinks logically and objectively", tended "to banish God and the supernatural from the universe."[10] Although scientific method and knowledge are closely connected to rationalism, the practice of science generally avoids asking the kind of questions that would construct a complete worldview, while rationalism has been free to raise and answer questions of value and metaphysical questions which would ultimately undermine and displace the rigid medieval cosmology. By the time of the Enlightenment, rationalism had provided alternative answers to the "Great Questions" regarding the meaning of life and reality to the extent that Christian belief was radically altered. Also, the growing prestige of science further set the stage for a new rationalist worldview after 1700.

Stating the new Enlightenment cosmology in an extreme form (as it might be understood by the *end* of the eighteenth century), Brinton describes it as "the belief that all human beings can attain here on this earth a state of perfection hitherto in the West thought to be possible only for Christians in a state of grace, and for them only after death."[11] Central to this belief was a new affirmation of Nature, the Nature of Newton, as orderly and essentially benign, indeed, as a normative model for the application of Reason to human affairs. "Reason will enable us to find human institutions, human relations that are 'natural'; once we find such institutions, we shall conform to them and be happy. Reason will clear up the mess that superstition, revelation, faith (the devils of the rationalists) have piled up here on earth."[12]

Brinton's claim that the worldview of modernity was born during the Enlightenment ("with the eighteenth century we are in many ways in modern times")[13] meant, more specifically, that the modern worldview emerged during the Enlightenment in three major stages to produce the essential presuppositions of European thought which have existed from the time of the American and French revolutions to the late twentieth century. During the early stage or generation, represented by Voltaire, Montesquieu, Pope, and the English deists, the *philosophes* generally were moderate humanists who still believed in restraint and the need to maintain social equilibrium, both of which were manifested in their deistic religious compromise. After 1750, however, most of the *philosophes* were radicals, which, in the realm of religion, led them from deism to the materialism and atheism exemplified in the thought of LaMettrie and Holbach. During this second generation, Brinton recognized Adam Smith and the Scottish *philosophes* in general as exceptions in being closer to the first generation in their moderation. Smith, for example, was no "doctrinaire believer in absolutely free economic competition; it was his followers who simplified his doctrines into 'rugged individualism.' "[14] The third generation of the Enlightenment is seen by Brinton as having adopted both rationalist-classical and sentimental-romantic elements prior to the American and French revolutions. Jefferson, Paine, Danton, and Robespierre exemplified late eighteenth century attitudes toward society and human relations, but hardly exhausted the range of attitudes passed on to the nineteenth century.

For Brinton the most important shift in the Western worldview which spread during the eighteenth and on into the nineteenth century was "the change from the Christian supernatural heaven after death to the rationalist natural heaven on this earth,"[15] in short, secularization. The doctrine of human progress would be the new secular faith leading towards long-term amelioration of the human condition on earth, and progress itself was based upon the spread of reason, which enables men to better control their environment. Here we see again the idea of human dominion over nature as an unmitigated good. Reason, applied to the understanding of nature by the great scientists from Copernicus to Newton, also produced by this time "the historic association of scientific and technological improvement with the idea of progress in the moral and cultural sense."[16] Also, by 1750 the practical implications of material improvements based upon reason were becoming clear even to the common man.

On the other hand, destruction of the traditional Christian cosmology as a side-effect of rationalism and scientific reasoning was not really understood or accepted by most Europeans by this time. Even the intellectuals of the day, led by the *philosophes*, were forced to deism as the typical compromise between traditional belief and the metaphysical implications of reason derived from relentless application of the new scientific method. The Enlightenment worldview also differed profoundly from that of the Middle Ages in its optimistic interpretation of human nature, which opened the way to viewing social change as progressive. If humanity was born free of original sin and instead the individual mind was essentially a *tabula rasa* (blank slate) upon which good laws could be inscribed by education and benign institutions, then the good life was readily achievable for all in the not too distant future.[17] If the *philosophes'* hopes that enlightened despots would carry out their program were dashed by oppression and censorship, then through revolution majority rule and democratic governments were soon to emerge from the theories of Locke and Rousseau to redeem their progressive vision.

Brinton's analysis of the origins of the modern worldview simultaneously pays homage to the insights of Carl Becker's *Heavenly City of the Eighteenth Century Philosophers* and anticipates the late twentieth century controversy over the meaning of the Enlightenment reflected in strongly differing interpretations of the meaning of progress (e.g. in the writing of Robert Nisbet, Hans Blumenberg and others, as discussed in Chapter Six of this book).[18] He ardently agrees with Becker "that the faith of the Enlightenment has an eschatology as definite as that of the Christians, a heaven that stands ahead as the goal of our earthly struggle."[19] However, the heavenly city of the *philosophes* lies in the future, a future not too distant in time in which material enjoyment and happiness would reign on earth. Observing that both Christianity and the faith of the Enlightenment share the belief "that man is no misfit in this world," Brinton also recognized their enormous differences. "If the faith of the Enlightenment is a kind of Christianity, a development out of Christianity, it is from the point of view of the historical Christianity of the Middle Ages a heresy, a distortion of Christianity; and from the point of view of Calvinism, a blasphemy."[20] Particularly in rejecting the Enlightenment doctrine of the natural goodness of man as the essential modern heresy, the conservative heirs of traditional Christianity have been left with the difficult task of encouraging self-denial and individual discipline in the behavior of the half-emancipated children of the Enlightenment whose worldviews remain stubbornly anchored in Christian religious belief systems.

Crane Brinton was one of the leading intellectual historians of the post-war era and, fortunately, he felt compelled to educate Americans concerning the origins of the modern worldview. In style and rigor, historian Franklin L. Baumer has been Brinton's most impressive successor in inculcating Enlightenment attitudes during the nineteen eighties and nineties. Baumer's major innovation in analyzing the origins of modernism was to approach the worldview concept from the perspective of what he called "the perennial questions," which is a more focused approach to what Crane Brinton called "the Great Questions".[21] I shall discuss Baumer's ideas in the analysis of nineteenth century evolutionary thought later on in this chapter.

In *The Passion of the Western Mind: Understanding the Ideas That Have Shaped Our World View* (1991), Richard Tarnas traced the classical and medieval worldviews to the period of ferment that gave rise to modernity from roots in the Renaissance, Reformation, and Scientific Revolution. In a chapter titled "Foundations of the Modern World View," Tarnas attempts to distill the essential elements of the modern worldview which emerged out of the three great historical movements of the sixteenth and seventeenth centuries. Tarnas sees the Enlightenment as the age when the modern worldview was established, with science as the most important underpinning of modernism. "During the two centuries following the Cartesian-Newtonian formulation, the modern mind continued to disengage itself from its medieval matrix. The writers and scholars of the Enlightenment...philosophically elaborated, broadly disseminated, and culturally established the new world view."[22] Regarding the importance of the Scientific Revolution, Tarnas writes that, in addition to offering the possibility of epistemological certainty and objective agreement, science suddenly stood forth as mankind's liberation" and "presented itself as the saving grace of the modern mind."[23]

In Tarnas' summary of the major tenets of the modern worldview, the influence of science is ubiquitous. In contrast to the Christian cosmos of the Middle Ages, the modern universe is perceived of being governed by comprehensible and quantifiable natural laws. Christian dualism between spirit and matter has been transformed gradually into a mind-matter dualism. "Science replaced religion as preeminent intellectual authority, as definer, judge, and guardian of the cultural worldview."[24] The modern mind perceived order in nature only through empirical studies which empowered humanity to control and dominate nature. Aspects of human nature other than rationality and empiricism were devalued because they failed to contribute to the modern scientific understanding of nature. The new heliocentric and mechanistic paradigms in science removed spiritual significance from the modern universe. By the nineteenth century, evolutionary theory brought even human nature and human origins into the arena of scientific explanation: "Darwinian evolution presented a continuation, a seemingly final vindication, of the intellectual impulse established in the Scientific Revolution."[25] Moreover, the spatial dimension established by Newtonian theory was both complemented and challenged by the temporal dimension introduced by Darwinian evolutionary theory. That Darwin banished Newton's God is clear, as Tarnas writes: "The early modern concept of an impersonal deistic Creator who had initiated and then left to itself a fully formed and eternally ordered world, the last cosmological compromise between Judeo-Christian revelation and modern science, now receded in the face of an evolutionary theory that provided a dynamic naturalistic explanation for the origin of species and all other natural phenomena."[26]

The final tenet of the modern worldview, according to Tarnas, is modern man's increased independence from both religion and nature: "the modern goal was to create the greatest possible freedom for man, from nature; from oppressive political, social, or economic structures; from restrictive metaphysical or religious beliefs; from the Church; from the Judeo-Christian God; from the static and finite Aristotelian-Christian cosmos; from medieval Scholasticism; from the ancient Greek authorities; from all primitive conceptions of the world."[27]

Abandoning all tradition, modern man might now create a brave new world of his own making through application of his recently acquired methods of systematic rationality and empiricism. Henceforth, an idea of positive, self-induced change, of progress, would dominate humanity's perception of both the past and the future, and traditional values would gradually be subsumed into a science-based technological society. The demise of traditional values and institutions has, of course, not conformed to the optimistic vision of the future held by some of the *philosophes* and their nineteenth century followers, and, throughout the nineteenth and twentieth centuries, with increasing intensity, the wisdom of abandoning tradition for "progress" has been seriously questioned.

Tarnas' emphasis upon science as essential to modernity is complemented by Warren W. Wagar's argument that secularization is at the heart of modernity. The cultural world of our Western traditions, of the Mediterranean basin and Western Europe, has seen myriad shifts in worldview over the millennia, but two transformations of worldview or metanoia stand out like bold peaks above surrounding mountains. The shift from the paganism and secularism of antiquity to the otherworldliness of late antiquity and the Middle Ages is one peak (discussed in detail in Chapters Three and Four), and the metanoia associated with the transformation of the medieval worldview into the secular worldview of modern Europe and its colonies is the other (discussed in Chapters Five, Six, and Seven). Gilbert Murray set the tone for understanding the first profound metanoia in *The Five Stages of Greek Religion*, and in more recent works Franklin Baumer, Warren Wagar, and others have documented the shift to modernism. Common to all of these interpretations of the medieval and modern metanoia is the premise that one of the dominant themes, if not *the* dominant theme of Western history is the conflict between faith and unbelief or skepticism.

Compared with the civilization of the Middle Ages, "modern" civilization is distinctly secular, materialistic, and this-worldly, although the earlier Christian worldview is often encased within the secular attitudes held by individuals living in the modern age. The consensus of European historians appears to be that prior to the eighteenth century and its Age of Enlightenment, thinking Western Europeans could still comfortably maintain the core beliefs of the Christian cosmology of the Middle Ages. However, during and after the intellectual controversies of the Enlightenment, continually fueled by advances in scientific knowledge, the medieval worldview became increasingly untenable to rational, open-minded individuals. More and more the secular equivalents of religious belief, ideologies or cryptoreligions (secular belief systems) such as humanism, scientism, and nationalism, attracted the interest of knowledgeable Europeans. Often enough, cryptoreligious beliefs were held simultaneously with watered down or mollified versions of Christianity. Whereas only an intellectually elite minority abandoned the old cosmology during the Enlightenment, in the nineteenth century a more general transition to secular attitudes occurred. As Wagar writes: "Martin E. Marty prefers to speak of the 'Modern Schism,' a time between 1830 and 1870 when the Western nations went over the hump of transition to a secular way of life and belief that demanded not 'the disappearance of religion so much as its relocation.' Marty refers to the displacement of traditional Christian or Jewish faith by systems of belief grounded in the values of this world, for example, nationalism.[28]

Wagar analyzes a major semantic problem in attempting to explain secularization, the problem of disagreement among scholars regarding the meaning of religion itself. He distinguishes between those who understand religion in a broad sense that includes non-transcendental belief systems, i.e., cryptoreligions, such as nationalism and scientism, and those who insist that religion must include supernaturalism, meaning "acceptance of a level of reality beyond the observable world known to science, to which are ascribed meanings and purposes completing and transcending those of the purely human realm."[29] Concluding that the mainstream of Christianity accepts supernaturalism as its hallmark, Wagar states that secularization is best understood as "the process by which the supernatural has been lost, an irrecoverable loss despite all the sophistries of so-called secular theologians, but one that man can learn to accept and even turn to his advantage, as Camus does in his postreligious masterpiece *The Myth of Sisyphus*."[30]

Wagar also emphasizes the importance of understanding secularization in the historical context of the intellectual battles "between those who see the rationalism of the Enlightenment or the idea of progress as essentially 'modern' and those who see them as reworkings of medieval beliefs, rooted in the Christian worldview."[31] This controversy over what modernity owes to its medieval cultural antecedents runs through Carl Becker's *Heavenly City of the Eighteenth Century Philosophers*, Karl Löwith's *Meaning in History*, and Hans Blumenberg's *The Legitimacy of the Modern Age*.

Whereas many historians have noted the beginnings of secularization in the sixteenth century, others have emphasized the secularizing influence of the Scientific Revolution and the Enlightenment. Acknowledging these antecedents in the experience of occasional individuals during the Renaissance and among some of the *philosophes*, Wagar stresses the importance of secularization of society at large, when the worldview of large numbers of people, if not the majority, was changed. Viewing secularization as but one facet of the modernization which occurred in the nineteenth century, displacing the pre-modern traditional rural economy and its antiquated quasi-feudal institutions and ubiquitous Christianity, Wagar writes:

> By contrast in a modern society most people live in towns and cities. The economy is predominantly industrial and commercial, the system of "social relations of production" is capitalism or state capitalism, the ruling classes are the masters of capital and their allies throughout the society, the national state with its quasi-democratic institutions of government has replaced the feudal-monarchical politics of the past, and the non-Western peoples have been fully incorporated into the Western world economy and state system. The majority of citizens are not practicing or believing Christians. The links between church and state have been mostly or entirely severed....[32]

The institutions described above can only be understood as organically related and as leading to the diminution of Christianity along with traditional, rural culture. However, the United States does not conform to the pattern established in modern Europe, perhaps because in America religion has joined, rather than opposed, secular trends. The ecclesiastical pluralism which has characterized America from its beginnings has allowed the attachment of religion to social causes as varied as ethnic identity, racism, imperialism, and entrepreneurial capitalism. Perhaps, given the renascence of the many predominantly Protestant Christian sects that spread to the regions that became the United States and Canada, the clock was reset for fundamentalist Christianity in the seventeenth and eighteenth centuries, followed by later revivals down to the present associated with a greater need for Christianity as a unifying force on the western frontier. Thus, America has to a considerable degree avoided the fate of secularization and cultural homage to the Enlightenment which characterizes the core, if not the periphery of contemporary European civilization. Of course, there are many liberal, scientifically informed American Protestants, but the rise of evangelical Christianity in recent decades has led to numerous challenges to acceptance of evolutionary theory and birth control, especially in conservative states such as Kansas, but also in Ohio, Michigan, and Pennsylvania.

The reader may well reflect at this point, what is the significance of the foregoing cultural analysis of modernity for environmental history and the contemporary environmental crisis? The answer, in a nutshell, is that the most ecologically destructive force in the modern world, the Industrial Revolution (and its offspring, the Technological Revolution), would not have occurred except for the antecedent changes in population, agricultural practices, political and economic ideas, and a broad transformation of the Western worldview during the period ca. 1500-1800. The shift in attitudes towards the relationships between man, nature, God, society, and history reflected the enormous changes which had taken place since the discovery of America and created the material and intellectual infrastructure favorable to the Industrial Revolution in England.

The Industrial Revolution

Modernity, the culture of Western European civilization and its colonies during the nineteenth and twentieth centuries, consists, as I have just elaborated, in far more than industrialization. At the same time, historical change in the West after 1500 introduced such a complex of new ideas and institutions that increasingly rapid historical change became inevitable. By the middle to late eighteenth century both the theoretical and technical foundations of the Industrial Revolution had been constructed, from Adam Smith's *Wealth of Nations* to the scientific and technological apparatus described in the *Encyclopedia* of Diderot and d'Alembert. It was by the second half of the eighteenth century largely a question of which modern nation-state would be capable of creating the prerequisites of an industrial revolution, rather than whether such a development would likely occur. The early modern centuries which witnessed the Reformation, Renaissance, Scientific Revolution, Great Discoveries, colonialism and imperialism, and the intellectual Enlightenment of the late seventeenth and eighteenth centuries gave birth to a modified Western worldview by the late eighteenth century, all the while witnessing the gradual manipulation and domination of nature as Europe and parts of North America and other colonial lands were transformed from native ecosystems and indigenous cultures into "civilized" landscapes, the new "Edens" of Western civilization. Secular thought and innovative ideas had altered Western consciousness irrevocably prior to the American and French Revolutions, which were contemporaneous with the beginnings of the Industrial Revolution. Radical as the changes from 1500-1800 were, they were comparatively glacial in their pace in comparison to the mad whirl, the "tourbillon" of the nineteenth and twentieth centuries, for the reason that once the Industrial and Technological revolutions began, the pace of change increased almost exponentially over the course of the nineteenth and twentieth centuries, with no respite in sight for citizens of the twenty-first century.

William McNeill, the leading "diffusionist" historian[33] in the United States (and probably in the world), wrote in *The Rise of the West* (1962) that the Industrial Revolution of the late eighteenth and nineteenth centuries "constituted a mutation in the economic and social life of mankind comparable in magnitude with the Neolithic transition from predation to agriculture and husbandry."[34] McNeill makes this judgment on the basis of the fact that both the Neolithic and Industrial revolutions resulted in the multiplication of human food and power resources many times over, in each case allowing radical increases in human numbers. He also explains the major reasons why the Industrial Revolution began in the British Isles rather than elsewhere: "Certainly Britain was favored by easy availability of beds of coal and iron ore, a labor force that could be constrained to accept new routines of work, and a class of innovators and entrepreneurs willing to develop new ideas and able to acquire the money or credit needed to embody them in new machines and procedures."[35] Decades later, in *The Human Web* (2003) McNeill and his son, J.R. McNeill, reaffirm the view that the Industrial Revolution was the single most important development in the eighteenth and nineteenth centuries by virtue of allowing humankind to increase its numbers, food supply, economic output, and global trade and communication.

Crucial to the Industrial Revolution was the shift to coal as England's forests disappeared. In further explaining why the Industrial Revolution happened first in England, the McNeills state that, in addition to English coal and iron, "tightening of the web both within Britain (roads, canals, railways, postal service) and worldwide (overseas trade and colonies, and population growth)"[36] created the necessary conditions for industrialization. The increased freedom and incentive to innovate under these new conditions produced several clusters of major innovations, categorized by the McNeills as follows: (1) 1780-1830, involving the textile and iron industries, and including such innovations as the flying shuttle, spinning jenny and the power loom in cotton production, and coking furnaces in the iron industry; (2) ca. 1820-1870, concerned with iron, coal, and steam engines and the rise of joint-stock companies as a major boost to capitalism; and (3) ca. 1850-1920, which witnessed major innovations in the coal and steel industries, railways and telegraphs, chemicals, and electricity, along with the rise of giant firms which operated internationally and on an enormous scale. Also, after 1860 science programs in universities became increasingly involved in the furthering of industrialization.[37]

Within this broad context of the Industrial Revolution, recent research by environmental historians shows more precisely how the transition was made from a traditional agrarian economy to the unique capitalist economy engendered by the Industrial Revolution, and even more important, how a gradual shift in the political economy of England initiated a profound transformation of the worldview of Western Europe and its colonies. In *The Great Transformation: The Political and Economic origins of Our Time* (1944), Karl Polanyi wrote: "What we call land is an element of nature inextricably interwoven with man's institutions. To isolate it and form a market out of it was perhaps the weirdest of all undertakings of our ancestors."[38] Environmental historian Matthew Osborn examines this "weirdest of all undertakings" during the early Industrial Revolution in Oldham, England and surrounding locales. That landscapes or "the land" and the socioeconomic practices of traditional societies associated with it should be subordinated to the higher order of a market economy is the idea at the heart of his study. Osborn observes that "the example of Oldham clearly shows that market links and coal reserves are not enough to 'industrialize' a society. The third essential component is the control of the land by an entrepreneurial elite committed to this kind of society, economy, and relationship with the land."[39]

The earliest development leading towards the rise of a market economy in Oldham was the increase in the felling of timber during the late eighteenth century. Only one-eighth of England was forested by 1700, and a century later the high cost of transporting timber was drastically lowered by the spread of a turnpike network and system of canals which opened new areas of woodland to exploitation. Timber prices had also been driven up by the American Revolutionary War, and London had long since been forced to begin substituting coal for wood as fuel. Even by the 1780s trees along or near watercourses or on the steep hillsides of relatively treeless farms were sold off without replacement by new plantings. Simultaneous with the late eighteenth century clearing of trees there was a surge in the building of mills and tenements to exploit the labor of a growing population. "In the latter part of the eighteenth century, timber scarcity was rationalized in much the same way as pollution from coal smoke was in the mid-nineteenth century, as a sign of progress."[40] Furthermore, the extraction of timber, through its use of the new transportation infrastructure, helped to set the stage for a new conception of the nature of land, even though it was still a part of the old pre-industrial society. "The completion of the revolutionary change to a system of market capitalism would be finalized through a much more intensive and long term endeavor, the mining of coal."[41]

Osborn concurs with other historians of the Industrial Revolution that the basis of sustained industrial growth in the late eighteenth and early nineteenth centuries was the substitution of coal for wood as a source of fuel. In the Oldham region southeast of Manchester expansion of the coal industry damaged and transformed the pastoral landscape, laid the foundations for a new elite's wealth and control of rural lands, and provided an impetus to the building of canals and turnpikes. Some early uses of coal in Lancashire were salt-making, drying malt, and the manufacture of dye-stuffs, lime, bricks, and glass. By the 1770s coal was widely used for heating homes in Oldham. The coal mines themselves also created increasing demands for specialized laborers to provide the coal miners with tools, baskets, candles, timber, and even masonry to wall off abandoned workings. During the late eighteenth century partnerships of men with larger financial resources replaced local owners and groups of workers, gradually consolidating mine ownership into fewer hands and displacing the local gentry from their lands. "Once the land was under the control of these men there was no need for leases, no need to mitigate mining damage, and the land was given over fully to this industry. These men were much less interested in the long-term viability of an estate, or agriculture, or the old social structures than were their predecessors."[42]

Osborn refers to the growing system of canals and turnpikes as the "conduits of commerce" without which industrialization would have been impossible. Improved transportation not only linked previously self-contained ecological and social systems to regional urban centers such as Manchester, but to colonial and other distant markets as well. The heyday of turnpike construction and reconstruction for Oldham was ca.1750-1825, after which this part of the Pennine Hills found its niche in the emerging industrial order in which different regions provided specialized resources and services to a complex network centered in the growing cities. Osborn recognizes, perhaps more strongly, the subtle and gradual transformation of the older agrarian worldview through the role of the canals (in expediting the growing trade in coal, wool, timber, corn, lime, and other goods), which he sees as "potent symbols of a much less accommodating view of the environment in that instead of working with or around the Pennines, they sought to change and bend the hills to the requirements of *their* economic necessities."[43]

The existence of coal in the Oldham area was crucial to its industrialization. The Parliamentary Act of Enclosure of 1802-1803 transferred ownership of common lands to nearby land owners and helped to turn farm laborers and herders into wage earners employed in the mines and manufacturing. Crucial to the shift towards an increasingly centralized market economy was the rise of a landowning elite not dependent upon the land, but allied to Manchester and its extensive market connections. "The newly established market linkages facilitated this group's increasing wealth and power, resulting in the concentration of land and capital under their control, and a 'great transformation' in the whole community's relationship with the environment."[44]

While coal, iron ore, the proximity of turnpikes and canals, and other local conditions determined the pace of industrialization of different rural areas, the city of London was at the center of the Industrial Revolution throughout the nineteenth century, although later in the century competition arose in continental Europe and North America. The Industrial Revolution developed during the period ca. 1770-1830, and gathered force during the Victorian era (1837-1901), when industries located in London, Manchester, and other centers of activity increasingly used machines powered by steam engines. Whereas the eighteenth century had depended on wood and charcoal, the machines of the Victorian Age were powered by coal and gas, taking some pressure off the dwindling forests. The use of coal turned London into a center of air and water pollution, which increased as the population grew from about one million in 1801 to 2.3 million in 1854 and 6.6 million in 1901.[45] The smog and dust of industry permeated the greater London area, killing lichens and other organisms and selecting for darker forms of butterflies and moths.[46] Liquid and solid wastes were dumped into the Thames, only to be swept back by the tides entering its estuary, killing virtually all fish and sea mammals in the vicinity, and creating a vast area of polluted, foul-smelling mudflats at low tide. Related pollution of drinking water caused thousands of deaths during the mid-nineteenth century.[47]

J. Donald Hughes has shown that the ecological fate of the English countryside was not much better than that of the London megalopolis. At the beginning of the nineteenth century England had proportionately less woodland than any other European country. The royal forests had been depleted for ships' timber and fuel, and disafforestation (change in the legal status of a royal forest to ordinary land) also contributed to deforestation and the loss of deer and other wildlife as well.[48] Loss of habitat and wildlife also resulted from the removal of hedgerows as small parcels of land were consolidated. We saw in Chapter Seven that most of the large mammals of the British Isles had been destroyed during the late Middle Ages and the sixteenth and seventeenth centuries, leaving only small mammals, birds, and enclosed deer herds by the Age of Enlightenment. This remnant fauna was nevertheless seriously depleted in the nineteenth century by commercial hunters who brought their bounty to the cities, especially London, and by the upper classes, who hunted their estates for birds and foxes. Demographically, England changed from three-quarters of the population living in the countryside in 1800, to a similar proportion living in the cities by 1900. Within only a century a largely rural country had been transformed into a nation of industrial cities. Simultaneously, the United Kingdom had become dependent upon foreign imports for almost half of its food. "In the process of industrialization, the UK began to draw raw materials from ecosystems abroad, subjecting them to monoculture, simplification, and deterioration. This was part of the reason for the tenacity with which the UK defended and extended its empire in the Victorian Age."[49]

Clive Ponting documents the English propensity for destroying animals considered to be pests or having some appeal to the human appetite. Traditionally, since the Middle Ages, large hunts were carried out to try to exterminate various animals, such as foxes, polecats, weasels, stoats, otters, hedgehogs, rats, mice, moles, hawks, buzzards, ospreys, jays, ravens and kingfishers. The nineteenth century witnessed the continuation of these practices. The range of birds eaten then was much wider than today, and included curlews, plovers, blackbirds, larks, thrushes and gannets, as well as the eggs of wild birds, and others such as goldfinches were captured in large numbers to be kept in cages.[50] However, the greatest ecological damage inflicted by the British and other European empires during the Victorian Age occurred in the lands colonized by these nations, and that destructiveness was gradually intensified by the application of new industrial techniques and machinery in the colonies. As Sing C. Chew writes: "the geographic limits of ecological decay now became global due to the political-economic empires constructed by Europeans that spanned the world at the dawn of the nineteenth century."[51]

Chew focuses on global deforestation as an index of ecological devastation in the nineteenth century. In addition to depletion of resources such as timber and minerals, the clearing of land for cash-crop plantations was a major factor in the process of deforestation. Despite the difficulties involved in estimating forest loss on a global basis, Chew surmises that global deforestation between 1700 and 1850 amounted to about 128 million hectares, chiefly the result of clearing for crop land. Europe, which had seen a doubling of its population to 145 million between 1600 and 1800, was only about 5-10 percent forested by 1800. England's dependency upon wood from northeastern Europe, the Baltic countries, and Russia shifted to North America in response to Napoleon's Continental Blockade in 1807, as a result of which heavy cutting of forests in New England, New York, and southeastern Canada ensued. European immigration to the United States during the first sixty years of the nineteenth century boosted land clearing in the Midwestern states, 120 million hectares having been cut during the entire century. Also, by 1856, over 80 percent of 560 iron furnaces operating in the northeast were still using charcoal for fuel, putting even more pressure on diminishing forests.[52]

We saw in Osborn's study of the early Industrial Revolution in Oldham that the transition from the use of charcoal to the use of coal was gradual. Ponting documents the world output of coal in 1800 at about 15 million tons, reaching 132 million tons by 1860 and over 700 million tons by 1900. In heavily industrialized Britain, coal production increased from 10 million tons in 1800 to 60 million tons in 1850. By 1900 95 percent of the world's energy consumption came from coal. In the United States, however, industrialization followed a different course because of the greater availability of wood and water power. Wood still accounted for 90 percent of United States fuel supplies in 1850, and steam power was not widely adopted for industrial use until the 1880s, when coal finally became the principal source of energy. As in Europe, this did not happen until easily available sources of energy such as wood and water power had been exhausted or outstripped by increasing demand for energy.

Environmental historian John Opie summarizes the history of industrialization in the United States and how it differed from that of England in *Nature's Nation: An Environmental History of the United States* (1998).[53] "The factory system found its most favorable climate in America. On American soil, economic advancement was understood to be a personal duty, part of a work ethic that goes back to Puritan rigorousness. A middle-class dogma stated that the virtues of hard work and self-discipline in a factory system would reap enormous rewards. Americans made the consumption of goods one of their life's goals."[54] Opie leaves no doubt that there has been an American predilection for turning nature, it's rocks, soils, waters, and biota, into utilitarian goods or resources from early on in its history. We have witnessed how, during and after the Colonial period, the British demand for timber followed the precedent for treating nature as resources on a large scale which was set by the early settlers and their planting of tobacco and other crops for export. Joseph Petulla, one of the leading scholars focusing on resource development in American environmental history, tells the story of Colonial resource extraction during the mercantilist, pre-Industrial era of the American colonies, when American attitudes toward nature were overwhelmingly exploitive, carrying to the new continent the detached utilitarian values characteristic of Western Europeans since the High Middle Ages and the European frontier.[55] Thus, by the beginning of the nineteenth century, when Thomas Jefferson was preparing to send an exploration party into the newly acquired Louisiana Territory, Americans were mentally prepared to develop the resources of western territories much the same way that they had converted the Atlantic frontier from a frightening and howling wilderness into New England's new pastoral Edens reminiscent of the farmlands of Europe (along with the slave-based plantation culture which had grown up in the South in response to mercantilist demands for tobacco, indigo, rice, and cotton).

In the early decades of nineteenth century America a limited industrialism produced cheap cotton goods, milled flour, small arms, and hand tools like axes. Coal and iron would not dominate industrialization until after the Civil War. However, as farmers migrated westward into the rich soils of the Midwest, the demand for iron and steel plows, seed drills, reapers, and harvesters boosted domestic production after the War of 1812. Only after 1830 did coal begin to replace charcoal in smelting iron. The appearance of steam locomotives in the 1820s greatly increased the demand for both coal and iron, and coal production jumped from 50,000 tons in 1820 to 14 million tons by 1860.[56] Another crucial factor contributing to industrialization was the immigration of more than 16 million Europeans between 1840 and 1900. Opie observes that it was not possible for these newcomers to foment a Marxist proletarian revolution because American "blue-collar workers believed, through hard work and pluck, they would attain middle-class prosperity."[57] Fortunately for the immigrants, American industry needed an inexpensive labor force.

At the heart of Opie's account of industrialization in America is his emphasis upon the rise of the corporation and the role of government in facilitating its concentration of economic wealth and power in a relatively few hands. "By the end of the nineteenth century, some Americans began to conclude that the nation went through a transition from a nation of independent property holders to a troubling dependence on a few big private corporations."[58] This process of radical change had created an impressive superstructure of new transportation and communication technologies integrated with financial and management institutions, the work of those aggressive nineteenth century entrepreneurs appropriately christened "the robber barons." Down to the present day many Americans have looked favorably on the robber barons as heroic captains of industry who championed the extension of the American right of free enterprise in a laissez-faire socioeconomic state, with the business corporation as its most praiseworthy institution. As Opie writes:

> As large enterprises, corporations function better than the individual small business owner because they can better amass great sums of capital to spend over long periods of time that modern industrial operations require. ...Directors might come and go, shareholders might die and sell out, but the corporation could go on raising millions of dollars in capital, building new factories and shutting down obsolete operations, hiring and firing generations of workers. The corporation became the enduring foundation for industrialization, a force stronger than natural resources or a labor force, with the capacity to survive the swings of business cycles, changing presidential terms, and even warfare.[59]

How well would this miraculous new and distinctly modern institution have succeeded without the assistance of the state, of American government? Although conceding that corporations received privileged status in the United States, Opie denies that the Constitution was a "capitalist tool." Subsequently, however, legislation favored business from early on, beginning with charters to corporations to build bridges, canals, and turnpikes in the late eighteenth century, and exploding in a spate of pro-business legislation during the Civil War. Economic historians of the left and right, Robert Heilbroner and Kevin Phillips, concur that, in fact, the state played a major role in the rise of corporate power. Heilbroner asserts that it is a profound mistake to conceive of capitalism as being fundamentally a "private" economic system, and that without the protecting, socializing and stimulating activities of the state to further the regime of capital, the normal operation of the market system would be unsustainable. "Remove the regime of capital and the state would remain, although it might change dramatically; remove the state and the regime of capital would not last a day. In this sense politics is prior to economics in that domination must precede exploitation....All this mocks the conventional economic view that the public realm is somehow secondary or ancillary to the private realm...."[60] Kevin Phillips factually supports Heilbroner's generalization with examples of corporate influence over the federal judiciary and domination of state legislatures and the U.S. Senate during the Civil War era, when railroads and other giant corporations "were big enough to methodically take over and dominate state legislatures, thereby taking control of the U.S. Senate and much of the federal and state judiciary."[61] Phillips is convinced that corporate growth and momentum after the Civil War was not the result of laissez-faire, but of pressure on the government to change statutes or create new ones favorable to the interests of big business. Phillips avers that from the late 1860s to the early 1900s the parallel ascent of the U.S. Senate and the corporations was the result of a symbiotic relationship between the two. One contemporary reformist commented that people wanted men in office who did not steal, but who would not interfere with those in private enterprise who do.[62]

The nineteenth century inevitably was one in which business interests engaged in rapid development of resources gained substantial control over government at the national, state and local levels. It was due to the seemingly heroic and promethean conquest of nature that resource exploitation was regarded by the general public as progressive. In the Age of Progress the small chorus of dissent in defense of nature coming from the Wordsworths of England and the Thoreaus of America was drowned out by the sheer abundance of goods and increased standards of living stemming from the Industrial Revolution.

In this context, the growing realization of the importance of corporations to industrial growth led to their being treated as individuals under the Fourteenth Amendment, protecting them against invasion of their rights to private property. "Not surprisingly, the major corporations exercised this unusual independence to establish monopolies of basic railroad, steel, petroleum, and coal industries through pools, trusts, and holding companies."[63] Subsequently, the majority of environmental impacts upon the land, from lumbering and mining to agribusiness and recreational development, came from the activities of corporations. Ecologically destructive corporate practices have continued to the present day under cover of the popular myth that laissez-faire capitalism had emerged from a Darwinian struggle for survival, when in fact the "free market" had been kept open by manipulation of governmental policies in favor of corporate growth and wealth. Only in the twentieth century did government begin to regulate industrial corporations with moderate rigor. "Through most of American history no human, social, ethical, or environmental needs seemed capable of overriding corporate self-interest and profit making. When environmental regulations took hold in the 1970s, they were an abrupt turnaround away from the historic premise that corporate profitability was the primary means to serve the public good."[64]

Once the modern industrial capitalist system was fully established during the middle and late nineteenth century in Western Europe and North America, with Russia and Japan not far behind, extraction of natural resources with all the attendant environmental consequences pursued a steady course of growth on down to the present day. However, the historically inordinate physical transformation of the globe during the past two centuries was not without enormous changes in the social and intellectual life of Western civilization as well. Capitalist ideology and practice were rooted in Calvinism, the Renaissance, and the general characteristics of modernity described earlier in this chapter. Once established and rationalized by Adam Smith, David Ricardo, Thomas Malthus, and others, the brave new laissez-faire world of industrial capitalism engulfed and transformed the traditional Christian belief system within a modified Western worldview. The institutionalization of the Scientific and Industrial revolutions as the dominant foundations of modern ideology gave rise to a schizophrenic perception of reality allowing for adherence to traditional religion and associated values simultaneously with the new mechanistic interpretation of nature as a font of resources conveniently available for human consumption through the techniques of science and industrial technology. The modern ideology of economic growth also claimed to represent an idea of progress built upon the ruins of the world's diverse ecosystems and species diversity. Since the nineteenth century the West has viewed the world through the lens of a strong dualism. Economic activity and religious beliefs each allow for highly individualistic behavior which is often contradictory. The accepted rule of modern society is that the two are basically unrelated, although capitalism owes something to the Protestant ethic as Max Weber has demonstrated. Undoubtedly, this convenient compartmentalization of ideas owes much to the dualism of René Descartes, who gave scientists philosophical permission to explore and exploit the natural world as a spiritless mechanism set in motion by a deity transcendentally isolated from his creation. Whatever one's beliefs concerning God's relation to nature, religion, once the keystone of Western culture, has been displaced by economic expediency as the primary arbiter of social behavior.[65]

Thus, economic and technological forces have been the primary cause of change in the Western cultural consciousness. The modern Industrial Revolution has mechanized the production of food, clothing and transportation in the centralized factory, but underlying these physical changes and mass production and consumption there have been shifts in our habits of thought and perception which facilitate our ability to rationalize the capitalist system of consumerism as a superior mode of civilization. "It is hard to exaggerate how far industrialism has gone in breaking down all the old notions of stability, community, and order. Our entire world-view has been transformed profoundly by this force. It has, among other consequences, led us to think that it is necessary and acceptable to ravage the landscape in the pursuit of maximum economic production."[66] In Osborn's study of the early Industrial Revolution in England we could see this shift in attitude occurring, soon to spread to the United States. The modernity described by Berman at the beginning of this chapter has been dubbed by Worster "the culture of industrial capitalism." They are one and the same in that constant innovation in science, technology, and industry play out as relentless change and adaptation in our social behavior. We have come to accept the chaos, the *tourbillon* of modern existence as *normal* to the point that we fear the notion of restoring or maintaining order in nature. "This may be the greatest revolution in outlook that has ever taken place. Traditional societies tended to see and value the order in nature; we of the modern industrial era have tended to deny it. And therein lies the deepest source of our contemporary environmental destructiveness."[67]

Some examples of modern industrial environmental destructiveness have been documented in this chapter, and pre-modern destructive practices have been described in some detail in Chapters Two, Four, Six, and Seven. What needs to be emphasized here is the distinctively greater pace and breadth of modern ecological devastation. The Spenglerian cycle of civilization of about a thousand years spanning the rise and fall of a culture was indicative of how long it took most civilizations to destroy the ecosystems and resources of the region in which a culture developed, flourished, declined, and collapsed. Modern Western civilization has simultaneously expanded the geographic parameters of ecological despoliation while increasing the rate of ecological transformation exponentially.

J.R. McNeil and William McNeill argue that the spread of trade, investment, and migration that was in place by 1800 created a tightening global web with devastating ecological consequences as a growing human population demanded more food and fiber from the world's farms and plantations. The area under crops roughly tripled between 1750 and 1910, as did pastureland. During the same period the world's forests shrank by approximately 10 percent, chiefly in North America. Wildlife habitat declined drastically during this period, and massive irrigation projects were undertaken in India, Russia, China, and the United States to support the surge in agricultural production. The "tightening web" of globalization also spread exotic organisms, plants, animals, and disease organisms, from continent to continent, facilitated by the machines of the new industrial age which powered steamships and railroads.[68]

The Industrial Revolution of the late eighteenth and early nineteenth centuries was in part the offspring of the scientific revolutions in astronomy and physics (and to a lesser degree, biology and chemistry) during the sixteenth, seventeenth, and eighteenth centuries. The spirit of inquisitiveness and inventiveness cultivated by the early scientists must have inspired mechanics and tinkerers, more practical men of high intelligence such as John Kay, inventor of the flying shuttle (1733), and Richard Arkwright, who invented and patented the water frame (1769) and introduced the steam engine in the 1780s to drive his spinning machinery, nearly a century after Thomas Newcomen built a steam engine to drive pumps in the coal mines in 1702.

Darwin and the Rise of Evolutionary Biology

Even before the rise of ancient civilization humans had speculated over the religious and philosophical implications of the fourth dimension, time. Heraclitus's famous utterance, "Whirl is king, having deposed Zeus," is suggestive of a developing cognizance of historical change on the part of the Ionian Greek philosophers. As we saw in Chapter Two, Plato was well aware of significant changes in nature during the centuries that preceded him. However, to his mind, changes in the natural and civilized world were merely evanescent shadows reflecting the timeless realm of ideal types or forms which existed in an ideal transcendental realm beyond the ordinary realm of human perception. Aristotle, the greatest natural historian of antiquity, rejected Plato's transcendentalism and its universal types for his own secular typology based upon observation of the properties and behavior of material things. He also rejected Plato's creationism, his conception of the entire universe materializing out of nothing. Nevertheless, as a student of Plato, Aristotle was destined to see nature as fixed or static in regard to individual species or types. Arranging animals and plants in terms of increasing complexity, they seemed to comprise a great chain or ladder of organic beings, in which those lower down were useful, especially as food, to those higher up, thereby placing humankind at the top of the ladder. Aristotle also believed that the universe was eternal, and that organic species were fixed in nature. "Through the constancy of the species, indeed, each individual could share something of the perfection and immortality with which those celestial divinities, the stars, were blessed."[69] Aristotle also replaced Plato's creation hypothesis with a cyclical philosophy of history in which the Sun, Moon, and planets periodically returned to the same relative positions, causing historical repetition of history under influence of the cosmic cycle (not unlike Plato's cyclical hypothesis explained in Chapter One, except that Plato's 72,000 year cycles [see Chapter Thirteen] fall within the creation and duration of the cosmos). Lucretius (ca. 94-51 B.C.), the leading Epicurean philosopher, attempted to explain the natural world in terms of the atomic theory (actually a hypothesis) developed by Democritus and Leucippus. This early atomic theory became the foundation of the Epicurean philosophical system, established with the purpose of freeing the classical mind from superstition and the fear of death. The fifth chapter of Lucretius's *On Nature* explains the origins of the world as well as its laws. After explaining the origins of the cosmos and the earth, Lucretius turns to organic life and attempts to prove that it has not always existed, but rather, was formed not long ago by the earth herself. "Thus then the passage of time alters the nature of the whole world and the earth enters upon one state after another, so that it is not able to produce what it once did, and can produce what it once could not."[70] Lucretius was also aware of the phenomenon of extinction of species, but he believed that species were fixed in form and did not change to new species over time. Although he thought that there was a kind of natural selection, its function was not to change the species, "but simply to eliminate monsters and weaklings thrown up spontaneously at the beginning...."[71]

There was little progress in the classical world's attempt to understand nature after Lucretius. This was largely the result of the gradual change in worldview which occurred from the Hellenistic era (approximately 323 B.C. - 14 A.D., the centuries from the death of Alexander the Great to the Roman Empire under Octavius Augustus) through the Roman Empire and its decline and fall. This *metanoia* or change in worldview (See Chapter Three) shifted the interests of intellectuals from a focus on this-worldly things to preoccupation with a transcendental realm beyond this world. The otherworldliness of the mystery religions and eventually triumphant Christianity essentially terminated attempts to understand nature until the late Middle Ages in the west. When late medieval "pre-scientists" studied nature, it was to understand how God's creation works, considered as knowledge complementary to the absolute knowledge of revelation contained in holy scripture. Western scientific achievement was accomplished almost entirely within this sacred context through the early modern centuries, when appropriately bowdlerized Aristotelian science was taken as the accepted framework of truth within which students of organic nature functioned. Although seventeenth century scientific discoveries opened new vistas for physics, astronomy, and cosmology, study of the earth itself and the organisms living upon it was greatly constrained by the Aristotelian science which was approved by the Christian church. Continued acceptance of the biblical time scale and the adoption of mathematics as the foundation of physical thought together strengthened the conception of a fixed order of nature in the late seventeenth and early eighteenth centuries.[72]

Modern geology, the time-scale of which would be essential to Darwin's theory of evolution, had no articulated scientific paradigm until the late eighteenth century, when James Hutton(1726-1797) established his uniformitarian theory of the earth in 1788. With the publication of Charles Lyell's *Principles of Geology* in 1830, Darwin soon possessed the geological piece of the evolutionary puzzle, but it took the ground-breaking thought of earlier evolutionary theorists and the economic speculations of Thomas Malthus to provide him with the other major components of his theory. Even by the mid-eighteenth century, however, a powerfully suggestive, essentially evolutionary schema had been proposed by the prolific French scientist George Louis LeClerc, Comte de Buffon. The first three volumes of his *Natural History* were published in 1749, in which he went far beyond Lucretius in developing an all-encompassing explanation of nature, from the solar system to the history of the earth and its inhabitants (Buffon and the eighteenth century evolutionary theorists, known as transformists, are discussed in Chapter Seven). Buffon linked the cooling of the earth to the genesis of various organic "stocks" which degenerated to later forms. His theory had little in common with Darwin's and like those of Lord Monboddo and Erasmus Darwin, was founded on ideas derived from Empedocles and Lucretius.[73] What is most important is the fact that these pre-Darwinian evolutionary theorists contributed to a growing scientific climate of opinion that *somehow* the evolution of species had occurred, but how? The notion that species were in fact fixed rather than changing over time was still reinforced by the general belief in the original Creation. However, the dissonant voices of the eighteenth and early nineteenth century scientists kept the idea of organic evolution "in the air" until a truly scientific explanation of the process could be established. Even during the Middle Ages philosophers debated the possibility of organic transmutation (why not, when their alchemists were seeking the means to transform lead into gold), but they had no clear criteria for distinguishing one species from another. Subsequently, the rise of modern taxonomy led not to evolutionary speculation, but to a strongly held doctrine of absolute fixity epitomized by John Ray's *The Wisdom of God* in the 1690s.[74] Thus, despite the speculations of Buffon and others, an acceptable scientific theory of evolution was held at bay until the appearance of *The Origin of Species* in 1859.

Reading *The Voyage of the Beagle*, Darwin's account of his voyage on the HMS Beagle from December, 1831 to October, 1836, one is immediately impressed by the mix of natural historical (i.e., biological) and geological observations which he made in South America and other parts of the globe. Thanks to Darwin's careful study of the copy of Charles Lyell's *Principles of Geology* that he had brought along on his extensive voyage, he began to make interdisciplinary connections between living and fossilized species of organisms within a geological context. In fact, his sense of the complex interrelationships of organisms over time suggests his grasp of the ecosystem concept long before the word ecology was coined in the 1870s. What was of greatest significance for his evolutionary theory was his experience, especially in South America, of finding fossils of organisms similar to living species in the same region. For example, reflecting on the geological history of Patagonia, Darwin wrote:

> At Port St. Julian, in some red mud capping the gravel on the 90-feet plain, I found half the skeleton of the Macrauchenia Patachonica, a remarkable quadruped, full as large as a camel. It belongs to the same division of the Pachydermata with the

rhinoceras, tapir, and palaeotherium; but in the structure of the bones of its long neck it shows a clear relation to the camel, or rather to the guanaco and llama. From recent seashells being found on two of the higher step-formed plains, which must have been modelled and upraised before the mud was deposited in which the Macrauchenia was entombed, it is certain that this curious quadruped lived long after the sea was inhabited by its present shells.[75]

Having recognized that a guanaco-like herbivore had been fossilized on top of a recently uplifted wave-cut platform, Darwin went on to compare its eating habits, i.e., its trophic niche, with those of the guanaco, after which he made similar comparisons of extinct South American species with contemporary capybaras, sloths, anteaters, and armadillos. Noting that the extinct species are far more numerous than those now living, Darwin concluded: "This wonderful relationship in the same continent between the dead and the living will, I do not doubt, hereafter throw more light on the appearance of organic beings on our earth, and their disappearance from it, than any other class of facts."[76] Shortly thereafter he speculated as to whether the arrival of humanity might have caused the extinction of some of the larger South American species. Elsewhere in *The Voyage of the Beagle* Darwin speculated upon the role of land bridges at the Bering Strait and between North and South America in determining the distribution and extinctions of mammalian species, suggesting that the arrival of early humans might have been a major cause of extinctions in North America. In many ways Darwin's travel journal covers the same ground as contemporary environmental histories such as Alfred Crosby's *Ecological Imperialism*.

Darwin's intellectual drive further to understand the origin and historical transformation of organic life was given additional impetus by his observations in the Galápagos Archipelago (*Galapago* is a Spanish word for turtle) towards the end of his voyage on the Beagle. Two sets of biological variations, on species of finches and land tortoises on different islands, provided Darwin with powerful field evidence for his theory of natural selection developed over the years after his experiences gained from his travels on the Beagle. Regarding the finches, all species were peculiar to the Galápagos Archipelago, having in common the general structure of their beaks, short tails, body form, and plumage. Upon observing the gradation in the size of the beaks of different species, Darwin wrote: "Seeing this gradation and diversity of structure in one small, intimately related group of birds, one might really fancy that from an original paucity of birds in this archipelago, one species had been taken and modified for different ends."[77] The tortoise, *Testudo nigra*, inhabits most if not all of the islands of the archipelago, usually frequenting the damp parts of the islands at higher elevations, but also found in the lower, arid areas. They tend to drink large volumes of fresh water when available and to survive drought by retaining water in the bladder as a reservoir. Darwin was surprised to hear from the Vice-Governor of the Galápagos that the tortoises differed noticeably from one island to another, and went about proving it for himself. He concluded that finches and tortoises alike appeared to be isolated on the different islands by powerful currents of the sea running westerly.[78] Thus, geographic isolation appeared to be responsible for the varieties of finches and turtles.

During the century and a half since Darwin made his observations on the Galápagos, the islands have gradually accumulated a sizeable human population (about 16,000 today) and undergone extensive ecological degradation. J. Donald Hughes, visiting the "shrine" of Darwin's finches in 1996, attests to the undesirable changes which have occurred despite legislation to protect the ecology of the islands and designation as a UNESCO World Heritage Site and Biosphere Reserve. His own study of the ecology of the Galápagos Archipelago and of Darwin's later writings led Hughes to remark: "It would be incorrect to suggest that Darwin built his system of evolution only on the observations he made in the Galápagos. He spent much of the rest of his life observing and collecting information on the ways in which breeders of domestic species produce the amazing varieties of form one sees in pigeons, for example. But the Galápagos offered the crucial stimulus, a fact he often acknowledged."[79]

Nearly a quarter of a century was to pass between Darwin's return home from the Beagle in October, 1836 and the publication of *The Origin of Species* in 1859. However, two years after his return he read by chance the treatise on population by Thomas Malthus, originally published anonymously in 1798. The essence of Malthus's thesis is fundamentally simple, although Malthus develops sophisticated arguments to prove it. Acknowledging the great variety and abundance of organic life and its tendency to reproduce prolifically, Malthus points out the limited space and food supply available for its sustenance. Without some kind of control upon numbers of individual organisms, the earth quickly would be overrun, but this does not happen. This is so because population is generally constrained by various checks such as predation, disease, and severe climate. This "imperious all pervading law," necessity, keeps population in balance in the natural world. The reproductive powers of mankind, however, having evaded the natural checks on population through civilized means, have produced geometric or exponential rates of population growth which must eventually outstrip arithmetic increases in food production.[80] In the natural world reproductive fecundity causes massive die-offs of the newborn and young. In civilized humanity the consequences have been misery and vice.

Malthus's famous essay most definitely acted as a catalyst to Darwin's search for a mechanism that would explain the evolutionary processes he saw in action in South America and the Galápagos, but which he could not explain. Many years after having read Malthus, Darwin wrote, in his *Autobiography*:

> In October 1838, that is, fifteen months after I had begun my systematic inquiry, I happened to read for amusement Malthus on *Population*, and being well prepared to appreciate the struggle for

existence which everywhere goes on, from long-continued observation of animals and plants, it at once struck me that under these circumstances favourable variations would tend to be preserved, and unfavourable ones to be destroyed. The result of this would be the formation of a new species. Here, then, I had at last got a theory by which to work; but I was so anxious to avoid prejudice that I determined not for some time to write even the briefest sketch of it.[81]

One must keep in mind that at the time Darwin read Malthus creationism was already locked in combat with evolutionary theory, especially Jean Baptiste Lamarck's theory of speciation by inheritance of acquired characteristics caused by such mechanisms as "use and disuse, adaptive variation and the influence of the environment...." However, Lamarck was rejected by his fellow scientists, who considered his arguments unconvincing.[82] Even Darwin's scientific associate Charles Lyell, the foremost authority on geology, had rejected evolution in the 1830s. So much that Darwin had seen in his travels was suggestive of the change from one species to another, but no one had ever observed such a thing because it was such a slow process, as Darwin had learned from his observations of geologic processes and fossils, much as Lamarck had. Now, at last, however, he had in mind a hypothesis based upon the mechanism of natural selection. But how might this hypothesis be tested to either support or undermine the concept of evolution? Let us consider Darwin's own arguments presented in *The Origin of Species* after briefly summarizing the theory in its simplest form, much of it foreshadowed in the foregoing pages.

First of all, animals and plants manifest observable differences or variation amongst the individual members of a given species. Humans have created varieties of domesticated animals and plants on the basis of selective breeding to enhance certain variable characteristics and to diminish others. Variation of individual characteristics is also observable in nature. Whereas variation and selection of domesticated species to create observable varieties, which are essentially incipient species, is directly observable, natural selection is not because it occurs relatively slowly over long periods of time. However, evidence of the mutability of species in nature exists in the fossil record, within which antecedent species similar to and genealogically prior to contemporary species are often found. The scarcity of intermediate species is a function of both the complex nature of natural selection and the imperfections of the geologic record. Furthermore, following Malthus, organisms generally reproduce numbers of offspring far in excess of the food supply and habitable space available to them, resulting in a struggle for existence both between species and between the individual members of a species. As a result, those individuals who vary in such a manner that they gain some kind of advantage in the struggle for existence are more likely to survive and produce offspring which pass on the advantageous variation. Eventually over geologic time, less competitive members of the species lacking the desired variation will diminish in numbers and possibly become extinct. Extinction is a common result of the transmutation of species and is well represented in the geologic record.

Darwin's theory of evolution as presented in abstract form above is not difficult to grasp, but the devil is in the details, which are not easily summarized. Returning to the mechanism of natural selection, Darwin states that natural selection "leads to divergence of character and to much extinction of the less improved and intermediate forms of life. On these principles, I believe, the nature of the affinities of all organic beings may be explained."[83] He goes on to explicate these affinities in terms of how biological classification, in terms of relating species by genus, families, orders, and classes, makes sense only within an evolutionary framework involving inheritance and divergence of character through natural selection. No such intrinsic relationships can be reasoned from the theory of independent creation of species. The sequential order of species recorded as fossils in the geologic record, with species becoming more complex over time, also suggest a tree-like branching pattern of divergence from common ancestors. The connectedness and continuity of organic life, comprehensible only through the process of evolution, suggested to Darwin the metaphor of "the Great Tree of Life, which fills with its dead and broken branches the crust of the earth, and covers the surface with its ever branching and beautiful ramifications."[84]

Darwin further substantiates his core theory of the mutability of species through variation, the struggle for existence, and natural selection with evidence from the study of animal instinct, hybridism, embryology, paleontology, and the geographic distribution of plants and animals. Throughout this long, carefully reasoned argument he implicitly suggests the larger ecological context within which the struggle for existence and the evolution and extinction of species takes place. He often uses terms such as "the economy of nature" or "the web of life" which provide an integrative perspective for his biological and geological observations and explanations. Donald Worster, who devotes several chapters to Darwin in *Nature's Economy*, appropriately recognizes Darwin as "The single most important figure in the history of ecology over the past two or three centuries...."[85] Before further pursuing Darwin's role as a founder of contemporary ecological biology, however, we should consider his achievement in the light of what he did not know about evolution when he wrote *The Origin of Species* and *The Descent of Man*.

Two ideas of enormous importance for understanding evolution (of both species and ecosystems) were beyond Darwin's grasp when he developed his evolutionary theory, which made it more of a hypothesis than a theory, and also made it difficult to defend against its detractors. As Darwin openly admitted, the mechanism of heredity was not well understood by him or anyone else when *The Origin of Species* was published. The hidden hereditary code was discovered by the Austrian monk Gregor Mendel (1822-1884) in the 1860s. Unfortunately, it was ignored until the early 1900s, when later biologists independently arrived at similar results in their own studies of the inheritance of single alternative characteristics in plants before discovering Mendel's earlier work. Crossing varieties of the common garden pea, Mendel theorized that visible alternative traits such as flower color, tallness and shortness, etc. were controlled by paired units of heredity which came to be called genes. In 1900, two scientists, Hugo de Vries and Carl Correns, were the first to understand what Mendel had achieved in his experiments, thereby initiating modern genetics as a branch of biology. Shortly after Darwin's publication of *The Origin of Species*, Mendel began his experiments with the hope of "reaching the solution to a question whose significance for the evolutionary history of organic forms must not be underestimated."[86] Despite unanswered questions regarding heredity and natural selection, Darwin's overwhelming evidence for evolution in the *Origin* had already gained a considerable following shortly after 1859. The virtual revolution in genetics during the twentieth century has further supported Darwin's explanation of evolution as a theory rather than a hypothesis.

Darwin's mentor, geologist Charles Lyell, is best known for having elucidated and popularized James Hutton's uniformitarianism, the idea that the same physical and biological processes which we can observe at work today, such as erosion, deposition, and vulcanism, have been operative on our planet over vast periods of time to form the land forms, structures, rock types, and fossils we see today. Hutton and Lyell both believed that these gradual and uniform processes were responsible for all of the features, including fossils, preserved in planet Earth. Their uniformitarian perspective clashed, for many years, with the alternative theory of catastrophism, which fitted better with the belief that Earth and all of its animal and plant species had been divinely created a relatively short time ago (4004 B.C. according to one ecclesiastical chronicler). The leading proponent of catastrophism, George Cuvier (1769-1832), was an accomplished comparative anatomist and paleontologist, whose study of the geologic history of the Paris Basin suggested to him that periodic inundations or catastrophes had wiped out all life *en masse* at intervals, resulting in extinction of many life forms. Cuvier was also convinced that the Earth's history had been short, and that several thousand years ago a great flood, Noah's flood, had cleared the Earth of many large mammals, thereby preparing it for human habitation. (He never imagined that humanity itself had done much of the clearing, such as we encountered in chapters Four and Seven). Cuvier and other catastrophists also confused the extensive alluvial and other unconsolidated deposits of Western Europe left by Pleistocene glaciers with those of Noah's flood. However, catastrophism gradually gave way to Hutton and Lyell's uniformitarianism during the decades prior to Darwin's *Origin*, in which he, Darwin, displayed a complete acceptance of the uniformitarian theory in geology, and applied it to biological evolution.

There is, indeed, a wonderful irony in all of this. Darwin frequently referred to the gaps or "imperfections" in the geologic record to explain why most of the expected links between species were missing. Cuvier recognized the same breaks in the record of sedimentary deposition, which in the Paris Basin reflected seemingly sudden changes from fresh water to salt water environments and associated species. Gradually, geologists learned to explain these gaps or hiatuses in the sedimentary record as *unconformities*, buried erosion surfaces wherever the rocks, sedimentary, igneous, or metamorphic, had been uplifted, eroded, and subsequently inundated by an encroaching or transgressing sea over far longer periods of time than the catastrophists could imagine. Thus, all seemed to be explained by unconformities well past the middle of the twentieth century when suddenly an unexpected explanation of catastrophic events, catastrophic both geologically and biologically, burst upon the earth science community of scientists and sent shockwaves directly into the somewhat rigidified structure of Darwinian evolutionary theory. Darwin's conceptions of variation and natural selection had already been modified in the light of genetics while leaving his uniformitarian perspective in tact. Then, in 1980, a paper published in *Science* by L.W. Alvarez, W. Alvarez, F. Asaro, and H.V. Michel, "Extraterrestrial cause for the Cretaceous-Tertiary extinction," initiated the return of catastrophism to geology and evolutionary biology.[87]

Stephen J. Gould thoroughly explains the context of Darwinian evolutionary theory and its uniformitarian context prior to the return of catastrophism, in the form of collisions of asteroids or bolides from outer space, in his scientific masterpiece, *The Structure of Evolutionary Theory* (2002), published only a few months before his untimely death.[88] Gould's enormous volume of 1433 pages is a kind of "summa biologica," the equivalent of Saint Thomas Aquinas's *Summa Theologica* but for evolutionary biology. To summarize Gould's observations briefly, Darwin believed that global mass extinctions appeared to take place suddenly, i.e., all at once, because of the imperfection of the geologic record (because sediments that were correlated by the same fossils could still actually be separated by long enough intervals of geologic time to allow favored new species to replace and cause extinctions of other species worldwide), and also because he thought that the dominant pattern for the evolution of new species was the separation of geographic regions by barriers such as mountain ranges and bodies of water. As he had observed in the Galápagos, such isolation allows separate populations to vary in different environmental settings to produce new species. Darwin was able to make the geologic argument (of imprecise correlation of extinctions) even for the demise of the trilobites at the end of the Paleozoic era, and the ammonites at the end of the Cretaceous period and Mesozoic era. Gould concludes that Darwin's clever argumentation had placed a road block in front of evolutionary theory for well over a century:

> I have discussed Darwin's defense of uniformitarian extrapolation in detail because his argument, in this case, proved so successful in directing more than a century of research away from any consideration of truly catastrophic mass extinction, and towards a virtually unchallenged effort to spread the deaths over sufficient time to warrant an ordinary gradualistic explanation in

conventional Darwinian terms, with any environmentally triggered acceleration of rate only serving to intensify the effects of ordinary competition, species by species.[89]

Darwin's authority in nineteenth and twentieth century evolutionary biology and paleontology was so overwhelming (comparable to that of Aristotle in the Late Middle Ages and early modern centuries) that research was skewed to fit his uniformitarian evolutionary model. Thus, many of the initial responses to the new extraterrestrially-caused catastrophism proposed by Alvarez et al were highly critical. Nevertheless, as the scientific evidence mounted during the 1980s, particularly in regard to increasing examples around the world of high concentrations of the rare element iridium (signature element of bolides, i.e., giant meteorites) at the Cretaceous-Tertiary or K-T boundary, when both the dinosaurs and ammonites became extinct worldwide, the hypothesis of extraterrestrial "neo-catastrophism" became a well-established fact.[90] Thus, today we are left with a modified Darwinian theory in which the uniformitarian and catastrophic principles are combined, producing over the long term what Gould has referred to as "punctuated equilibrium." In a book-length chapter of 279 pages, a book within a book, titled, "Punctuated Equilibrium and the Validation of Macroevolutionary Theory," Gould demonstrates that the fundamentals of Darwinian theory modified by post-1980 neo-catastrophism and modern genetics, are still essentially sound, much like the remaining walls and vaults of a Gothic English cathedral bombed out but not totally destroyed by a German blitzkrieg during World War II. Long periods of genetic variation and natural selection followed the uniformitarian processes of Hutton, Lyell, and Darwin until, every several tens of million years, bolides of various sizes caused sudden waves of mass extinction on planet Earth. In this simple but unanticipated manner have cosmology, astronomy, physics, chemistry, geology, and biology been integrated as parts of the planet's physical and biological history.[91]

   This bare summary of the essential links between evolution and ecology may serve as a prolegomenon to kaleidoscopically shifting conceptions of nature and ecology during the nineteenth and twentieth centuries. The evolutionary implications of the relatively new late twentieth century uniformitarian-catastrophist compromise are enormous. However, in the nineteenth century one took one's Darwinism neat, and it served quite adequately to revolutionize and redefine the way that Western Europeans looked at nature. On the heels of Darwin's *Origin* came Ernst Haeckel's (1834-1919) ecology concept, the history of which has been explicated in great detail by Donald Worster in *Nature's Economy*. Although we have already defined ecosystems (Chapter One, pp.3-5) and considered them in various contexts throughout this book, we must consider ecology and ecosystems once more in the specific setting of the nineteenth century and the triumph of Darwinian evolutionary theory. The preceding discussion of Darwin's geological and biological observations during his voyage on the Beagle clearly suggests that Darwin had already grasped the complex interactions of the physical and biological environments, what we would today call ecosystems, by that time, before the word and concept were in common usage. He displays that understanding more explicitly in the *Origin*, beginning with the introduction, when he rejects both the idea of individual creation and the notion that external conditions are the only possible cause of variation by reference to "the case of the co-adaptations of organic beings to each other and to their physical conditions of life, untouched and unexplained..." by such causes.

> In the case of the mistletoe, which draws its nourishment from certain trees, which has seeds that must be transported by certain birds, and which has flowers with separate sexes absolutely requiring the agency of certain insects to bring pollen from one flower to the other, it is equally preposterous to account for the structure of this parasite, with its relations to several distinct organic beings, by the effects of external conditions, or of habit, or of the volition of the plant itself.[92]

Much of what Darwin has to say about ecology, the complex network of interactions between organisms and their environment, can be found in his chapter on the struggle for existence, where he gives examples of "how plants and animals, most remote in the scale of nature, are bound together by a web of complex relations." He goes on to describe the relations of different types of bees and plants to field mice, which destroy the combs and nests of bees. However, the bees' nests survive better when cats are numerous and control mouse populations. "Hence it is quite credible that the presence of a feline animal in large numbers in a district might determine, through the intervention first of mice and then of bees, the frequency of certain flowers in that district."[93] His next example suggests the concepts of climax ecosystems and the succession of plant communities. Observing that very different plants grow up after an American forest is cut down, but that ancient Indian ruins in the southern United States that were once cleared of trees have re-grown forests with the same diversity and proportions of species, he then remarks on the struggle which must have gone on between species of trees, insects, snails, and larger animals, "all striving to increase, and all feeding on each other or on the trees or their seeds and seedlings, or on the other plants which first clothed the ground and thus checked the growth of the trees!"[94] Shortly thereafter Darwin states that the most severe competition for survival must be between organisms of similar species "which fill nearly the same place in the economy of nature; but probably in no one case could we precisely say why one species has been victorious over another in the great battle of life." From these examples he draws a corollary which he claims to be of the highest importance: "that the structure of every organic being is related, in the most essential yet often hidden manner, to that of all other organic beings, with which it comes into competition for food or residence, or from which it has to escape, or on which it preys. This is obvious in the structure of the teeth and talons of the tiger; and in that of the legs and claws of the parasite that clings to the hair on the tiger's body."[95]

Donald Worster is convinced "that Darwin's theory of evolution was grounded in ecology."[96] Recounting examples of co-adaptations much as I have above, Worster makes the assertion that Darwin broke away from the traditional, Platonic view of "places" or ecological niches in nature, in which places were ideas existing in the mind of God prior to his creation of plant and animal inhabitants to fill them. In Darwin's account the availability of places for different species gave rise to the ceaseless competition or struggle for existence that enabled natural selection to improve the adaptability of organisms and create new species. However, divergence into new types could also "open a more peaceful route and a well-rewarded one; the organism that was born different and found a way to use its uniqueness might establish itself without the need for competition."[97] Despite his awareness of this alternative, Darwin clung to his belief that the essential mechanism of natural selection was competitive replacement to fill ecological niches. Worster points out that Darwin, near the end of his life, "was as strongly impressed as ever with the universality of violence in man and nature." Darwin even wrote of the triumph of the Caucasian races over the Turkish in the struggle for existence, and observed that higher civilized races must wipe out numerous lower races under the law of history and progress.[98] Thus Darwin himself had become a follower of "Darwinism," a dubious conflation of evolutionary ideas with historicism and the idea of progress.

The idea of an endless struggle for the survival of the fittest was "in the air" during the first half of the nineteenth century, along with the concept of evolution, with its roots in eighteenth century transformism. Darwin's scientific explication of these ideas was preceded not only by Malthus and his depiction of the dismal human struggle involving ultimate overpopulation and famine, but by Herbert Spencer, who wrote an essay defending evolution against believers in special creation in 1852, seven years before the *Origin*. Spencer carried the idea of evolution explicitly into the realm of human society, and is the best known of the "social Darwinists." There is no little irony that Malthus's emphasis on the social struggle for existence reappeared later as "Darwin's idea" in liberal doctrine, but it was Spencer who popularized social Darwinism as an idea of human progress by recognizing the struggle between individuals and among nations, but only until the struggle lost its social utility, after which societies would evolve to an "industrial" phase, when peace and cooperation would reign. This "law" of evolution and human progress was well-received in Europe until its newly industrialized war machines began to blow one another to pieces after 1914, the end of the "century of progress" that began with Napoleon's defeat at Waterloo.[99]

This chapter began by emphasizing the devastating social and environmental impacts of the industrialization and urbanization that picked up speed during the early nineteenth century in tandem with imperialism and increasing resource extraction in the tightening global network dominated by

European nations. A demographic explosion in Europe accompanied this transformation, with the European population increasing from about 200,000,000 in 1815 to 460,000,000 by 1914. During the same period approximately 40,000,000 European emigrants settled in other continents. Significantly, Europeans differed from the colonial world in that their numbers increased by lowering their death rate rather than raising their birth rate. Thus, although uncountable millions were forced or lured off of their farms and grazing lands, either by acts of enclosure, economic necessity, or hope for a better life in the city, and millions suffered the privations, suffering, humiliation, and the ills of alcoholism and tuberculosis as a consequence of their exploited labor in the satanic mines and mills of industrializing Europe and its neo-European colonies, the overall trajectory of the human species in the European world was upward. Hence, despite the sad and deplorable fates of so many individuals, caught up in the vortex of the industrial-urban *tourbillon*, Europeans as a whole prospered increasingly between 1815 and 1914. "Each generation enjoyed an increase in wealth and comfort, a widening of economic opportunity, an improvement in the standards of nutrition, health, and sanitation."[100]

If the whirlwind of modernity in the socioeconomic realm wreaked havoc with all too many lives of the lower classes, at the intellectual level the ideas and climate of opinion which gave rise to Darwinian evolutionary science and Spencerian social Darwinism were ultimately devastating to the Western Christian worldview, or, more precisely to the numbers of individuals who continued to accept the medieval legacy of ideas and myths. A virtually new worldview grew up to challenge the traditional worldview, a belief system still largely intact since the early Middle Ages. This dichotomy of two Western worldviews, the Christian, transcendental belief system on one hand, and a science-based, naturalistic and evolutionary perspective on the other, have persisted as two camps in a schizoid Western civilization down to the present day. Intellectual historian John C. Greene has performed the commendable task of carefully delineating the foundations of the mature modern secular perspective, referring to it specifically as the "Darwinian worldview." He writes that "*Darwinism* should be used to designate a world view that seems to have been arrived at more or less independently by Spencer, Darwin, Huxley, and Wallace in the late 1850s and early 1860s...."[101]

It is because these intellectuals of the Victorian Age held similar views regarding the nature of God, man, history, science, and society that Greene recognizes them as expressing a commonly held worldview. Each of these men "had fused several streams of Western thought into a unitary view of nature-history as a continuum undergoing progressive change in accordance with fixed laws discoverable by science."[102] The antecedent ideas and postulates upon which this new worldview is based are summarized by Greene as follows: (1) that the world consists of a wisely ordained, law-bound system of matter in motion; (2) that the present structures of the physical universe are derived from a previous, more homogeneous system of matter through the operation of natural laws; (3)

that Earth and its surface have undergone perpetual geological change, thereby continually modifying the conditions of existence for living organisms; (4) that this system of nature produced, in addition to physical changes, the generation of simple organisms capable of changing in structure and faculties in response to a changing environment; (5) that a social science of historical development would discover the necessary laws and stages of human progress; (6) that a constant struggle for existence was generated by the tendency of human populations to multiply beyond the means of subsistence; and (7) that humanity's sole means of acquiring knowledge of reality was to be found in the methods of natural science.[103]

Modernity itself is a *tourbillon*, a whirlwind of ceaseless change which is not easy to identify and define. However, visualizing modernity from the perspectives of such brilliant intellectual historians as Crane Brinton, Richard Tarnas, and Warren Wagar at the beginning of this chapter, we have acquired enough knowledge of its essential nature to realize that Greene's "Darwinian worldview" which arose midway through the nineteenth century, essentially integrated and completed modernism as a worldview and pragmatic foundation for our present way of life. This *weltanschauung* or world outlook has spread widely and persisted for a century and a half down to the present day with minor modifications based upon new scientific knowledge such as the evidence for catastrophic interruptions of evolutionary processes by extraterrestrial objects or bolides. The essentials of the modernist worldview remain intact despite the vast increases in scientific knowledge over a period of one hundred and fifty years! Nevertheless, in the United States the modernist worldview or at least some of its essential postulates, is rejected by somewhere between one-third and one-half of the population. Wagar's secularization has engulfed us overall, but in terms of belief in the meaning of existence, i.e., of religion and philosophy, Americans have retained the medieval Christian worldview to a remarkable degree, as have nations such as Ireland and Poland, on the periphery of Europe. At the core of Europe, however, in France, Germany, England, and in Scandinavia as well, secularization and acceptance of the "Darwinian" worldview is dominant. In this sense, the heart of Europe is more modern than the United States. This is hardly surprising, given that the nations at the center of the eighteenth century Enlightenment appear to be the ones which have taken its epistemology and philosophy more seriously, its destructive twentieth century consequences not withstanding. Although the religious-secular split in the modern Western worldview has not entirely disappeared in contemporary Europe, it has not engendered the profound schism and sociopolitical turmoil which rages today in the United States. Europeans have socioeconomic problems enough with the consequences of having absorbed large numbers of the inhabitants of their nineteenth century empires into the mother countries: traditional colonial inhabitants swept into the vortex of the modern tourbillon, the whirlwind of change.

The schizoid nature of contemporary modern Western culture has already spelled very bad news for the protection of our natural environment. The failure of twentieth century environmentalism, to be discussed in the following two chapters, is directly related to the religious-secular schism. This is so because a mandate for good environmental laws which are well enforced cannot come about without a widespread, general acceptance of the Darwinian version of the modern worldview. Such a worldview entails an understanding of the *evolutionary* character of ecosystems. Lacking such understanding among the majority of citizens, attempts to implement environmental ethics as environmental law must be misguided, and poorly implemented and supported. If only a few brave and insightful souls and their followers laid the foundations of modern environmentalism in nineteenth century America, by the 1960s there appeared to be a full-blown environmental revolution in ideas and action which created the Environmental Movement. However, as we shall see in the following chapter, the sense of an American environmentalist mandate for positive reconstruction to preserve nature was actually the manifestation of individual concern over the hazards of pollution for *human* life rather than a growing acceptance of our responsibility for the health and continuation of Earth's evolutionary ecosystems. Instead, human domination of global ecosystems through the application of mechanistic science was conflated with the Environmental Movement and its goals to produce a misleading conception, *sustainability* of industrial development and population growth while maintaining environmental quality.

NOTES

[1]. Berman, Marshall, *All That is Solid Melts Into Air: The Experience of Modernity* (New York: Penguin Books, 1988), p.16.

[2]. Ibid., p.17.

[3]. Ibid., p.19.

[4]. Ibid., p.20.

[5]. Ibid., pp. 345-6.

[6]. Ibid., p.24.

[7]. Eric Hobsbawm, *The Age of Extremes: A History of the World, 1914-1991* (New York: Pantheon Books, 1994), pp.2-17.

[8]. Brinton, Crane, *The Shaping of Modern Thought* (Englewood Cliffs, N.J.: Prentice Hall, Inc., 1963), p.22.

[9]. Ibid., p.107.

[10]. Ibid., p.82.

[11]. Ibid., p.109.

[12]. Ibid., p.110.

[13]. Ibid., p.113.

[14]. Ibid., p.111.

[15]. Ibid., p.114.

[16]. Ibid., p.116.

[17]. The blank slate fallacy is considered at length in Steven Pinker, *The Blank Slate: The Modern Denial of Human Nature* (New York: Viking, 2002).

[18]. See Robert Nisbet, *History of The Idea of Progress (New York: Basic Books, 1980).*

[19]. Crane Brinton, *The Shaping of Modern Thought*, p.136.

[20]. Ibid., p.137.

[21]. Ibid., p.109.

[22]. Tarnas, Richard, *The Passion of the Western Mind* (New York: Ballantine Books, 1991), p.284.

[23]. Ibid., p.282.

[24]. Ibid., p.286.

[25]. Ibid., p.288.

[26]. Ibid., pp.288-289.

[27]. Ibid., p.290.

[28]. Wagar, Warren, ed. *The Secular Mind* (New York: Holmes J. Meier Publishers, Inc., 1982), p.3.

[29]. Ibid., quoting David Martin, pp.4-5.

[30]. Ibid., p.5.

[31]. Ibid.

[32]. Ibid., p.6.

[33]. See Chapter 1, pp.10-11.

[34]. William McNeill, *The Rise of the West* (Chicago: The University of Chicago Press, 1963), p.732.

[35]. Ibid., p.733.

[36]. J.R. McNeill and William McNeill, *The Human Web: A Bird's-Eye View of World History* (New York: W.W. Norton and Company, 2003), p.234.

[37]. Ibid., p.235.

[38]. Karl Polanyi, *The Great Transformation: The Political and Economic Origins of Our Time.* (Boston: Beacon Press, 1944, 1957), p.130.

[39]. Matthew Osborn, "The Weirdest of all Undertakings: The Land and the Early Industrial Revolution in Oldham, England," *Environmental History* V. 8 (April, 2003), pp.246-269, p.248.

[40]. Ibid., p.253.

[41]. Ibid.

[42]. Ibid., p.259.

[43]. Ibid., p.264.

[44]. Ibid., 266.

[45]. J. Donald Hughes, *An Environmental History of the World* (London: Routledge, 2001), p.120.

[46]. Ibid., p.119.

[47]. Ibid., pp.120-121.

[48]. Ibid., p.123.

[49]. Ibid., p.124.

[50]. Clive Ponting, *A Green History of the World* (New York: Penguin Books, 1991).

[51]. Sing C. Chew, *World Ecological Degradation: Accumulation, Urbanization, and Deforestation* (Walnut Creek, CA: Alta Mira Press, 2001), p.131.

[52]. Ibid., pp.132-134.

[53]. John Opie, *Nature's Nation: An Environmental History of the United States* (Fort Worth, Texas: Harcourt Brace College Publishers, 1998), 216-241.

[54]. Ibid., p.216.

[55]. Joseph Petulla, *American Environmental History* (Columbus, Ohio: Merrill Publishing Company, 2nd Ed., 1988).

[56]. John Opie, *Nature's Nation*, p.224.

[57]. Ibid., p.228.

[58]. Ibid., p.229.

[59]. Ibid., p.231.

[60]. Robert Heilbroner, *The Nature and Logic of Capitalism* (New York: W.W. Norton and Company, 1985), pp.104-106.

[61]. Kevin Phillips, *Wealth and Democracy* (New York: Broadway Books, 2002), pp.412-413.

[62]. Ibid., pp.238-239.

[63]. John Opie, *Nature's Nation*, p.232.

[64]. Ibid., p.234.

[65]. Gilbert F. LaFreniere, "World Views and Environmental Ethics," *Environmental Review*, V. 9, No. 4, (Winter 1985), pp.307-322, p.316.

[66]. Donald Worster, *The Wealth of Nature: Environmental History and the Ecological Imagination* (Oxford: Oxford University Press, 1993), p.178.

[67]. Ibid., p.180.

[68]. J.R. McNeill and William McNeill, *The Human Web*, pp.264-267.

[69]. Stephen Toulmin and June Goodfield, *The Discovery of Time* (New York: Harper and Row, Publishers, 1965), p.45.

[70]. Lucretius, *On Nature* (Indianapolis: The Bobbs-Merrill company, 1965), p.184.

[71]. Toulmin and Goodfield, *The Discovery of Time*, p.48.

[72]. Ibid., p.75.

[73]. Ibid., p.176.

[74]. Ibid., p.172.

[75]. Charles Darwin, *The Voyage of the Beagle* (New York: Bantam Books, 1958), p.148.

[76]. Ibid., p.149.

[77]. Ibid., p.328.

[78]. Ibid., pp.330-344.

[79]. J. Donald Hughes, *An Environmental History of the World*, p.131.

[80]. Thomas Malthus, *An Essay on the Principle of Population* (Harmondsworth, Middlesex, England: Penguin Books Ltd., 1970), pp. 70-72.

[81]. Darwin quoted by Anthony Flew in introduction to *Malthus, An Essay on the Principle of Population* pp.49-50.

[82]. Toulmin and Goodfield, *The Discovery of Time*, p.186.

[83]. Charles Darwin, *The Origin of Species* (Oxford: Oxford University Press, 1996), p.105.

[84]. Ibid., pp.106-107.

[85]. Donald Worster, *Nature's Economy*, p.114.

[86]. Mendel quoted in John A. Moore, *Science As a Way of Knowing: The Foundations of Modern Biology* (Cambridge, Massachusetts: Harvard University Press, 1993), pp.286-287.

[87]. Alvarez, L.W., W. Alvarez, F. Asaro, and H.V. Michel., "Extraterrestrial cause for the Cretaceous-Tertiary extinction." *Science*, 1980, No. 208: 1095-1108.

[88]. Stephen J. Gould, *The Structure of Evolutionary Theory* (Cambridge, Massachusetts: The Belknap Press of Harvard University Press, 2002).

[89]. Ibid., p.1303.

[90]. Ibid., pp.1303-1308.

[91]. Ibid. See pp.745-758 for an introduction to "punctuated equilibrium."

[92]. Charles Darwin, *The Origin of Species*, p.5.

[93]. Ibid., pp.61-62.

[94]. Ibid., pp.62-63.

[95]. Ibid., p.64.

[96]. Donald Worster, *Nature's Economy*, p.155.

[97]. Ibid., p.161.

[98]. Ibid., p.165.

[99]. Franklin L. Baumer, *Modern European Thought: Continuity and Change in Ideas, 1600-1950* (New York: Macmillan Publishing Company, Inc., 1977), pp.363-364.

[100]. Geoffrey Bruun, *Nineteenth Century European Civilization* (New York: Oxford University Press, 1960), pp.2-3.

[101]. John C. Greene, *Science, Ideology, and World View: Essays in the History of Evolutionary Ideas* (Berkeley: University of California Press, 1981), p.130.

[102]. Ibid., p.148.

[103]. Ibid., pp.148-149.

CHAPTER TEN: THE TWENTIETH CENTURY ROAD TO RUIN

An Overview of the Twentieth Century

The periodization of history does not generally lend itself to neat compartments of millennia, centuries, and decades. The nineteenth century, for example, better fits the years 1815-1914 than 1800-1900, and the twentieth century better fits 1914-1991 or 1914-2001 than 1900-2000. Waterloo and the beginning of the First World War meaningfully demarcate a nineteenth century of relative peace and general progress in Western civilization just as the beginning of full-scale modern technological warfare in 1914 and perestroika or the Islamist terrorist attack on New York in September 11, 2001 best encapsulate the turmoil of the twentieth century. Historians of Europe and its colonies often emphasize a transitional period, 1871-1914, which links the nineteenth century early industrial world to the explosive development of population, science and technology, energy use, industry, and consumption in the twentieth century. Some sense of the magnitude of intensified human activity on the planet during the twentieth century is suggested by environmental historian J.R. McNeill's observation that more energy probably has been deployed in the twentieth century world than in all of human history prior to 1900.[1]

Thus, we should consider the transitional era of 1871-1914 to better understand the major substructural or nature-related currents (i.e., increases in human population, the impact of science and technology upon drastically increased energy use, industrialization, and human consumption of resources) underlying the *events* of the twentieth century (World Wars I and II, the Great Depression, the rise of nationalistic communism, American leadership in science, technology, industrialization and mass production, the Cold War, and resource wars in the Middle East). Without such a framework linking the two centuries it would be very difficult to grasp the deeper causes of twentieth century historical events. However, before considering this transitional period, some general observations should be made regarding the judgments of contemporary thinkers upon the twentieth century experience as a whole.

In *The Age of Extremes* (1994) British historian Eric Hobsbawm summarized an opinion survey of European intellectuals concerning their individual, subjective opinions about the predominant characteristics of the twentieth century. Their responses ranged from " the most terrible century in Western history" (philosopher Isaiah Berlin) through "I see it only as a century of massacres and wars" (French agronomist and ecologist René Dumont) and "I can't help thinking that this has been the most violent century in human history" (British writer and Nobel Laureate William Golding), to "the chief characteristic of the twentieth century is the terrible multiplication of the world's population. It is a catastrophe, a disaster. We don't know what to do about it," and "the most fundamental thing is the progress of science, which has been truly extraordinary" (Spanish scientist and Nobel laureate Severo Ochoa). British anthropologist Raymond Firth also stressed the development of electronics as one of the most significant events of the century.[2]

I might add that the vast majority of those who lived in the twentieth century saw its horrors and its threats only occasionally and for relatively short periods of their lives. Of course, we owe our own present good fortune to the sacrifices of those serving in the military in behalf of democratic causes. However, concerning losses in human population directly or indirectly caused by twentieth century warfare, they were small in comparison to the planet's exploding human population. Whereas tens of millions died as the result of modern warfare in the twentieth century, world population grew from 1.65 billion in 1900 to 6.07 billion in 2000.[3] This constitutes a gain of 4.42 *billion* human beings in a single century, more than two and a half times the population in 1900! Most of this enormous population growth took place during the second half of the twentieth century, during which 3.58 billion of us have been added to humanity's mushrooming numbers. Looking ahead into the twenty-first century, one wonders to what degree enormous population increases will generate friction in the political and socioeconomic realms as burgeoning populations and diminishing resources generate uncomfortable situations for the planet's increasing billions. And looking backwards to the late nineteenth and twentieth centuries, to what extent did rapid population growth in Europe, along with diminishing resources, contribute to its unprecedented destructive wars? Sorting these factors out from the political friction engendered by competition between newly created national polities engaged in imperialistic adventures around the globe is no easy task, but, there is little doubt that they played an important role in generating military conflict. Such factors very likely will only be compounded globally in the twenty-first century.

Fortunately for western Europe, the self-proclaimed heart of the "civilized world" in the nineteenth century, population growth began to decline during the nineteenth century, as early as 1830 in France, partly as a result of the Code Napoleon, which required that inheritances be divided among all sons and daughters. According to R.R. Palmer it was "economic security and the possession of a social standard that led to the reduced birth rate in France."[4] The repeal of child labor, increasing use of contraception, and the social desirability of small families all contributed to decreased birthrates, and the rest of western Europe eventually followed this demographic model, which was contemporary with the growth of industrializing cities, leading to a diminution in the rural population. Simultaneously, especially during 1850-1940, Europeans were migrating to new lands in the Americas, Australia, and New Zealand, the regions dubbed "neo-Europes" by Alfred Crosby. As many as 60 million people left Europe between 1850 and 1940.[5] However, since World War II Europe has been inundated with an influx of population from Asia and Africa.

Despite the decreasing birthrate, Europe's population grew from 263 million in 1850 to 396 million in 1900, although several tens of these millions emigrated. These demographic changes were contemporary with a shift in the Industrial Revolution to a new phase ca. 1870. Whereas the early nineteenth century was characterized by the use of steam power in mines, railroads, and ships, as well as the growth of the textile and metallurgical industries, the "New Industrial Revolution" began to tap petroleum and to expand global transportation, communications, trading systems, and the organization of business, finance, and labor.[6] It was during this period that rapid growth occurred in the coal and steel industries, railways, telegraphs, chemicals, and electricity, leading to the development of giant firms "that took advantage of economies of scale and operated internationally. For the first time, most of the innovations came not in Britain, but in Germany and the United States, especially the large-scale bureaucratic business organization, pioneered by American railroads."[7] The Industrial Revolution also became increasingly global as science and technology spurred its growing complexity. Thus, during the period 1871-1914 global economic hegemony shifted from Europe to the United States.

A thorough and lucid account of the rise of American resource industrialism is presented in Joseph Petulla's *American Environmental History* (2nd edition, 1988). A major theme threading through Petulla's history is that resource utilization has intensified following every American war: Revolutionary, Civil, World War I, World War II, and Vietnam. After the Civil War, new methods of manufacturing iron and steel led to the search for additional iron ore deposits as the demand for iron and steel for railroads, bridges, and high rise buildings increased. Rapidly expanding railroads allowed economic titans like Carnegie and Rockefeller, the "Robber Barons," to integrate mining production, processing, and marketing. The beginning of the twentieth century witnessed increasing monopolization as capitalists with enormous financial resources acquired smaller companies, leading to limited business reforms during the Progressive Era of the early twentieth century. Especially after World War I the development of highways and suburbs was powered by the booming automobile, steel, and oil industries.[8] The internal combustion engine that revolutionized modern warfare in World War I also revolutionized the American landscape, a process of almost uncontrolled development on to the present day.

Petulla describes the 1920s as "a preview of the modern era of growth, affluence, economic concentration, conspicuous consumption, and the rise of science and technology, ... a prelude to social and economic disaster."[9] Unfortunately, the economic booms and reckless pursuit of overnight fortunes during the jazz era ended in the stock market crash of 1929 and a depression which lingered until World War II. The stock market crash which initiated the Great Depression was the consequence of easy credit, mass speculation in the stock market, and skyrocketing stock prices. "The flimsiest of rumors, expansions, dividends, mergers, would push up stock prices to undreamed-of levels. The average person was encouraged to buy on margin, paying for only half the stock price and borrowing the rest, because buying on margin allowed the purchaser to buy twice as much stock."[10] At the same time, holding companies which produced nothing watched their corporate assets burgeon as the savings of private citizens were collected. When the American "cardboard economy" finally collapsed in 1929 the top 5% of the nation's population received more than a third of the nation's income, and between 1929 and 1933 investments in capital goods dropped 88%. Also in the 1920s trade wars among the world's major exporters led to prohibitive tariffs in America and elsewhere, and as the tariffs remained into the 1930s up to approximately 40% of American wage earners were unemployed. Just as the excesses of the "Robber Barons" led to the weak reforms of the Progressive Era, the excesses of unrestrained capitalism during the Roaring Twenties gave rise to a desperate attempt to salvage the United States from depression and massive unemployment through "welfare capitalism." Franklin Delano Roosevelt's New Dealers "supported the system of private profits but wanted to engage government in the task of unionization and welfare programs. New Dealers generally accepted the dominant position of monopoly and oligopoly...."[11] They argued that the manipulation of prices and production were necessary to stabilize an economy dominated by giant corporations. They followed the ideas of Thorstein Veblen, particularly that the emphasis of the national economy on business profits needed to be reoriented towards the common good of the nation and the needs of its citizens. Such a shift in priorities would require national planning by the federal government.

In Europe and other parts of the world the boom and bust cycles of Western capitalism likewise posed the constant threat of unemployment in industry and devastatingly low prices for farmers. These chronic problems led to the search for political and socioeconomic alternatives such as fascism and communism, thereby contributing to the causes of the Second World War. Both Italian fascists and Russian Soviet communists sought a solution to the ills of capitalism in the pursuit of autarky or regional self-sufficiency, an idea which became popular again later on in the century during the American Environmental Movement, even though its opposite, globalization, became the dominant pattern of historical change as the movement weakened. During the 1930s Japan also attempted to become self-sufficient in food and military armaments while extending its empire in Southeast Asia. The pursuit of autarky by Germany, Italy, and Japan also drove these nations towards imperialist expansion in pursuit of mineral and fossil fuel resources. In 1931 Japan seized Chinese territory which possessed abundant coal and iron ore, and in the late 1930s attempted to gain control of the oil fields of the Dutch East Indies. This resulted in the establishment of a United States embargo on oil exports to Japan, which convinced Japanese military strategists that war with the United States was imminent. Thus, their pre-emptive strike on Pearl Harbor was partly motivated by their need for oil for military purposes. Nazi Germany's 1941 invasion of Russia was rooted in the Nazi ideology of race and Lebensraum but also similarly motivated by the attempt to gain control of the major oil fields at Baku (in what is now Azerbaijan).[12]

According to J.R. McNeill and William McNeill 3 percent of the world's 1940 population, about 60 million people, were killed during World War II. The greatest carnage occurred in eastern China and the western U.S.S.R., where as many as 25 million died. The United States, whose highly productive factories contributed so much to victory, lost a total of 400,000. Out of World War II, the United States emerged as the first global superpower in the history of the world. Its military prominence was based upon: (1) a monopoly on nuclear weapons; (2) the world's largest navy; and (3) about half of the world's industrial capacity. The key to American power "was the efficiency of its heavy industry, symbolized by the moving assembly line, which made production faster and cheaper."[13] In addition, mobilization for World War II revitalized the American economy and laid the foundation for the extraordinary economic growth and global trade which have characterized the second half of the twentieth century, thereby creating the affluent but wasteful consumer society within which we function today. As Joseph Petulla points out, postwar growth involved federal cooperation with the automobile boom which produced the "automobilization" of the American economy and culture to the point that by the 1970s there were over 100,000,000 American cars, and one out of six jobs in the United States were connected to the production or use of the automobile.[14] Capturing the essence of the post-World War II era down to the present day, Petulla writes:

> The economic source of the affluent society resulted from the war. It increased both government expenditures and economic concentration. Appendages of the new society were high-pitch advertising competing for higher incomes; bigger automobiles to be driven over hundreds of new highways; slick supermarkets for the suburbs; dozens of new credit devices for thousands of new consumer items. Inflation came in the wake of a transformed America, and a growing class of the poor and the elderly were left out of the mainstream of abundance.[15]

Obviously, it required a sudden surge in energy production and use during and after the Second World War to put the global marketplace back on track. Even as early as the 1920s, however, some resource analysts had warned of the exhaustion of the world's crude oil supplies, but improved technology soon doubled the yield of gasoline from crude oil, so that gasoline sold at only 10 cents a gallon in 1927. By the 1930s the United States government began to set aside oil reserves for future military use. World War II demanded unlimited oil production in the United States as supplies were cut off from the Dutch East Indies and other parts of the world. Despite the high demand for oil during the war, postwar prices remained at 20 to 30 cents per gallon during the 1950s, which encouraged growth in the automobile industry. Oil also began to replace coal for household heating during this time, as well as the fuel for expanding fleets of ships and diesel trains. The new plastics industry also added to the demand for oil. Thus, the United States gradually became addicted to imported crude oil from the Middle East.[16] One might have expected that the manipulated "oil crisis" of 1973 would have shaken the American public and its government into energy sobriety, which did happen to the extent that small Japanese and German cars became popular and mileage standards were considered important to American automobile manufacturers for awhile, but by the 1990s Americans, more affluent than ever, were hopelessly addicted once more.

To place twentieth century energy demands in perspective, between 1900 and 1990 the demand for coal increased five times while the demand for oil increased 150 times. By 1990 the world produced 5,000,000,000 tons of coal and 3,000,000,000 tons of oil. Overall, energy production during the twentieth century increased sixteen-fold over what it had been in the nineteenth century. In terms of energy inequality, during the 1990s, the average American used 50-100 times as much energy as the average Bangladeshi. The energy-powered world economy grew fourteen-fold in the twentieth century despite the slump during the Great Depression. The environmental and social consequences of this immense economic growth have been overwhelming. "Intellectually, politically, and in every other way, adjusting to a world of rapid growth and shifting status was hard to do. Turmoil of every sort abounded. The preferred policy solution after 1950 was yet faster economic growth and rising living standards...."[17]

Contemporaneous with the rapid economic growth of the post-World War II half of the twentieth century, world history was dominated by the Cold War. Tensions between the United States and Western Europe, and the communist block of the Soviet Union and its satellites escalated after Soviet scientists provided Stalin with the atomic bomb. Although the U.S.S.R. and China pursued cooperation as fellow communist states, their armies clashed along their border by 1968-1969, and over the following several years China established warmer diplomatic relations with the United States. Meanwhile, because the Soviet economic system tended to discourage technical and managerial innovation, their agriculture fell into decline, so that by the late 1970s they could not feed their own population. Development of their Siberian oil and gas fields, however, allowed the Soviets to sell oil to the West while importing American grain, technological innovations, and consumer goods. Along with the moderating effects of trade upon the Cold War, after the 1970s Soviet state control of information regarding the high material standards of living in Western nations gradually broke down, consequently creating a more open society in the U.S.S.R. and allowing a flood of Western consumer goods and values, particularly the ideas of liberal democracy, to penetrate the "iron curtain." By 1989 the last of the communist buffer states fell from Soviet control, and in 1991 the Soviet Union itself collapsed as fourteen republics declared their independence from mother

Russia. "The Soviets lost the Cold War for the same reason that the Axis lost World War II: they could not create an interactive, cooperative, innovative, international economy to match the American-led one. They remained too wedded to the autarkic economy that Stalin had built...."[18]

The second half of the twentieth century also witnessed the continued decolonization of the Eurasian and colonial empires that had been building up during the nineteenth and earlier centuries. The end of the imperial period was signaled by the breakup of the Austro-Hungarian, Ottoman, and Russian Empires at the end of World War I, when Ireland also won independence from England. The major era of decolonization occurred between 1943 and 1975, when India, Indochina, Libya, Syria, Lebanon, Algeria, Korea and other colonial regions broke away from the grip of England, France, Portugal, Holland, and Japan, cutting the roots of empire which reached back to the sixteenth century. Decolonization played into the hands of United States corporations and national strategic planning, as countries such as South Korea and Taiwan were aided by American loans and trade in order to build barriers against the spread of communism. The disintegration of the U.S.S.R. into myriad successor states capped off a half century of decolonization in 1991, followed by economic stagnation in the successor states as well as in Russia itself.[19] Boosted by innovations in electronics and the worldwide adoption of computer technology, the twentieth century ended on the positive note of extraordinary economic growth, especially in the United States, but also on the negative signals of corruption in large corporations and the stock market. Gigantic corporate wealth and deception of stockholders caused a moderate crash and exposed the ubiquity of corruption in corporate capitalism. As John Opie observes regarding corporations in the United States: "Government regulation of industrial corporations began to build up in the twentieth century, but when compared to business operations in other industrialized nations, corporate freedom remains essentially untouched."[20]

In *The Human Web*, J.R. McNeill and William McNeill assess the sporadic process of globalization throughout history, and they see the twentieth century as a successful culmination of the process, with urbanization and population growth as the dominant social changes of the time. Culturally, the twentieth century also witnessed greater homogenization around the globe, with a final surge in this direction after 1980, when "technology and policy combined to produce faster globalization, this time combined with rapidly growing inequalities. The cosmopolitan web encompassed the habitable globe, all the world's peoples and ecosystems, in a swirl of kaleidoscopic interaction."[21] They express concern that this surge of population growth, based upon technological improvements in as well as the spread of modern agriculture, has created pathogens and crop pests, and observe that " to date the institutionalization of science, and of technological research and development, has kept humankind ahead in this ecological arms race."[22] I found myself somewhat surprised and disappointed at this conclusion. After all the local, regional, and finally, global destruction and degradation of ecosystems by *Homo sapiens*, is it not unabashedly anthropocentric to focus on the hazards of agricultural pests and parasites at this time rather than on the ecological consequences of the agricultural and urban activities of humanity itself? A single primate species has now exploded the limits to its reproductive capacity and placed insupportable, unsustainable demands upon all the other species and resources of the planet. Shouldn't this have been the main concern voiced at the end of *The Human Web*? As historians, the McNeills have been cautious about making judgments concerning the impact of humankind upon nature. William McNeill, in particular, has been a major booster of the virtues of globalization in his historical works. Globalization is the heroic protagonist of the late nineteenth and twentieth centuries, world trade having made great strides from 1871-1914, then having contracted from 1914-1945 under the pressures of two wars, economic depression, and the autarkic experiments of communism and fascism, and finally developing with hurricane force from 1945 to the present. From the time that William McNeill published *The Rise of the West* in 1963, in part to refute Oswald Spengler's historical depiction of distinct, essentially isolated high cultures or civilizations, he has been elaborating the theme of globalization through trade (along with the disease transfers and the spread of technological innovation that accompany it) not only as the dominant pattern of world history, but as by far the only possible course which world history could take. There is, of course, a secular idea of progress implicit in McNeill's historical writing. In the course of describing globalization over time, he has become a powerful advocate of globalization, a process which to many environmentalists is one of the main causes, if not *the* main cause of today's environmental crisis.

Fortunately, McNeill's son, J.R. McNeill, followed his father's interest in world history in his own academic career, but with a forward-looking specialization in environmental history. World historians are hard to find nowadays, particularly given the staggering complexity of the past two centuries. Nevertheless, J.J. McNeill, in the grand synthesizing tradition of his father, has written the first environmental history of the twentieth century. Although wary of prognostication in *Something New under the Sun*, McNeill junior does a thorough job of summarizing human impact upon the environment before drawing brief conclusions about the historical and ecological significance of the twentieth century experience. He writes that: "It is impossible to know whether humankind has entered a genuine ecological crisis. It is clear enough that our current ways are ecologically unsustainable, but we cannot know for how long we may yet sustain them, or what might happen if we do." He goes on to point out that there have been many unsustainable societies in the past and that they have either disappeared or transformed themselves into something new, a topic discussed in the earlier chapters of this book. Then, citing China's 3000 years of unsustainable development, McNeill states that "unsustainable development on the global scale may be another matter entirely, and what China did for millennia the whole world perhaps cannot do for so long. If so, then collapse looms, as prophets of the ecological apocalypse so often warn."[23] Thus, even J.R. McNeill, a prudent historian and admirer of the global web constructed by humankind, displays uneasiness over the future of both mankind and the natural world, moving a step beyond his distinguished father towards historical pessimism rooted in our treatment of nature. In the pages that follow I will attempt to summarize the extent of environmental damage inflicted upon our planet during the twentieth century, focusing in particular upon ecosystems, especially forests, and agricultural lands created out of forests and grasslands. Then, in the second half of this chapter I will discuss changes in the Western worldview related to modern environmental damage, including the roots and antecedents of the Environmental Movement.

Environmental Impacts of Late Nineteenth and Twentieth Century Development

I prefer to think of the context of environmental impacts from the ground up, from the minerals, fossil fuels, and groundwater that lie beneath the surface of the earth, to the thin, fragile layer of soil that covers consolidated and unconsolidated rocks over much of the planet's surface, to the systems of organisms, i.e., ecosystems, composites of plants and animals interacting with soil, water, and air, and existing at or near the interface of soil or rock and the atmosphere on land, and from sea floor to the ocean surface and atmosphere. As terrestrial mammals we have most severely damaged or changed terrestrial ecosystems or biomes, but as we became more sea-faring in recent millennia, we have greatly changed aquatic ecosystems as well. In earlier chapters the gradual devastation of Mediterranean and European ecosystems, as well as those of their colonies, has been described. In this chapter I will focus upon the effects of industrialization upon environmental degradation in the United States and other regions where, in pursuit of a society of material abundance (which became the affluent or consumer society of today) we have inflicted ecological damage upon the North American and other continents over a relatively short period of time, in the United States moving westward and disrupting nature with increasing intensity during the late nineteenth and twentieth centuries.

In addition to J.R. McNeill's *Something New Under the Sun*, a number of other recent environmental histories have examined the environmental history of the twentieth century, notably Paul R. Josephson's study of *Industrialized Nature*, J. Donald Hughes' *An Environmental History of the World*, Clive Ponting's *A Green History of the World*, and Sing C. Chew's *World Ecological Degradation*. Environmental history has arrived at an age of synthesis after more than three decades as a sub-discipline of history. The amount of descriptive history covering all periods of environmental history in all parts of the world today is overwhelming, benefiting from the rapid increase in numbers of publishing scholars in the information age. A small number of historians have integrated and synthesized all of this research in books like the aforementioned titles. Before discussing the particular contributions of these authors and others to a general understanding of ecological degradation in the twentieth century, I would like to sketch out a broad outline of typical environmental impacts and the ways in which they impinged upon my own life

In the United States and Canada massive deforestation of southern mixed deciduous, and northern mixed and coniferous forests along the Atlantic seaboard during the seventeenth , eighteenth, and nineteenth centuries subsequently spread westward to the great Pacific Cordilleran forests of California and the Pacific Northwest, where the last stand to preserve small remnants of old growth forests is occurring today. The late nineteenth and twentieth centuries have witnessed increased mechanization of logging techniques simultaneous with extensive road building into private, state, and federal forests. Loss of biodiversity, eroding watersheds, and increased siltation and flooding in stream systems have all increased rapidly during this period, consequently reducing the biodiversity of anadromous fish and other species. Deforestation on fertile lands in the eastern United States opened up the great bread basket of loess (wind-blown dust and silt from glacial deposits) soils which extend from the interior lowlands and Mississippi Valley to the eastern Great Plains. Further west on the Great Plains, the combination of a semi-arid climate and the removal of native grass cover by "sodbusting" farmers created the unsustainable farming which gave rise to the Dust bowl during the Great Depression in the 1930s. Gradually, the Missouri and Mississippi rivers became choked with silt. During the great dam-building era of the 1930s-1960s, impacts from dam building under the Army Corps of Engineers further destroyed riparian habitat. Mining operations followed the westward flow of trappers, traders, and farming from the Rocky Mountains to the Sierra Nevada, leaving in their wake a chain of ruined riparian ecosystems and streams loaded with toxic heavy metals. The Bureau of Reclamation, established in 1902, began the process of filling in "worthless swamps" along floodplains and shorelines, destroying about half of these valuable breeding grounds for fish, invertebrates, birds, and mammals in the twentieth century.

Deforestation in the United States during the second half of the nineteenth century led to a growing awareness on the part of federal officials and politicians that some kind of national forest management policy would be necessary. Partly inspired by George Perkins Marsh's *Man and Nature* after 1864, the state of New York was first to act in behalf of forest conservation in establishing the Adirondack Forest Preserve in 1885. In 1891 the Forest Reserve Act gave the President power to set aside forest reserves, and in 1897 the Secretary of Interior was awarded jurisdiction over national forest reserves. A year later Gifford Pinchot became head of the Division of Forestry under the United States Department of Agriculture, and in 1905 the first chief of the United States Forest Service. Pinchot committed the national forests to "sustained yield" management in order to produce a regular flow of wood to consumers without damaging the long-term productivity of the forests. Unfortunately, sustained yield was compromised during World War II and subsequently, down to the present day.

Clearcutting was the established mode of logging the national forests in the 1930s and 1940s just as it was during the colonial period, and the policy of "multiple use," which allowed mining, stock grazing, outdoor recreation, hunting, and fishing along with logging also contributed to ecological degradation within the forests.[24]

Contemporary with the exploitation of forests in the United States, political liberation of European colonies, especially after World War II, led to increased exploitation of forests worldwide, because "indigenous governments, spurred on by the call of nation-building, combined with neocolonial policies on the part of the West resulted in a modernization trajectory that has meant the emulation of the development strategies of the West."[25] Multinational companies based in the United States, Japan, Europe, Canada, and Malaysia set about harvesting global forests in highly unsustainable ways. Towards the end of the twentieth century the world's forests had shrunk by nearly half of what they had been eight thousand years ago. It has been estimated that an increase in global population to 9.5 billion by 2050 will cause the destruction of much of the remaining forests in tropical and temperate regions, threatening 270,000 plant species and 75 percent of the world's mammals.[26]

In the United States the widespread use of the automobile after World War I increased the demand for roads, highways, bridges, and, above all, oil. As we passed through the 1929 crash, the Great Depression and into World War II, trucks and highways more and more dominated the landscape and reached deeper into forests and other uninhabited regions. While coal gave way to oil for home heating and other uses, oil became the great necessity of increasingly industrialized warfare, benefiting the Allies and helping to bring the Axis to its knees by 1945. Development of the federal interstate highway system during the post-war era linked states and commerce more tightly, and federally funded dams for hydroelectric power flourished in the 1950s and 1960s, with the promise of nuclear power fulfilled by 1957. It was within this milieu of hope and optimism after winning the war in Japan and Europe that I grew up in New York City. Even before the war there are memories that link industrial development and modern technological warfare. I remember reading the Sunday comic strips on the floor when the attack on Pearl Harbor was announced on our radio. I was seven years old, and eighteen days later I received a train set for Christmas, an "army supply train" with a removable tank that shot out sparks, a functioning search light on one car, and a spring artillery piece on another that shot rubber rounds across the living room. "Made in Japan" was stamped on the sides of the low-quality steel made from Chinese iron. From then on it was war toys and war games until somehow, I discovered nature in the Big Apple.

Living in Manhattan on the lower eastside until I was eight years old, my parents rarely went to Central Park, and I almost never saw a tree for all those years. The "Third Avenue El," the elevated train, ran by our cheap apartment, and I began solo trips to the movies at the age of five, walking over dark cobblestoned streets beneath the El. Happily, when I was eight we left Manhattan for Staten

Island, a city borough full of old neighborhoods with tree-lined streets and a good bus system. The extensive, several square mile hilltop woodland of the New York City boy Scout Camp was the magical place at the center of the island which turned my mind towards a fascination with wildness and wild animals somewhere in distant wildernesses. I thought of becoming a forest ranger when I grew up, but as you know, I settled for geologist, at least for awhile. However, rather suddenly my parents fled this greener part of New York City, partly due to the fear of nuclear war between the United States and Russia in 1950. Bomb shelter drills were already common in the grade schools, and Cold War paranoia was intensifying. When I left Staten Island at age 15, my interest in natural history and geology were based upon sheer intellectual curiosity, with no thought whatsoever about what I might be employed for except somehow to work in behalf of nature. No sooner did I embark upon my undergraduate career in geology and biology that I was reminded of the professional tracks for geologists. One would either become a petroleum geologist to keep the oil flowing, or a mining geologist to develop coal and mineral resources. Somewhere along the way my ideal of working for the preservation of nature was compromised. On through graduate school I became a mining geologist, and upon graduation engaged in exploration for lead, zinc, iron, dolomite (for smelting iron), uranium, boron (for the space industry), and even gold in Central America where I earned enough money in a short time to buy myself the freedom of travel which began the process of gradually breaking away from the treadmill of working as a resource-development scientist.

Naively, in the early 1960s I still believed that working for the federal government would be more altruistic, less self-serving, and more on the side of nature than working for private mining companies. Smokey the Bear had a good reputation at this time. I took the federal professional geologist examination and became a ground-water geologist, assigned to a ground-water study in Santa Barbara County, California. My colleagues at the office enjoyed nature for hunting and fishing. They were also comfortable with recommendations, in many of our reports, to destroy riparian vegetation along streams in order to increase the long-term safe yield (the amount of ground-water which could be pumped in perpetuity without harming the basin) of ground-water basins for urban and agricultural uses. These attitudes, which blithely accepted the destruction of riparian wildlife habitat for utilitarian purposes, were also held by the Army Corps of Engineers who were lining the area's stream channels with concrete to maximize flood control efficiency. During the mid-1960s the citizens of Santa Barbara began to raise their voices in protest against the practice of concretization of waterways. About the same time many of us were reading Rachel Carson's *Silent Spring*, and the Wilderness Act was passed in 1964. Something new was in the air, and when the USGS closed their ground-water research program in Santa Barbara, I declined the transfer to Long Beach, took a half-time job as a geologist for the Santa Barbara Museum of Natural History, and returned to graduate school to study European history. At the same time I became an environmental geologist writing ground water studies and environmental impact reports for Santa Barbara County while finishing the Ph.D. in History. My early interests in and concerns for nature as a boy had merged with a growing nationwide interest in environmental protection and management.

The point of this brief account of my own personal experience in relation to environmental impacts is that during the first half of the twentieth century there were many citizens like myself concerned with slowing and constraining the modern industrial assault upon nature, but the general public was not listening, much as today it is not listening in the first decade of the twenty-first century. However, as the environmental impacts of modern industrial civilization reached the threshold of negatively affecting a large percentage of the United States population, a popular mandate developed in the 1960s for protecting citizens from air pollution, water pollution, the rise in cancer, and other unanticipated environmental impacts of twentieth century industrial society.

Before considering the Environmental Movement to which this shift in consciousness gave rise, we must develop a larger perspective of twentieth century industrial development. There is no better place to begin to grasp the magnitude of twentieth century industrial expansion and its ecological consequences than with Josephson's *Industrialized Nature*, because his book emphasizes the truly monstrous power of twentieth century technology, what he calls "brute force technologies," even more strongly than J.R. McNeill does in *Something New Under the Sun*. Less comprehensive than McNeill's environmental history of the twentieth century, Josephson's history demonstrates the ecological destructiveness of the mega-technological triumphs of the twentieth century, the gigantic dams, efficient timber harvesting machines and earth movers, floating fish-factories, and the network of "corridors of modernization", highways, railroads, canals, bridges, power stations and power lines, and enormous extensive irrigation networks which facilitate rapid resource development. "What I term the corridors of modernization make it possible to move people and machines rapidly and aggressively into the interior of a land, establish outposts of technological civilization, and begin to harness natural resources *to the demands of urban residents*." [27] I emphasize the demand of urban residents because city-dwellers are for the most part particularly disconnected from the facts regarding where their food, water, fiber, and fuel come from, and what ecological damage is done in the processes of extracting, refining and transporting these goods to them. Recall from Chapter One that Spengler identified this estrangement of urban man from nature in the late stage of all civilizations.

In Chapter One I emphasized the distinction that should be made between ecosystems on the one hand and urban and other humanized systems, i.e. *dominantly* artifactual systems, on the other. Although environmental history itself has contributed to the blurring of these distinctions, they nevertheless suggest the *degree* of human intervention with natural processes on Earth. Even J.R. McNeill's use of the term "agroecosystems" is misleading in that modern twentieth and twenty-first century agricultural practices have created managed monocultural systems for the most part, simplified systems which humans have constructed *in place of ecosystems*, which are far less modified by human intervention. Josephson's research supports this crucial distinction by calling our attention to the vast disjuncture separating these two kinds of systems, especially since the twentieth century. Around the globe, Josephson demonstrates the gargantuan scale of our modern projects, from the transformation of Amazonian tropical forests into grazing lands (and unsustainable ones at that), rubber plantations, mining districts extensively removing soil and vegetation, to the "Stalinist Plan for the Transformation of Nature," which took giant steps towards transforming central Russia and the Ukraine into " a mighty agricultural and industrial machine," and which included the largest hydropower station in the world.[28] Documenting comparable mega-projects in China, Norway, and the United States, Josephson calls our attention to the fact that these Earth-transforming systems consist of far more than merely large technologies. They also include government bureaucracies and world trade organizations proselytizing and financially supporting Western-style development, legions of scientific researchers, engineering firms that design specific technologies, construction firms that build them, and the armies of workers who manipulate the resources.

   More insidiously, the brute force technology established in the twentieth century has gradually transformed our perception of humankind's relationship to nature through the changing language of science. In "resource management" journals industrial metaphors have supplanted biological ones, beginning in the 1920s and 1930s. "From that point on, nature was industrial--in machine metaphors that supplanted biological explanations for vital functions of plants and animals and in the view of rivers, fields, and forests as closed systems that would operate as humans specified."[29] Nature had become "industrial" in the sense that Taylorism (the scientific management of human labor) and Fordism (mass production on an assembly line) would now be extended to nature herself, as forests were converted *en masse* into lumber and wood pulp, and water, fish, and other resources, especially food plants, were mass-produced. Brute force technology and its mass processing of nature-as-resources into commodities developed into a modern norm which has transcended differences in political ideology to become a global practice. Brute force technology spread so widely in the twentieth century because it is highly efficient as practiced under all ideologies, it has been held up as a symbol of human dominion over nature associated with worldwide "progress", the enormity of its achievements has been used to glorify both capitalist and socialist states, it has been used to justify neocolonial development on the Western model in so-called "backward" or undeveloped nations, and, related to achievements in material progress, industrial technology has been used to glorify the heroic onward march of science and technology in behalf of the good of humankind.

In the long run, perhaps the potentially most devastating of twentieth century brute force technologies involved the conversion of long-established labor-intensive agricultural practices into "agribusiness" or industrial agriculture. The mechanization of agriculture began in the United States with the introduction of horse-drawn threshers and reapers in the 1830s. Given the cumbersome nature and consequent limitations of steam engines, power for farmers arrived in the United States in 1892 in the form of the gasoline-powered tractor. "The United States led the way in tractor adoption because of high American labor costs and large farm size. The spread of gas stations, repair shops, and mechanics provided farmers with the needed support system. The American conversion to the tractor took place between 1920 and 1955." The U.S.S.R. became a tractor enthusiast in the 1930s, and tractors proliferated in the United Kingdom and Europe after 1950.[30] In France, for example, where prior to World War II agriculture was labor-intensive, economically inefficient, and based upon the small family farm, the American model of machine agriculture spread rapidly after 1945. Tractors in France increased from 35,000 in 1946 to 230,000 by 1954, whereas in the United States in 1945 there were 2,354,000 tractors. The mechanization of French agriculture disassembled the nation's peasant communities and initiated a great exodus to new jobs in the towns and cities.[31] A major ecological impact of this transformation throughout postwar Europe was the removal of hedgerow and treeline property boundaries to increase field size for machine harvest of monocultural crops such as wheat, thereby removing limited but precious "edge habitat" so important to the continued viability of populations of small mammals, reptiles, birds, and insects in dominantly agrarian environments.

Worldwide, the mechanization of agriculture drove agrarian populations out of traditional farmlands and into the towns and cities. In the United States, where half of the population worked as farmers in 1920, by 1990 the farm population had dropped to 2-3 percent. This widespread demographic abandonment of farmland in many parts of the world continued throughout the twentieth century, as breakthroughs in plant breeding and the Green Revolution increased crop yields while making farm laborers redundant. "The Green Revolution was a technical and managerial package exported from the First World to the Third beginning in the 1940s but making its major impact in the 1960s and 1970s. It featured new high-yielding strains of staple crops, mainly wheat, maize, and rice."[32] Beginning with the development of hybrid corn in the United States in the 1920s, plant breeding in Mexico under the direction of Norman Borlaug and his associates in the 1940s produced new wheat strains which produced high-yield crops in response to heavy applications of nitrogen and irrigation water. Comparable new breeds of rice followed in the 1960s, and these breakthroughs promised the hope of feeding the world's growing populations. Ecologically, however, the Green Revolution combined new plant hybrids with mechanization to promote the farming of monocultures. Farmers hereafter would have to purchase seed, fertilizers, and pesticides as well as mechanized equipment. Because monocultures are prone to pest problems, application of large dosages of fertilizers and pesticides throughout the world has poisoned air, surface water, soil, and groundwater. By 1990 20,000 people were dying each year along with wildlife that ingested the widespread pollutants associated with industrial agriculture.[33] Rachel Carson's *Silent Spring* blew the whistle on this increasing hazard in 1962, but the problem has continued almost unabated to the present day. Increasing incidences of cancer in humans, as well as other serious illnesses, have been traced to the rise of agribusiness. J.R. McNeill observes that the twentieth century antibiotic and agricultural revolutions both represent "a drift toward ever greater complexity, and potential vulnerability to disruption, in the systems that underpin modern life."[34] Further, Clive Ponting places the agricultural revolution in a global context that portends future struggles between developed and underdeveloped nations: "The result of all these changes in food production, trade and population growth has been to create an unbalanced world agricultural system. It reflects the distribution of political and economic power between the industrialized countries and the Third World that emerged in the period following the great expansion of Europe after 1500."[35]

The purpose of this section has been to give some idea of the range of environmental impacts associated with modern scientific and technological development in the twentieth century. It is beyond the scope of this work to extend this discussion into the realm of global impacts related to massive energy production since 1900. However, these global environmental impacts, such as acid rain, depletion of the ozone layer, atmospheric and oceanic warming, are described not only in thousands of books and scientific articles but in the daily newspapers and popular magazines of our time. We have become so used to this background information concerning global environmental degradation that, like the Cold War and nuclear threat of the mid-50s and 1960s, we have relegated these portentous changes to the level of something like background noise, so little important to daily life that they are generally ignored by the vast majority of citizens worldwide. "Life goes on," as the saying goes, but if most of us truly understood and deliberated upon our predicament, we might actually begin to initiate serious thought and action in defense of a nature under brutal assault worldwide. Fortunately, at least a small number of us have realized the gravity and extent of human-caused environmental degradation intermittently since classical antiquity, and with the onset of modern Western civilization and its mechanistic science and technology, increasing numbers of us have taken up the cause of nature's defense against human depredation. In the section that follows I will focus on the origins of this defense in the United States during the nineteenth and early twentieth centuries, but from a different perspective from that taken in chapters Six and Seven, which focused more on changes in ecosystems in Europe and consequent changes in European attitudes toward nature. Because the twentieth century Environmental Movement, the most important manifestation of changing attitudes towards nature in the Western worldview, had its immediate origins in the United States and Canada, I shall focus on North America in pursuit of its origins.

Pre-Environmentalism in the United States in the Nineteenth and Twentieth Centuries

Viewed from the twenty-first century, European and North American industrialization in the nineteenth century probably does not appear so devastating as recent ecological impacts described in the previous pages, particularly in North America. A century that began with the Lewis and Clark expedition to unexplored Indian territory, experienced the Civil War and completion of the transcontinental railroad in the 1860s, fought the western Indian Wars into the 1890s, and built brownstone churches and libraries in large cities and towns before the century ended appeared to all but a few to be a "century of progress." During the same century, across the Atlantic Europeans struggled through the Napoleonic wars, attempted to reinstall decadent ancient regimes, fought revolutions and counterrevolutions in the streets of Paris in 1848 and 1870, celebrated the triumphs of industry and technology at the London Crystal Palace exhibition of 1851 (eight years before Darwin's *Origin* created a cultural disturbance), maintained imperial control of resource bases in India, Asia, Africa, and Latin America, and expanded trade and industry vigorously from 1870-1914. Unfortunately, greater killing technologies and fire power were major creations of the European and American experience of progressive change. Less noticed along the road to rapidly expanding commerce, industry, increased longevity, and higher living standards for the majority of Western humanity was the massive ecological degradation that resulted from so much frenetic activity. Struggling to surpass one's peers or simply to stay afloat at any level of economic activity, citizens of both Europe and North America generally understood the conversion of forests, grasslands, floodplains, and marshes into productive agricultural, grazing, industrial, commercial, and urban uses as a wonderful thing, particularly in America where nature's abundance appeared to be inexhaustible for most of the nineteenth century.

As Carolyn Merchant and numerous other scholars have demonstrated, Americans, i.e., Euro-Americans, saw themselves as a new wave of the Christian God's chosen people, destined to sweep the barbarous, benighted indigenes aside to create a new Eden for the people of the book. Beginning with Columbus, the Native Americans were relentlessly slaughtered, enslaved, embroiled in nationalistic European wars of empire, and both purposely and inadvertently decimated by European diseases, as described in Chapter Eight. Much of this devastation of indigenous culture happened in the nineteenth century, in the footsteps of Lewis and Clark. The grasslands of the central lowlands and Great Plains of North America provided a relatively easy access to the mountains and mineral deposits of the great western Cordillera, from Colorado to California. Thus, while Europeans consolidated their industrial revolution, and large cities grew with newcomers from the depopulating countryside, Americans struggled and fought to build a giant nation out of most of the habitable part of the North

American continent. Given the grand dynamic of "manifest destiny," how many would reflect upon the impact of all this positive activity upon the natural world, particularly when it was usually inhabited, if only sparsely, by "savages." To convert nature into culture was a progressive, civilizing process.

In Chapter Eleven we explored the rise of "pre-environmentalism" in the United States, documenting Rousseau's direct influence upon Ralph Waldo Emerson (1803-1882) and indirect influence upon Henry David Thoreau (1817-1862). Emerson and Thoreau were giants among the early and mid-nineteenth century Americans who began to reflect upon the relationship of mankind to Nature, and they were much influenced by European intellectual reflection on that relationship. However, as the frontier moved rapidly westward after Lewis and Clark, such ideas were cultivated in the highly civilized rural-urban setting inhabited by the transcendental nature philosophers in "old" states like Massachusetts. Vermont's George Perkins Marsh developed a more pragmatic argument for improving humanity's relationship with nature at the beginning of the Civil War, but it remained for John Muir (1838-1914) to invent a western, more purely American foundation for a benign human interaction with nature. Although Thoreau and the artist George Catlin (1796-1872) both had conceived of preserving nature, albeit on very different scales, and Yellowstone National Park was already established without Muir's assistance, it was his heartfelt religious experience of nature, particularly the mountains, of the western United States that gave impetus to an ideology of preservation that has formed the most powerful strand of pre-environmentalism. From Muir's role in founding the Sierra Club in 1892 to the Wilderness act of 1964, preservationism gained in stature. One might say that what Rousseau did for increasing appreciation of the pastoral and mountainous landscape of eighteenth century Europe, Muir did for the wilderness of nineteenth century America.

Chapter Six demonstrated that respect for the environment in the post-medieval Western tradition was closely related to religious belief and worship. Rousseau's deistic nature worship continued the seventeenth and eighteenth century focus upon God's creation as an indirect way of venerating the deity. Wordsworth and other European Romantics, Emerson, Thoreau, and John Muir continued this tendency, which over time concentrated more attention upon nature itself and less upon the creator. Simultaneously, mechanistic scientists maintained a Cartesian dualism which allowed them to devour nature-as-resources while attending church services with a clear conscience on Sundays. Wild nature having been all but eradicated in Europe, the American "wilderness," as the North American continent was perceived in the wake of depopulation of the indigenes, readily attracted the attention of the minority of aesthetically and spiritually hungry neo-Europeans seeking purity in wild nature. While the Industrial Revolution took hold in nineteenth century America, however, the vast majority of those engaged in taming the western frontier saw only forests to be cut, plains to be farmed, animals to be hunted for food and furs, and mineral ores to be converted into valuable metals. Worst of all, western settlers resented recalcitrant

Indians resisting the white, Christian conversion of wilderness into civilization. While industry flourished in the eastern states, albeit nonchalantly polluting soil, water, and air, trading posts became towns, and a few towns along major waterways and coasts became new cities in the west. Only during the second half of the nineteenth century, as the limits of empire were defined and cut-and-run timber companies sought new virgin forests in the western mountains, did the precursors of environmentalism become well-known to the larger American public. Also, while most Americans had accepted the environmental degradation associated with expanding industrialization for most of the nineteenth century, by 1890 some individuals were seeking ways to regulate industry and restore livability to cities which had grown up alongside industrial development.

While the urban pollution and unhealthy physical and social conditions associated with rapidly growing industrial cities was largely ignored throughout most of the nineteenth century in America, there had been, since the early decades of the century, a minority culture of individuals sensitive to the beauty and worth of nature. Certainly the Hudson River School of landscape painters (ca. 1825-1870) did a great deal to popularize the aesthetics of wild, rugged landscapes, beginning with the paintings of the Catskill and Adirondack mountains in New York, the White Mountains of New Hampshire, and even the hills surrounding the Connecticut River in Massachusetts and Connecticut. Thomas Cole (1801-1848) and his protegé Frederick Church (1826-1900) were the leading figures among American artists who attempted to represent aspects of the divine presence in their landscapes. The transatlantic European-American culture which had transmitted the foundations of American Transcendentalism from Europe to America also provided early mid-nineteenth American landscape painters with attitudes reflecting the European Romantic Movement (ca. 1770-1830). "Turner, Claude, and perhaps German romantics aided Cole's quest during his first trip abroad, out of which grew a more 'majestic' style of landscape painting and an urge to illustrate grand themes from history."[36]

Cole was born into a middle-class family in a Lancashire town in 1801. As a youth growing up during the early Industrial Revolution he was directly affected by the social pressures and upheavals brought about by industrialization. "Like other romantic artists, Cole developed a powerful antipathy to industrialization; however, unlike many of the romantics, he did so from immediate experience."[37] Cole learned engraving in England after displaying artistic ability as a youth, and, after moving to the United States with his family in 1818, continued his self-education and taught drawing and painting in Steubenville, Ohio in 1819. Cole developed an interest in landscape painting while supporting himself as a portrait painter in Steubenville, and also tried history and genre painting, copying, and painted transparencies. Simultaneously, he studied the work of landscape painters Thomas Birch and Thomas Doughty. While Cole gained experience during the early nineteenth century, Americans were developing a taste for scenery. Tourists were drawn to scenes of awesome grandeur, particularly waterfalls. Such tourist destinations as Niagara Falls, the

Catskill Mountain House with its views of the Hudson River Valley, the view of the Connecticut Valley from Mount Holyoke, Massachusetts, and the natural Bridge of Virginia were becoming well known through guidebooks, travelers' descriptions, and paintings, prints, and drawings of natural features. Although small numbers of tourists had visited the spectacular view site of the Catskill Mountain House decades earlier, once the Mountain House Hotel and a good road were built, the site was made famous in newspaper accounts and in James Fenimore Cooper's 1823 novel, *The Pioneers*.[38] Thus, the stage was set for the arrival of Thomas Cole seeking employment as a painter in New York City, soon after which he discovered the view site and its spectacular surroundings of canyons and waterfalls for himself. Beginning in 1825, Cole's magnificent and inspiring canvases began to capture the attention of artists and aesthetes in New York City. Cole, of course, is also well-known for his religiously inspired *The Voyage of Life* series of paintings and his didactic set of historical paintings, *The Course of Empire*. The important point here is that he was not isolated in popularizing a taste for wild nature in the 1820s and 1830s, but, rather, was one of the geniuses who stood out as some Americans began to recognize the value of their natural environment. While a majority of Americans idealized an Edenic, pastoral nature at this time, a pre-environmentalist minority culture gained adherents particularly in the eastern states where the threat of predators and savage men had disappeared from all but the remotest corners of that region. Environmental historian Alfred Runte affirms the importance of Cole and the Hudson River School for a growing "party of nature" in America:

> With the rise of the Hudson River School of landscape painting, cultural nationalists found their first vindication. Prior to evolution of the genre during the 1820s and 1830s, its predecessors usually did little more than imitate European styles and subject matter. In contrast the Hudson River School broke the bonds of tradition and looked directly to nature for guidance and inspiration. For the first time American artists disdained merely reinterpreting Old World buildings and ruins for the hundredth or thousandth time. Instead the Hudson River School searched for truth and realism in the natural world, confident that only the unchanging laws of the universe contained real wisdom and meaning for mankind. Artists were advised to depict mountains, forests, river valleys, and seacoasts, where, despite random human interruptions, the hidden but ever-consistent laws of nature could still be deciphered.[39]

Frederick Church was the pupil and protegé of Thomas Cole, and became famous in the 1850s and 1860s for the expansive canvases which spectacularly represented the sublime in nature, from Niagara Falls to the Andes Mountains of South America and icebergs of the North Atlantic, all of which he experienced personally, sometimes at the risk of his life. Church, following two years of study under Cole, was already successful at the age of 21, and painted a memorial to his mentor after Cole's premature death in 1848. Church's stunning landscapes and seascapes, and especially the skies associated with them, outdid even Cole's considerable technical skill, and his work was often compared with that of Caspar David Friedrich. "Whether one compares Church to Friedrich, Bryant to William Wordsworth, or Ralph Waldo Emerson to Friedrich Wilhelm von Schelling, the fundamental point remains the same: the investigation of the complex relationships between man and the natural world dominated the cultural life of the mid-nineteenth century in a way perhaps unmatched by any other single concern." [40] As John Opie writes, there was an " extraordinary influence of romantic artists like Albert Bierstadt, Thomas Moran, and Frederick Church on an American public that wanted instruction as to how to see its wild places. A visit to a national park like Yellowstone, Glacier, the Grand Canyon, or Yosemite was soon sought as a once-in-a-lifetime adventure, not just another summer vacation trip."[41]

The nineteenth century roots of environmentalism consist of an evolving interplay of Christianity and Western nature worship, the former an ancient but protean foundational mythology, the other a new, modern phenomenon reflecting an increasing preoccupation with nature in the emancipated, post-medieval Western psyche. The religious context of the developing nature philosophy in America during the nineteenth and early twentieth centuries is seminal to our understanding of profound shifts in our attitudes toward nature down to the present day. A brief sketch of this powerful Christian context is necessary at this point, particularly in order to grasp the idiosyncratic way in which the American belief system assimilated nature worship into its reinvigorated Christian worldview since the period of colonization.

The early settlers in the Massachusetts Bay colony and in Virginia established what was essentially a state church, in which all people were required to attend the church services of a particular religious denomination (Congregationalism in the case of the Massachusetts Bay colony, Anglicanism in Virginia). These official religions were extremely intolerant of other denominations. When, following the American Revolution, the constitution was adopted, the implicit intolerance of the early state churches was repudiated, not because of a strong tendency towards secularism, but rather to legitimate a balance of power between existing denominations, none of which was strong enough to impose a single state religion upon the new nation.[42] Protestant domination of American religious belief was forced to loosen its grip, however. "The United States' original Protestant-Christian bias was challenged and broadened by waves of Roman Catholic and Jewish immigrants in the nineteenth and early twentieth centuries and later by agnostics and secularists."[43] Following the constitution's disestablishment of religious authority, religious denominations needed to innovate in the struggle for the minds of believers, and they found their best opportunities, particularly new sects such as the Mormons and Disciples of Christ, on the American frontier, where sparse populations in small, widely separated settlements found comfort in frontier congregations. Baptists and

Methodists were especially successful in gaining converts on the western frontier, the Methodists becoming the largest religious group in the country by 1850, with the Baptists not far behind. These developments offer part of the answer to a question that has often vexed my mind. Why is American Christianity so vibrant in comparison to that of the Western European nations from which it came? Churchgoers in France, England, and Germany today are no more than 15-20%, whereas perhaps nearly 50% of Americans attend services regularly. What is it that has kept Christianity alive in America despite the nation's vulgar materialism and barbarous popular culture? One might suggest that Christianity was renewed and the clock of history reset with each new beginning: with colonization and settlement, in the brutal struggles with native Americans; with the French and Indian wars of the middle eighteenth century, with expansion and colonization of frontier lands; with the challenge of the imported ideas of freethinkers, atheists, secularists, communists and socialists, and; perhaps most challenging of all, with the evolutionary biology of Charles Darwin which crossed the Atlantic in the minds of some late nineteenth century immigrants. Only on the periphery of Europe, in countries like Ireland and Poland, has Christianity held out against the challenges of dissonant new scientific ideas undermining to Christian belief, and the lure of a secular life style which grew out of Enlightenment unorthodoxy.

Nevertheless, despite the secularizing and demythologizing inroads of modernity which assaulted educated Americans in the city and undermined their Christian beliefs, new waves of Christians, ready to restart their denominations relatively free of persecution in a new land, arrived later in the nineteenth century, as Lutherans from Germany and the Scandinavian countries settled the farmlands of the Midwest and the north-central part of the United States. Catholics also immigrated in substantial numbers beginning in the 1830s, causing uneasiness and intolerance in a dominantly Protestant nation. With this brief sketch of American Christianity in mind we can better understand the shift in attitudes toward nature which run through Transcendentalism and the Hudson River School to John Muir and Aldo Leopold during the century and a half that preceded the Environmental Movement of the 1960s and 1970s.

The Transcendental nature philosophers, Emerson and Thoreau, and the Hudson River School landscape painters Cole and Church all perceived nature as the creation of a transcendent deity who designed the intricate complexity and beauty of the natural world. Their perception of nature generally reflects the "argument-from-design" which was accepted by most Enlightenment deists as well as liberal clergymen in the eighteenth and nineteenth centuries, a view of nature which persists in the minds of many Christians today. Whereas Thomas Cole passed away contentedly without any challenge to this belief, his younger pupil, Frederick Church, very possibly experienced doubt over the argument-from-design as a result of his encounter with Darwin's evolutionary thought. Church was an admirer and a careful reader of Alexander von Humboldt's encyclopedic descriptive works on nature, including *Cosmos: A Sketch of a Physical Description of the Universe* (1849). Church even modeled his painting expeditions to the Andes on the earlier expeditions of Humboldt. Humboldt was inspired by the belief that the external phenomena of nature represented " one great whole, moved and animated by internal forces.... Nature is a unity in diversity of phenomena; a harmony, blending together all created things, however dissimilar in form and attributes; one great whole animated by the breath of life."[44] Humboldt's vision of a harmonious universe was a pseudo-scientific pronouncement of the argument-from-design, the internal forces identified with the handiwork of the creator. "It also embodied the guiding principles that animated Church, and that Darwin would tear down with a theory of conflict and balance between internal and external (largely random) forces."[45] Thus does evolutionary theorist Stephen J. Gould apprise us of his intention to demonstrate Darwin's influence upon Church. This deconstruction began with Darwin's theory of natural selection in *On the Origin of Species* in 1859. Perhaps fortunately, Humboldt died that year, when Darwin's theory (described in detail in chapter Nine) "and the radical philosophical context of its presentation drove Humboldt's pleasant image to oblivion."[46] Recapitulating three crucial aspects of the new Darwinian worldview: (1) nature is a scene of competition and struggle, not higher harmony; (2) evolutionary lineages have no intrinsic direction toward higher states or greater unification; and (3) evolutionary changes are not propelled by an internal and harmonious force, Gould observes that Church was but one of many humanists whose worldview was shattered by this new and heartless view of nature. Late nineteenth and early twentieth century "weldschmerz" (world sorrow) undoubtedly owed more than a little to Darwin's obliteration "of a world lovingly constructed with intrinsic harmony among all its constituent parts."[47] Gould thereafter speculates upon the devastation wreaked upon Church's psyche and creative will to paint benign and harmonious landscapes, unfortunately with little evidence at hand to prove his point. Later in life Church suffered severe inflammatory rheumatism and lost the use of his painting hand. He was also much engaged in the construction, furnishing, and enjoyment of his Hudson River Valley mansion, Olana, after 1859. His extensive library, replete with the works of Humboldt, Wallace, and post-Darwinian Christian evolutionists, and even some works of Darwin, including *The Voyage of the Beagle*, nevertheless contained no copies of *On the Origin of species* or *The Descent of Man*. Perhaps the aging Church clung to the argument-by-design after all.

American environmental historians often give the Hudson River School a certain amount of credit for contributing to increased interest in nature on the part of the general public in America between the 1820s and 1870s. Through the 1860s paintings of the eastern United States by Cole and Church gained a wide audience of admirers, and Albert Bierstadt and Thomas Moran in the 1860s and 1870s led the "Rocky Mountain School" of landscape painters with their large and spectacular canvasses of Yellowstone, Yosemite, Rocky Mountain peaks, and the Grand Canyon. Environmental historian Alfred Runte gives these artists the full credit they deserve by relating their inspirational paintings to the origin of American national parks.[48]

Although paintings of the Catskills, Adirondacks, and White Mountains and unusual features such as Niagara Falls and Natural Bridge, Virginia by Cole, Church, and others were certainly dramatic for the most part, they would be overshadowed by the grand vistas of the far west after 1859, the year that church completed and displayed his grand masterpiece, "Heart of the Andes," a South American sequel to "Niagara" (1857). As Runte remarks, "The modern discovery of Yosemite Valley and the Sierra redwoods, in 1851 and 1852, respectively, provided the first believable evidence since Niagara Falls that the United States had a valid claim to cultural recognition through natural wonders."[49] Niagara Falls, the only great scenic wonder that Americans could hold up to Europeans as superior in scale, had suffered severe exploitation at the hands of private promoters. Suddenly, with the opening of the "wild west" at mid-century, the United States could claim possession of grandiose landscapes comparable and even superior, it was claimed, to the grandeur of the Alps and Pyrenees. As early as 1857 colonel A.V. Kautz proclaimed the Cascade Range around Mount Rainier to be " mountain scenery in quantity and quality sufficient to make half a dozen Switzerlands."[50] The waterfalls of the Yosemite came in for the highest praise beginning in 1859 when Horace Greeley, owner and editor of the *New York Tribune*, after visiting the Yosemite Valley, proclaimed it "the most unique and majestic of nature's marvels."[51] Thus began a virtual campaign of announcing to the civilized world, notably Europe, the superiority of American mountain scenery. Runte encapsulates this nature-based nationalism as follows:

> These claims, however trivial from today's perspective, then filled an important intellectual need. For the first time in almost a century Americans argued with confidence that the United States had something of value in its own right to contribute to world culture. Although Europe's castles, ruins, and abbeys would never be eclipsed, the United States had "earth monuments" and giant redwoods that had stood long before the birth of Christ. Thus the natural marvels of the West compensated for America's lack of old cities, aristocratic traditions, and similar reminders of old world accomplishments.[52]

Albert Bierstadt and Thomas Moran took to painting these "natural monuments" with gusto, thereby contributing to a developing interest in preservation that led to the creation of the great national parks of the West. Bierstadt and Moran's works were even purchased by the federal government to be displayed in the nation's capitol in the 1870s. Thus, the stage was set for the preservation of monumental landscapes and the beginning of the National Park system. There was no thought of preserving wilderness, ecosystems (the word ecosystem was coined ca. 1870-1875), or species diversity, but sizeable tracts of land became unavailable to exploitation and destructive development beginning with Yosemite in 1864 and Yellowstone in 1872. Yosemite and the Mariposa

Grove of Sierra redwoods were turned over to California for administration at first, but in 1905 these lands were ceded back to the federal government and joined with a larger area to create Yosemite National Park. Therefore, Yosemite was the first national park in practice, if not in name. Exploration of the Yellowstone region in the early 1870s quickly led to proselytization in behalf of its extraordinary features (geysers, hot springs, waterfalls, mountain scenery, and spectacular wildlife) as worthy of federal ownership for public enjoyment and recreation, and the park was created in 1872.

Thus the stage was set for a major step to be taken in behalf of wilderness itself when John Muir (1838-1914) arrived at Yosemite Valley in 1868, several years after Albert Bierstadt had made his sketches for *Valley of the Yosemite* (1864).[53] Muir would establish and promote the basis for preserving wilderness for the value of its ecosystems, although that term was not yet in use. However, a broad public acceptance of such an idea "awaited an age receptive to the life-giving properties and aesthetic beauty of all ecosystems. Well into the twentieth century, Americans valued the natural wonders of the West almost exclusively for their scenic impact."[54] The wisdom Muir gained during his many wilderness treks and mountain climbs in the Sierra Nevada Mountains and elsewhere generally fell upon the deaf ears of an American public steeped in traditional Christian religion (in all of its abundant manifestations), which assured them of personal salvation and respect within civil society while their related utilitarian values rationalized the conversion of wild nature into farms, towns, cities, and factories at a dizzying pace by past historical standards. Nevertheless, Muir's insights into the intrinsic value of nature were heeded by a growing constituency of emancipated minds, which paid off in the creation of additional national parks and a National Park system by 1916, and in the founding of the Sierra Club in 1892. He may have lost the battle to save the scenic Hetch Hetchy Valley from becoming a water-supply reservoir for San Francisco (approved in 1913), but the struggle gained such publicity that public appreciation of the value of wild places was much enhanced. As Runte observed, "no defeat so forced the issue of how best to guard the national parks in an urban, industrial age."[55] Muir also lost a titanic struggle with utilitarian forester Gifford Pinchot (1865-1946) over the future of our national forests, a struggle which was rekindled within the U.S. Forest Service during and after the Environmental Movement of the 1960s and 1970s, and which sputters on to the present day.

John Muir and Gifford Pinchot became good friends beginning in 1892, when they met in the Adirondacks and began the habit of hiking and camping together as opportunities allowed. Only a year earlier congress had passed the General Land Law Revision Act, which became known as the Forest Reserves Act of 1891 and, which laid the basis for the future national forests by withdrawing land from the public domain under federal protection. At the time, Muir surmised that the protected forests would not be utilized for lumber, but rather would serve as watershed and undeveloped wilderness. Pinchot, on the other hand, recently trained in European forestry methods and having preferred clear cutting to select cutting, envisioned managing these forests for a sustained yield.[56] The degree of grazing and mining to be allowed in the federal forests was also a major issue. In 1897, the subsequent Forest Management Act "provided the opening wedge for the rational development which Pinchot preferred."[57] That same year Pinchot was appointed as a special forest agent for the Department of Interior. Unlike some of his preservationist friends, Muir still believed that Pinchot would support preservationist causes from his governmental position. Even by this time, Muir himself thought that national forests could be preserved and used, but in 1898 he shifted to the belief that forestry and wilderness preservation were incompatible. Pinchot and Muir were finally on opposite sides of the fence when in 1905 Pinchot was appointed head of the Division of Forestry in the Department of Agriculture. The schism between conservationism (as rational, efficient use or resource utilitarianism) and preservationism (as preservation of unutilized wilderness for its own sake and human enjoyment of it) was out in the open, soon to be exacerbated by the rift between Muir and his preservationist followers, and Pinchot's growing community of government scientists and politicians acting in behalf of their conception of the public welfare.[58]

The spectacular Hetch Hetchy Valley became part of Yosemite National Park in 1890 and was designated a "wilderness preserve." However, even as early as the 1880s San Francisco's water board and politicians had coveted the valley as a reservoir site for the city. A protracted struggle to acquire this land for dam construction continued from 1903 until 1913, when Congress finally approved the dam. This decision in favor of utilitarian development over wilderness preservation resulted in numerous protests " from women's groups, outing and sportsmen's clubs, scientific societies, and the faculties of colleges and universities as well as from individuals. The American wilderness had never been so popular before."[59] Bitter with the taste of defeat in his mouth, John Muir died a year later. Although preservationists won the next round in the struggle for setting wild lands aside with approval of the National Park Service in 1916, the paradigmatic model for the dominant attitudes and practices of twentieth century Americans was set by the "gospel of efficiency" preached by Gifford Pinchot and growing numbers of utilitarian conservationists within the resource agencies of the federal government and state governments in the United States. As Samuel P. Hays has observed: "Conservation, above all, was a scientific movement, and its role in history arises from the implications of science and technology in modern society.... Its essence was rational planning to promote efficient development and use of all natural resources."[60] Thus, despite the growing interest in monumental landscapes linked to railroads and automobiles in the new system of national parks after 1916, John Muir's dream of a wilderness experience for multitudes of Americans was compromised from the start by a utilitarian worldview rooted in the Christian idea of dominion over nature, the centuries-long practice of implementing dominion over nature in America, and the rising hubris among early twentieth century scientists and technologists that nature-as-resources could be managed in perpetuity under the scientific laws and principles of the modern mechanistic scientific paradigm. After Darwin, the radically different perception of nature based upon evolutionary biology and the related biological sub-discipline of ecology developed in the minds of a minority group of scientists. However, it was not until the "brute force" technologies of the twentieth century had spun off severe environmental impacts which were called to the attention of the American public in the 1960s that the conflict between the two scientific paradigms (mechanistic and ecological) became an open public controversy.

The Twentieth Century Worldview in America before the Environmental Movement

For the remainder of this chapter I shall consider the rising tension within the Western worldview precipitated by events leading to the Environmental Movement as the industrial-technological society gained momentum on both sides of the Atlantic. However, whereas Americans looked with envy at the historic landscapes and architecture of Europe in the eighteenth and nineteenth centuries, by the twentieth century Europeans were taking note of American advances in technology and industrialization. While Europeans admired these developments, a seer like John Muir was exploring the Sierra Nevada and seeking respite from the accelerating pace of ordinary life in the United States. His insights regarding the growing rift between Americans and a vulnerable natural world may throw some light on changes occurring in the American version of the Western worldview early in the twentieth century.

The great paradox of American civilization, the most successful colonial outpost of Western civilization, is its consistent and often strident Christianity on the one hand, and its extreme materialism and cultural vulgarity on the other. Contemporary televangelists appear to reconcile these seeming polar opposites effortlessly to produce an effect at once stunning and grotesque. Although highly critical of American electronic barbarism, Europeans and Asians impressively mimic American culture themselves. Thus, when one attempts to depict the American character or spirit, problems immediately arise in the face of its ambivalence. However, if one poses the question of the American character in terms of our perception of and treatment of nature, the muddy waters of our syncretistic culture definitely clarify somewhat. Like our European forefathers, we are relentlessly utilitarian and exploitative, with a small minority of dissenters such as Thoreau and Muir in the nineteenth and early twentieth centuries. The entire American historical era of 1865-1962, from the end of the Civil War to the beginning of the Environmental Movement, was one of fascination with what some historians have dubbed "the technological sublime," which involved the worship of technology as representing an awe-inspiring power, much as the natural sublime (discussed in relation to deism and Romantic nature worship in Chapter Six) implied the power of the deity represented in the argument-from-design. If the monumental sublime fascinated the Hudson River and Rocky

Mountain painters, alongside their nationalistic nature worship there arose a worship of technology. "The appearance of large-scale industrial systems was hailed as 'the second creation of the world' and 'the second discovery of America.' "[61] The world's greatest steel production and automobile manufacture also drew European admiration early in the twentieth century, and by the 1940s tourists were flocking to see the Hoover and Grand Coulee dams, along with steel skyscrapers and gigantic bridges, the greatest manifestations of the technological sublime.

Lewis Mumford has attempted to explain the American fascination with invention, technology, and the giant structures associated with brute force technology:

> Given an old culture in ruins, and a new culture *in vacuo*, this externalizing of interest, this ruthless exploitation of the physical environment was, it would seem, inevitable. Protestantism, science, invention, political democracy, all of these institutions denied the old values; all of them, by denial or by precept or by actual absorption, furthered the new activities. Thus in America the new order of Europe came quickly into being. If the nineteenth century found us more raw and rude, it was not because we had settled in a new territory; it was rather because our minds were not buoyed up by all those memorials of a great past that floated over the surface of Europe. The American was thus a stripped European; and the colonization of America can, with justice, be called the dispersion of Europe, a movement carried on by people incapable of sharing or continuing its past.[62]

No one in America was unaffected by the progress of invention, Mumford tells us. Invention, in turn, propelled the utilitarian conquest of the American environment. This implicit ethic in American society was made fully explicit in the gospel of efficiency of Gifford Pinchot, Theodore Roosevelt, and the American Progressive Movement. "Technical prowess also laid an ethical framework on an otherwise brutal and amoral nature because it could serve a higher good. The Progressives repudiated the Transcendentalists: Society should not emulate nature but instead become a finely balanced, well-oiled, and efficiently humming machine."[63] Thus, the upshot of the Progressive Movement of 1890-1920 was to create and implement a belief in the efficacy of the machine as a model for both the proper utilization of nature-as-resources, and the appropriate management of mass democratic society in the twentieth century and on into a benign technology-based future. All of this would seem to leave the neo-protestant, ex-Protestant John Muir and his followers as a chorus crying in the wilderness. However, Muir's discontent with the triumph of efficient management of resources and society produced a profound critique of modernism which resonates with the protests of late twentieth and early twenty-first century environmentalists. Oelschlaeger recognizes this deeper meaning of Muir's writings since 1867 as confirming "his dismantlement of a Judeo-Christian-based anthropocentrism and an unmistakably clear grasp of a biocentric perspective on wild nature."[64]

Within Muir's biocentric perspective, humans should be empathetic participants within natural systems rather than scientific observers separated from nature by their anthropocentric objectivism. Oelschlaeger argues that Muir passed through the stages of theism, panentheism,[65] and on to pantheism during his life. Muir's mature wilderness theology/gospel is "a remarkable post-Darwinian pantheism." Muir, therefore, differed from the Transcendentalists, who were panentheists because they recognized a separation of the deity from its divine creation. Muir's search for humanity's place in nature led him to eschew the Christian construction of "Lord Man" for the diminished but eminently saner role of biotic citizen, an idea which anticipates the thought of Aldo Leopold. Muir is usually depicted as a critic of modernism, that "complicated concatenation of ideological presuppositions, including ideas that progress is inevitable, that the power of science and technology is unlimited, that humankind represents the apex of creation, and that the natural and cultural worlds can be understood on the basis of a machine metaphor."[66] Oelschlaeger insists, however, that Muir is not an antimodernist but a precursor of constructive postmodernism. For example, Muir does not reject science but scientism, the use of traditional, classical science to support the mechanistic view of nature which underlies the modernist worldview. The basis for re-evaluating Muir in postmodern context is to understand him as a "proto-ecologist" struggling to assimilate the implications of Darwinian evolutionary biology in the context of modern destructiveness toward nature. Muir's arch-nemesis, as well as that of contemporary environmentalists, is *resource conservation*, which "uses the bandages and palliatives of mainstream ecology to ensure that advanced industrial societies extract the last measure of value of the natural world, and accordingly does not question the underlying assumptions of that culture."[67]

The American variant of the Western worldview reflected a society made up of displaced Europeans whose worldview was already under the influence of scientific and technological change imbibed within the framework of Protestant Christian dogma. Beginning in the age of colonization and continuing into the twentieth century, the new American society necessarily subsumed its increasingly powerful utilitarian capability to exploit nature within the context of Christian belief. "Manifest destiny" was the American expression of the Christian idea of providence adapted to the reality of modern rational thought in politics as well as in science and technology. The natural sublime of the American landscape was thus transformed into the modern technological sublime, i.e., increased human control over nature as a manifestation of modern progress. In Chapter Six I explained this blend of Christian providentialism and modern progressivism as an idea of progress termed "millenarian progressivism." Carolyn Merchant has explained this mixture of supernaturalistic and secular interpretations of American history in her account of the technological transformation of America as a recovery narrative in which the displaced European Christians of the Old World build a new Christian Eden out of a godless wilderness, thereby uniting the ideas of providence and progress in a unique American worldview.[68] This worldview allows members of American society, down to the present day, to "have their cake and eat it too" by rationalizing Christian materialism in an increasingly secularized world dominated by European ideas. America became more isolated from its European motherland, however, with the growing intellectual dominance of the Darwinian worldview which grew up concurrently with the settlement of the American west. The unwanted Darwinian worldview (described at the end of the last chapter) spawned by modern science threatened to undermine the moral integrity of the technological sublime created within the framework of the Christian millenarian (providential-progressive) worldview which has dominated the American mind for several centuries. Around 1920 H.L. Mencken remarked, "Heave an egg out of a Pullman car anywhere in the United States today and you're bound to hit a fundamentalist." Plus ça change, et rien ne change pas (The more things change, the more they remain the same).

The interplay between Darwinism and Christian fundamentalism in America symbolizes the cultural schism which divides not only the United States, but also the objectives of modernity and those of traditional societies worldwide. Before examining this fissure in greater detail in the following chapter, we must examine the decades immediately preceding the American Environmental Movement.

Prelude to the Environmental Movement

The Progressive Movement died out as World War I took center stage in America as well as in Europe. It was followed by the "Roaring Twenties," a decade of prohibition (1920-1933), bootleg liquor, jazz clubs, automobiles, gangsters, and wildly escalating stock-market investments. "This carefree attitude followed the prosperity that arose from tax cuts and the government's 'hands-off' approach to business, during which they allowed progressive regulations to lapse or simply be ignored."[69] Environmental devastation during the twenties, as described earlier in this chapter, was generally ignored under presidents Warren G. Harding and Calvin Coolidge. Herbert Hoover, however, somewhat re-ignited the federal interest in gigantic public works projects, authorizing the Boulder Canyon (Hoover Dam) project, flood control, oil conservation, and other federal projects. The twenties also witnessed the proliferation of automobiles and electrically-powered consumer products such as refrigerators and radios, but it was the automobile which truly revolutionized American life by causing construction of a network of roads across the landscape. Supported by a growing infrastructure of gas stations, hotels, and restaurants, tourism gradually transformed the objectives of the National Park system created in 1916.

Following the stock-market crash of 1929 and the early years of the Great Depression, American conservationism was integrated into the New Deal of Franklin Delano Roosevelt after 1933. Under the New Deal, which greatly expanded the power of the president, Roosevelt, who displayed an interest in nature and conservation from early on in his political career, was able to develop partial solutions to the problems of unemployment, business stagnation, and maldistribution of wealth through creation of such agencies as the Civilian Conservation Corps (CCC), the Soil Conservation Service (SCS), and the Tennessee Valley Authority (TVA). The CCC hired young men to build roads, firebreaks, trails, and lookout towers in National Forests, the SCS taught and assisted with soil management throughout the country, and the TVA undertook construction of a hydroelectric network designed to rejuvenate a devastated southern economy through the construction of dams, flood-control structures, a soil conservation program, and recreational facilities. Conservation under the New Deal generally had little to do with benefitting wildlife or habitat, but was, rather, a continuation of the utilitarian policies of Gifford Pinchot. An exception was the addition of new national parks (Olympic in Washington, Kings Canyon in California, and Shenandoah in Virginia), although the following discussion will explain how the parks were compromised by utilitarian objectives.[70]

The ideals of wilderness preservation and, even more so, of protection of scenic landscapes, were realized in the National Parks, but the National Parks were soon invaded by roads and automobiles in increasing numbers during the 1920s and 1930s. During the early 1930s the New Deal public works projects, including those of the Works Progress Administration (WPA), made disturbing inroads into the heart of many national parks and national forests. Among the men who responded to these developments by founding the Wilderness Society in 1935, Aldo Leopold (1886-1948) is certainly the best known. Beginning his professional career in the United States Forest Service (USFS) in 1909 after completing a Master's degree in forestry at Yale, Leopold passed through many changes in his attitudes toward nature as he moved from utilitarian forestry to game management and wilderness preservation, and finally, to laying the ecological and philosophical foundations for sustainable land-use practices. By 1924 Leopold had published a wilderness proposal for the Gila National Forest which was successfully implemented by the USFS. Although years later he would articulate a wilderness ideal in keeping with that of the Wilderness Act of 1964 (in which humans were to be present as short-term visitors in large, undeveloped tracts of undisturbed land), at the time that he conceived the Gila Wilderness as a District Manager for the USFS he envisioned an area open to wilderness hunters, and which also allowed grazing and local use of timber, provided that there were no permanent roads to allow commercial timber extraction. His main objective was "to save public landscapes from the popularity of the automobile, outdoor recreation, and the improvements that accompanied both."[71] Recreational use in national forests, along with the creation of the National Park system, coming at the same time that automobile use for recreation exploded in the United States, drove Leopold, Robert Marshall and other wilderness proponents to create the Wilderness Society as a bulwark against the destruction of wild nature by automobiles, roads, and related development.

In addition to his role as wilderness advocate, Leopold is perhaps the foremost pre-environmental figure in the twentieth century for his environmental philosophy, presented succinctly in the final section of *A Sand County Almanac* (1949) as "The Land Ethic." *A Sand County Almanac* was published a year after Leopold's death. Also, in 1948, Fairfield Osborn's *Our Plundered Planet* was published, these two books powerfully foreshadowing the coming Environmental Movement of the 1960s and 1970s. If Osborn unequivocally portrayed modern humanity's destruction of Earth's ecosystems, Leopold provided the basis for a possible solution to many of our environmental problems. During his mature years Leopold gained a general knowledge of geology, evolutionary biology, and ecology, particularly the prevailing ecological ideas of F.E. Clements and Arthur Tansley during the early decades of the twentieth century. "Although Elton's community paradigm (later modified... by Arthur Tansley's ecosystem idea) is the principal and morally fertile ecological concept of 'The Land Ethic,' the more radically holistic superorganism paradigm of Clements and Forbes resonates in "The Land Ethic" as an audible overtone."[72] If these early ecological concepts are compromised by contemporary chaos theory into a conception of ecosystems as representing mere aggregates of organisms rather than community-like, integrated systems, Leopold also developed a "land aesthetic" which might serve as a fall-back position in defense of ecosystems. Whatever the future may bring in the way of epistemological and logical challenges to the viability of the ecosystem concept, one thing is certain. Ecosystems, however we define and analyze them, are giving way to artifactual human systems at an alarming rate. Leopold was well aware of this even in his time, which led to his attempt at reform. At this point I wish to represent "The Land Ethic" in the context of the 1930s and 1940s when *A Sand County Almanac* was written and revised several times before publication by Oxford University Press.

In 1864 George Perkins Marsh warned Western humankind of their highly destructive treatment of nature, but his argument was for utilitarian, wise use of the land. Nearly a century later Leopold argued against ecologically destructive practices, whether on forests, marshes, or farmland, in defense of the intrinsic value of ecosystems themselves; i.e., in terms of the integrity of "biological communities," arriving at the moral precept that: "A thing is right when it tends to preserve the integrity, stability, and beauty of the biotic community. It is wrong when it tends otherwise."[73] Leopold builds an argument for a land ethic which views it as the third stage in the evolution of human ethics, from relationships between individuals to relationships between individuals and society, to an "ethic dealing with man's relationship to land and to the animals and plants which grow upon it."[74] All of these ethics rest upon the premise that the individual belongs to a community of interdependent parts. "The land ethic simply enlarges the boundaries of the community to include soils, waters, plants, and animals, or collectively, the land."[75]

After further elaborating the nature of the land ethic Leopold discusses the "ecological conscience" of the time, which was the 1930s era of the New Deal, when conservation was receiving strong financial support from the federal government. Defining conservation as "a state of harmony between men and land," Leopold observes that something is lacking in the conservation education of his day, and that farmers and other users of the land do not take their obligations for care of soil and water seriously because they lack an extended social conscience concerning the land. In a later section called "The Outlook," he wrote:

> Perhaps the most serious obstacle impeding the evolution of a land ethic is the fact that our educational and economic system is headed away from, rather than toward, an intense consciousness of land. Your true modern is separated from the land by many middlemen, and by innumerable physical gadgets. He has no vital relation to it; to him it is the space between cities on which crops grow. Turn him loose for a day on the land, and if the spot does not happen to be a golf links or a "scenic" area, he is bored stiff.[76]

Returning to Leopold's concern that something was lacking in the content of conservation education, he discusses the New Deal conservation effort in Wisconsin in the 1930s, particularly the federal offer of CCC labor to adopt remedial practices for improving and saving the soil, which was rapidly eroding away at the time. "The offer was widely accepted, but the practices were widely forgotten when the five-year contract period was up. The farmers continued only those practices that yielded an immediate and visible economic gain for themselves."[77] The Wisconsin Legislature passed the Soil Conservation District Law in 1937, which offered free technical service and specialized machinery from public funds if farmers would write their own land-use rules, which would have the force of law. A decade later, no county had yet written a single rule. Such rules might have protected woodlots from grazing and steep slopes from erosion, which would have helped to prevent increased flooding. Leopold concluded that education mentioning obligations to land should have preceded the rules. Why is it, he asked himself, that obligations over and above self-interest are taken for granted in rural community enterprises such as the improvement of roads, schools, churches, and even sports teams, and not in protecting soil and water, let alone preserving the beauty and diversity of the farm landscape? In other words, why are land-use ethics governed entirely by economic self-interest? A new land ethic is needed in America, he thought, because obligations have no meaning without a social conscience. "The proof that conservation has not yet touched these things lies in the fact that philosophy and religion have not yet heard of it. In our attempt to make conservation easy, we have made it trivial."[78]

Leopold's frustrated speculations concerning the difficulty of implementing a land stewardship ethic raise some questions regarding the roles of human nature and human culture, respectively, in influencing human attitudes toward and treatment of nature. In the earlier chapters of this book I made the argument that all human cultures have tended to exploit their surrounding environment, and Western civilization most of all. At two levels we are exploitative: (1) in order to survive and earn a living from the earth, we must necessarily cause environmental impacts, and our *self-interest* motivates us in our endeavors; (2) *culture* theoretically can reinforce or constrain our self-interested practices in gaining a livelihood from the land. Pre-civilized cultures often developed benign relationships with nature in their capacity as hunter-gatherers. Was this the result of cultural constraint upon self-interest as hunter-gatherers reached the carrying capacity of their lands? Alternatively, when some hunter-gatherers solved their regional population problem through the invention of agriculture, there was a shift to managed landscapes under strong authoritarian rule, as in ancient Egyptian, Sumerian, or Aztec-Mayan civilizations. Concurrently, a powerful cultural worldview evolved to rationalize agricultural practices. In Western civilization a comparable agrarian-based worldview developed which conceived of land use as a God-given right of Jews and Christians to exploit the Earth as God's created home for humankind. Leopold pays homage to the Christian foundations of Western farming in his references to Ezekiel, Isaiah, and Abraham in *The Land Ethic*. What surprised him in his Wisconsin experience of federal conservation and its failure was the power of individual self-interest combined with the profit motive in resisting changes in behavioral practice. The unanswered question for him was that of the degree to which cultural, i.e., Western Christian and modern cultural attitudes, reinforced self-interest in resisting the passage to an acceptable land ethic. This issue will be addressed in the following two chapters. Aldo Leopold would have been more than pleased, had he lived to a ripe old age, to witness the beginnings of environmental philosophy (environmental ethics) and environmental history two decades after publication of *A Sand County Almanac*. The question for us, to be answered in the next chapter, is: Has this revolution in environmental education produced sufficient change in agricultural and other land-use practices of modern civilization to prevent an ecological crisis in the twenty-first century?

Christianity, Secularism and the Environment to 1960

Looming war clouds over Europe during the late 1930s added to the gloom of global depression just before World War II. The New Deal and conservationism had not bailed the United States out of its economic woes when Pearl Harbor shocked the country into action in 1941. Conservation immediately gave way to mass industrial production during the war. The 1940s and 1950s were thoroughly unconducive to positive activity in behalf of nature. There were exceptions, such as the Echo Park Dam controversy, in which the Sierra Club,

Wilderness Society, and other conservation-preservation groups prevented construction of a dam which would have flooded the Dinosaur National Monument in Utah and Colorado. However, for the most part, having led the victors in World War II, "Americans thought they deserved improved living standards and material comforts. Conservationists struggled as the nation focused on these expectations rather than on environmental concerns. The public at large did not begin to comprehend the environmental danger caused by two hundred years of uncontrolled industrial expansion until the mid-1960s."[79]

After 1945 the consumer society restructured itself, and industry retooled from the production of tanks, bombers, and guns to turn out automobiles, television sets and the electronic gadgets that filled the modern home. With 6% of the world's population in 1955, the United States was producing almost half of the world goods. After the Korean War (1950-1953) GIs attended colleges, suburbs expanded, and the Federal Highway Acts of 1956 and 1958 projected construction of 40,000 miles of roads financed by federal gasoline taxes. New housing tracts in the suburbs tied in with the "automobilization" of the American economy and culture, creating almost unimaginable land-use sprawl patterns around major cities. Petulla observes that the affluent society of the post-war era was indeed the result of the war, which had increased both government expenditures and economic concentration in growing corporations. American society was also in general disharmony during the 1950s, as the Cold War, the threat of a nuclear attack, McCarthy witch hunts for traitorous American communists, and the struggles of the civil rights movement took center stage. As a result, national parks and other federal lands were ignored, poorly managed, and degraded by overuse, while rural lands were gobbled up by suburban sprawl. Simultaneously, pesticides such as DDT came into regular use, the United States producing 124 million pounds in 1947 and 637 million pounds by 1960. Air pollution problems also intensified from the increased use of automobiles, and litter became a common sight on the highways. Books such as Leopold's *A Sand County Almanac* (1949), Osborn's *Our Plundered Planet* (1948) and *The Limits of the Earth* (1953), and William Vogt's *Road to Survival* (1948) sounded an ecological alarm during this period, but the American public was not listening. Only a decade or so later Rachel Carson's *Silent Spring* finally ignited the Environmental Movement and gradually forced the American political system to take action in defense of nature.[80]

Another development of importance in laying the foundation of environmentalism during the first half of the twentieth century in America was the increased secularization of its inhabitants. Christian religion dominated our sub-culture of Western European civilization throughout the seventeenth through the nineteenth and early twentieth centuries. However, several historical developments, including increasing cultural literacy and awareness of European ideas, many of which strongly challenged the Christian belief system, the spread of scientific thought in geology, evolutionary biology, and ecology in American colleges and universities, and contact with European secular thought during and in between World Wars I and II paved the way to the alternative culture of secular humanism. Its influence was far less than in Western Europe, but it did pose a challenge to the dominant Christian worldview. By the 1950s, as an undergraduate and graduate student in eastern universities, my perception of the United States was that it was a predominantly secular nation with Christianity on the wane as an anachronistic "cultural appendix," a redundant set of beliefs which would be removed during several more decades of secularization. French existentialism and Zen Buddhism appeared to be the preferred philosophies of the time, and I became a follower of one and then the other of these alternatives to Christianity. All these decades later, however, traditional religion is alive and well, and the nineteenth and twentieth century prophets of a secularized Western civilization, and a secularized world in the long run, have been sadly disappointed, with the exception of western Europe, where church attendance has dropped to only 9-12% in the Scandinavian countries and 17% in France.

I recently did a personal survey of religious attitudes and their relationship to perceptions of the ecological crisis, making inquiries among friends, acquaintances, students, and people in the workplace. One question I asked concerned the degree to which pastors and priests mention environmental issues during sermons and religious services. All respondents answered with an emphatic "never." One Baptist person added that "ecology is a left-wing issue which we wouldn't think of dealing with." I was surprised and disappointed by the results of my mini-survey. It led me to inquire into the sociology of contemporary American religion and its historical antecedents. Among the relevant knowledge that I acquired was the fact that church/synagogue attendance in the United States has not changed significantly over the past sixty years, although there have been noticeable variations during this interval from the 1940s to 2000. Most striking to me was the post-World War II revival, when church membership rose from 49 to 63.6 percent between 1940 and 1959. The "secular age" that I thought I was living in as a student had simply reflected an island of secularization in a rising sea of religious belief, or at least membership. Also, along the Potomac River during the Eisenhower presidency, Bible breakfasts and congressional prayer groups flourished in the nation's capital, we inserted the phrase "under God" into the pledge of Allegiance, and "In God We Trust" appeared on postage stamps and paper money. Sociologists warily recognized an upsurge of interest in religion. Given that the United States was reconstructing itself after a terrible depression and world war, more people were undoubtedly seeking stable moorings in an unstable world living under the shadow of potential nuclear warfare and the "menace" of global communism.[81] This led one observer to interpret the religious revival as a "cult of reassurance," which he described as "a flocking to religion, especially in middle-class circles, for a renewal of confidence and optimism at a time when these are in short supply. It is a turning to the priest for encouragement to believe that, despite everything that has happened in this dismaying century, the world is good, life is good, and the human story makes sense and comes out where we want it to come out."[82]

To some extent the post-war religious revival can be interpreted as a commentary on the enfeebled state of the belief in progress and the retreat to the idea of providence after World War II. And with the Vietnam War to follow, good times were ahead for the church just when the Environmental Movement arrived in the early 1960s. Considering the previous quotation in this context, what is the relationship between religious revivalism and environmentalism since the 1960s? A turning to ministers and priests for reassurance that the human story, i.e., history, is sensible and benign places the Environmental Movement within a worldview which, following a failure of nerve at the horrors of the twentieth century, essentially denies the truth of historical process in order to put the common mind at ease. Thus, the ecological crisis, like the nuclear, communist, and population (identified by Osborn and others in the 1940s) crises, shrank in the popular individual mind in the light of lives guaranteed salvation by a predominantly Christian religious belief system. This conclusion appears to be applicable today on the basis of my mini-survey. However, even more threatening than the threat of megacrises in general was the more immediate threat, by the early 1960s and thereafter, that any individual in American society might face death from cancer caused by various pollutants, including nuclear radiation, pesticides on food and in water, and lead-contaminated air along the freeways connecting cities and their post-war suburbs. This concern of the American public for the health of individual humans was the most powerful force initiating the Environmental Movement. Church services were inefficacious against the spread of pollution-caused cancer and other diseases of modern industrial civilization. Out of this realization there arose a popular political mandate for protection against newly discovered threats to human life and health.

NOTES

[1]. J.R. McNeill, *Something New Under the Sun: An Environmental History of the Twentieth-Century World* (New York: W.W. Norton and Company, 2000), p.15.

[2]. Eric Hobsbawm, *The Age of Extremes: A History of the World, 1914-1991* (New York: Pantheon Books, 1994), pp. 1-2.

[3]. R.R. Palmer, Joel Colton, and Lloyd Kramer, *A History of the Modern World*, 9th ed. (New York: Alfred A. Knopf, 2002), p.556. Between 1650 and 2000, world population grew from 545 million to 6073 million, or an increase of 11 to 12 times in 350 years. During this 350 year period, the European percentage of world population dropped from 18.3% to 12% as population growth in Asia accelerated relative to that of Europe.

[4]. Ibid., p.557.

[5]. Ibid., p.560.

[6]. Ibid., p.564.

[7]. J.R. McNeill and William McNeill, *The Human Web*, p.235.

[8]. Joseph M. Petulla, *American Environmental History* (Columbus, Ohio: Merrill Publishing Company, Second Edition, 1988), pp. IX-XV. Also pp. 175-196, 275-301, 325-349, 373-405.

[9]. Ibid., p .351.

[10]. Ibid.

[11]. Ibid., p.354.

[12]. Michael T. Klare, *Resource Wars: The New Landscape of Global Conflict* (New York: Henry Holt and Company, 2001), p.31.
[13]. J.R. McNeill and William McNeill, *The Human Web: A Birdseye View of World History* (New York: W.W. Norton and Company, 2003), p.298.
[14]. Joseph M. Petulla, *American Environmental History*, p.326.
[15]. Ibid., p.377.
[16]. John Opie, *Nature's Nation: An Environmental History of the United States* (Fort Worth: Harcourt Brace Publishers, 1998), pp. 226-227.
[17]. John R. McNeill, *Something New Under the Sun*, p.17.
[18]. J.R. McNeill and William McNeill, *The Human Web*, p.304.
[19]. Ibid., pp.305-309.
[20]. John Opie, Nature's Nation, p.233.
[21]. J.R. McNeill and William McNeill, *The Human Web*, p.318.
[22]. Ibid.
[23]. J.R. McNeill, *Something New Under the Sun*, p.358.
[24]. J. Donald Hughes, *An Environmental History of the World: Humankind's Changing Role in the Community of Life* (London: Routledge, 2001), pp.189-190.
[25]. Sing C. Chew, *World Ecological Degradation: Accumulation, Urbanization, and Deforestation* (Walnut Creek, California : Altamira Press, 2001), p.140.
[26]. Ibid., p.141.
[27]. Josephson, *Industrialized Nature* (Washington: Island Press, 2002), p.132. Even in ancient Rome, the demands of urban residents required the ecologically destructive resource extraction of metals from Spain and England, and crops from the North African littoral.
[28]. Ibid., p.1.
[29]. Ibid., p.9.
[30]. J.R. McNeill, *Something New Under the Sun*, p.216.
[31]. Gilbert F. LaFreniere, "Greenline Parks in France: Les Parcs Naturels Régionaux." *Agriculture and Human Values*, Vol. 14, No. 4, Dec. 1997, p.339.
[32]. J.R. McNeill, *Something New Under the Sun*, p.219.
[33]. Ibid., p.224.
[34]. Ibid., p.227.
[35]. Clive Ponting, *A Green History of the World: The Environment and the Collapse of Great Civilizations* (New York: Penguin Books, 1991), pp. 252-253.
[36]. William H. Truettner and Alan Wallach, eds., *Thomas Cole: Landscape into History* (New Haven: Yale University Press, and Washington D.C.: National Museum of American Art, Smithsonian Institution, 1994), p.153.
[37]. Ibid., p.25.
[38]. Ibid., p.31.
[39]. Alfred Runte, National Parks: *The American Experience* (Lincoln: University of Nebraska Press, 1987), p.23.
[40]. Franklin Kelly (with Stephen Jay Gould, James Anthony Ryan, Debora Rindge), *Frederic Edwin Church* (Washington: National Gallery of Art and Smithsonian Institution Press, 1989).
[41]. John Opie, *Nature's Nation*, p.400.
[42]. Ronald L. Johnstone, *Religion in Society: A Sociology of Religion* (Upper Saddle River, New Jersey: Prentice-Hall, 2001), 258-263.
[43]. Ibid., p.263.
[44]. Stephen Jay Gould, "Church, Humboldt, and Darwin: The Tension and Harmony of Art and Science" in Franklin Kelly, *Frederic Edwin Church* (Washington: Smithsonian Institution Press, 1989), pp.94-107, p.97.
[45]. Ibid., p.98.
[46]. Ibid., p.104.
[47]. Ibid., p.104-105.

[48]. Alfred Runte, *National Parks: The American Experience*, pp.22-32.

[49]. Ibid., p.19.

[50]. Ibid.

[51]. Ibid., p.20.

[52]. Ibid., p.22.

[53]. Ibid., p.24.

[54]. Ibid., p.31.

[55]. Ibid., p.82.

[56]. Roderick Nash, *Wilderness and the American Mind,* Third Ed. (New Haven: Yale University Press, 1982), pp.133-134; and Samuel P. Hays, *Conservation and the Gospel of Efficiency: The Progressive conservation Movement, 1890-1920* (New York: Atheneum, 1969), p.36

[57]. Samuel P. Hays, *Conservation and the Gospel of Efficiency*, p.36.

[58]. Char Miller, *Gifford Pinchot and the Making of Modern Environmentalism* (Washington: Island Press, 2002), pp.136-139.

[59]. Roderick Nash, *Wilderness and the American Mind*, pp.176-177.

[60]. Samuel P. Hays, *Conservation and the Gospel of Efficiency*, p.2.

[61]. John Opie, *Nature's Nation*, p.234.

[62]. Lewis Mumford, *Interpretations and Forecasts* (New York: Harcourt, Brace, Jovanovich, Inc., 1973), p.16.

[63]. John Opie, *Nature's Nation*, p.235.

[64]. Max Oelschlaeger, *The Idea of Wilderness* (New Haven: Yale University Press, 1991), p.173.

[65]. Panentheism regards the external world, the cosmos, as divine, but as a manifestation of a transcendent being separate from the cosmos, whereas pantheism sees the cosmos as directly the body of God.

[66]. Max Oelschlaeger, *The Idea of Wilderness*, p.202.

[67]. Ibid., pp.201-204.

[68]. Carolyn Merchant, *Reinventing Nature: The Fate of Nature in Western Culture* (New York: Routledge, 2003).

[69]. Benjamin Kline, *First along the River: A Brief History of the Environmental Movement* (San Francisco: Acada Books, Second Edition, 2000), p.60.

[70]. Ibid., pp.61-63.

[71]. Paul S. Sutter, *Driven Wild: How the Fight against Automobiles Launched the Modern Wilderness Movement* (Seattle: University of Washington Press, 2002), p.71.

[72]. J. Baird Callicott, "The conceptual Foundations of the Land Ethic" in J. Baird Callicott, ed. *Companion to A Sand County Almanac: Interpretive and Critical Essays* (Madison: The University of Wisconsin Press, 1987), p.200.

[73]. Aldo Leopold, *A Sand County Almanac, with Essays on Conservation from Round River* (New York: Ballantine Books; San Francisco: Sierra Club, 1974), p.262.

[74]. Ibid., p.238.

[75]. Ibid., p.239.

[76]. Ibid., p.261.

[77]. Ibid., p.244.

[78]. Ibid., p.246.

[79]. Benjamin Kline, *First along the River*, p.70.

[80]. Ibid., pp.71-73.

[81]. Ronald L. Johnstone, *Religion in Society: A Sociology of Religion* (Upper Saddle River, New Jersey: Prentice Hall, 6th Ed., 2001), pp.274-275.

[82]. Paul Hutchinson quoted in Ronald L. Johnstone, *Religion in Society*, p.275.

CHAPTER ELEVEN: THE ENVIRONMENTAL MOVEMENT

Introduction

In Chapter Ten I attempted to summarize some of the most important antecedents of the Environmental Movement of the 1960s and subsequent decades. Some scholars see the movement as a child of the 1960s and 1970s, and most think that it continues down to the present day. Others, like Samuel P. Hays, consider "environmental movement" a misnomer because its formal ideas are overshadowed by concrete situations, and that "environmental engagement" and "environmental culture" better represent what has gone on since Rachel Carson's *Silent Spring.*[1] Nevertheless, the term "environmental movement" is so thoroughly ingrained in the literature of environmentalism that it is here to stay. Personally, I prefer a capitalized "Environmental Movement" in order to demarcate it as a particular set of events which occurred in the 1960s and 1970s and afterwards, just as the Romantic Movement is defined in European history. To my mind, much confusion is insinuated into the collective consciousness of the general public by including the roots of modern post-1962 environmentalism under the term "environmental movement," but this is an unfortunate practice in books too numerous to bother documenting. The term pre-environmentalism, as in pre-romanticism, helps to distinguish the discussion of the roots or antecedents of the Environmental Movement *per se.* As for the "pre-rational impulse" that has motivated modern man to join this movement, Peter Hay writes "that it is a deep-felt consternation at the scale of the destruction wrought, in the second half of the twentieth century, and in the name of a transcendent human progression, upon the increasingly embattled life forms with which we share the planet."[2] I fully agree with Hay, who complements this statement with the acknowledgment that many people are more concerned with the effects of twentieth century industrial-technological society upon humans and human environments than they are with species and ecosystems. These complementary emphases represent two major themes of the Environmental Movement: wildlife and ecosystem preservation; and concern for human health and welfare. I would suggest that the latter concern for *human* welfare has co-opted the Environmental Movement, and that preservationism, despite its impressive genealogy of pre-environmental preservationists, has lost its ability to focus the attention of Americans on the proper center of gravity of the Environmental Movement: the biospheric processes upon which all life depends, including humans. Our failure to recognize, collectively, the primacy of nature in structuring human societies is due in large part to the assimilation of liberal economic theory into the modern Western worldview, in which ecosystems (nature) are viewed as *resources*, i.e., as merely a subset within the economic system that rules the contemporary world. In the long run we will be forced to acknowledge that the reverse is true, that economics is only a part of the biosphere and must conform to its needs or the ecological system will collapse and carry its human economic parasite down with it.

Before developing these ideas I would like to introduce one of Garrett Hardin's favorite analogies in thinking about the challenges to modern environmentalism.[3] Envision a basement laundry room in which an irreparable plugged drain has caused a sink to overflow. There are two ways in which one might deal with the problem. The first, admittedly stupid, would be to mop up the overflow, wring it out into buckets, and pour the water down a functional drain. The obviously preferable solution would be to turn off the taps which are causing the sink to overflow. Hardin proposes that we see the overflow as the symptoms or environmental impacts of modern industrial civilization and the taps as pouring out virtually unlimited population growth and consumption of goods which extract natural resources. What modern societies have done is to run for the mop to clean up the accumulating environmental messes resulting from industrial-technological development rather than turning off the taps of overpopulation and overconsumption. "Running for the mop" has created the National Environmental Policy Act (NEPA) and Environmental Protection Agency (EPA) to mop up these messes, but, not surprisingly, they are piling up much faster than anyone can clean them up. Deep ecologists, who believe that a general change towards an ecologically benign worldview (that respects the biosphere and our place in it) is necessary in order to shift to an ecologically stable system of global polities, refer to the "running for the mop" syndrome as "shallow ecology." The United States and other developed countries have legislated and implemented numerous shallow ecology programs, leaving the taps running full blast, and lacking any popular intellectual grasp of the need for a profound shift in worldview in order to facilitate sustainable practices. During the 1960s, however, considerable thought was applied to a serious critique of the modern technological-industrial civilization and its impacts upon both humankind and nature, with an emphasis on the former. Intellectuals such as Norman O. Brown, Herbert Marcuse, Paul Goodman, Theodore Roszak and many others functioned as mentors of the 1960s counterculture, which was a seedbed of ideas and activism which contributed to the rise of the Environmental Movement. Before analyzing the countercultural-environmental connection, however, let me first generalize the accepted interpretation of the Environmental Movement's beginnings made by most environmental historians and philosophers.

An inventory of opinions on the subject suggests to me that the primary cause of the Environmental Movement was a general panic on the part of the American public in the light of accumulating evidence that industrial activity was polluting soil, water, air, and food with a variety of dangerous substances, many of which caused cancer. There is a general consensus that Rachel Carson's *Silent Spring*, which enjoyed 31 weeks on the *New York Times* best-seller list when published in 1962, was the literary fire alarm that set off the Environmental Movement, acting in consort with the spread of ecological ideas in the scientific community, which was facilitated by the publication of Eugene P. Odum's textbook, *Fundamentals of Ecology* (1953). Carson's popularization of the fact that synthetic pesticides, including chlorinated hydrocarbons such as DDT, were poisoning humans and other organisms around the globe, shocked large numbers of Americans. Until she blew the whistle, "the Department of Agriculture, the Public Health Service, and the Food and Drug Administration has repeatedly proclaimed that synthetic pesticides were harmless."[4] *Silent Spring* was also excerpted in *The New Yorker* in 1962, and its ideas presented on a CBS prime-time television special in 1963, a year before Carson died of cancer. For the first time in their history, Americans learned that industry and technology, megaheroes of the American Dream, had poisoned their air, rivers, streams, and ground water, as well as soils and crops, with health- and life-threatening chemicals while governmental agencies responsible for their health and safety were "whistling in the wind."[5]

Although there is a general consensus among scholars regarding the seminal importance of Rachel Carson's *Silent Spring* as a major immediate impetus to the rise of the Environmental Movement, there is some disagreement regarding the relative importance of wilderness preservation on the one hand and the public fear of harmful pollution on the other. Important developments on the wilderness agenda occurred simultaneously with Rachel Carson's warning in 1962. By that time congress was receiving more mail about a wilderness bill (originally proposed in 1957) than on any other proposed legislation. Unfortunately, when the bill was finally passed in 1964, it was a substantially gutted version which allowed mining, dams, and power plants in wilderness areas, which constituted only a fraction of those originally proposed. Thus, these two strands, of preservationism and concern over pollution, began to intertwine during the early 1960s. While the wilderness strand of environmentalism could be traced to nineteenth century preservationists and the Progressive Era, the pollution strand was something shockingly new. The greater influence of the pollution strand upon the character of the Environmental Movement is indicated by Kirkpatrick Sale in describing the 1962-1970 years when the movement began:

> In its first eight years the environmental movement compiled a record of accomplishment that no one could have calculated when it began, but even more important, it achieved a broadened scope that no one could have imagined. The concern was no longer just the impact of the human society on the wilderness and its species; it now included the impact of human society on humanity as well. And though some important victories would be won on the former

front in the decade to come, it was primarily the latter that was the preoccupation of movement and public alike.[6]

The majority of environmental historians and philosophers appear to support this view, namely that the pollution and other negative side effects of industrial development were found to be intolerable by the American people, resulting in a popular outcry and a national political mandate that demanded responsible action on the part of government. "The environmental movement, which erupted from the social changes of the 1960s, was not merely an expression of organized conservation groups but also a manifestation of growing anger among the public."[7] Between 1965 and 1970 polls showed an increase from 17 to 53 percent of respondents who ranked "reducing pollution of air and water" as one of the three top national priorities. Other major defenders of the environment chimed in following the Wilderness Act of 1964, including Barry Commoner, Garrett Hardin, Lynn White Jr., and Paul Ehrlich. The theme of overpopulation was also at the forefront of concern, which gave rise to a pessimistic, doomsday climate of opinion during the late 1960s and early 1970s. "The mood of the 1970s, set in place by predictions of imminent, ecologically wrought doom, was grim and desperate. The environment movement was Hobbesian, deeply despairing and anti-democratic. It judged human nature harshly. That environmentalism could be a positive force for refurbishing the quality of life on earth was scarcely considered."[8] Major exceptions, among others, to the preceding generalization, were Ernst Callenbach's *Ecotopia* (1973) and Theodore Roszak's *Where the Wasteland Ends* (1972), both books carrying positive values as well as criticism from the 1960s counterculture into the 1970s and merging their ideas with those of the growing Environmental Movement. This is a very important point, for the countercultural critique of the modern technological society was at the heart of what idealists of the Environmental Movement would re-invent in the form of environmental philosophy, generally known as "environmental ethics." Deep ecology, social ecology, and ecofeminism all contributed to a profound and potentially subversive attack upon the dominant social paradigm, i.e., the technological-industrial society propelled by capitalist expansionism through unrelenting economic growth. In order, let us consider the contributions of the 1960s counterculture, the major events of the 1960s, and the radical environmental philosophies, to the Environmental Movement.

The Counterculture and the Environment

The countercultural critique of modern technological-industrial society was anticipated by such earlier cultural critics as Thomas More, Jean-Jacques Rousseau and the Romantics, the Luddites in England, Thomas Jefferson, Ralph Waldo Emerson, Henry David Thoreau, Herman Melville, Nathaniel Hawthorne and James Fenimore Cooper in the American "pastoral" tradition, to name some major rebels against over-commercialization and over-development. During the nineteenth and twentieth centuries the modern "tourbillon," that maelstrom of ceaseless change propelled by the Industrial and Technological revolutions, has generated legions of alienated humans too numerous to name. People uprooted from their traditional ways of life have suffered mass alienation (e.g., displaced farmers working for marginal wages in coal mines and factories) throughout Western civilization and its colonies, and modern intellectuals have documented the physical and mental suffering endured by humans in a condition of social and cultural alienation. Marxism and socialism have attempted to respond to this ubiquitous social condition produced by ceaseless change by attempting to restructure the sociopolitical context of the modern tourbillon. Concern over the problem of mass and individual alienation within modern technological-industrial society was re-ignited in the 1950s and 1960s, producing numerous books on alienation, which included the neo-Marxian and neo-Freudian works of Herbert Marcuse and Norman O. Brown. Their books became seminal documents in the intellectual arsenal of that loosely aggregated congeries of ideas and practices which came to be known as the counterculture, and whose most eloquent synthesizer and spokesperson was the historian Theodore Roszak.

To those of us who experienced the hippy culture and intellectual counterculture which flourished in the late 1960s, looking back several decades it takes on a quaint, superficial aura. Its music still persists remarkably well, but its quirky costumes, soft and hard drugs, spiritual gurus, carefree disregard for conventional customs and employment, and sexual promiscuity have not fared so well, as the established culture has regained a degree of control over its youthful progeny. In contrast to the Environmental Movement, the countercultural movement was ephemeral, after the late 1960s continuing on largely by imitation devoid of substance. Perhaps it can best be understood, retrospectively, as a general social outburst of collective alienation resulting from the crushing onrush of modernity, including the excesses of post-war capitalism combined with permissive liberal governments. At the beginning of the twentieth century the sociologist Max Weber predicted that the incessant change of modernity would cause a "disenchantment of the world" involving human mechanization, bureaucratization, and routinization of life engendered by science, technology, and relentless industrialization. Furthermore, Weber envisioned "the possibility of a human revolt against the 'iron cage' that would represent a return to spontaneity, mystery, charismatic personality, and other preconscious impulses." [9] More immediately, social changes during and after World War II responded to a relatively sudden shift from deprivation to affluence and abundance, creating the first generation of American children who took endless material satisfaction for granted, and who were not overly concerned with money, employment, or housing. This sense of freedom gave rise to the first generation of American youth free of the fear of scarcity.[10] Simultaneously, this generation reflected upon a society which, despite its Christian pronouncements, had produced and used the atomic bomb, was fighting an unnecessary and brutal war in Vietnam in the name of capitalism and Christianity, and which continued to oppress Native Americans, blacks, and women. Historian W. Warren Wagar has attempted to encapsulate the general attitudes of the youthful counterculture:

> Whatever we choose to call them, adherents of the counterculturalist paradigm aspire to be more radical than the radicals. What is wrong with modern man and woman, they say, is not this or that institution or trend, but modernity itself. The rape of the environment is just one example, to which most of us have only recently awakened, of how the values of modernity, so seductive at first blush, are in fact demonic. Thus, the great question for late twentieth-century society is not whether to limit growth, as technoliberals imagine, but whether to replace the values of the dominant culture that make us *want* to grow.[11]

"Something is happening, but you don't know what it is, do you Mr. Jones?" Thus did Bob Dylan mock the typical establishment citizen that the triumphs of modernity were leading us to unforeseen disasters, and that only the hip members of the counterculture could recognize the destructive potential and materialistic perversity of modern values in America. In the above quotation of Wagar's perception of the underlying intellectual essence of the counterculture, we see evidence of the connection between hippie culture and the developing Environmental Movement. Joni Mitchell's "Pave paradise, put up a parking lot" also come's to mind. It is hardly surprising that hippies and incipient environmentalists read similar books and expressed a common interest in saving the Earth from extreme overdevelopment. Callenbach's *Ecotopia* envisions an emancipated piece of the United States (the Pacific Northwest and northern California) in which ecologically sustainable industry, preservation of extensive areas which have been returned to nature, mass transit, and de-automobilization have been implemented by groovy citizens kept happy by marijuana, wine, and sexual promiscuity. The deeper level of congruence between the counterculture and a growing number of environmental advocates, however, is their common harsh critique of modernity, its values, and its despised life style of wasteful superabundance and designer consumerism. If the counterculture acknowledged the hopelessness of changing this dominant social paradigm into a benign society comparable to the one suggested by Callenbach, and instead sought refuge in Big Sur and escape into drugs, sex, and exotic mysticism, the Environmental Movement took the major strengths of modernity, rationalism and science, and turned them against the technological-industrial society. Along the way, environmentalism built a sophisticated philosophical critique of this society, but it has been relegated to a limited audience of committed radical environmentalists out of reach from the general public consciousness and the channels of political power. The Environmental Movement would be forced to settle for Hardin's "running for the mop," with little hope of turning off the taps of population and consumption. Nevertheless, it will be useful to compare the criticisms of the modernist paradigm articulated by both the counterculture and the Environmental Movement. Almost singlehandedly, historian Theodore Roszak has made the countercultural critique available to a wide audience of American intellectuals.

In *The Making of a Counterculture: Reflections on the Technocratic Society and its Youthful Opposition* Roszak synthesized the ideas of Herbert Marcuse, Norman Brown, Allen Ginsberg, Alan Watts, Paul Goodman, and others into a powerful critique of the modern technological society. He defines the technocracy as "that social form in which an industrial society reaches the peak of its organizational integration. It is the ideal men usually have in mind when they speak of modernizing, up-dating, rationalizing, planning."[12] The technological society is one in which the political structure, be it capitalist, socialist, or communist, is influenced by bureaucracies of technical experts who provide governing bodies with the latest scientific and technological knowledge. Rendering itself ideologically invisible, the technocracy provides the larger society with its own values and presuppositions about the nature of reality, where they become an essential part of our worldview, "as unobtrusively pervasive as the air we breathe." The technocracy acts as a grand cultural imperative which exists beyond the realm of public debate and which underlies the common perception of human history as progress through science, technology, and industrialization. Furthermore, "while possessing ample power to coerce, it prefers to charm conformity from us by exploiting our deep-seated commitment to the scientific world-view and by manipulating the securities and creature comforts of the industrial affluence which science has given us."[13] Roszak sees the scientific worldview as our "unimpeachable mythology," which has been used to reduce modern humanity to the rationalist-mechanist model of seventeenth and eighteenth century European philosophers. Like the Beat Generation of the 1950s and most countercultural intellectuals of the 1960s he is almost obsessed with the problem of alienation in modern society, and he finds this widespread alienation to be a consequence of the broad, unexamined influence of the scientific worldview. Roszak refers to this component of our modern worldview as "the myth of objective consciousness" which, to his mind, is the major root cause of most of the problems of modern industrial societies, from psychological and social alienation to the hidden connection between science and totalitarian social control, to the ecological destructiveness of an affluent consumer society dependent upon relentless industrialization.

> ... the natural environment must be conquered and subjected to forceful improvement. Climate and landscape must be redesigned. Waste space must be made livable, meaning covered over with an urban expansion into which nothing that is not man-made or man-arranged will intrude itself. Similarly, the social environment--the body politic--must be brought as completely under centralized, deliberative control as the physical body has been brought under the domination of the cerebrum.... An objective, meaning an

alienated, attitude toward the natural environment comes easier these days to a population largely born and raised in the almost totally man-made world of the metropolis.[14]

One might think that there is nothing new here, that the critique of the technological society has been around at least since the nineteenth century, foreshadowed by Thoreau and other critics of the "machine in the garden" and carried on by Henry Adams towards the *fin de siècle*, and Lewis Mumford for most of the twentieth century. What is new in Roszak is that he participated in an actual rebellion, a movement against the oppressive technocracy that showed promise of positive change for the future. However, in collating and representing the spirit or essence of the countercultural movement, Roszak revived the supernaturalist and escapist tendencies of the nineteenth century Romantic Movement. According to cultural historian Leo Marx, Roszak "portrays the contemporary world as the scene of an all-encompassing Manichean struggle between opposed views of reality, each marked by an ideal type of knowing: scientific naturalism and gnosis. One is reductive, partial, analytic; the other augmentative, holistic, synthetic."[15] Roszak appears to view history in terms of struggles between ruling epistemologies or worldviews, as though they functioned separately from the groups that adhered to them. This leads him, according to Marx, to the erroneous conclusion that a science grounded in instrumental reason necessarily produces evil outcomes due to its epistemological inadequacy. "No accommodation between science as we know it and Roszak's conception of an adequate epistemology is conceivable."[16]

In *Where the Wasteland Ends* Roszak attempts to build a new religious alternative to the scientific myth of objective consciousness which has built the "air-conditioned nightmare," Henry Miller's term for the arid, dehumanized artifactual environment. He forthrightly turns to the Romantic sensibility for inspiration. "Since Rousseau and the Romantics, hostility toward the artificial environment has run through our culture like a soft, lyrical counterpoint to the swelling cacophony of the machine."[17] Roszak goes on to express his intention to revive the Romantic critique of technological society, with the hope of reinforcing countercultural resistance against it in order to prevent environmental collapse, nuclear Armageddon, or a political descent into neo-fascism. He also foresees our present social predicament with remarkable acuity. "When public education collapses under the weight of its own coercions and futility, the systems teams will step forward to propose that the schools invest in electronicized-individualized-computerized-audio-visual-multi-instructional consoles."[18] Thus, with each new technological innovation the dissenters are disoriented and the culture of technique is insinuated more deeply into the collective mind. In *the Technological Society* (1959) Jacques Ellul defined technique as "more and more sophisticated means to poorly examined ends."[19] Roszak scrutinizes these technological means and their consequences in detail before proposing his

Romantic counterpoint in the second half of *Where the Wasteland Ends*, in which the ends or purposes of life are re-evaluated in the light of the Romantic legacy.

Roszak contends that the secular humanism and scientism, i.e., the cult of science and technology which grew out of Baconian-Cartesian dualism, have produced the myth of objective consciousness or "single vision," the one-dimensional perception of reality which has expunged poetic vision, religiosity in its deeper sense of mystical experience, the sense of the sacredness of nature and human life embedded in it, and, above all, the sheer experience of the joyous exuberance of life from the modern consciousness. William Blake, Goethe, Wordsworth and Shelley, and even Rimbaud are called upon by Roszak to defend the richness of a naturally endowed, full human consciousness of reality which has been so sadly diminished over the past three modern centuries. Enlightenment "Reason" has gradually blinded the perception of modern humanity to the larger reality that lies beyond our limited single vision, leading us towards demonic outcomes. Heed the warning of William Blake, Roszak exhorts his reader, that mankind must and will have some kind of religion, and this need can only be misguided into self-destructive channels by the single vision generated in our minds by reductionist science and the pursuit of power through technology.

Although his neo-Romantic and mystical alternatives have been poorly received by most academics, Roszak has developed a limited following, and his more recent books have constituted a one-man defense of nature through expanded consciousness and awareness. In *The Voice of the Earth* (1992) for example, Roszak attempts to effect a solution to the environmental crisis through development of an "ecopsychology," a psychological perspective within which the human psyche is formally recognized as being embedded in the history and processes of the Earth itself. "Nothing less than an altered sensibility is needed, a radically new, holistic and ecological standard of sanity that undercuts scientific rationality and uproots the fundamental assumptions of industrial life."[20] In other words, Roszak understands that a change in worldview or metanoia is absolutely necessary in order to shift from the technological society to a culture responsible for the ecological health of the planet. However, when Roszak describes humanity and its intellect, particularly its ecological insights, as the evolved "mind" of *gaia* (the planet as an enspirited superorganism) environmental philosophers and historians retreat from what sounds like "New Age" nonsense not far removed from astrology and crystal healing. Nevertheless, rigorous academic philosophers such as Michael Zimmerman are respectful of the often brilliant insights thrown off by Roszak in his quest for mystical or quasi-mystical foundations for a neo-romantic counterculture. Such a culture would be responsible for inculcating a metanoia capable of finally placing humankind in a transformed society in which a new sensibility towards nature would complement sustainable ways of making a living from Earth. Roszak is not easy to classify, but Zimmerman observes "that in the 1980s, New Age began giving way to New Paradigm, whose leading figures include Roszak, Capra, and others linked with deep ecology."[21]

Ecology was "in the air" during the late 1960s, and, superficially or otherwise, the flower children of the counterculture were major proselytizers of the value of nature and the need to change our industrial civilization to something less destructive. With this in mind, I will summarize the major events of the Environmental Movement during the 1960s and 1970s before sorting out the major strands of *environmentalism*, ranging from resource conservationism to the varieties of "radical ecology" which have profoundly influenced attitudes toward nature in academic communities, turning loose upon the conservative, established technological society a segment of our youth full of enthusiasm for saving the planet.

The Environmental Movement during the 1960s and 1970s

In 1991 a poll taken of members of the North American Association of Environmental Education asked them to rank the ten most significant environmental events of the twentieth century. The result showed a ranking as follows: (1) *Silent Spring* by Rachel Carson (1962); (2) Earth Day, 1970; (3) the National Environmental Policy Act (1969); (4) *A Sand County Almanac* by Aldo Leopold (1949); (5) the Clean Air Act of 1970; (6) the Endangered Species Act of 1966; (7) the World Commission on Environment and Development (1983); (8) the Clean Water Act of 1972; (9) the Wilderness Act of 1964; and (10) the Stockholm Conference on the Environment (1972). Laying aside the fact that such a ranking is somewhat subjective, the results do make it clear that most significant legislation occurred between 1964 and 1972, when seven of the ten most significant events took place.[22] The mid to late 1960s and early 1970s were the most active period in the history of the Environmental Movement, which gave the 1970s the label of "the environmental decade." In the 1960s, symbolically, the Wilderness Act of 1964 was probably the single most important piece of environmental legislation, but in terms of challenging the status quo in our utilization of nature the National Environmental Policy Act of 1969 (NEPA) was probably the most threatening to vested interests in unlimited development. Environmental organizations, many at the local level, came into existence, including the Population Crisis Committee in 1965, The Chesapeake Bay Foundation in 1966, the Freshwater Foundation in 1968, and the Friends of the Everglades and Union of Concerned Scientists in 1969, the same year that NEPA was legislated. Also in 1969 newspaper headlines told of the Santa Barbara Channel oil spill, the chemical burning of the Cuyahoga River near Cleveland, and the shocking fact that Lake Erie was dying from chemical pollution and sewage effluent. On the national level, the Environmental Defense Fund was created in 1967, Zero Population Growth, inspired by Paul Ehrlich's *The Population Bomb* (1968), in 1969, and Friends of the Earth splintered off from the Sierra Club the same year. Simultaneously, memberships in environmental organizations were exploding during the 1960s. The National Wildlife Federation doubled its membership from 271,900 in 1966 to 540,000 in 1970, and Sierra

Club membership increased from a mere 20,000 in 1959 to 113,000 by 1970. The Wilderness Society grew from 27,000 in 1964 to 54,000 by 1970, and the Audubon Society from 41,000 in 1962 to 81,500 in 1970.[23]

The greatest year in the history of the Environmental Movement was 1970. In January, 1970 NEPA became federal law and in December the EPA was instituted. In between, on April 22 the first Earth Day was celebrated, the result of Wisconsin senator Gaylord Nelson's efforts to organize nationwide "teach-ins" on college campuses, following the model of anti-Vietnam War teach-ins during the 1960s. Although the Nixon White House was opposed to Nelson's proselytization and political activism on behalf of the event, April 22 witnessed a turnout greater than that of any event in the 1960s, involving 1500 colleges and 10,000 K-12 schools. Large rallies were also held in New York, Washington, and San Francisco, altogether involving an estimated 20,000,000 people or roughly 10 percent of the American public. However, some conservative organizations, such as the Daughters of the American Revolution, denounced Earth Day as a Communist plot, and anti-war and civil rights radicals saw it as an establishment ploy to direct youthful energy away from the war and civil rights. Although it did not receive support from the mainstream majority, Earth Day did demonstrate a large degree of support for environmentalism in 1970, enough to send a message to the Nixon administration to take action in defense of nature.[24]

Although President Richard Nixon somewhat reluctantly signed the National Environmental Policy Act (NEPA) into law on January 1, 1970, which was a major victory for environmentalists, it also turned out to be a compromise that essentially maintained the status quo through a bureaucratic process that focused the attention of the Environmental Movement upon Hardin's "mop," cleaning up the messes created by rapid industrial growth and brute force technologies, rather than focusing on the long-term causes of environmental problems, i.e., the gushing taps of overpopulation and overconsumption, not to say the lack of ecology-based land-use planning. However, on the positive side a spate of environmental legislation followed NEPA, beginning with the revised Clean Air Act of 1970, and followed by the Clean Water Act of 1972, which regulated the release of pollutants into storm sewers, streams, rivers, and reservoirs, and began the reclamation of recreational waters for swimming and fishing. The pesticide DDT was finally banned in 1972 and the Ocean Dumping Act was established to prevent marine pollution. A year later, Congress passed the Endangered Species Act (ESA), which began listing endangered or threatened species and carrying out plans for their recovery. Despite these salutary developments, it took only several years after the establishment of NEPA for the Environmental Movement to lose its initial impetus. "Despite the plethora of new environmental legislation, public support for environmental issues began to decline after the first flush of enthusiasm. The troubles that befell the Environmental Protection Agency demonstrate this. With strong congressional support, the "EPA rode the crest of the environmental wave of the late 60s and early 70s, right into the first oil shock of 1973."[25]

In the following pages, two separate sections discuss: (1) the effects of environmental legislation upon private land and managed public lands, which together comprise approximately 95 percent of the land area of the United States; and (2) wilderness areas affected by the Wilderness Act of 1964 and the Endangered Species Act (ESA) of 1973, including areas within Bureau of Land Management lands, national wildlife refuges, and national parks, which altogether comprise approximately 5 percent of the land areas of the United States.

Private Land and Publicly Managed Land

The early history of NEPA and the EPA provide a foreshadowing of the vacillating future of United States environmental legislation and regulation. Under NEPA there was a declaration of national environmental goals which sounded idealistic and promising as follows: (1) Fulfill the responsibilities of each generation as a trustee of the environment for succeeding generations; (2) Assure for all Americans safe, healthful, productive, and aesthetically and culturally pleasing surroundings; (3) Attain the widest range of beneficial uses of the environment without degradation, risk to health or safety, or other undesirable and unintended consequences; (4) Preserve important historical, cultural, and natural aspects of our national heritage and maintain, where possible, an environment that supports diversity and variety of individual choice; (5) Achieve a balance between population and resource use that will permit high standards of living and a wide sharing of life's amenities; (6) Enhance the quality of renewable resources and approach the maximum attainable recycling of depletable resources.[26] Briefly reflecting upon our successes and failures in attempting to implement these goals for a third of a century, we clearly have not been good trustees of the environment but we have been more successful in producing a safe and healthful environment. We have produced a wide range of beneficial uses, but with the consequences of massive degradation of natural habit nationwide and the world's greatest contribution to global warming, acid rain, and ozone depletion, whereas we have done a fairly good job of historical and cultural preservation. Worst of all, we have maximized both population and consumption, and enjoyed high standards of living without any sustainable balance whatsoever, and our overcutting of old growth and other forests, as well as soil-eroding and polluting agribusiness, i.e., machine-fertilizer-pesticide agriculture, have devoured and poisoned natural resources like there's no tomorrow.

Despite our failure to live up to the idealistic goals established by NEPA, our environment would be much worse off without that seminal legislation. Section 102 of NEPA established the environmental impact statement (EIS) process, which has done considerable good in forcing environmental review of all federal projects, from large and small dams to highway and airport construction and the dredging and channelization of rivers and harbors. Some of the nation's worst pork barrel projects have been exposed as ecologically harmful enough to be turned down. All EISs require: (1) an environmental impact assessment (EIA) of the positive and negative changes which would apply to projects, chiefly "major actions significantly affecting the quality of the human environment"; (2) the writing of and public input into an EIS if the assessment finds the project to have the potential to cause significant adverse environmental effects; and (3) public hearings and submissions of comments from public and private agencies, institutions, and individuals before the EIS is sent on to congress as an informational document designed to provide adequate information in behalf of environmental concerns related to a proposed federal project. An EIS demonstrating significant environmental impacts does not mean that a project will necessarily be denied, but a number of particularly egregious projects have been rejected or slowed down. The best thing about NEPA for environmentalists and the concerned public has been to bring the development activities of the federal government out into the open for public scrutiny and, occasionally, even denial or major modification through adjustments referred to as mitigation measures, such as decreasing the size of a project or preserving wildlife corridors.

The original proposed legislation that became NEPA did not include requirements for preparing EISs on proposed federal projects. These requirements under Section 102 of NEPA were added late in the legislative review process, shortly before action on the part of congress, and are referred to as the "action-forcing mechanism" of NEPA, "indicating that agencies must prepare a draft statement, which is then subject to review and critique by other federal agencies as well as state and local governmental and private groups."[27] Section II of NEPA also established the Council on Environmental Quality (CEQ) in the office of the President of the United States, the agency being responsible for preparing general guidelines for the writing of EISs by all federal agencies, for reviewing EISs on controversial projects, and for developing comparative analyses on the EIS process. Unhappily, it was within the politically manipulated CEQ that the weakening of the EIS process would take place. Before explaining this development, the relationship of the EPA to NEPA should be mentioned. The EPA was established in December of 1970 as the environmental regulatory agency of the United States. The EPA is not responsible for the environmental impact analysis process, but does review EISs submitted by other federal agencies in which environmental impacts related to the EPA's concerns, such as air and water pollution, toxic wastes, pesticides, radiation, and solid waste management are involved.

At first, federal agencies undertook their responsibilities for reviewing their own work from an environmental standpoint with some trepidation. Many of the early EISs written were too brief or contained major omissions and were sent back for further study, thereby slowing down proposed projects such as Army Corps of Engineers dams, canals, and deepening of harbors, or a United States Forest Service Plan (USFS) for clearcutting old growth in the Pacific Northwest. Worst of all for the process of development at the federal level, every project, every dam, highway widening, or clearcut was now open to public scrutiny since each federal project was sited in someone's state and county, and usually near a town or city, and the full-disclosure EIS process necessarily had to inform the private citizens and local governments wherever the project might be. This sudden openness of public development projects to individuals concerned for their well being (NIMBYS, i.e., not in my backyard), and to members of the Sierra Club, the Wilderness Society, and Audubon Society, as well as city, county, and state governmental agencies, brought challenges and lawsuits into existence for many federal projects, and terminated a number of them, including the large Elk Creek Dam on the Rogue River in my own state of Oregon. There was a growing fear among federal development agencies immediately after the creation of NEPA that development would crawl to a halt and the bottom would fall out of the economy. A solution was readily forthcoming through the CEQ which as a result of its being located in the Office of the President, could change its position depending upon whether the President was an environmentalist Democrat, anti-environmentalist Republican, or somewhere in between. The CEQ solution, which effectively resulted in gutting the potential enforcement capability of the NEPA EIS process, was the addition of a new section to be included in all EISs which would call for "an indication of what other interests and considerations of federal policy *are thought to offset the adverse environmental effects of the proposed action.* (Italics my own). This section is oriented to a discussion of other decision factors that the agency feels tend to counterbalance any adverse environmental effects."[28] Cost-benefit analyses were to be included in this section of the extended EIA format.

What this simple piece of legal *legerdemain* meant, referred to thereafter as "overriding considerations," was that even though an EIS might demonstrate significant adverse effects to wildlife, or air or water quality, the "other interests and considerations of federal policy," i.e., economic development interests, would be acceptable as more important and more economically needed and beneficial than the loss of wildlife or other ecosystemic components or health. Generally, federal developers could thereafter breathe a sigh of relief that most projects would be approved by Congress despite the significant adverse environmental impacts identified in EISs. Before this weakening of NEPA occurred, however, there were some significant victories. In 1970 the Supersonic Transport (SST) project was terminated because of potential environmental impacts, and the $50 million Cross-Florida Barge Canal was shut down, leaving the Army Corps of Engineers in shock.[29] The greatest environmental brawl involving the NEPA EIS process in its early years was over the Trans-Alaska pipeline, which was proposed by real estate developer Walter J. "Wally" Hickel after he (barely) became governor of Alaska in 1966. When Hickel was nominated Secretary of Interior by incoming President Richard Nixon, he announced plans to build a 400-mile pipeline from north of Fairbanks to the oil fields of Prudhoe Bay across an ecologically fragile arctic tundra. The application for a permit to build the pipeline preceded NEPA, but was nevertheless required to pass through the NEPA EIS process. "The battle over the pipeline pitted development in the United States' last wilderness against growing national sentiment that favored preserving nature."[30] In 1971 the pipeline EIS was made public in Washington, D.C., setting off an unanticipated public furor that resulted in the Department of Interior's sending the document back for revision. In 1972, after a cost of $13 million, a Final Environmental Impact Statement was released, and after extensive debate the pipeline was approved by Congress in 1973, the same year that the CEQ added the section of "overriding considerations" to the EIS format. The pipeline was subsequently challenged in court, but the OPEC oil embargo of that year was considered a national emergency as the price of oil soared and Americans fumed at the long gasoline lines, resulting in congress's swift approval of the pipeline. It was soon constructed and by 1977 oil was flowing to the port of Valdez, eight hundred miles from where it was pumped from the sedimentary rocks below the surface. "The legacy of the pipeline controversy became not environmental regulation but the fierce allegiance of Americans to their favorite personal transportation device--the automobile."[31]

A beleaguered President Nixon entangled in Watergate tapes was responsible, along with his lawyers and advisers, for pulling the teeth from the EIS process, which was designed as an informational tool to lead federal legislators towards environmentally benign, or more benign development than was taking place. Due to the 1973 formulation of "overriding considerations," NEPA has caused an enormous amount of paper shuffling while negligibly changing or shutting down few ecologically destructive projects. It would be difficult to imagine that the fierce struggle over the Trans-Alaska pipeline did not play a role in gutting the efficacy of the intended EIS process, NEPA's once-powerful action-forcing mechanism. Another major setback during the early 1970s was the congressional defeat of a plan to provide state governments with federal funds in order that local county and city governments " would undertake adequate planning and in so doing would observe ordinary precautions about dealing with sensitive land areas and sensitive land uses."[32] In other words, our federal government, including the Nixon administration, was actually considering approval of a federal land management program which came to be known as the National Land Use Planning and Policy Act. It was submitted to Congress during the early 70s and was almost passed, having been passed by the Senate but failing in the House of Representatives. As we shall see, this was a terrible defeat for the Environmental Movement. Roughly from 1973-1976 the land-use planning bill was re-introduced several times, but was beaten back by a conservative coalition of real estate developers, investors in property, farm and ranch organizations, and small business proprietors. "By 1976 almost everyone agreed that there would be no such thing as a national land-use policy in the United States of America. And there isn't to this day--which makes us unlike virtually every other industrial democracy in the world."[33] Rather than being dealt with through land-use planning, such issues as preserving and managing natural wilderness areas, maintaining the integrity of urban and suburban neighborhoods and rural communities, and preserving prime agricultural land would have to be handled piecemeal under different policies in different states and local jurisdictions. Thus, Americans as a whole came very close to establishing much-needed national land-use planning. Fortunately, some of the void left by this political defeat at the hands of special interests was filled by new state land-use programs in California, Oregon, and a number of other states.

The California and Oregon land-use planning systems were established in 1972 and 1973, respectively, and became models for the planning programs of other states. A number of states still have minimal land-use planning today. Prior to explaining the California and Oregon systems, one must understand the difference between general plans and comprehensive plans, and also the difference between planning by individual projects, as in the case of the NEPA EIS process, and planning by design, or comprehensive planning. There was no planning *per se* when Western towns were built on the edge of the frontier, but common sense suggested the location of commercial and industrial uses along the major mode of transportation, whether river or railroad, the latter soon following gentle stream gradients into the interior of the continent. Houses were generally built further back from the river than commercial and industrial buildings, and a cemetery and even a park and wealthy mansions further back from the din and dust of the town. Thus, something like planning, imperfect as it was, gave rise to the idea of the *general plan*, in which towns and cities laid out more precisely where industrial, commercial, residential, and agricultural uses should go, along with special zones for schools and hospitals. However, the general plan was a pre-ecological tool which took no regard for nature, or, in utilitarian terms, for the natural resources upon which towns, cities, and counties would develop and grow. The ecology-based plan which made the hydrology, geology, soil and vegetative cover, wildlife, and human historical monuments and artifacts an important, if not the most important, aspect of where different zones of human use would be located, was first popularized by University of Pennsylvania planner Ian McHarg, and has come to be known as the *comprehensive plan*. McHarg's well-known book, *Design with Nature* (1968), suggests that the planner first inventory the natural resources of a town or city site, or of county lands, and then design the planning zones to the natural parameters of the area.

Planning by "planning," or by one project at a time, is what the NEPA EIS process did, but it also incorporated the basic concept of comprehensive planning, the *environmental inventory* of the site of the proposed project--the geology, hydrology, wildlife, etc.. The state of California took this NEPA concept and ran with it, creating an Office of Planning and Research (OPR) in Sacramento and in 1972 requiring state equivalents of EISs on all *public* projects, including sections of the California Aqueduct. Under the new California Environmental Quality Act (CEQA), these reports would be called *environmental impact reports* (EIRs) and would serve the same functions as NEPA EISs: full disclosure of information regarding the environmental resources of the proposed project site and an assessment of potentially significant adverse environmental effects if the project was implemented. That same year a California Supreme Court decision known as "Friends of Mammoth" extended CEQA to include all private as well as public projects for environmental review. Assembly Bill 889 was an outcome of the

State Supreme Court decision in favor of the "Friends of Mammoth," a group of property owners and environmentalists who sought to prevent valuable open space and wildlife habitat on the edge of the town of Mammoth Lakes from conversion to a sprawling condominium development. The court requirement of an environmental report set the precedent for the state requirement of EIRs on *all projects*, private and public, under Assembly Bill 889, including small subdivisions, zoning changes, and special use permits. On April 1, 1973 the CEQA EIR process began throughout the state, much to the consternation of land developers, stockholders, and other property owners.

I worked as a geologist and land-use planner under CEQA for seven years, from 1972-1979, and was impressed with the fierceness with which land-use battles were fought in California, with so much land wealth at stake. San Diego County planners had a special framework for CEQA review called Integrated Regional Environmental Management, and hopes were high throughout the state that environmentally sound land-use planning could be enforced under the CEQA EIR process. Also, in California comprehensive planning, which inventories all aspects of the land, from geology and soils to wildlife in a given county or city, was recommended by the OPR, but not required. Forward- looking, ecologically interested counties and cities, such as Santa Barbara County and its main city, quickly developed sophisticated comprehensive plans as a data base to support the CEQA EIR process. Overall, there were major land-use successes resulting from "rezones" of mountains and foothills, and especially, agricultural land, in Santa Barbara County during the 1970s, and I worked on most of the CEQA EIRs used in making rezone decisions at the Board of Supervisors, the elected local governing body in California counties. Under the old general plans, mountain foothills and farmlands had been zoned into small parcels of 1, 3, 5, or 10 acres for large-lot sprawl development, long before chronic wildfires and the Environmental Movement changed our perception of appropriate land use. During the mid 1970s these badly zoned lands--low density housing developments on steep slopes covered with wildfire prone chaparral (an ecosystem of small oaks and brushy plant species), and low density developments sprawling away concentrically from higher density town centers in places like the Santa Ynez Valley, were eating up prime agricultural land and wildlife habitat. Both a Santa Ynez Mountain foothill rezone to larger parcels of 5, 10, and 20 acres, and a Santa Ynez Valley agricultural rezone from parcel sizes of 1, 3, and 5 to 10, 20, and 100 acres were approved by the Santa Barbara County Board of Supervisors. Environmentalists, including myself, were optimistic that development could be rationalized and made ecologically less damaging in the years ahead. We definitely seemed to have saved the agricultural use of the Santa Ynez Valley from destruction by urban sprawl.

Gradually, land developers and property owners invented ways of developing lands piecemeal, covering farmland and open space with sequences of small developments which cumulatively ate away the countryside outside the areas which had been rezoned. Also, following revision of the NEPA EIS process to include "overriding considerations," or the need for economic development despite significant environmental impacts, the California OPR incorporated "overriding considerations" into the CEQA EIR process, in the long run effectively gutting it as boards of supervisors began approving almost every project on the basis of need for economic development. By the time I left CEQA and Santa Barbara to begin a teaching career, county morale was low and some environmental planners had joined development firms to use their insider expertise on behalf of further development. At this point I described the CEQA EIR process as "processing the disaster." I remember a friend saying, "You thought that you could change the world," meaning that he saw the cause of environmental planning as a debacle.

In Oregon I encountered a different approach to environmental planning, one which ignored the EIR project-by-project approach for a state requirement of environmentally-based comprehensive planning required for all 278 counties and cities in the state. In 1973, under Assembly Bill 100, Oregon established the Land Conservation and Development Commission (LCDC) and a staff of several dozen planners to see that every local government use its planners to develop and submit to the state a comprehensive plan for their area. The greater Portland area and other large cities responded very positively to this new requirement, backed as it was by the ecology-oriented climate of opinion which still flourished in the early 1970s, when Republican Governor Tom McCall said "come and visit, but don't stay." Conservatives in generally Republican rural counties were outraged, however, that governmental restrictions were going to be placed on their liberty to use land in any manner which pleased them, and a tug of war began in order to extract comprehensive plans from rural areas. An attempt by LCDC to convince Polk County (which includes part of Salem, the state capitol) in the Willamette Valley to rezone large tracts of low density parcels of 5 acres (amounting in all to about 40,000 acres or 60 square miles) created by the old general plan to large parcels of 40-80 acres resulted in a stalemate and compromise parcels of 15 acres. Resistance to comprehensive planning, it turned out, was stronger than it had been in California, and as the 1970s passed by, that resistance increased. In California most counties have voluntarily developed recommended comprehensive plans in order to facilitate project-by-project environmental analysis under the CEQA EIR process. The two approaches re-enforced one another in what was the best land-use planning program in the nation; but pressures exerted in various ways by developers, increasing property values which doomed farmland and other open space to suburban sprawl, and weakening of the EIR process by "overriding considerations" has gradually overrun the best possible environmental management, which essentially acted as a rear-guard action against overwhelming pressures for development in a state of 30 million people. The Oregon LCDC and its comprehensive planning approach (without EIRs) has fared somewhat better, although breakdown occurs at the level of local government because county and city planners traditionally have facilitated development rather than slowed it down. Also, with only 3.5 million or so in population, pressures for development have been far less than in California, to the point that in 2004 the governor of Oregon called for an emphasis on the *development* aspect of the Land Conservation and Development Commission, inviting growth in population to fuel a hoped-for economic boom, the opposite of Tom McCall's "visit but don't stay." After three decades of land-use battles, NEPA, CEQA, and LCDC are still legally in place, but they have lost their teeth and the short-lived public mandate for ecologically sound land development. Furthermore, Oregon's LCDC was disempowered by a 2004 initiative based upon the arguments of frustrated landowners and real estate developers. A majority of 61% of Oregonians rejected the state's model land-use planning program (which had been supported against similar past initiatives by 55%). By 2005 this crucial land-use struggle was headed for the Oregon Supreme Court to resolve the "taking issue," i.e., the extent to which citizens should be compensated for loss of the potential value of land affected by land-use planning. Nevertheless, without these institutions produced by the Environmental Movement, the aesthetics and ecology of much of the American landscape would be far more disagreeable than it is today.[34]

Aldo Leopold would be proud of the efforts undertaken in behalf of the land ethic he proselytized in the 1930s and 1940s. However, the displacement of small farms by the large holdings of agribusiness corporations since the Second World War has destroyed the livelihoods of many individual farmers. In most instances these people were generally conservative in outlook and not particularly interested in saving nature on their land. Agribusiness corporations are, with a vengeance, even less interested in the ecology of streams and scraps of "wasteland" contiguous to their highly efficient operations. Consequently, there really has been little room in the American mind for anything like a land ethic, save in the minds of a minority of individuals and groups associated with the Environmental Movement. Fortunately, before environmentalism, in 1885 the state of New York set the precedent for preserving large tracts of public lands which were neither national parks nor wilderness, but a mix of forest lands sold off by cut-and-run timber barons, along with farmlands, grazing lands and small towns, much of which became the Adirondack State Park, the original "greenline park." A greenline park consists of "coherent landscape areas with outstanding public values that were (or could be) partially owned by public and quasipublic agencies, but for the most part would consist of unspoiled land still in private ownership."[35] The Environmental Movement helped to inspire legislation creating other greenline parks such as the New Jersey Pinelands Reserve and the Columbia Gorge National Scenic Area in Oregon and Washington. Given the rate of loss of similar noteworthy rural or pastoral landscapes in the United States, if we possessed enough citizens with the moral sense and the political will to preserve such a "middle landscape" in accord with Leopoldan land-use ideals, we would lobby to create a national system of such parks. National systems of pastoral greenline parks already exist in France and the British Isles.[36] The development of greenline park systems is partly the result of the failure of human nature to live up to Aldo Leopold's expectations in his desire to establish a land ethic in American farm communities. The hard realities of agribusiness expansion and the economic bottom line which has become more oppressive for small, individual farmers competing against large corporations left the adoption of the land ethic to a small number of altruistic idealists and environmental advocacy groups whose lobbying has paid off in the creation of a few American greenline parks, but not nearly enough. Just as in the case of forests and wilderness, we have placed boundaries prohibiting ecologically destructive uses around a few small regions amounting to about five percent of the United States in national parks, wilderness, and greenline parks. In France today, more than ten percent of pastoral lands, i.e. mixed forests, farming, and grazing lands, are managed under special greenline land-use management plans, and those lands are in addition to their wilder national parks in remote mountains like the Alps and Pyrenees. American environmentalists can take some satisfaction in the fact that the French national park system was inspired by the creation of American national parks. Having established their own national parks, the French realized the need to preserve the best of their pastoral landscape, "la campagne," under the relatively strict land-use plans (special comprehensive plans) which are the basis for creating greenline parks.[37]

Wilderness: Less than Five Percent of Our Land

The preceding explanations of NEPA, CEQA, LCDC, EISs, EIRs, comprehensive plans, and greenline parks should suggest the high level of practical accomplishment achieved as a result of the Environmental Movement. The United States and many other countries would be disastrously worse off without these institutions in place along with the better known environmental legislation produced by the Environmental Movement, such as the Clean Air Act, Clean Water Act, Safe Drinking Water Act, and Endangered Species Act. Taken altogether, these achievements are impressive, but the history of their success is compromised by the weakening of the Environmental Movement after the 1970s.

Environmental legislation related to land-use planning in the foregoing section is of great importance because approximately 95% of the land area of the United States is either in private ownership or intensively managed public lands such as national forests and Bureau of Land Management (BLM) grazing lands. All of these lands have been affected by the creation, through federal legislation, of NEPA and the EPA, and legislation by the states creating environmental planning agencies. In sharp contrast, wilderness in the United States, including Alaska, comprises only 4.67 percent of all land in the United States, as of April, 2004 consisting of 662 wilderness areas totaling 105.7 million acres. This land area includes relatively pristine portions of federal lands, including 43.6 million acres in national parks, 34.8 million acres in national forests, 20.7 million acres in national wildlife refuges and 6.5 million acres on western lands of the Bureau of Land Management. Over half (57.5 million acres) of all designated wilderness is in the extensive forests, national parks, and wildlife refuges of Alaska. This means that within the contiguous forty-eight states, less than 3 percent of the total land area is wilderness as part of the National Wilderness Preservation System created by the Wilderness Act of 1964. [38]

Thus, wilderness takes on a disproportionate significance for the preservation of native ecosystems and endangered species if, on 95% of the American landscape, ecosystems have been fragmented and partially or entirely transformed into managed human systems of urban, agricultural, grazing, mining and timber lands. In fact, as we shall see, wilderness itself is delimited and fragmented to such a degree that the long-term survival of large predators such as grizzly bears and wolves in the forty-eight contiguous states of the United States is doubtful under present circumstances. Thus, the Endangered Species Act of 1973 is closely linked to the future of wilderness in America.

The fact that 13.1 percent and 9.1 percent of the land areas of the United States and Canada, respectively, are protected by national parks and wilderness areas suggests a more promising future for endangered species than the percentage of land in wilderness alone. Nevertheless, human use of national parks and wilderness areas has increased dramatically in recent decades as population and the desire for recreational areas have multiplied. Also, many preserved areas are in mountain or arid environments, where the effects of human use are disproportionately destructive. In Pacific borderland parks and wildernesses that I have hiked, backpacked, and mountain climbed in since the 1950s, in California, Oregon, and Washington, eutrophication of lakes and ponds, and the spread of organisms such as *Giardia* have resulted from the increase of human waste. Simultaneously, air pollution from distant cities and highways has contributed to the acidification of lakes and streams. The once-pristine environments of our national parks and wildernesses have undergone obvious as well as unperceived changes in the physical and biological parameters which threaten both ecosystems and individual species. Overlaid upon local impacts are the predictable and unpredictable consequences of long-term climate change related to global warming. All of these hazards to wilderness suggest the need to expand wilderness areas to the greatest extent possible, along with necessary animal corridors and the linking of habitat areas to allow survival of ecosystems and their component species as human-induced changes occur.

The primary cause of the loss of species leading towards extinction is the loss of habitat; i.e., the displacement of ecosystems by conversion to human use. In the United States, by the late 1980s 85 percent of the original pre-European "virgin" forest had been destroyed by urban development, timber harvesting, and agricultural development. Canada today, in contrast, still retains about 90 percent of its original forest cover, although half of the old-growth forests in British Columbia have been cut down. In addition, 60% of wetlands in the United States have been lost to urban and agricultural development, with comparable losses occurring in Canada. The devastating effect of these changes on species inhabiting the North American continent, threatening about 1000 species in the United States and 52 species in Canada, was partially mitigated by preservation of habitats beginning with Yellowstone National Park in 1872, and the focus on individual species in the U.S. Endangered Species Act of 1973. Canada, on the other hand, has no federal legislation covering endangered species.[39]

The Endangered Species Act (ESA) of 1973 was reauthorized in 1988, and was scheduled for reauthorization in 1992. However, opposition to the act from development, timber, and mining interests has forced a political stalemate and operation of the ESA on year-to-year budget extensions. Anti-environmental groups and Republicans in Congress continually lobby for weakening or abolishing the ESA. Private environmental groups such as Defenders of Wildlife have challenged the agencies responsible for endangered species, the Fish and Wildlife Service (FWS) for terrestrial and freshwater species, and the National Oceanic and Atmospheric Administration (NOAA) for marine and anadromous species, to act on species which are endangered or threatened, eventually resulting in a moratorium by the overwhemed FWS on listing activities in 2000, at the beginning of the George W. Bush presidency. Listing, critical habitat, and recovery plans comprise the three essential elements requisite to designating a species as endangered or threatened. Listing of species directly by FWS or NOAA is complemented by listings based upon petitions from environmental groups and state agencies. The responsible agency must also designate areas where the species is currently found or where it may spread, if it recovers, as critical habitat. To facilitate preservation of individual species, the agency is also required to develop a recovery plan. Recently, the Bush Administration asked a federal judge to invalidate critical habitat protection for the California gnatcatcher, arguing that the FWS "did not carry out an adequate economic analysis of the critical habitat designation, as required by law."[40] Thus, the same reasoning that undermined NEPA, the doctrine of "overriding considerations," namely, that the need for economic development should override the possibility of closing down or preventing a project due to potentially significant environmental impacts, was applied to thwart the objectives of the ESA. In short, no matter what the environmental damage, such as not providing critical habitat protection for an endangered or threatened species, economic business-as-usual must be carried out. Neither environmental impact statements nor habitat protection requirements should blunt the onward rush of economic development.

The northern spotted owl represents a high-profile case in point for the ESA. An icon for the preservation of remaining old-growth forests in the Pacific Northwest, the northern spotted owl population had dwindled to between 4,000 and 6,000 by 1990, leading the FWS to list the owl as a threatened species. Several years of bitter controversy followed, which pitted environmentalists against the timber industry, which was already shedding many jobs due to increased mechanization. In 1994 a compromise, the Northwest Forest Plan, was worked out by the Clinton administration. The United States Forest Service (USFS), employing its recently adopted "ecosystem management approach," set aside 7.4 million acres of land in the plan, where logging of tree stands older than 80 years would be prohibited (in California, Oregon, and Washington). In 2003 the Bush administration settled a lawsuit by a timber industry group (the

American Forest Resource Council) by agreeing to review spotted owl protection in order to re-asses the amount of timber which needed to be protected, with the expectation that it would be reduced.[41]

As Democratic and Republican administrations come and go, the environment becomes a political football kicked left and right. Laws and plans can be undone not too long after they have been formulated, leaving the practice of rapid economic growth as a constant in the long-term. As the United States has drifted to the political right, and religious and related social issues have transformed the Republican and Democratic parties, the legacy of the environmental movement has fallen to the responsibility of a minority referred to as "environmentalists" by the larger American public. As further attrition of environmental gains occurs (e.g., rejection of the Kyoto Treaty and the near opening of the Arctic National Wildlife Reserve (ANWR) in 2005 under the fiercely anti-environmentalist Bush administration), the American public sinks deeper into the pursuit of wealth, electronic entertainment, soteriological religion, and environmental apathy.

In recent years my wife and I have made annual pilgrimages to the great national parks and wilderness areas of the Canadian Rockies, from Banff and Jasper to Yoho and Waterton Lakes, as well as Glacier in Montana. We have been amazed and impressed at the abundance of wildlife that we have encountered each year: black bears, grizzly bears, elk, moose, deer, mountain goats, bighorn sheep, wolverine, and smaller mammals and birds. I had seen comparable wildlife in Yellowstone in 1959, but in the 40 years spent in California and the Pacific Northwest I have only seen a few deer, elk on two game reserves, and three black bears against a starry sky in the Tioga Pass area of the Sierra Nevada, when they leaned over my sleeping bag, decided not to eat me, and headed off for a nearby campground to scavenge for food.

At first glance, the abundance of game in the Canadian national parks would seem to suggest long-term ecosystem stability. One measure of stable ecosystems is the condition of the top predators in the system; in the case of the Canadian Rockies this means grizzlies, wolves, and cougars, as well as black bears, although vegetation constitutes about 90 percent of their food intake. Ecologists have learned that most wilderness areas in the United States are missing their top predators, especially grizzly bear and wolves, although cougars have made a spectacular comeback in Pacific Borderland states in recent decades, since hunting them with dogs was outlawed. Looking northwards to Canada and Alaska for true wilderness, Americans generally do not realize the threatened status of grizzlies and wolves in Banff, Jasper and other parks of the Canadian Rockies. Concern for preserving adequate habitat (including linking corridors of public and private lands) for these predators led, in the 1990s, to the Yellowstone to Yukon Conservation Initiative, or Y2Y. Y2Y grew out of the speculations of conservation biologists in the late 1980s, including Michael Soulé and Reed Noss. They argued that what the future required was not giant "superparks" but, rather, the linking of existing protected areas by wildlife corridors. "A network of corridors could provide what none of the existing reserves could do on their own: enough room and freedom for wildlife to flourish." [42]

By the early 1990s greenways and wildlife corridors were being established in Florida, Oregon, and California, in each case connecting isolated populations of animals to prevent local extinction. During the 1980s and 1990s wildlife ecologists had been surprised at learning of the wide geographic ranges of wolves, grizzlies, and even trout. Because existing wilderness areas have been carved out of national parks, national forests, national wildlife refuges and Bureau of Land Management lands in the United States, corridors in these public lands in conjunction with conservation easements on private lands, a popular land-use tool in Massachusetts and other preservation-conscious states, would allow for the spread or continued existence of large mammals even as more urbanized lands developed in the future. In Massachusetts, a large area in state parks combined with conservation easements has allowed the return of moose and black bear to a heavily populated state.

One brave soul in Canada decided to test the validity of the corridor approach to wildlife preservation by walking the Y2Y region along or near the continental divide from Mammoth Hot springs, Wyoming in Yellowstone National Park to the southern Yukon in Canada, a journey of 2,200 miles largely by foot, but also by ski and canoe. Karsten Heuer, a wildlife biologist, found the Y2Y corridor presently open to the wanderings of grizzlies and wolves, but threatened by new roads, oil and mineral exploration, legal and illegal hunting, forest destruction by clearcutting and extensive road networks followed by hunters and snowmobilers, expanding recreational development, and a knot of overdevelopment within Banff National Park itself, compounded by fast traffic linking it to nearby Calgary, only an hour and a half away. Near the end of an adventure that nearly cost him his life, Heuer wrote:

> I will need to remember how this stillness and simplicity feeds me, how the humility and awareness evoked by bears and wild swims feeds into my everyday life. I will need to remember the good that inevitably follows the bad, the warmth that follows the cold, the calm that returns after the storm, and the quiet patience of animals that endure these cycles day after day, year after year. I need to remember that everything, including life, is only temporary. [43]

Heuer completed his epic journey in 1999, and since that time several encouraging developments include: new wildlife corridors in British Columbia; embracing of the Y2Y ethic by the Jackson Hole, Wyoming Chamber of Commerce; a 2003 plan by the Canadian federal government for a new national park on the British Columbia-Yukon border, expansion of Waterton lakes National Park into British Columbia's Flathead River Valley, and new protected areas in the Northwest Territories; President Clinton's "Roadless Rule" that would have protected 16.5 million acres of the Y2Y region in Idaho, Montana, and Wyoming were it not for the Bush administration's attempts to keep the area open to road building, logging, and motorized access; and awards from the Canadian Geographic Society and the World Conservation Union for the foresight of the Y2Y vision.

Like the Swords of Damocles of possible global nuclear warfare and actual global warming that hang over the heads of business-and consumption-oriented citizens in their daily pursuits of the good life and higher standards of living, the threats to wilderness resulting from the spread of global capitalism fail to stir most of Earth's citizens to action. Otherwise, anti-environmentalist, backward-looking conservative politicians like George W. Bush and his ilk would not be elected. Instead of being led towards sustainable practices that would leave room for wilderness and long-term biodiversity, it has been left to a relatively small minority of ecologists and other academics, the staffs of numerous private environmental institutions, and heroic activists like Heuer to move the ideals of the Environmental Movement forward into a future dominated by explosive economic growth and human population expansion.

The 3 percent of wilderness in the contiguous forty-eight states and the larger but gradually fragmenting wilderness of Canada together take on enormous significance while the Old World continents approach total humanization, with Africa not far behind. Implicitly, the global technological society which spread from Western European, Faustian civilization is leading us toward the effacement of the natural world and the separation of humanity from its natural origins. "When wilderness has been consumed, our understanding of what is natural can be changed as required, and no facet of the human psyche or biology will be left invulnerable to revision. Reason, and only Reason, will prevail." [44] Not reason, but the tyranny of reason is what we will be threatened with in a post-wilderness world. Ancient and medieval civilizations substituted agrarian societies cemented by soteriological religious belief systems for eradicated wilderness. Modernity complements such traditional beliefs with endless novelty. The idea of the tyranny of reason makes no sense to contemporary urban man, immersed in new technologies and entertainments.

> What is reason if not consensus? And how can any tyranny exist in such a proliferation of choices, such an unprecedented prosperity and scope for self-expression? Already for millions of such men the rationale of the technocracy has become absolute, and the highest use of intelligence consists in maintaining their position in it.
>
> Only in wilderness is it possible to escape this tyranny. In wilderness a man or woman has physically left behind the milieu of conditioning--the pervasive sociability, the endless "information" from the mass media, and so on. To some extent, the wilderness traveler will be reminded of his animal nature, and share again the profound, irrational correctness of trees, lakes,

birds, and beasts. For urban men this can be a subverting experience. Some must react violently in an attempt to debase or destroy the source of their disturbance, and to bring ancient terrors to heel.[45]

Environmental Ethics and the Western Worldview

Before summarizing the challenges to the movement since the 1970s, I will discuss one more important contribution of the Environmental Movement to our civilization during its early stages, which was the rise of environmental philosophy, also known as environmental ethics, synchronous with the beginnings of environmental history. Environmental ethics are especially important as potential progenitors of new elements to be incorporated into a postmodern, ecologically responsible worldview. Environmental philosophies have arisen and spread much more quickly than philosophical thought in classical antiquity, or the competing mystery religions which flourished during the Hellenistic and early Christian eras of the Roman Empire, laying the foundations for a new transcendental worldview out of which the Western civilization of the Middle Ages was constructed. Here I will briefly sketch out a simple classification of competing environmental ethics which applies to the decades reaching from ca. 1970 to the present.

Traditional Western philosophy has its roots in classical antiquity, dominated by the genius of Plato and Aristotle. Plato's transcendentalism and Aristotle's teleological orientation, manifested in his conception of the "great chain of being," were integrated into Christian theology by Plotinus (A.D. 205-270) and Saint Thomas of Aquinas (ca. 1224-1274), respectively. For almost the entirety of the history of philosophy in Western civilization, philosophy has been practiced under the influence of Christian supernaturalism. As discussed in Chapters Five and Six, the various strands leading towards the substitution of a modern, secular worldview for the supernaturalistic medieval worldview included the Great Discoveries, the Renaissance and Reformation, the rise of Western science to pre-eminence during the 16th and 17th centuries, philosophical skepticism in the seventeenth century and later, and the integration of eighteenth century philosophy and continuing Christian religion into a compromise philosophical perspective which arose during the Enlightenment (in the broad sense including late 17th century antecedents). While Voltaire, Rousseau and the majority of the philosophes adhered to this compromise called deism, other more radical philosophers forged ahead toward the logical culminating terminus of philosophical speculation set in turbulent motion by the onrush of modern history (and its implicit blasphemies from the medieval or Reformation perspectives). In other words, philosophy began to detach itself from the age-encrusted thought of the medieval philosopher-theologians. Despite the gradual emancipation and secularization of philosophy, which by the 19th century could finally express itself with impunity in the more liberal parts of Europe, it hardly occurred to these

Enlightenment moderns that any harm could come from all the modern wonders of science, technology, industry, and commerce. That is, until Thomas Malthus came along (See Chapter 9). Malthus, with his mathematical proof of long-term population increase and (implicitly) ecological disaster, might compete with Aldo Leopold for the title of "father" of environmental ethics. Whatever the merits of his candidacy for that honor, Malthus did introduce a secular approach to human population-environmental problems, although as a Protestant clergyman he did allow the Almighty to participate by making population die-offs, even of humans, all part of the grand providential design. If his speculations are undeserving of the term "philosophy," he still focused the nineteenth century mind on distinctly environmental issues. Leopold did the same for the pre-environmentalist twentieth century, but with particular reference to humanity's moral responsibility for nature, or, more precisely, nature's health, as expressed in the "land ethic." Thus, it is Leopold who has been acknowledged as the true founder of environmental philosophy.

Environmental philosopher J. Baird Callicott has constructed his own system of environmental ethics upon Leopoldan foundations, and has generated a valuable philosophical controversy in the process.[46] The nature of this controversy, focused as it is upon the nature of the ecosystem concept as the lynchpin of all environmental ethics, is beyond the scope of this book, but from the perspective of environmental history and the problematic Western worldview as regards our attitudes toward and treatment of nature, environmental ethics is possibly the key to any possible metanoia or change in worldview. My purpose here is to lay out a framework, a simple classification of currently existing environmental ethics, including negative or anti-environmental ethics such as the Christian idea of human dominion over nature, in order to assess the potential for metanoia, and also, of great importance, to better understand the *problem of development or economic growth* in relation to our contemporary worldview as it relates to nature.

A classification of *environmental positions or kinds of environmentalism* (which can be extrapolated from different environmental ethics) that is simple enough for my purposes is presented in Robyn Eckersley's *Environmentalism and Political Theory: Toward an Ecocentric Approach* (1992). Eckersley relates five predominant types of environmentalism to the philosophical reasoning that underlies each type. Considered on a scale which ranges from extreme anthropocentrism on one hand to ecocentrism on the other, we might envision these five types as also roughly paralleling the political spectrum of right to centrist to left or radical. So envisioned, we begin on the anthropocentric right with the most conservative category of environmentalism (considered anti-environmentalist from the perspective of all four other types), which is the idea of *resource conservation*, also referred to as utilitarianism, scientific management, or "wise use," with a history traceable from Plato and the Old and New Testaments to the modern Scientific Revolution and its refinement in the thought and practices of Gifford Pinchot and natural resource development agencies such as the USFS, the BLM, the USBR, and the Army Corps of Engineers. For Pinchot and other resource conservationists, *development* is the first principle of conservation, "for the benefit of the many, and not merely for the profit of the few." Moving from the anthropocentric right towards the center we encounter *human welfare ecology*, a product of the largely urban and suburban environmental concerns of the 1960s, with antecedents in the Marxist and other health and labor movements of the nineteenth century. Human welfare ecology has been concerned with "toxic chemicals or 'intractable wastes'; the intensification of ground, air, and water pollution generally; the growth in new 'diseases of affluence' (e.g., heart disease, cancer); the growth in urban and coastal high rise development; the dangers of nuclear plants and nuclear wastes; the growth in the nuclear arsenal; and the problem of global warming and the thinning of the ozone layer ...."[47] Clearly, these problems peculiar to the twentieth century and its exponential industrial growth are no small stakes, but there is no grandiose vision for reforming an ecologically destructive system in human welfare ecology. Rather, it expresses the health and safety concerns of individuals and their families, although it "has been highly critical of economic growth and the idea that science and technology alone can deliver us from the ecological crisis."[48] Nevertheless, although some proponents of human welfare ecology have called for a stewardship ethic involving a shift to appropriate technologies, organic agriculture, and soft energy such as wind and solar power, they are generally anthropocentric in their primary concern for *human* welfare. In contrast to the "wise-use" orientation of resource conservationism and the environmental quality issues of human welfare ecology, the essence of *preservationism*, a politically centrist position, is reverence for wilderness, particularly as a source of aesthetic and spiritual inspiration for humanity. Eckersley argues that wilderness campaigns "have generated the most radical philosophical challenges to stock assumptions regarding our place in the scheme of things, thereby forcing theorists to confront the question of the moral standing of the nonhuman world."[49] While acknowledging that John Muir's pantheistic view of nature was largely anthropocentric in its emphasis on the aesthetic and spiritual response of humans to wild nature, preservationism has gradually become more ecologically informed, thereby shifting its focus to the preservation of ecosystems for their intrinsic value and not just for their awe-inspiring beauty. Eckersley is also well aware of the problem I have emphasized earlier in this chapter in terms of Garrett Hardin's metaphor of "running for the mop" as opposed to "turning off the tap" of population and consumption, when she writes "from an ecological point of view, it is self-defeating to focus exclusively on setting aside pockets of pristine wilderness while ignoring the growing problems of overpopulation and pollution, since these problems will sooner or later impact upon the remaining fragments of wild nature."[50]

    From the relatively conservative environmental viewpoints of resource conservationism and human welfare ecology, and the moderate stance of preservationism, Eckersley moves to the more radical left of *animal liberation* and *ecocentrism.* The animal liberation movement has its roots in the modern utilitarian school of moral philosophy epitomized by the thought of Jeremy Bentham (1748-1832) and brought up to date by the contemporary philosopher Peter Singer. The essence of animal rights environmental ethics is that because animals can suffer, i.e., because they are sentient or feeling creatures, at least down to a certain level (perhaps just above the oyster), they are worthy of moral standing and should be treated accordingly. Particularly, such inhumane practices as "factory farming" and vivisection should be abandoned if we take animal rights theory seriously. The theory also implies the need to protect the habitat of sentient fauna, but at the same time leaves non-sentient fauna and vegetation in a moral limbo except for their instrumental value to sentient animals. It is also inadequate in dealing with the human role towards predators which inflict pain upon their prey, and is atomistic rather than holistic in locating intrinsic value in individual organisms and not in whole communities or ecosystems. In contrast, *ecocentrism,* which can be partly understood as a more ecologically informed variant of preservationism, is Eckersley's preferred type of environmentalism. Ecocentric environmentalists strongly support the idea of leaving or reclaiming large tracts of wilderness or wild areas in order to support species and ecosystem diversity. Whereas regions with little or no wilderness such as Europe have emphasized human welfare ecology, regions with large wilderness areas have focused on preservation and given rise to ecocentrism as *deep ecology* or *ecofeminism.* These environmental ethical positions are also referred to as *radical ecology,* and generally adopt a holistic rather than an atomistic perspective, which values ecosystems and species as well as individual organisms.[51]

With these five broad categories of the types of environmentalism which developed during the 1970s in mind, I will briefly continue the history of the Environmental Movement and its institutionalized offspring, collectively referred to as *environmentalism* (including : 1. Resource conservationism; 2. Human welfare ecology; 3. Preservationism; 4. Animal liberation; and 5. Ecocentrism or radical ecology). Whether or not it is more appropriate to speak of an Environmental Movement after 1980, when its personal antithesis, Ronald Reagan, was elected President, or whether it makes more sense to refer to the post-Reagan era as simply "environmentalism" because it has not reflected a national mandate since the 1970s is a moot point, although most historians and media journalists still write of the Environmental Movement as though it were alive and well with strong public support. In the latter view, one would at least have to admit that the movement is presently moribund, or the Bush administration would not be running rampant over environmental laws today, in 2007.

The conservative backlash of the 1980s "marked a reversal in the move toward environmental protection. The energy crisis of the mid-1970s and the Reagan revolution of the 1980s demonstrated that environmental issues, though important, were subordinate in the public mind to material living standards and economic security."[52] It was the purpose of the Reagan administration to roll back the environmental accomplishments of the 70s, notably the large regulatory institutions like the EPA and legislation such as NEPA and the ESA. These objectives were undertaken in the manner described by Garrett Hardin as "Quis Custodiet Ipsos Custodes?" (Who will watch the watchers themselves?). In his famous "Tragedy of the Commons" essay a crisis occurs and legislation responds with new laws and agencies, creating reassurance and eventually apathy among the general citizenry. In the final stage, administrative positions on policing agencies such as the EPA are staffed by those responsible for the crisis, in environmental matters these being CEOs of polluting corporations and other anti-environmental individuals. Thus, we wind up with the fox guarding the henhouse, and the likes of James Watt (Department of Interior) and Ann Gorsuch (EPA) in charge of environmental agencies in order to weaken or destroy them. These anti-environmental appointments by the Reagan administration created a firestorm of angry criticism from individual citizens, the Sierra Club, and other private environmental institutions, and Watt and Gorsuch were replaced, but the Quis Custodiet syndrome continued, including the staffing of the CEQ with anti-environmentalists under the Reagan and later Republican administrations. As a rule, Democratic administrations tend to support environmental goals and Republican administrations generally undermine them. Between 1980 and 1992 we experienced federal apathy to the point that a number of states adopted stronger environmental legislation of their own. A moderating influence was the strong environmentalist response, supported by much of the public, to the Reagan era of deregulation. In 1985 about 80 percent of the public supported existing environmental regulations.[53] Also, environmental lobbyists gained enough of a footing in Washington to somewhat offset the power of the corporations. Environmentalist writer Kirkpatrick Sale goes so far as to remark that in 1989 the Environmental Movement was stronger than it had ever been.[54] And yet in 1988 George Bush had been elected the (pseudo)"environmental" President (environmentalists did not believe it) with the following results:

> On the international front, the Bush Administration refused to sign a treaty on carbon dioxide emissions in the atmosphere, accepted by all the other nations of the world, until it was watered down to meaninglessness, and at the 1992 Rio "earth summit" it was alone in opposing and refusing to sign international accords on preserving endangered species and protecting forests and wetlands. At home Bush made development of the Arctic National Wildlife Refuge for oil drilling the centerpiece of his energy policy (the measure was shelved by Congress in 1990), while taking no action on reducing energy consumption or increasing automobile efficiency.[55]

If the Environmental Movement was so strong, and the majority of Americans were behind it, why were they electing Republican presidents and apathetically accepting anti-environmental policies? Obviously, the polls were misleading, and we had other priorities. Kline refers to the early 1990s as a period of government retrenchment and public apathy, a syndrome which has perpetuated itself right on to the present, when Bush junior's policies are worse than a clone of his father's. Ever since Ronald Reagan, the economy and multiplying human needs have overshadowed environmental concerns in the United States while, simultaneously, the high technology computer revolution has transformed the business world and its corporations into a global network. All the same, there was good reason for environmental optimism when the Clinton-Gore administration took power, and treaties turned down by Bush at the Rio Conference in 1992 were signed. Clinton, however, was as much focused on economic growth as his predecessor, and the anti-environmental *wise-use movement* gained momentum during his eight years as President. Clinton also may have opened a Pandora's (environmental) box when he lobbied for passage of the North American Free Trade Agreement (NAFTA) which passed in 1993. "Whatever Clinton may have believed, NAFTA has proved to be a detriment to the environment but a boon for business."[56] During his first term, Clinton also alienated environmentalists by approving a "salvage logging" rider that increased timber cutting in national forests, but later in his first term he earned their praise for vetoing bills that would have slashed the EPA's budget, opened the Arctic national Wildlife Refuge to oil drilling, and allowed clear-cutting in Alaska's Tongass National Forest. The balancing act between advocating rapid economic growth and protecting the environment was difficult to perform, and during Clinton's second term he displayed a moderate, disinterested approach to environmental issues. His administration vacillated over strict implementation of the Kyoto accords designed to limit man-made emissions of greenhouse gasses, and gave little support to a United Nations proposal to create a worldwide fund to deal with global environmental crises. The United States has remained, down to the present day, the most economically conservative voice when it comes to supporting solutions to global environmental problems. We are ruled by the need to maintain the treadmill of production in an economically inegalitarian society in which the lower economic orders would suffer egregiously from an economic slowdown, much as they did in the Great Depression of the 1930s.[57]

The environmental institutions created in the 1970s (and subsequent environmental legislation) have survived strong political attempts at deregulation from the early 1980s to the early 2000s. Has the institutionalization of the movement falsely reassured Americans that environmental problems have been taken care of so that it is safe to ratchet up the pace of economic development? Kline asks himself whether the institutionalization of the Environmental Movement has really accomplished anything, and answers that, like democracy, it is governed by numerous conflicting opinions and compromises. Democracy survives this process, he opines, but what of the Environmental Movement? "We can only hope that eventually the environmental movement will work. Unfortunately, unlike the workings of governments, there is a time limit to solving our domestic and global environmental problems."[58] I agree with historian

Kline that the achievements of the Environmental Movement are impressive but nonetheless compromised by our inability to slow down or escape the escalating speed of the economic treadmill.

The Death of Environmentalism?

Even worse than Kline's lukewarm optimism, has the explosive global economic development of the late twentieth and early twenty-first centuries sounded the death knell to what remained of the Environmental Movement in 2005? In an article titled "The Death of Environmentalism: Global warming politics in a post-environmental world" [59] the authors argue that contemporary environmentalism is incapable of dealing with the problem of global warming, the worlds most serious ecological crisis, and that this is so because "the environmental community's narrow definition of its self-interest leads to a kind of policy literalism that undermines its power." [60] The proposed antidote to the environmental gridlock that has resulted from rigidification of the Environmental Movement might be new institutions growing out of a new vision for environmental problem-solving in the face of the conservative shift in American values since the 1980s and 1990s.

Environmentalism today is seen by many citizens as just another special interest, so caught up in its pursuit of individual issues, from old growth-forest preservation to global warming legislation, that it has lost sight of a larger framework of values of enough worth to American citizens to gain their support. As environmentalists have carved out a multitude of institutions they have become so involved with financial support and political lobbying over single issues that they have lost touch with the majority of Americans. Meanwhile, over the past several decades, right-wing conservatives have carefully delineated the "family values" of the "contract with America" and built the powerful coalition which has dominated early twenty-first century American politics. Nevertheless, American values are still a mix of traditional religious and modern Enlightenment values, which together inform the opinions and behavior of American citizens. The authors of "The Death of Environmentalism" accuse liberals of having approached politics with the objective of winning one issue at a time while ignoring the shift in values which has driven the United States to the right. "If environmentalists hope to become more than a special interest we must start framing our proposals around core American values and start seeing our own values as central to what motivates and guides our politics. Doing so is crucial if we are to build the political momentum--a sustaining movement--to pass and implement the legislation that will achieve action on global warming and other issues." [61] Curiously, in the above quotation the authors seem to miss the point that the most environmentally destructive core American values, such as utilitarianism, resourcism, and fundamentalist or quasi-fundamentalist religious belief are alive and well, and even spreading as part of the neo-conservative political revolution. This should be cause for pessimism rather than optimism regarding the future of American environmentalism as it relates to "core American values." Environmentalists lack any explicit core set of values comparable to those of right-wing conservatives, but rather are comprised of a coalition including deep ecologists, ecofeminists, tree-hugging activists, agency bureaucrats, ecologists, and soft technology businessmen.

There is as much disagreement as agreement among "environmentalists" as to who they are and what their collective goals and policies ought to be. These fractures within the environmental community are suggested by the authors of "The Death of Environmentalism" themselves when they write: "Environmentalists need to tap into the creative worlds of myth-making, even religion, not to better sell narrow and technical policy proposals but rather to figure out who we are and who we need to be." [62] If "environmentalists" are in need of defining themselves as a collectivity, they are also in need of better understanding the inertial power of the core values of Western, Faustian civilization and its unique colonial variant, America. A major purpose of this book has been to elucidate these deeply-rooted values in order to assess the barriers of traditional belief and behavior which stand in the way of metanoia, a profound transformation of values such as we need to shift from societies of utilitarian, nature-destroying "takers" [63] to one seriously responsible for human activities within the larger framework of ecosystems and the biosphere. I am convinced that at least part of the key to solving the problem of American resistance to major environmental reforms leading towards a sustainable society is to be found in the religious component of our worldview, and the same is probably true of other traditional societies. This is not to say that secularization is the answer to our problems. The recent history of Western Europe and the Soviet Union suggests that modern, secular society is, if anything, even more destructive of nature than Christian America. What is needed in both cases is something tantamount to a new religion as part of a shift in worldview based upon our recently acquired evolutionary and ecological knowledge. A naive, fundamentalist element in Christianity strongly opposes such knowledge, and this resistance to intellectual progress is manifested in the present domination of American politics by right-wing ideologues typically adhering to fundamentalist Christian beliefs. [64]

More than a few people are alarmed by the drift to right-wing politics rooted in Christian fundamentalism. Bill Moyers, the distinguished Public Broadcasting System (PBS) journalist and commentator, himself a practicing Christian, reflects that " the delusional is no longer marginal. It has come in from the fringe, to sit in the seat of power in the Oval Office and in Congress. For the first time in our history, ideology and theology hold a monopoly of power in Washington." [65] According to a recent Gallup poll, one-third of the American electorate believe that the Bible is literally true. Many of these citizens believe in the "rapture index," the literal belief of an imminent millennium, the Antichrist, Armageddon, and the lifting of true believers directly to heaven to sit on the right hand of God.

What is the significance of these beliefs for public policy and the environment when nearly half of the U.S. congress is under the influence of Christian right advocacy groups and a Time-CNN pool showed that 59 percent of Americans believe that the prophecies of the Book of Revelation are going to come true?[66] Obviously, despoiling the environment must take a back seat if such momentous events, mandated by an irate God, are expected just around the corner. How can fundamentalist believers in the rapture, claiming the Creation of a mere six millennia ago as their first premise of logical argumentation, take seriously the felling of an old-growth forest, the loss of an endangered species, or the future environmental hazards resulting from global warming? True belief in the Christian philosophy of history, in a predetermined world programmed by the Judeo-Christian God from start to finish, is absolutely obliterating to the Darwinian evolutionary and geological idea of time. Nothing could be more remote from the mind of a prosyletizing religious fundamentalist. To the extent that Christianity in its literal mode has captured the American mind, it has had and will continue to have an undermining effect on the ideals, goals, and policies of environmentalism.

The second Bush administration flagrantly carried the banner of fundamentalist Christianity into the Oval Office, and has subsequently worked to undermine, weaken, and subvert environmental laws, from the Clear Air and Clean Water acts to NEPA and the ESA. Through the EPA, numerous pollution laws are under assault, and the Arctic National Wildlife Refuge has been in grave danger of being opened to drilling. In addition, the Bush administration has denied the reality of global warming and its long-term consequences to the point of American embarrassment on the international political stage. Why is it that no majority public outcry is heard in the United States? Bill Moyers is perplexed at our lack of outrage at the environmental injustices of our time. Is it greed or apathy he asks, but for him there is no obvious answer. What he fails to see clearly is that, in addition to rampant corporate political power, grotesque consumerism, and population growth, the outmoded worldview that sees nature as little more than the home and resources that the Christian God provided to his chosen people is even more profoundly destructive to our ability to cope with environmental problems than most of us can imagine.

The Western worldview locates God's divine creation, Lord Man, at the center of the physical universe and above the natural world. Within this worldview the aspect of religion has long dominated all of the others, including nature, history, man, and society, as discussed in Chapters Three and Four. Nature and man are the Christian God's creation, history is the interlude of God's testing of mankind between the creation and the millennium, and society is the hierarchical structure of clergy herding humanity to heaven for some and hell for others. Modern Western man has perpetuated the myth that this Christian worldview and modern, secular scientific and rational thought are compatible. However, the modern model of nature, as ecosystems produced by evolutionary processes, is totally incompatible with the Christian conception of nature, which itself is *not a viable model of reality or basis for an acceptable explanation of human nature.* The contradictions of the Christian and other mythopoeic, transcendentalist worldviews alloyed with capitalist ideology are outwardly manifested by unrestrained and uncontrollable ecological destruction, and inwardly displayed by the fractured modern Christian humanity so well described by Dostoevsky and Freud. Furthermore, whereas Christianity has provided the *ends* of our behavior towards nature, modernity has increased our *means* for manipulating nature a hundred fold.

Solutions to the devastating ecological problems resulting from this lethal Christian-modern capitalist amalgamation have included attempts to create a "green" Christianity as well as to proselytize the superiority of secular, religion-free, rational and science-based societies.[67] However, in the long run a green Christianity only perpetuates the fallacious and destructive Judeo-Christian religious myths about man and nature which have contributed to the ecological crisis. On the other hand, is it realistic to expect to displace traditional religious myths with a secular worldview in the twenty-first century? A European would answer that it is almost a *fait accompli* already, but Marxist societies in Russia and China suggest a deep-seated need in many humans to maintain some kind of traditional or other religious belief system. For many environmentalists nature provides the basis for either secular or mystical interpretations of reality. For the vast majority of humans, however, it does not appear likely that a new nature-based worldview is attainable in the modern world.

A sophisticated recent explanation of the demise of environmentalism is the theory of "post-ecologist politics" advanced by Ingolfur Blühdorn. This approach focuses on processes of cultural change in recent decades that have given rise to "value orientations and social concerns which devalidate or overlay those advanced by ecologists."[68] Blühdorn argues that American and European liberal democracies have failed to absorb the environmental agenda, finding it irreconcilable with the capitalist consumer economy. Simultaneously, scientific research and technological advances have helped to "disarm the ecological time bomb" for awhile, and although health and safety issues remain important, "a certain deradicalization of the ecological debate and eco-politics is undeniable. Environmental issues have indeed lost much of their ideological explosiveness."[69]

The decline of "ecologist" patterns of thinking and the transitions to "post-ecologist" politics is based upon the premise that modern liberal democratic societies have convinced themselves that environmental issues have been dealt with and pacified. This self-delusional stance requires strategies of "simulation," which describe a "set of social practices which function to preserve and regenerate a societal self-description which is rapidly losing its sociological foundation."[70] Thus, by means of simulative politics the illusion of completing the unfinished project of modernity is sustained. "As the logic of the market conquers all formerly autonomous social systems and the individual itself, the modernist project of the all-integrating systemic coherence is coming closer to completion."[71] Individuals functioning within late-modern, twenty-first century society continue to conceive of themselves as distinct from the all-embracing system of the market, and as possessing the normative standards which direct the course of society. Unknown to the individual is the fact that "the conceptual instruments which had once served the emancipation and autonomy of the exclusively defined human self are being turned into instruments for the reproduction and stabilization of the economic system."[72] Thus, the Kantian Enlightenment ideal of the "autonomous identical subject," capable of benefiting the human species as a whole, has been subliminally deconstructed and sacrificed to economic efficiency through the politics of simulation.

> In the formation of the new patterns of identity construction, the dynamics of the capitalist system has an important role to play. Economic thinking and the logic of the market have invaded all societal sub-systems, and marginalized all alternative ways of thinking. The permeation of all social relations and activities by the code of payment and profitability implies that identity formation becomes a primarily economic matter, and takes first and foremost the form of material accumulation and consumption.[73]

As identity construction becomes increasingly one-dimensional within the economically dominated Western worldview (having spread throughout the world), competition for scarce resources and development opportunities increases at the expense of what remains of the natural world. A post-ecologist, thoroughly capitalist politics of nature has arisen as part of the new political economy of uncertainty in which the normative certainties of traditional modernity are replaced by new constructed certitudes, such as the spread of liberal democracy and industrialization. Furthermore, the new patterns of identity formation "have reshaped the entire political agenda placing unprecedented emphasis on issues such as tax cuts, public expenditure, welfare parasites, border control and the protection against crime and terrorism. Contemporary political debates are fuelled by neo-materialism, social envy and the fear for security."[74] Against such a setting, ecological concerns fall into the background of political debate.

Blühdorn's sociopolitical analysis explains and essentially substantiates the generalization made by Oswald Spengler early in the twentieth century (see the end of Chapter I) that Western, Faustian civilization, like other great civilizations, has reached the final, winter phase of its development in which *money* itself takes on a life of its own, as megalopolitan humanity worships its power almost exclusively and devours the countryside (what is left of nature) in a frenzy of unrestrained inorganic growth. Although the universality of Spengler's cyclical explanation of the history of complex civilizations remains questionable, his identification of ecologically destructive shifts in attitudes toward nature and resources in the later stages of historical development is substantiated by recent developments in environmental history in works such as J. Donald Hughes's *Environmental History of the World*, Clive Ponting's *A Green History of the World*, and Jared Diamond's *Collapse*.

NOTES

[1]. Samuel P. Hays, *A History of Environmental Politics Since 1945* (Pittsburgh: University of Pittsburgh Press, 2000), p.94.

[2]. Peter Hay, *Main Currents in Western Environmental Thought* (Bloomington: Indiana University Press, 2002), p.3.

[3]. See Garrett Hardin, "The tragedy of the commons." *Science* 162:1243-1248, 1968.

[4]. John Opie, *Nature's Nation: an Environmental History of the United States* (Fort Worth: Harcourt Brace College Publishers, 1998), p.413.

[5]. Ibid., pp.414-415.

[6]. Kirkpatrick Sale, *The Green Revolution: The American Environmental Movement 1962-1992* (New York: Hill and Wang, 1993), p.27.

[7]. Benjamin Kline, *First Along the River: A Brief History of the U.S. Environmental Movement* (San Francisco: Acada Books, 2000), p.80.

[8]. Peter Hay, *Main Currents in Western Environmental Thought*, pp.174-175.

[9]. John Patrick Diggins, *The Rise and Fall of the American Left* (New York: W.W. Norton and Company, 1992), p.245.

[10]. Ibid., p.246.

[11]. W. Warren Wagar, *The Three Futures: Paradigms of Things to Come* (New York: Praeger, 1991), p.42.

[12]. Theodore Roszak, *The Making of a Counterculture: Reflections on the Technocratic Society and its Youthful Opposition* (Garden City, New York: Anchor Books, Doubleday and Company, 1969), p.5.

[13]. Ibid., p.9.

[14]. Ibid., pp.225-226.

[15]. Leo Marx, *The Pilot and the Passenger: Essays on Literature, Technology, and Culture in the United States* (Oxford: Oxford University Press, 1988), p.172.

[16]. Ibid., p.174.

[17]. Theodore Roszak, *Where the Wasteland Ends: Politics and Transcendence in Postindustrial Society* (New York: Doubleday and Company, 1972), p.13.

[18]. Ibid., p.65.

[19]. Jacques Ellul, *The Technological Society* (New York: Vintage Books, 1964).

[20]. Theodore Roszak, *The Voice of the Earth* (New York: Simon and Schuster, 1992), p.232.

[21]. See Michael E. Zimmerman, *Contesting Earth's Future: Radical Ecology and Postmodernity* (Berkeley: University of California Press, 1994), pp.79-85.

[22]. Frank Knight, "Environmental Celebrations: The Ten Most Significant Environmental Events of the 20th Century," *Nature Study*, Vol. 44, Nos. 2 and 3, Feb., 1991, p.1.

[23]. Kirkpatrick Sale, *The Green Revolution*, pp.11-23.

[24]. Ibid., pp.24-25.

[25]. Benjamin Kline, *First along the River*, p.94 and pp.92-94.

[26]. Larry Canter, *Environmental Impact Assessment* (New York: McGraw-Hill, 1977), p.6.

[27]. Ibid., p.4.

[28]. Ibid., p.8.

[29]. Hal K. Rothman, *Saving the Planet: The American Response to the Environment in the Twentieth Century* (Chicago: Ivan R. Dee, 2000), p.140.

[30]. Ibid., p.143.

[31]. Ibid., p.146.

[32]. Charles E. Little, *Hope for the Land* (New Brunswick: Rutgers University Press, 1992), p.vii.

[33]. Ibid., p.viii.

[34]. See Gilbert F. LaFreniere, "Land-Use Planning and the Land Ethic," *The Trumpeter*, Vol. 10, No. 2, 1993, pp.59-62.

[35]. Gilbert F. LaFreniere, "Greenline Parks in France: Les Parcs Naturels Régionaux." *Agriculture and Human Values*, Vol. 14, No. 4, 1997, pp.337-352, p.337.

[36]. See Gilbert F. LaFreniere, "Greenline Parks in France" for a detailed description of the French greenline park system (Les Parcs Naturals Regionaux) as well as an in-depth comparison of these parks with British National Parks, which are actually pastoral, greenline parks totally different from American National Parks.

[37]. Ibid., pp.337-341.

[38]. The Wilderness Society (Ben Beach, Bart Koehler, Leslie Jones, and Jay Watson, Eds.), *The Wilderness Act Handbook* (40th Anniversary Edition) (Washington, D.C.: The Wilderness Society, 2004), p.3.

[39]. Char Miller, Editor, *The Atlas of U.S. and Canadian Environmental History* (New York: Routledge, 2003), pp.176-177.

[40]. Richard T. Wright, *Environmental Science: Toward a Sustainable Future* (Upper Saddle River, New Jersey: Pearson-Prentice Hall, 2005), p.271.

[41]. Ibid., pp.269-273.

[42]. Karsten Heuer, *Walking the Big Wild: from Yellowstone to the Yukon on the Grizzly Bear's Trail* (Seattle: The Mountaineer's Books, 2004), p.xii.

[43]. Ibid., p.228.

[44]. Ibid., pp.234-235.

[45]. Wayland Drew, "Killing Wilderness," *The Trumpeter* Vol. 3, no. 1, Winter, 1986, p.20. A deep ecology journal, *The Trumpeter* can now be found online.

[46]. See J. Baird Callicott, *In Defense of the Land Ethic* (Albany: State University of New York Press, 1989).

[47]. Robyn Eckersley, *Environmentalism and Political Theory: Toward an Ecocentric Approach* (Albany: State University of New York Press, 1992), pp.36-37.

[48]. Ibid., p.37.

[49]. Ibid., p.39.

[50]. Ibid., p.41.

[51]. Ibid., pp.42-47.

[52]. Benjamin Kline, *First along the River*, p.101.

[53]. Kirkpatrick Sale, *The Green Revolution*, p.52.

[54]. Ibid., p.69.

[55]. Ibid., p.75.

[56]. Benjamin Kline, *First Along the River*, pp.122.

[57]. Ibid., pp.133-152.

[58]. Ibid., p.152.

[59]. Michael Shellenberger and Ted Nordhaus, "The Death of Environmentalism: Global warming politics in a post-environmental world." *Grist Magazine*, 3/22/2005, pp.1-25. http://www.grist.org/cgi-bin.

[60]. Ibid., p.3.

[61]. Ibid., p.22.

[62]. Ibid., p.23.

[63]. See Daniel Quinn, *Ishmael: An Adventure of the Mind and Spirit* (New York: Bantam/Turner, 1992). Quinn tells the story of the rise of the "takers," modern agrarian-based civilizations, in contrast to the "leavers" or hunting and gathering societies that inflicted far less damage upon the natural world.

[64]. See Susan Jacoby, *Freethinkers: A History of American Secularism* (New York: Henry Holt and company, 2004) for a detailed analysis of America's resistance to Darwinian evolutionary

biology, Chapter Five, pp.124-148. Southern resistance to Darwin's ideas in the nineteenth century laid the foundations for twentieth and twenty-first century right-wing religious extremism which contributes to strong anti-regulatory and anti-environmental attitudes particularly in the southern and southwestern states today.

[65]. Bill Moyers, "There is no Tomorrow," *Free Inquiry*, Vol. 25, No. 4, June-July 2005, pp.21-23, p.21.

[66]. Ibid., p.22.

[67]. For opposing views on the value of traditional religious belief for modern society, see Max Oelschlaeger, *Caring for Creation: An Ecumenical Approach to the Environmental Crisis* (New Haven: Yale University Press, 1994), and Sam Harris, *The End of Faith: Religion, Terror, and the Future of Reason* (New York: W.W. Norton and Company, 2004). Oelschlaeger argues that the Environmental Movement can be empowered through discursive interaction with traditional religion. Harris sees no possibility of solving modern problems, political, international, or environmental, unless contemporary societies abandon their traditional religious beliefs.

[68]. Marcel Wissenburg and Yoram Levy, eds., *Liberal Democracy and Environmentalism: The End of Environmentalism?* (London and New York: Routledge, 2004), p.36.

[69]. Ibid., p.35.

[70]. Ibid., p.44.

[71]. Ibid.

[72]. Ibid., p.45.

[73]. Ibid., p.39.

[74]. Ibid., p.40.

CHAPTER TWELVE: THE PROBLEM OF ECONOMIC GROWTH AND THE WESTERN WORLDVIEW

Introduction

Layers of unconscious, subliminal destructive attitudes toward nature permeate the Western mind, and especially the American mind. The first two chapters of this book demonstrate how *all* civilizations have severely degraded the ecological context of civilizational development, typically leading to the extinction, severe impairment, or displacement of agrarian-based societies. The fundamental layer or structure responsible for human abuse of nature which is present in all *Homo sapiens* is not much different from that possessed by other creatures seeking sustenance from Earth. We might call this fundamental stratum *natural utilitarianism*, the biological need to utilize the ecosystems of the planet for food, water, and shelter. Pre-civilized cultures found that when nature was over-utilized by human populations, starvation, disease, and death followed, and over long periods of time learned to live in relative balance and harmony with their surroundings. Their natural utilitarianism was, in many if not most cases, gradually constrained by the establishment of rules, values, and myths which maintained hunter-gatherers in approximate equilibrium with somewhat human-modified ecosystems over the millennia. (In the long term, however, Pleistocene shifts from glacial to interglacial periods during the past two million years have subjected hominid bands to natural selection leading to the evolution of human intelligence and the rise of Homo sapiens to dominance over competing hominids.)[1] Therefore, it is hardly surprising that environmental historians have expressed surprise, rather recently, at finding that Native Americans, Australian aborigines, and other pre-civilized societies have caused animal extinctions and other ecological damage. On the positive side of the ledger, environmental philosophers such as J. Baird Callicott have recognized the moderating worldviews of hunting and gathering societies to include something similar to what we call an environmental ethic in modern developed societies. This ethical component of pre-civilized worldviews helped to prevent the abuse of ecosystems resulting from natural utilitarian ways of earning a living from nature. However, gradual changes such as long term population increases, and climate changes like the drying and warming related to the shift from the Pleistocene Ice Age to the present (Holocene) interglacial climate, eventually pushed hunting and gathering cultures beyond their carrying capacity, forcing them to invent alternative modes of survival such as herding and agriculture.

Once pastoralism and agriculture (with its fixed communities) were established, a rationalization of their more intensive utilization of nature had to be constructed, giving rise to the several distinctive worldviews of early agrarian civilizations, and the enormous difference between the Mesopotamian and Egyptian worldviews described in Chapter Two. The Judeo-Christian culture manifested another distinctive worldview rationalizing its increased utilization of nature in the arid lands of the eastern Mediterranean. All of these early civilizations, then, had constructed values and myths peculiar to their individual ecological circumstances, thereby creating a secondary explanation of the utilization of nature in agriculturally conditioned civilizations. This second level of understanding human interaction with nature we might term *rational utilitarianism*, which is the civilized equivalent of the values and myths created by pre-civilized hunter-gatherers in order to maintain a degree of ecological equilibrium. The ecological consequences of rational utilitarianism, on the other hand, have been degradation of ecosystems by imperialistic expansion of civilizations. In the long run, the basis for long-term historical change towards human disequilibrium with nature has resulted from the accumulation of technics and knowledge allowing long-term improvements in agriculture and other aspects of civilization. Thus, rational utilitarianism magnifies the impacts upon nature resulting from simple natural utilitarianism. Consequently, partly by trial and error, civilizations have developed, overshot carrying capacity, and disintegrated or been subsumed into other civilizations, or expanded at the expense of primitive cultures and wild nature.

Clarence Glacken, in *Traces on the Rhodian Shore*, throws light on the subject of how pre-civilized natural utilitarianism was rationalized by early civilization, first by myth in Mesopotamian (Sumerian) civilization, and later by rational argument in ancient Greco-Roman civilization. Glacken's book is an exegesis of the relationship between nature and culture expressed in Western thought from ancient times to the eighteenth century. Focusing upon three fundamental ideas, of a designed earth, of the influence of the environment on man, and of man as a modifier of the environment, Glacken traces these three themes through the literature of ancient, medieval, and modern civilization. It is apparent from his survey of ancient and Western thought that rationalization of and belief in the rightness of human utilization of nature is a venerable and morally respectable tradition, closely allied with religion and philosophy from its very beginnings in ancient Mesopotamia.

In Sumerian theology it was assumed that order was designed into the cosmos at its creation by superhuman and immortal beings, humanlike in form, who governed the cosmic order which they initially created. This theology was apparently based upon an analogy of human society and the artisanship which had created cities, temples, palaces, and cultivated lands. "Therefore the cosmos must also be controlled by living beings, but they are stronger and more effective because their tasks are far more complex."[2] Glacken thinks that later, Western ideas of a divinely designed earth, including the idea that divine power is inseperable from the order of nature, may well have been influenced by Sumerian theology. The Sumerians, however, had no sense of history, assuming that their civilization, its farms, cities, and institutions, had been essentially the same from the time of their creation. The shift to rational thought and an awakening sense of history occurred in Greek civilization, shifting the understanding of human utilization of nature from the mythological towards the historical speculations of Greek historians and philosophers, from Herodotus to Plato and Aristotle.

Linked to his three seminal ideas of a designed earth, the influence of the environment on man, and of man as modifier of the environment, Glacken examines four secondary ideas throughout Western history, including the principle of plentitude, interpretations of cultural history, the effects of human institutions such as religion and government upon man-nature relationships, and the organic analogy applied to the rise and fall of nations and peoples and to the earth. Within this framework of ideas, the first step in making the environment useful to humanity was the acceptance of the primordial *ordering* of nature by the gods.

> In ancient and modern times alike, theology and geography have often been closely related studies because they meet at crucial points of human curiosity. If we seek after the nature of God, we must consider the nature of man and the earth, and if we look at the earth, questions of divine purpose in its creation and of the role of mankind on it inevitably arise. The conception of a designed world, in both classical and Christian thought, has transcended personal piety. In Western thought the idea of a deity and the idea of nature often have had a parallel history; in Stoic pantheism they were one, and in Christian theology they have supplemented and reinforced one another. Whether a God or the gods had a share in the life of men spent in their beautiful, earthly home, or whether, as the Epicureans believed, the order of nature was not to be ascribed to divine causes, the interpretations which arose out of these arguments have had a dominant place in molding the conception of the earth as a suitable environment for the support of life.[3]

This passage succinctly encapsulates Glacken's major thesis and also emphasizes the ubiquitous utilitarian assumptions regarding the relationship between humankind and nature in Western civilization and its antecedents. Glacken's history of the idea of a designed earth provides a necessary perspective for the modern ecologist, evolutionary biologist, and environmental historian when assessing the enormous inertia and resistance to acceptance of the Darwinian evolutionary perspective. This is so partly because the idea of a designed earth is itself subsumed within the broader idea of the natural world. The idea of a designed earth reconciled the plentitude, the sheer variety of the natural world with the achievement of an artisan-creator (based upon human utilization of the earth), thereby providing the pre-Darwinian world with a holistic conception of nature that was compatible with prevailing religious doctrine (as well as with heretical religious and philosophical speculations). The validity of the artisan analogy was challenged in the ancient world, and more seriously in the seventeenth and eighteenth centuries by Spinoza, Buffon, Hume, and Kant, long before the challenge of Darwinian evolutionary biology provided the *coup-de-grace* to the argument-from-design. This venerable tradition of thought still finds a large following in contemporary versions of Christian and other traditional religious belief systems.[4]

Western European civilization is perhaps foremost in the degree to which it has added to, and expanded upon, its natural and rational utilitarianism. It has developed a set of ideas, values, and beliefs which have carried Western European civilization (and its global imitations) to a new and profoundly destructive level of rationalizing the utilization of nature. This new synthesis, which we call *modernity*, is overlaid upon older Christian ideas of dominion over and stewardship of nature. This novel modern cryptoreligious element in our worldview is properly an *ideology* or secular belief system possessing the power to move people to acts of sacrifice and aggressiveness towards nature and other civilized and pre-civilized cultures. Thus, the *problem of development* (economic growth) ultimately exists within the unique worldview impressed upon the mind of modern man, setting in motion the whirlwind or "tourbillon" which has turned world history into an explosion of change increasing at exponential rates, leaving in its wake the wreckage of the biosphere and traditional cultures.

Nothing comparable to modern economic development has ever occurred in the past. It is true, as demonstrated in Chapter Two, that a number of civilizations have destroyed their lands and ecology over long periods of time. But never before has humankind witnessed a virtual blitzkrieg of economic growth and land development comparable to what we are seeing in the Western world and its former colonies today. Europe itself, having devoured nature there, has been on a rampage of development for *half a millennium*! It is no wonder that so little of wild nature or wilderness remains in the world, generally located in regions of extreme climate, be it arctic or equatorial. It would not be unreasonable to suggest that all this rapid change is just the usual matter of humans making a living from the earth, but that there are simply more of them. Unfortunately, this is not the case. Contemporary humans are impelled to utilize and destroy nature in a peculiarly exuberant and self-satisfied way. What is it that makes them unique in history? Our study of Western civilization in relation to nature has provided most of the evidence needed to move towards a conclusion. Development (economic growth) at increasingly rapid rates is a distinctly modern phenomenon, but, more important, *the problem of modern development is an extrapolation of the worldview of modern humanity*, a problem which became global as western ideas and techniques spread with colonization.

Traditional   societies   typically   develop   uniform,   well-integrated worldviews which are held by the vast majority of individuals belonging to those societies. Depending upon the relative degree of isolation of a culture, its worldview may be relatively simple and uniform, as in the case of Mesopotamian, Egyptian, and Aztec-Mayan civilizations. The Greco-Roman civilization and its worldview became more complex by the blending of Greek and Roman religion and philosophy, later assimilating additional elements, include Egyptian mystery religions, into a complex or syncretistic worldview, so that individuals could pick and choose religious or philosophical beliefs according to taste. The modern worldview is highly syncretistic because it has swept up elements of other cultural belief systems from around the world in the course of its expansionist period of imperialistic adventures. Despite this agglomeration of ideas and religions, history and sociology allow us to reconstruct the core values of Western civilization prior to dilution by exotic elements. Throughout the Western Middle Ages, for example, our core Christian religious belief system remained largely unchanged for a thousand years (ca.500-1500). Then, during the Late Middle Ages, Renaissance, and Reformation a ferment of change transmuted the core medieval worldview into something we call modernity. Prior to modernity Western Christians had essentially used up the natural resources of western Europe under a worldview which explained humanity as God's favorite creature inhabiting Earth, designed for humanity, at the center of the cosmos. To be fruitful, multiply, and dominate Earth's other creatures expressed God's will and command. During the Middle Ages the natural human utilitarian impulse to make a living from the Earth was reinforced by the Judeo-Christian ideal of dominion over nature. Thus, there were two levels of motivation compelling Europeans to turn its wilderness into a garden. By 1300, this task was essentially completed, and thanks to Europe's temperate climate, the result was a garden rather than a wasteland such as much of the Mediterranean basin had become during classical antiquity.

The modern period, broadly speaking, 1500-2000, has added additional components to the Western worldview, leading towards peculiarly Western destructiveness towards nature, overlaid upon the natural utilitarian and Christian dominion components. First, the rise of modern science and associated technological achievements integrated the model of a perfect, God-designed mechanical cosmos, the key to more masterful manipulation of nature. The Scientific Revolution of the 16th and 17th centuries and gradual technological improvements also facilitated the commercial capitalism of the early modern centuries and gave rise to a growing sense of optimism among the literate townsmen of growing cities. Separately, in England and France, ideas of human progress were expressed, and in the English tradition blended with the belief in divine providence in the thought of John Locke and, perhaps, Adam Smith.[5] In France a more secular progressive vision, described in Chapter VI, grew out of the seventeenth century "quarrel of ancients and moderns." Protestant sects leaving for America in the seventeenth century were not initially believers in progress, but by the late eighteenth and nineteenth centuries they would adopt the English providential-progressive attitude which I call *millenarian progressivism*, as compared to the secular, science-based idea of progress in France which I term *scientific progressivism.* Thus, over time, both mechanistic science and its manageable clock-like universe, and a millenarian belief in progress became part of the uniquely American variant of the European worldview.

Another crucial development in the history of Western European civilization tended to intensify American overutilization of nature. In the course of increasing commerce and expanding towns and cities in Europe during the period of colonization, a bourgeois culture arose which became increasingly skeptical and open-minded during the 17th and 18th centuries. French skeptics were followed by the *philosophes*, popularizers of science, liberal political ideas, and increasing criticism of aristocratic and ecclesiastical beliefs and social position. This Enlightenment bourgeois "café culture" was to some degree slowed down in the American colonies by the task of subduing nature and Native American Indians, as well as by the social influence of revived Protestant religious communities. During the second half of the eighteenth century secularism and liberal thought began to flourish in America as well as in Europe, creating the transatlantic civilization of the late Enlightenment which produced such American *philosophes* as Thomas Jefferson and Benjamin Franklin. The Enlightenment and its secular thought never fully took hold in the United States, however, apparently due to the freshness and social vigor of Protestant churches in a frontier environment or not far removed from such surroundings. Whatever complex combination of causes was responsible for the French Revolution, in Europe that conflagration initiated a sputtering, intermittent progression towards increased secularism whereas the American Revolution did so on a much smaller scale. In America, the growth of industrial cities and a high culture in cities like New York and Boston accomplished a growing secularization while outside on the farm and in the towns and villages Protestantism reigned supreme. Rural religiosity in turn reinforced the utilitarian-Christian worldview. During the nineteenth century this built-in aggressiveness toward nature would be reinforced by an American variant of the idea of progress intertwined with the liberal economic thought of Adam Smith and David Ricardo, further motivating the rapid utilization of *nature-as-natural-resources.*

With this brief historical background in mind, we may now turn to a more detailed analysis of the components of the modern Western worldview which motivated our virtually demonic obsession with multiplying material wealth (at the expense of nature and traditional societies) through the processes of economic growth or development. The ideology of classical liberal economic thought and its neo-classical variant are a product of modern Western European civilization and although only two centuries old, already so deeply ingrained in the Western mind that it would seem almost unthinkable for most Americans, Europeans, other Neo-Europeans, and the rest of the "developed" world to imagine an alternative way of interpreting how we ought to live in the world. This certainty that the present form of economic development is a natural and progressive turn of events gave rise to the hubris of Francis Fukuyama's *The End of History and the Last Man* described at the end of Chapter Seven. It has also given rise to a sophisticated critique of modern economic thought from the perspective of the Environmental Movement, beginning with E.F. Schumacher's *Small is Beautiful: A Study of Economics as if People Mattered* (1974) and H.E. Daly's *Towards a Steady State Economy* (1973). In 1977 William Ophuls linked modern economics to pre-environmentalist politics in *Ecology and the Politics of Scarcity* (revised, 1992), and in 1997 mounted a full-scale critical assault upon the entire modernist system in *Requiem for Modern Politics: The Tragedy of the Enlightenment and the Challenge of the New Millennium*. The essence of his general theory of civilization was summarized briefly in Chapter II. To my mind, Ophuls's *Requiem* is the most cogent detailed explanation of the modernist foundations of economic development available to environmentalist critics of the contemporary growth paradigm, and I will therefore begin my assessment of the problem of development (as a faulty and inevitably self-destructive model for modern society in our time) with a summation of Ophuls's thesis, a particularly lucid framework for further discussion.

Recalling Ophuls's "four great ills of civilization" (over-utilization of nature; violence against civilized and uncivilized outsiders through conquest and imperial expansion; political and religious despotism over insiders; and sociopolitical inequality, including slavery), Ophuls offers a powerful argument that the modern Western European attempt to overcome the ills of despotism and socioeconomic in inegalitarianism during the Enlightenment came at the expense of enormously increasing our over-utilization of nature and increasing Western violence against outsiders through imperialistic colonialism (which in turn increased European and neo-European over-utilization of nature). *Requiem* elaborates this process of diminishing two of civilization's "great ills" at the expense of greatly increasing two others. If the development of commerce and trade linked to the Age of Discovery, the Scientific Revolutions, technological innovation, and the rise of European nation-states laid the foundations of Western-style development, it was the political and economic theories of Thomas Hobbes, John Locke, and Adam Smith which morally validated the Western prototype of endless economic growth as the way the world should work. Ophuls demonstrates how economic, political, and ethical theory is inextricably linked in the modernist worldview which spread with the aid of English and Scottish political and economic theory to become the dogma of classical economics.

Long-term environmentalist criticism and reform of contemporary corporate capitalism and the state capitalism of socialist and communist nations is not likely to be very successful lacking an understanding of the broad historical perspective which Ophuls's prescient analysis provides. Ophuls inquires why Western economic development has been and remains so successful despite the two enormous side-effects of its implementation, first, the massive destruction of global ecosystems at local, regional, and global levels, and secondly, the rapid displacement of traditional cultures and the ecological and social knowledge and wisdom which they have acquired over millennia of pre-historical and historical development. The economic development that we have experienced in the modern world is an abnormal historical era which has occurred under extraordinary and *unrepeatable conditions*. These include, first of all, the discovery of the "Great Frontier" of the New World with its cornucopia of resources "available for the price of killing, enslaving, or colonizing the native inhabitants." Secondly, spurred on by the exigency of having arrived, for a second time, at carrying capacity during the era of the Great Discoveries, Europeans "invented technologies that gave them both the power to kill, enslave, and colonize on a global scale *and* the ability to exploit the resources more 'intensively' than the allegedly benighted natives could do." Thirdly, because these resources were available in a pristine and concentrated form, the ecological costs of the early stages of development were modest, as a consequence of which development appeared unquestionably beneficial to the pre-ecological observer.[6] Development during the late nineteenth and twentieth centuries, however, has spectacularly revealed the latent ecological and social costs of rapidly expanding industrial activity, and also the unpleasant fact that although these costs are considered by economists to be "external" to the market, and generally not reflected in the price mechanism, they nevertheless have to be paid by someone. For example, cheap corn or soy are produced at the "external" costs of soil erosion, stream sedimentation and flooding, pesticide and fertilizer pollution, and the displacement of small family farms, leading Ophuls to conclude that " this is the story of the Industrial Revolution in a nutshell: technological cleverness has allowed us to lower market prices by driving up ecological costs. In the end, therefore, economic development is intrinsically predatory and destructive."[7] Details of Ophuls's thesis are summarized in the endnotes.[8]

Adam Smith's theory of market economics completes the Enlightenment ideological template for the modern "growth society." I have discussed Smith's contribution in the context of the Enlightenment in Chapter Seven (pp.19-21), but here I will continue to summarize Ophuls's argument, within which Smith plays the culminating role of completing the revolution in political economy foreshadowed by Hobbes and elaborated by Locke in the seventeenth century. As I explained in Chapter Seven, the ambivalence in Smith's economic thought stems from his earlier intellectual pursuits as a moral philosopher, quite literally a believer in the classical model of virtue, which enjoined a heroic command of the self, to which he later opposed the self-indulgent values of economic man. So Smith denigrated riches and strongly recommended moderation: let a man be content with little, with a cottage instead of a mansion."[9] Yet by 1776 Smith reaffirmed the doctrine of the "invisible hand" as a justification of the market economy in which the pursuit of selfish individual economic interests contributed to the common good of the general economic welfare. Following his theory, the modern state would cooperate with rather than dominate the economic process, and the market system and free trade would produce riches beyond the imagination of the pre-modern mind. Ophuls reminds us that Smith, due to his earlier career as a moral philosopher, was paradoxically anti-economic in his personal values, his own moral beliefs anticipating later critiques of bourgeois political economy. "Those who use him to justify unrestrained greed and absolute economic freedom both misunderstand and misrepresent him--to the point of slander."[10] Ophuls is acutely aware that if Smith legitimated selfishness in the economic sphere, he intended that it be kept firmly in check by government regulation through just laws and social constraint through proselytization and practice of moral principles and civic virtue. Smith himself was well aware of the degradation of individual laborers in the mines and factories of an incipient Industrial Revolution. Yet, somewhat naively perhaps, he expected the moral constraints imposed by civil society to prevent excessive abuses resulting from the self-seeking actions of entrepreneurs. Like Hobbes and Locke before him, Adam Smith could not imagine that "endless" economic development would materially, as well as socially, impoverish us in the long run. Nor could he imagine that his theory of laissez-faire economics represented by the "hidden hand" would be thoroughly, even brutally disconnected from the Enlightenment moral context that mollified the idea of economic development in his own era. Ophuls also demonstrates that the free trade proselytized in *The Wealth of Nations* was fundamentally imperialist in its objectives. "As the title indicates, Adam Smith's philosophy of economic development is first and foremost about national wealth; next, it is about the enrichment of the trading and manufacturing classes; lastly, if at all, it is about improving the welfare of the common people (although Smith clearly intended this end)."[11]

Ophuls's *Requiem* fully elaborates the argument summarized briefly in the last few pages, arriving at a conclusion that the Enlightenment precepts of the modern worldview have led to consequences far more disastrous than simply rapid environmental degradation, which Ophuls acknowledges as anomalously destructive during modern history, but not essentially different in its end result from the ecological destructiveness of earlier civilizations. The political-economic-social tragedy of modernity is a testament to the thoroughgoing wrong-headedness of the Hobbesian-Lockian-Smithian legacy:

> Everything that does not work, all that we hate and fear about the modern way of life, is the logical or even foreordained consequence of the basic principles that we have chosen to embrace. Explosive population growth, widespread habitat destruction, disastrous pollution, and every other aspect of ecological devastation; increasing crime and violence, runaway addictions of every kind, the neglect or abuse of children, and every other form of social breakdown; antinomianism, nihilism, millenarianism, and every other variety of ideological madness; hyperpluralism, factionalism, administrative despotism, and every other manifestation of democratic decay; weapons of mass destruction, terrorism, the structural poverty of underdevelopment, and many other global pathologies--all are deeply rooted in Hobbesian politics, whose basic principles set up a vicious circle

*of power seeking and self-destruction. In other words, the most
intractable problems of our age are due not to human nature itself
but, instead, to the way in which the Enlightenment in general, and
Hobbesian politics in particular have encouraged the worst
tendencies of human nature to flourish in the modern era.*[12]

I agree, to some extent, with Ophuls's fundamental conclusion that it is
not so much the domination of human nature by its baser instincts and proclivities
that is responsible for the modern crisis that he describes, but rather the shift to
the Enlightenment worldview which has been shorn of traditional and rational
moral constraints to form societies tied to and manipulated by *the idea of
development* in a mode blind to the ecological and social destruction which it
must inevitably create. The "growth society" is a uniquely destructive culture run
amok from a Western civilization at war with itself; in short, a schizoid culture.
Perhaps nowhere in the world is this culturally schizophrenic behavior more
spectacularly manifested than in the United States, today the epicenter of
exponentially increasing global economic development and its environmental and
traditional social casualties. On the other hand, I would qualify Ophuls's
judgment here with the observation that in virtually all times and places,
humanity's worst behavior has been called forth by corrupt institutions and ideas,
whether in the Colosseum of Rome or the inquisitions held by an all-powerful
medieval Christian church. The Enlightenment causes of global environmental
degradation are superimposed upon the misconceptions of the natural world and
human interaction with it that grew up in ancient and medieval civilization. The
historical roots of the modern ecological crisis penetrate far deeper than Ophuls
will allow in his diatribe against the unintended evils that grew out of the Anglo-
American Enlightenment. However, Ophuls is correct in diagnosing the most
pernicious manifestation of rational utilitarianism in the history of civilization.

Ophuls's critique of modern polity as the root cause of environmental
crisis is not intended at all as an attempt to compare the economic practices of
liberal democracies with a more virtuous Marxist society. To the contrary, Ophuls
finds communist and socialist regimes even more self-destructive than the West,
because Karl Marx "criticizes the bourgeoisie in terms that are essentially
bourgeois: he makes economics the explanatory force and thus takes Hobbesian
materialism to its ultimate limit--surpassing not only Locke but even Smith!"[13]
However, Marx rejects the liberal Lockean, economically dominated polity for
having substituted the tyranny of property for the autocracy of feudal monarchs.
In the process, capitalism destroyed traditional community while enriching and
empowering the limited class of entrepreneurs while the market turned the
common populace into commodities whose labor could be bought cheaply. Such
property-less wage earners could gain social salvation only through the
communistic solution of universal proprietorship. "Under the aegis of the famous
'dictatorship of the proletariat,' socially directed production will eventually

abolish natural scarcity; and the resulting superabundance will cause the state to 'wither away' into a mere utility like the post office."[14]

In establishing his communist philosophy of history as a process which must inevitably culminate in the dictatorship of the proletariat, Marx also re-introduced the very religious (or cryptoreligious) zealotry that Hobbes had attempted to expunge from politics, thereby unleashing a new era of quasi-religious warfare which is not yet over. Thus, communist ideology created new states which sought to create equality through the domination of nature and material abundance, which would serve as a substitute for traditional morality in maintaining social harmony. It is hardly surprising, then, that liberal Western political philosophers and environmentalists found the communist states to be even more brutally degraded ecologically than those of the West when comparisons were made in the 1960s and 1970s.

We have learned that all human cultures, and particularly civilizations dependent upon agriculture, have affected the ecosystems within which they grew up to varying degrees, with regions such as the Mediterranean basin being almost completely transformed. Some level of economic development has always caused ecological damage. However, what makes modern Western European civilization (and its imitators) unique is its obsessive preoccupation with economic growth and its proselytizing, imperialistic imposition of its unique "growth society" upon the entire world. Moreover, this process of worldwide economic development through Western means of "brute-force" or "bulldozer" technology is viewed by the majority of contemporary humans as a positive, scientific, progressive, and even heroic transformation of "backward" or "undeveloped" regions and Third World countries. Very few of us gain the opportunity to analyze the modern Western worldview which underlies the historically unprecedented process of *globalization*, the transformation of Earth according to Western beliefs and practices. We have witnessed throughout this book a mass of evidence indicating the unusually powerful attitudes of human dominion over nature which have contributed to the modern Western worldview, from the Christian myth of human dominion, to early modern Western mechanistic science, the rise of industrial technology, and the obsessively growth-oriented political economy postulated by Hobbes, Locke, Smith, and Marx which dominates Western political and economic practices and their global proliferation. Given the venerable genealogy and general veneration of Western "progressive" institutions, only a small minority of individuals within our culture have been able to peer back in time through the window of history to the origins of our revered modern culture and apprehend the ecological and social dangers inherent in our culturally dominant worldview. Attempts to educate American citizens in this regard have been a dismal failure in the Environmental Movement. Perhaps we will only gain limited understanding of the problematical Western worldview and its implementation through globalization from the negative feedback that we are beginning to experience from the globalization process itself. Therefore, I will continue our analysis of the problem of development from the perspective of contemporary globalization and its social and environmental discontents.

Sustainable Development and Globalization

The term "sustainable development" has become the environmental catchphrase of the late twentieth and early twenty-first centuries, meaning many different things to environmentalists, politicians, economists, and sociologists. In light of Garrett Hardin's analogy of "running for the mop" of environmental cleanup while the taps of population growth and rampant consumerism are running full blast, sustainable development seems worse than ambiguous and oxymoronic; it seems downright absurd. It also serves as a convenient symbolic smokescreen representing the possibility of a future ecotopian world while business-as-usual entrepreneurs and technologists increase the speed of the global treadmill of development. I am not at all alone in this harsh-sounding judgment. "Sustainable development" was linked to the agenda of trade liberalization when it was recommended in the 1987 Brundtland Report published by the World Commission on Environment and Development. Most people active in the Environmental Movement now believe that the goal of "sustainable development" has been thoroughly subverted, even serving as " a cynical exercise to accord a fake legitimacy to full-on, business-as-usual, global environmental rapine."[15] The lack of definitional precision had been deliberately constructed to leave the concept open to manipulation by global development.

"Sustainable development" has been ill-defined: "Despite the fact that the term sustainable development has become common currency among many groups, it is a confused and sometimes contradictory idea, and there is no widespread agreement as to how it should work in practice."[16] "Sustainable development" is defined in the Brundtland Report as "development which meets the needs of the present without compromising the ability of future generations to meet their own needs." One must focus upon the ambiguity of the words *need* and *development* to demonstrate that "sustainable development" could represent either the extreme of the dominant economic worldview or the extreme of a deep ecology worldview. Under the dominant economic worldview sustainability would involve: dominance over nature; perception of the natural environment as a resource for human use; material and economic growth for increasing human populations; belief in ample reserves of resources; complex technological progress and solutions; consumerism and growth in consumption; and centralized communities. Under the deep ecology worldview a set of entirely antithetical principles would be implemented: harmony with nature; appreciation of the intrinsic worth of nature; simple material needs; awareness of limited Earth resources; appropriate science and technology; recycling and anti-consumerism, regional and local traditional community, and the shrinkage of human population over the long term. Obviously, as applied today, with Garrett Hardin's taps wide open, the dominant economic worldview thoroughly dominates the application of the concept of "sustainable development." This disparity between economic as opposed to ecological principles of development is further confused by application of the term "sustainable development" to both economic and ecological sustainability.[17] This has caused environmentalists to qualify "sustainable development" as "ecological sustainable development" to deny the legitimacy of simple economic development according to the dominant economic worldview. For economists, the environment is incorporated into economic calculations and the context of the economic system or market system. What is at stake here is enormous; i.e., nothing less than "rival claims over what is ontologically primary: the natural environment or the human economy. Is the environment a sub-set of the economy, or is it a prior and sustaining context within which, and dependent upon which, all human systems, including economic systems, reside?"[18]

That such a fundamental disagreement regarding the nature of ecological and historical reality can even be argued today is a testament to the inertial power of the dominant economic worldview and the woeful public ignorance of our relationship to nature which underlies our educational system. Morever, our attitudes toward nature in the syncretistic mix of religious belief and popular worldviews which predominate in modern societies today is a barrier to ecological understanding. It is within such a pervasive fog of intellectual confusion that the Environmental Movement has struggled unsuccessfully to substitute ecological knowledge for a centuries-old accumulation of misunderstanding of nature and of our relationship to it. The popular intellectual brew of the Christian idea of dominion, the Promethean indefatigability of mechanistic science and technology as intrinsically progressive in nature, the obsessive preoccupation with the market system as the only reasonable basis for structuring society, and the infatuation with proselytizing for and building the global network of technology and trade altogether leave any steady-state or deep ecological foundations for true "ecological sustainable development" beyond our mental grasp. Lacking metanoia (a general shift in worldview towards an ecological understanding of the appropriate goals for future development) the term "sustainable development" presently does more harm than good by creating the illusion of progress towards ecological goals while business-as-usual and all the associated environmental ills continue. We have witnessed "the apparently successful capture of 'sustainable development' by a neo-liberal worldview disinclined to make the slightest concession to ecological imperatives...."[19]

The environmentalist attempt to establish a successful working paradigm of sustainable development began in the 1960s and 1970s when Kenneth Boulding, E.J. Mishan, E.F. Schumacher, and Herman Daly formulated critiques of the growth paradigm of traditional mainstream economics. As Herman Daly has noted, after the mid-nineteenth century, controversy over the desirability of unlimited economic growth ended, and the notion of virtually endless economic growth has been taken for granted ever since. Given this established premise, economists down to the present day have elaborated theories which rationalize that the only solution to problems caused by growth is more growth. Daly and other environmental economists have demonstrated the inherent illogic in this perspective from a theoretical point of view, but with little pragmatic effect. To my mind, modern economics of the past 150 years is tantamount to a form of "collective madness" that social theorist Eric Hoffer characterized in *The True Believer* (1951) when he described the mass movements of Christianity, fascism, and communism. Ironically, he left out our own Western capitalistic delusion while reasoning from the soapbox of the modern liberal political economy and its quasi-maniacal obsession with endless economic growth! These liberal economic beliefs and actions assume infinite resources and unlimited waste-assimilating capacities when, in actuality, natural systems are finite, closed loops. As Carolyn Merchant writes, "human beings since the transition to an inorganic economy, have been living off non-renewable geological capital. This means that humans are no longer in equilibrium with the rest of nature, but are depleting and polluting it, overloading the natural cycles."[20] Unfortunately, when all is said and done, the ecological conception of "sustainable development" formulated by Schumacher, Daly, and other environmentalist scholars has been absorbed by mainstream economics and subsequently stood on its head to serve the economic status quo.

The topic of development is also discussed under "globalization" in a spate of books and articles published over the past decade. Joseph E. Stiglitz, a former cabinet member in the Clinton administration, chairman of the Council of Economic Advisors, and senior vice president and chief economist of the World Bank, defines globalization as " the closer integration of the countries and peoples of the world which has been brought about by the enormous reduction of costs of transportation and communication, and the breaking down of artificial barriers to the flows of goods, services, capital, knowledge, and (to a lesser extent) people across borders."[21] Globalization has also reactivated older institutions like the United Nations, the International Labor Organization, and the World Health Organization, while empowering the International Monetary Fund (IMF), World Trade Organization (WTO), and the World Bank. Stiglitz focuses on problems and controversies associated with the IMF and World Bank in particular, including global financial crises and the transition of former communist states to market economies, with the objective of seeking the means of ensuring global economic stability. He finds the IMF dominated by financial interests at the expense of real assistance, such as food subsidies, to Third World (developing) countries. The WTO is likewise under the thrall of financial interests, to the detriment of the environment and the infrastructures of traditional societies. "Those who seek to prohibit the use of nets that harvest shrimp but also catch and endanger turtles are told by the WTO that such regulation would be an unwarranted intrusion of free trade. They discover that trade considerations trump all others, including the environment!"[22] Meanwhile, the bureaucratic leaders of the IMF and World Bank genuinely believe that their agenda is in the general interest of global communities, proselytizing that all will eventually benefit from trade and capital market liberalization. Consequently, they support forcing Third World countries to accept the reforms associated with global trade, which include the promotion of democracy.

Globalization, with its increasingly interdependent national economies, financial markets, trade, corporation, and consumer marketing, are driven chiefly by technology, with its accelerated innovation in electronic communications, and their convergence with television, electronic trading in stocks, bonds, and futures options, and the global explosion of the Internet and e-commerce, and a twenty-year wave of deregulation, privatization, liberalization of capital flows, extension of global trade, and export-led growth policies since the early 1970s collapse of the Bretton Woods fixed currency-exchange regime. Add to these factors the demise of the state-commanded Soviet economy and the gradual deregulation of global markets known as "The Washington Consensus," i.e., "the dominant Western economic paradigm promoted by the United States, the World Bank, the International Monetary Fund, and their dominant schools of academic economists on both sides of the Atlantic."[23] The "global information age" has contributed to enabling a new world created since the 1960s and 1970s out of three independent processes, including the information technology revolution, the economic crises of both capitalism and statism and their subsequent restructuring, and evolving social movements including libertarianism, feminism, human rights, and environmentalism. Multiculturalism and postmodernism might be added to the list. The recent fermentation of all these ideas and processes has propelled the complexity of contemporary civilization a quantum leap beyond our previous understanding of modernity, and the modern "tourbillon" has exploded into "a new dominant social structure, the network society" which we recognize as the "worldwide web" described by McNeill and McNeill in Chapter Ten.

Whereas Stiglitz analyzed the problems and dysfunctions of the extended global market system itself, Hazel Henderson emphasizes the social and environmental effects of this brave new world of global economics, stating that "ever more problems and issues have become global--beyond the reach of national governments: from climate change, cross-border pollution, desertification, and loss of biodiversity to space junk. Proliferating weapons--trafficking, drugs trading, organized crime, nuclear and toxic wastes and epidemics spread by air travel, not to mention global terrorism...."[24] The list goes on to cloning and genetically modified organisms, migrant populations, and the disturbing growth of giant mega-cities, the Spenglerian megalopolis with a social and ecological vengeance. Even worse, the uneasy responses of traditional cultures in the Middle East and elsewhere to Western ideas and technologies have stirred a rise in fundamentalism. The United States itself, as I have noted, is anything but immune to this naive turn to the heavens in the face of seemingly intractable earthly difficulties. Despite her documentation of the elements of what would appear to be a dismal future, Henderson is still optimistic that enough time remains to create a sustainable global economy, and she has been prolific in writing about the possibilities of such a transition. However, as I will demonstrate in the remaining pages of this chapter, it is more probable that the current global economic paradigm will continue without serious regard for ecological sustainability. Furthermore, global tensions and warfare, especially terrorism, between cultures and nation-states, are likely to focus public attention upon more immediate threats rather than the long-term stability of the natural world as the global marketplace expands along with increased population and consumption.

Amy Chua's *World on Fire* (2003) makes a powerful argument that the exportation of free-market democracy by the developed countries, especially the United States, Europe, and Japan, has released vitriolic ethnic hatreds in Africa, Asia, Russia, and Latin America, not to mention the Middle East. In these largely Third World regions, Chua has found that free markets have concentrated wealth in the hands of resented ethnic minorities, leading to the rise of governments antagonistic to those minorities, as in the case of Milosevic in Serbia, for example. Thus, the democratization of many regions, in tandem with free trade, has bred unexpectedly negative social consequences, and a rising tide of anti-American sentiment around the world, recently exacerbated by the arrogant unilateral and internationally uncooperative foreign policy of the Bush-Cheney administration.

In order to understand this trend we must first grasp the fact that the United States is primarily responsible for the global spread of free markets. According to Thomas Friedman, in *The Lexus and the Olive Tree*, today's worldwide prescription of privatization, deregulation, and economic liberalization was produced by America and Great Britain. As Chua writes, "it was America, determined after the Second World War to promote capitalism and contain Communism, that drove the creation of the World Bank, the International Monetary Fund, GATT, and most recently the World Trade Organization as well as a host of other free-market-oriented international institutions."[25] The United States is the leading nation in pressuring other countries to open their markets for free trade and investment, with its global armed forces ready to keep these markets and sea lanes open for globalization. America's place today on top of the global economy is compared by Chua with the market-dominant Chinese in Southeast Asia in terms of the resentment that accrues against the minority in economic control. This form of cultural imperialism gives the regional indigenous majority the uneasy sense of being engulfed by the United States. "Just 4 percent of the world's population, America dominates every aspect--financial, cultural, technological--of the global free markets we have come to symbolize. From the Islamic world to China, from our NATO allies to the southern hemisphere, America is seen (not incorrectly) as the engine and principal beneficiary of global marketization."[26]

The specious argument commonly made by entrepreneurs and bureaucrats engaged in the economic globalization process is that once the traditional societies of Third World countries are enriched by industrialization and the shift takes place from subsistence farming to modern "green revolution" (agribusiness) agriculture to develop products for trade, their increased national incomes and "higher" standard of living will allow them the surplus wealth to pay for environmental regulation and management. Nothing could be further from the truth. Sing C. Chew, in his study of world ecological degradation, recognizes an "incessant drive for the maximal utilization of the bounties of Nature over world history." In the course of increasing materialistic acquisition and consumption, humans have largely become alienated from their natural beingness as well as from the ecosystems they have diminished. Over the course of ca. 5000 years, during which a number of civilizations have collapsed partly due to ecological degradation, humans appear not to have learned useful environmental lessons. Despite our recently acquired ecological knowledge, contemporary modern industrial humanity behaves with the same negligence toward the natural world. Chew concludes that the spread of the global marketplace through the remaining pristine and modified ecosystems of the world may be likened to a malignant or cancerous cellular process involving rapid, uncontrolled growth, invasion and destruction of adjacent tissues, and metastasis or distant colonization. "Paralleling this in the case of human communities as a whole are rapid, uncontrolled growth of global population; invasion and destruction of all planetary ecosystems... and increasing undifferentiation in structure and appearance as a consequence of cultural forms via the dynamics of the world economy."[27]

If Chew's metaphor for the consequences of American-led global transformation through free-market trade and democratization sounds too harsh, I have personally witnessed the abject poverty and massive ecological degradation associated with Latin American "development" of tropical forests into overgrazed, eroding wastelands, and lowland jungles transformed into banana plantation monocultures now so infested with parasites that half of the bananas in American supermarkets are infected with rot, along with polluted streams and coastal waters accumulating fertilizer and pesticides. This describes the real-world model of the environmental consequences of the expanding global marketplace. Moreover, the people displaced from their traditional ways of making a living have agglomerated into urban masses, many of them existing in dire poverty. Should they be thankful to NAFTA, GATT, WTA, IMF, the World Bank, and the USA?

Robert D. Kaplan attempts to answer this question in *The Coming Anarchy* when he tells us that the chaos in Africa that we have heard about and witnessed on the BBC and CNN in recent years can be considered a global model for what war, borders, and ethnic politics will be like in the decades ahead, and he considers environmental scarcity to be a major factor in these events. As for the future of sustainable development, he writes the following: "Mention 'the environment' or 'diminishing natural resources' in foreign-policy circles and you meet a brick wall of skepticism or boredom. To conservatives especially, the very terms seem flaky. Public-policy foundations have contributed to the lack of interest, by funding narrowly focused environmental studies replete with technical jargon which foreign affairs experts just let pile up on their desks."[28]

Kaplan finds the attitudes of foreign-policy experts particularly deplorable because he personally considers "the environment' to be the leading national-security issue of the twenty-first century. He has in mind such problems as the political and strategic impact of exploding populations in the Third World, disease transfer, deforestation and soil erosion, water shortages, and possible rises in global sea level which would inundate low-lying regions such as Bangladesh and the Nile Delta, which in turn would prompt mass migrations. Countries threatened by diminished resources, and which might shift to anti-democratic regimes include Indonesia, Brazil, and Nigeria. As Earth's human population soars from 6 to more than 9 billion over the next half-century, 95 percent of population increase will be in the poorest regions of the world, such as Africa, where governments are already in various stages of dysfunction. Environmental stress in the developing world is likely to present people with a choice of totalitarian, fascist, and road-warrior alternatives. The long-term security implications of Kaplan's scenarios are frightening to contemplate. Soil degradation alone bodes ill for West Africa, the Middle East, India, China, and Central America. Kaplan concludes: "We are entering a bifurcated world. Part of the globe is inhabited by Hegel's and Fukuyama's Last Man, healthy, well fed, and pampered by technology. The other, larger, part is inhabited by Hobbes's First Man, condemned to a life that is 'poor, nasty, brutish, and short.' "[29]

Marxist, American Green Party activist, and scholar Joel Kovel is essentially in agreement with Kaplan's dismal view of a Third World transformed by American-led globalization. He observes that the international market system and its development projects have had three decades to moderate the negative social and environmental consequences of growth, and "has failed so abjectly that even the idea of limiting growth has been banished from official discourse."[30] Further, the present system of "growth" amounts to increasing the wealth of the few in developed countries at the expense of the many. Like Chew, he compares contemporary 'growth' to a cancer on a living organism which, if not treated, will mean the destruction of human society, and even the possibility of extinction of the human species. Why, he asks, do we speak of capital "as though it has a life of its own, which rapidly surpasses its rational function and consumes ecosystems in order to grow cancerously."[31] No matter what the ideologues of capitalism say, monetization is not included in the laws of nature, but exists in the context of ecosystems "whose internal relations are violated by conversion to the money-form. Thus the ceaseless rendering into commodities, with its monetization and exchange, breaks down the specificity and intricacy of ecosystems." [32]

Kovel refers to globalization as the establishment of a planetary regime to supervise the growth process. It is nothing fundamentally new, since capitalism has always been a world system, but the present system of global capitalism has taken on the mission to convert the undeveloped portion of the world's economy into full participation: "to achieve new 'lean' ways of production utilizing dispersed locations, to take over the natural resources, to consume the labour power cheaply, and to keep commodities rolling so that the values embedded in them may be realized."[33] Particularly as capital has been dematerialized and can be moved about by electronic means, it has become fundamentally uncontrollable, giving it greater flexibility to overcome its major obstacles, ecological knowledge and informed democracy. Thus, "great waves of capital batter against and erode ecological defenses. Similarly, democracy, and not government, is the great victim of globalization. As global capitalism works its way, the popular will is increasingly disregarded in the effort to squeeze ever more capital out of the system."[34]

One could go on encyclopedically documenting the ecological devastation and social dislocation occurring at accelerating rates around the globe today, and indeed this is what one reads daily in newspapers and numerous periodicals. The American public and its somnambulent leaders have been deluged with such information to the point of largely ignoring it and continuing their ecologically destructive decision-making without substantial changes drawn from knowledge of an ongoing environmental crisis related to population growth, overconsumption, and an expanding, relatively unregulated global market system. What one rarely reads about, if at all, is the worldview responsible for our destructive behavior, rationalized as necessary and progressive despite the informational negative feedback. Actually, it is a *subset* of our variable, syncretistic worldview that has become the predominant economic belief system capable of absorbing or displacing other elements of the modern worldview. Essential aspects of the current economic belief system have been discussed *en passant* in this chapter, but with the preceding discussion of the Environmental Movement and the "problem of development" in mind, it is once more time to scrutinize the existing syncretistic worldview held by the American citizenry in order to isolate the economic component of our assumptions about reality which are the basis for our behavior and rationalizations of it, i.e., our rational utilitarianism.

The complex modern worldview gradually displaced the Christian medieval worldview in European and neo-European colonial societies during the past several centuries. What was lost was an integrated, "organic" traditional worldview in which Christianity permeated every aspect of political, social, and even economic life. God, nature, humanity, history, and the structure of society were all neatly and lucidly explained by Christian mythopoeic thought. The breakup of medieval society and its beliefs and institutions has bequeathed to us a far more complex, much less certain mix of beliefs and attitudes labeled as "modern," subject to incessant revision in response to the historical rollercoaster of endless revolutions in scientific thought, relentless innovations in technology, and, above all, the gradual emplacement of a cryptoreligious ideology, our modern economic systems of thought and practice. Out of these variables, *economics* has expanded its influence upon our lives to a degree that members of most traditional cultures find appalling and frightening. Traditional Muslims and Christians alike find secular humanism, materialism, hedonism, individualism, selfishness, even political democracy to be anathema to the true believer. In

America, as in Europe, traditional religious followers have for the most part managed a compromise between the wealth and freedoms enabled by modern society and the moral imperative of religious faith. In nations such as Sweden, France, and Germany secular humanism has largely displaced traditional religious belief. Whatever mix of worldviews predominates in the nations of the Western world and its westernized (modernized) followers, all of us as individuals find our lives increasingly controlled by technology and economic growth. As individuals, many, if not most, of us feel as tossed about by the modern tourbillon as the plant and animal species that are overrun and scattered by bulldozers. And the whirlwind of modernity whirls faster with every passing year. Our lives and the natural world are at the mercy of forces we scarcely understand because a system of economic ideas and practices has been emplaced without democratic decision-making, much as technology has produced more and more sophisticated means to unexamined ends exempt from democratic consensus. Even when politics are involved, our collective worldview has been captured by a revised theory of free-market economics which has stood poor Adam Smith on his head. This contemporary "neoliberal" theory may be described as follows:

> In the worldview in which we have all been trained, economic growth, also called "economic development" when applied to the Third World, is the central panacea for our problems. In such a view, free markets and free trade are essential tools for maximizing growth and development, since they both strive for business activity that is unfettered by regulations on environment or health or worker's rights, by tariffs, by protection of local businesses, or by democratic governance itself. If trade worldwide can be freed from all this, then a single global economy can be created that unifies all nations and peoples in the same activity under corporate guidance, supposedly leading to global prosperity.[35]

We have not *all* been brainwashed with the contemporary free-market theory of development or there would not be such an extensive literature opposed to the global economy. However, corporation heads and underling business bureaucrats, and the vast majority of our political leaders, Democratic and Republican alike, along with a preponderance of Internet and media communicators and bureaucrats, as well as Ivy League economic theorists, have imbibed this neoliberal doctrine with enthusiasm. It has been for several decades the seminal idea that is transforming and destroying our planet. Economist David C. Korten has challenged this conventional wisdom by arguing that right-wing economic ideologues have created a myth pleasing to American and European business and political elites by distorting the original ideas of Adam Smith beyond recognition. Korten maintains that most people accept free-market ideology without question much as they accept the basic doctrines of their religious faith and of democratic government. However, along with neoconservative economics they implicitly accept such underlying assumptions of economic liberalism as: the idea that humans are essentially selfish; that competitive behavior is more rational than cooperative behavior; that action leading to the highest financial returns to individuals or corporations are the most beneficial to society; and that human progress is best measured in terms of increases in the value of what members of society consume, and that this is a measure of progress.

The gist of Korten's critique of contemporary economic liberalism centers around Adam Smith's contention that government support and protectionism tended to distort the self-corrective mechanisms of a competitive market comprised of small buyers and sellers, without the presence of large business (read *corporations*) or monopolistic market powers. He also distinguished the "natural price" of products from the "market price," the latter eventually approximating the former through the "invisible hand" of the competitive market, without advertising or promotional costs. Korten emphasizes that Smith also assumed that capital would be rooted in a particular place. Smith "was most definitely not an advocate of gigantic corporations, detached from commitment to any place, funded through capital markets that separate the management of capital from its actual ownership."[36] To the contrary, he would have found this model inefficient and contrary to the public interest. Ideologues of contemporary economic liberalism such as Francis Fukuyama, like Marxist theorists, are historicists, proclaiming the historical inevitability of the forces advancing their cause. Furthermore, they offer Western publics the illusion of only two choices, between "free world" liberal economic policies and a centrally planned Soviet-style economy. "In defiance of history and logic, economic liberalism's worldview permits no possibility of supporting a market economy without supporting the *free* market or favoring trade without advocating free trade. In its extremist manifestations, economic liberalism seeks to paint its opponents into an equally extremist corner and denies the possibility of a middle ground...."[37] Korten locates the pragmatic foundations of present economic theory in the United States of the 1920s when the economic system closely approximated the free market idea. Then, financial interest had almost free reign, stock market speculation was maximized, large corporations endlessly merged, and the nation's wealth became concentrated in fewer hands. Does this sound like the late twentieth century? Korten thinks that the contemporary economic system based on the ideology of free market capitalism is destined to collapse for many of the same reasons that the Marxist economic system fell. These include: concentration of economic power in unaccountable, centralizing institutions (transnational corporations rather than the state); both are highly destructive of nature or natural capital in the name of economic progress; both venerate mega-institutions that politically disempower and socially weaken the populations they are supposed to service; and both take a narrow, economistic view of human needs and the connection to nature and community necessary to a healthy society. According to Korten, only democratic pluralism and a regulated market that maintains domestic competition and favors domestic enterprises that favor local workers paying local taxes can remain healthy in the long run. Such a model can integrate green market practices and environmental regulations far more efficiently than the present corporate system.

The problem of economic growth/development has been an insurmountable obstacle to the success of the Environmental Movement, particularly in the deregulated era of giant corporations and monopolies propelled by American government itself with increasing vigor since Ronald Reagan took office in 1980. It is the reason why some environmental historians, my self included, have dated the Environmental Movement 1962-1980. Despite the proliferation of laws and regulations which environmentalism has produced, there is today a sense among many of us that the Environmental Movement has failed. Environmental science majors may find the idea unpalatable, but ecological insults to the planet as a whole are becoming greater and more widespread. From a stack of related articles I pick up one entitled "A Planetary Defeat: The Failure of Global Environmental Reform."[38] The author, John Bellamy Foster, an experienced veteran of the 1992 Rio summit and the Johannesburg summit a decade later, summarizes a decade of intensified environmental degradation since the Rio Summit, with the United States leading the charge. He also assesses U.S. entrepreneur and environmentalist Paul Hawken's argument for "natural capitalism," capitalism that fully incorporates nature into its system of value, and rejects the idea of tinkering with corporations and institutions such as the WTO, IMF, etc. Instead, like Korten, Foster favors action in behalf of the principle of *localization* in order to challenge globalization and lay foundations for real sustainable development. He concludes: "The main lesson to be derived from the failure of global environmental reform associated with the Rio summit is that there is no possibility for an effective movement for social justice and sustainability separate from the struggle for an alternative society."[39] This judgment supports my own and other radical environmentalists' view that social reconstruction based upon our embeddedness in nature cannot occur without *metanoia*, a fundamental shift in worldview. Although this chapter in particular and the entire book in general have laid the foundations supporting such a perspective, it will be developed more fully in the following and final chapter within the context of the philosophy of history, which was introduced in Chapter I.

The contemporary reductionist, purely economic worldview in which politics, education, and even our historical view of reality, as well as illusions of a supertechnological, affluent future are all manipulated and controlled by business-dominated government as preceptor, *did not happen by accident or because it is the best approach to gaining a long-term living from the planet*. Based upon the preceding history of our environment and the Western worldview, it is the expectable, logical outcome of its historical antecedents. In the final chapter I will attempt to explain our dilemma and the possibility of its solution in the light of the historiographical revolution of the twentieth century and its relationship to speculative philosophies of history and the rise of environmental history.

NOTES

[1]. Personal communication, Max Oelschlaeger, Professor of Philosophy and Religion, Northern Arizona University.

[2]. Clarence Glacken: *Traces on the Rhodian Shore: Nature and Culture in Western Thought from Ancient Times to the End of the Eighteenth Centuy* (Berkeley: University of California Press, 1967), p.4.

[3]. Ibid., p.35.

[4]. Ibid., pp.707-709.

[5]. See Carolyn Merchant, *Reinventing Eden*, pp.77-84 and 88-89.

[6]. William Ophuls, *Requiem for Modern Politics: The Tragedy of the Enlightenment and the Challenge of the New Millennium* (Boulder, Colorado: Westview Press, 1997), p.11.

[7]. Ibid., p.12.

[8]. In a chapter of *Requiem* titled "moral entropy" Ophuls explains how Thomas Hobbes's philosophical revolution was combined with John Locke's bourgeois politics and Adam Smith's market economics to create an Anglo-American liberal political paradigm which "unleashed human will and appetite, but provided no countervailing source of moral principle strong enough to preserve society from their ravages over the long term." (William Ophuls, *Requiem for Modern Politics: The Tragedy of the Enlightenment and the Challenge of the New Millennium*, Boulder, colorado: Westview Press, 1997, P.30.) Thus, Ophuls finds liberalism, the foundation of the modern development paradigm, to be intrinsically self-destructive, and to continue today only by force of inertia deriving from its successes initiated by development of the Great Frontier (i.e., the colonies) and the global economy built upon Third-world exploitation. In the next several pages I will summarize Ophuls's major arguments supporting this thesis.

The Hobbesian political foundation of modern societies structured for the pursuit of unlimited economic development begins with Hobbes's rejection of Aristotle's assumption of the innate sociability of the human species manifested in the ancient polis, perceived as "a tribe writ large, a community given by nature that is united by common blood and shared ideals." (Ibid.) Against this traditional view of human political society, Hobbes posited the idea of the "state of nature" in which atomistic individuals struggled against their fellows in a war of all against all that was " solitary, poor, nasty, brutish and short." Hobbes reasoned that humanity could escape his postulated "state of nature" and the atomistic struggle for power once that, through reason, they adopted a "social contract" in which "the sovereign makes and enforces rules that permit and even encourage individuals to seek their own selfish ends--but now peacefully, within a framework of laws that preserve both public order and private rights." (Ibid., p.31) In an orderly commonwealth in which private pursuits were guided by laws acting as hedges to contain unsociable behavior, modern humanity would "progress in the arts and sciences, or what we would today call economic development." (It was precisely this process and its dehumanizing and inegalitarian consequences that Rousseau attacked in his discourses on the arts and sciences and on the rise of inequality, as described in Chapter Seven, and which made him, in the eyes of Francis Fukuyama and others, the father of modern environmentalism.) As Ophuls points out, it was Hobbes's philosophical genius that allowed him to redefine the pre-historical and historical conditions of human existence in terms of a new, modern political myth, the social contract, which justified the monarchical ruler or Leviathan in terms understandable to the Protestant mind once the myth of the divine right of kings had been destroyed.

Hobbes laid the political foundation of a modern society focused upon "progress" or economic development at the expense of overthrowing the familial and communal basis for traditional "organic" society. His theory of isolated, asocial individuals, having displaced

Aristotle's "political animal," helped to create *modern* society composed of largely amoral, self-seeking individual atoms deracinated from traditional communal roots. In the modern commonwealth or Leviathan " the motives for joining the commonwealth are purely negative: fear of violence on the one hand and a realization that 'the contentments of life' can be obtained in no alternative way on the other. The Hobbesian polity is therefore a necessary evil, to which individuals give only grudging consent." (Ibid., p.32.) In this view of politics, the American desire for minimal governmental regulation and maximum freedom to pursue selfish, individual ends makes perfect sense, and social behavior begins to approximate the war of all against all which Hobbes wrongly attributed to his hypothetical "state of nature." Modern anthropology and history have corrected Hobbes's model of pre-history with the model of small, integrated hunter-gathering societies, bound together by common traditions and values. In comparison, the Hobbesian polity is comprised of a collectivity of selfish and passionate individuals who exist as social atoms guided by merely prudential rather than traditionally moral laws. "In practice, therefore, one is largely free to do whatever one can get away with. Worse, although most individuals are reasonable and inclined to live at peace, a greedy, ambitious, and unreasonable few will always stand ready to embroil the whole society in a vicious circle of competitive self-seeking." (Ibid., p.33.) Herein are to be found the roots of American political conservatism and libertarianism. Ophuls further observes that Hobbes had, in effect, invented a politics appropriate to the coming machine age, a politics preoccupied with the machinery of constitutional government combined with Montesquieu's doctrine of checks and balances. The resulting modern administrative state has become "a governmental machine for satisfying the material aspirations of the people" and "Hobbes is indeed the 'father of us all'--and of American politics above all--to a much greater extent than is commonly realized." (Ibid., p.34.) In short, Hobbes terminated the classical conception of politics as founded upon natural community, with virtue as the end of politics, while dispensing with the medieval conception of the divine right of kings and reestablishing monarchy and future forms of the state on secular grounds.

Building upon Hobbes's theory of the social contract, John Locke revised Hobbes's assumptions: (1) of scarcity in the state of nature leading to a war of all against all; (2) that human passion tends to overwhelm reason; and (3) that the only alternatives are between absolute anarchy in nature and the strict order of the absolute state. To the first assumption, Locke responded with the New World, America, as his model for the state of nature, where vast and inexhaustible lands and resources awaited appropriation. Under such benign circumstances, only a sort of "night watchman" would be needed to referee fair play as economic development took place. To the second assumption Locke answered that except under extreme conditions the conscience and reason of most men keep their passions in check, thus requiring only limited government to maintain civil order. Thirdly, Locke rejected Hobbes's extreme alternatives of anarchy or autocracy for a more benign society developed under a social contract in two stages, the first involving the foundation of a civil society, and the second consisting of agreement by its members to establish a commonwealth of limited powers. As Ophuls explains, Locke was not arguing for weak government but for a basis for representative democracy: "thanks to his more sanguine view of both the state of nature and human nature, as well as his ingenious modification of the social contract, Locke was able to argue for government by continuing consent of the governed in place of Hobbes's one-time consent followed by autocracy." (Ibid., pp.34-36.)

The other fundamental element in Locke's theory of government, of paramount importance for modern humanity's obsession with rapid economic growth, is his theory of private property. Although Ophuls credits Hobbes with in effect inventing political economy "by making material satisfaction the end of politics and economic development the task of the sovereign," it was Locke who became "the founder of modern political economy by making private property the basis of liberal politics." (Ibid., p.36.) I am quoting Locke himself here, to emphasize his importance in transmitting the Christian idea of human dominion over nature. He writes, in *The Second Treatise of Government:*

God, who hath given the World to Men in common, hath also given them reason to make use of it to the best advantage of Life, and convenience. The Earth, and all that is therein, is given to Men for the Support and comfort of their being. And though all the Fruits it naturally produces, and Beasts it feeds, belong to Mankind in common, they are produced by the spontaneous hand of Nature; and no body has originally a private Dominion, exclusive of the rest of Mankind, in any of them, as they are thus in their natural state: yet being given for the use of Men, there must of necessity be a means *to appropriate* them some way or other before they can be of any use, or at all beneficial to any particular Man. (John Locke, *Two Treatises of Government*, New York: Cambridge University Press, 1960, p.328.)

Thus, beginning with the explicit pronouncement of God's gift of Earth to humankind, Locke builds his argument for rights to private property. With America as his model, Locke's "state of nature" is superabundant, and he extends Hobbes's right of appropriation with the key idea that the application of human labor to the "commons" makes it the property of those laboring upon the land. This is not harmful to others "since there was still enough, and as good left; and more than the yet unprovided could use." (Locke quoted in Ophuls, *Requiem for Modern Politics*, p.36.) Over time, the introduction of money (representing property) replaces the rough equality of the state of nature with the harsh inequality of civilization (the transformation that so vexed Rousseau and led him to write the *Discourses*). As Ophuls points out, this transformation does not challenge Locke's principles of private property because "the existence of 'America' makes the issue of inequality moot, for anyone not happy with the current distribution of property in Europe can light out for the New World and appropriate land from the American commons." (William Ophuls, *Requiem for Modern Politics*, p.36.) In actuality, of course, such European colonizers actually took or stole the land from Native Americans while offering Locke's justification of property rights in order to do so. Thus, European colonists and later pioneers could conquer or seize Indian lands and deprive the Native Americans of their livelihoods as hunter-gatherers or primitive farmers while maintaining a clear Christian conscience and sense of righteousness. Meanwhile in Europe, and especially in England, marginal farmers and herdsmen were expelled from the "commons" as they were appropriated into private property by the gentry and bourgeois who possessed the power and money to do so. Ophuls adds the observation that Locke's clever theory was also contradicted by the fact that much of American "property" was created by slave labor. Furthermore, "ecological scarcity renders Locke's argument untenable on its own terms: without the free commons of 'America' as a safety valve, those lacking property have nowhere left to turn and are likely to be pauperized in perpetuity." (Ibid., p.37.)

Ophuls clearly recognizes Locke's property doctrine as a political myth, one of critical importance to his general theory of the state because "economics in general and private property in particular are the basis of the Lockean political order." (Ibid.) Within this new order material satisfaction is the end of politics and the process of acquiring property displaces fighting over religion (the bane of sixteenth and seventeenth century Europe) with the collective conquest of nature and material enrichment. Government is needed only to manage the division of the spoils of conquest, and the drive for material wealth sublimates political ambition and potential social conflict under a state in which *economics becomes a surrogate for politics*. Under such a regime political participation is generally restricted to the propertied class within which the economically necessary traits of enlightened self-interest, enterprise, initiative, thrift, prudence, and responsibility are cultivated. Ophuls's reflections on Locke conclude:

For Locke, then, life, liberty, and estate are deeply interrelated, and the word *property* summarizes and encapsulates a whole range of important political and social goods. The security of property is therefore of paramount importance, and the primary duty of a Lockean sovereign is to replace the free-for-all of the state of nature with a well- ordered and well-enforced set of rules guaranteeing property rights: 'Government has no other end but the Preservation of Property.' Locke is thus the root philosopher of the emerging capitalist order: he completes the economization of politics begun by Hobbes, makes private property the dominant political principle, and lays the foundation for modern political economy. (Ibid., p.38.)

[9]. Ibid., p.39
[10]. Ibid.
[11]. Ibid., p.102.
[12]. Ibid., pp.267-268.
[13]. Ibid., p.40.
[14]. Ibid., p.41.
[15]. Peter Hay, *Main Currents in Environmental Thought*, p.213. See section titled All Hail "Sustainable Development," pp.212-219.
[16]. Nick Middleton, *The Global Casino: An Introduction to Environmental Issues*, 3rd Ed. (New York: Oxford University Press Inc., 2003), p.38.
[17]. Ibid., pp.39-40.
[18]. Peter Hay, *Main Currents in Environmental Thought*, p.215.
[19]. Ibid.
[20]. Carolyn Merchant, quoted in Peter Hay, *Main Currents in Western Environmental Thought*, p.207.
[21]. Joseph E. Stiglitz, *Globalization and its Discontents* (New York: W.W. Norton & Company, 2002), p.2.
[22]. Ibid., p.216.
[23]. Hazel Henderson, *Beyond Globalization: Shaping a Sustainable Global Economy* (West Hartford, Connecticut: Kumarian Press, Inc., 1999), p.1.
[24]. Ibid., p.2.
[25]. Amy Chua, *World on Fire: How Exporting Free Market Democracy Breeds Ethnic Hatred and Global Instability* (New York: Doubleday, 2003), p.232.
[26]. Ibid., p.230.
[27]. Sing C. Chew, *World Ecological Degradation: Accumulation, Urbanization, and Deforestation* (Walnut Creek, CA: Alta Mira Press, 2001).
[28]. Robert D. Kaplan, *The Coming Anarchy: Shattering the Dreams of the Post Cold War* (New York: Vintage Books, 2001), p.19.
[29]. Ibid., p.24.
[30]. Joel Kovel, *The Enemy of Nature: The End of Capitalism or the End of the World?* (Nova Scotia: Fernwood Publishing Ltd, and London: Zed Books Ltd, 2002), p.5.
[31]. Ibid., p.39.
[32]. Ibid., p.40.
[33]. Ibid.,, p.69.
[34]. Ibid., p.76.
[35]. Jerry Mander and Edward Goldsmith, eds., *The Case against the Global Economy; And for a Turn Toward the Local* (San Francisco: Sierra Club Books, 1996), p.181.
[36]. David C. Korten, "The Mythic Victory of Market Capitalism" in Jerry Mander and Edward Goldsmith, Eds., *The Case Against the Global Economy*, p.188.
[37]. Ibid., pp.188-189.
[38]. John Bellamy Foster, " Planetary Defeat: The Failure of Global Environmental Reform," *Monthly Review*, January, 2003, pp.1-9.
[39]. Ibid., p.8. Numerous books and articles strongly support the validity of the view that ecological damage is rapidly worsening as the global economy spreads and intensifies. See, for example: J. Donald Hughes, *An Environmental History of the World: Humankind's Changing Role in the Community of Life* (London: Routledge, 2001), pp.206-213 and 221-233; Michael N. Dobkowski and Isidor Wallimann, Eds., *The Coming Age of Scarcity: Preventing Mass Death and Genocide*

in the Twenty-first Century (Syracuse, New York: Syracuse University Press, 1998), pp.43-60, 83-100, and 253-257.

CHAPTER THIRTEEN: NATURE AND THE PHILOSOPHY OF HISTORY

Introduction

Nature and the philosophy of history? Not only to the popular mind, but even to many intellectual historians, a dissonance springs from the juxtaposition of these categories, even though the great philosophers have focused on the idea of nature for millennia. For most of us, scientific study of nature seems appropriate, but nature is not generally thought of as the realm of philosophers. Plato and Socrates certainly would agree, particularly because they redirected the nature philosophies of the Ionian, pre-Socratic philosophers towards an increasingly moralistic, transcendental, and anthropocentric emphasis. As for philosophies of history, few people are aware of such things, and yet they are experienced by virtually everyone, but subliminally, as part of dominant worldviews in any given time or place. Within the civilizations of the West, the cyclical speculative philosophy of history dominated the ancient Greco-Roman mind, the Christian idea of providence prevailed during the Middle Ages, and the idea of progress has conditioned modern thought, beginning in the late seventeenth century. In other civilizations traditional mythopoeic religious belief analogous to Western Christianity has usually prevailed. In the modern Western world, as in regions shaped by it, relatively uniform belief in a dominant religious myth has yielded to complex syncretism in which a multitude of worldviews compete for adherents. In the United States these include a kaleidoscopic range, from overwhelmingly dominant Christianity to Buddhism, Islam and a seething variety of sects and religious cults, to atheists, agnostics, deists, existentialists and other children of the Enlightenment. Traditional cultural uniformity has been shattered into factions of believers held loosely together by the political precepts of modern constitutional democracy and the practice of increasingly global economic development.

What has this to do with academic study of nature and the philosophy of history? First of all, despite our cultural syncretism and the quiet dissonance of competing worldviews within a mass democracy with an alleged common value system, American civilization is largely guided by two worldviews and the contained philosophies of history which grew out of our historical experience. The Christian worldview and the idea of providence, much discussed in earlier chapters, provides the majority of Americans with the assurance that all is well in the long run for individuals living in accord with the Christian notion of salvation. The idea of providence, i.e., the Christian philosophy or explanation of history, also promises a renewal of this tired, worn-out planet Earth when Christians have completed the paradise or *parousia* at the end of history in preparation for the second coming of Christ and the millennium of his reign on earth. In this view, economic development is the reclamation of the Garden of Eden from the thorny, predator-ridden wilderness into which mankind had been cast by a disgruntled God. For the Christian true believer *development* is not only a God-given right, but a responsibility on the part of serious, dedicated Christians. The historical forces leading to the displacement of medieval institutions and Christian belief by modern, science-based secularism substituted a new, mechanistic worldview which perceived history to be progressive on the basis of gradually accumulating knowledge, technology, and, especially, science. The scientist, much like the Christian, carries the responsibility of transforming the natural world into a "better" place, a more comfortable home for humankind, free of hunger, disease, and want. Science and technology have thus been harnessed to the project of economic development perceived as "progress." Unfortunately, this pursuit of "progress" has led to such rampant growth and overdevelopment that nature itself, and the natural processes that sustain us, have been devastated in recent centuries. In sum, the Christian providential and modern progressive interpretations of the meaning of history are largely subliminal philosophies of history, unexamined by the great majority of citizens participating in our grand project of economic development, and unaware of the power of these ideas and their implementation, and their contribution to the destruction of nature to the point of our own self-destruction. However, twentieth century challenges to the methods of writing history, and the doubtful credibility of speculative philosophies of history have led to the unmasking of modern historical writings as too often a rationalization of economic development in the name of "progress." This ongoing dialogue over the nature of history itself has significant implications for our attitudes toward and treatment of nature, if only intellectuals of our society as a whole could grasp them.

The Twentieth Century Historiographical Revolution

Historiography, the study of the history and method of historical writing, was not formalized in the modern world until the nineteenth century, when the scientific method was taken as a model worthy of imitation by historians, who, thereafter, through careful scrutiny of texts and other sources, would write "scientific history" undistorted by subjective opinion and prejudice. However, as we saw in Chapter I, inadvertent distortion of "what actually happened in history" always seemed to creep into historical writings, and it did not take long before "scientific" history would be undermined. One of the greatest of the new, self-assured nineteenth century historians, Leopold von Ranke (1795-1886), believed that if he scientifically managed the facts, divine providence would take care of the meaning of history, thereby demonstrating the persistence of the idea of providence as exerting an influence comparable to that of the idea of progress in his time. The two beliefs blended to produce "millenarian progressivism," the equivalent of a Christian belief in progress. Ranke's perception of the meaning of history as implicit and self-evident was also influenced by the economic doctrine of laissez-faire, whereby each individual pursuing his own ends was led by a hidden hand to create universal harmony. "The facts of history were themselves a demonstration of the supreme fact of a beneficent and apparently infinite progress towards higher things." [1] Thus does E. H. Carr identify the nineteenth century blending of the ideas of providence and progress, the two dominant Western philosophies of history, as being so thoroughly ingrained in the nineteenth century worldview that historians claiming to have dispatched the subjective element in historical writing were knee-deep in what we might call "providential progressivism," or what I termed millenarian progressivism in Chapter Six.

World War I shattered the secularized eschatological vision of fin-de-siècle historians, in the 1880s and 1890s Wilhelm Dilthey (1833-1911) and Benedetto Croce (1866-1952) launched an assault upon the nineteenth century historians' "doctrine of the primacy and autonomy of facts in history." [2] Croce argued that all history is contemporary history, by which he meant that the historian sees the past through the present and its problems, thereby distorting it. Therefore, the paramount task of the historian is less the collection of facts than their evaluation. Only by grasping the meaning of past thought and reconstituting it in the present can the historian meaningfully write history. "By and large, the historian will get the kinds of facts he wants. History means interpretation." [3] Hermeneutics, the scholarly interpretation of Biblical texts, would therefore be a better model for historical writing than the scientific method, alas, opening the door in the long run to postmodern criticism of the "Truth" found in sacred and other texts.

World War I further shattered the view of nineteenth century European historians that history is progress guided by divine providence, and once that meaning was gone, they rejected any belief suggesting that history has a meaning, even though belief in progress and providence persisted in the popular mind. This is why they strongly rejected Spengler's cyclical philosophy of history as well as Toynbee's attempt to reconcile the cycle with progress and providence. On to the present day they have accepted the alternative, non-committal view that there is no general pattern in history at all. [4] Historians in the United States likewise moved towards ignoring or rejecting any concern with ultimate meaning in history. To earlier twentieth century social and economic history, American historians added various sub-disciplines of "multicultural" history, from the history of black and Hispanic minorities, ethnic histories of other neglected peoples outside of the Western world, to feminist history and gay history. Some historians include environmental history in this spectrum of new histories which emerged in the 1960s and 1970s, but I think that this is a mistake. Whereas multicultural histories deal with various neglected groups of humans and their problems during and after the Civil Rights and Feminist movements, environmental history grew out of a concurrent but different milieu, the

Environmental Movement explained in Chapter Eleven. As a graduate student in history at the University of California at Santa Barbara during the late 60s and early 70s, I was particularly interested in the philosophy of history, having read Spengler and Toynbee. During those years, books on the speculative philosophy of history were still being published, but they disappeared after the early 1970s as multicultural history gained academic appointments in colleges and universities. (During this period, speculative philosophy of history has largely functioned at a subliminal level in American politics where the spread of capitalism and democracy have been understood as part of a grand providential design, although neoconservatives such as Francis Fukuyama have presented these tendencies in academic garb). American historians generally have ignored the speculative philosophy of history since that time, and the teaching of "Western Civilization" as a required or strongly recommended core course has diminished greatly.

Simultaneously with the rise of multiculturalism in history, a new field of philosophy and critical theory in the United States took its bearings from an eclectic group of French philosophers and critical literary theorists usually referred to as "postmodernists." The term "postmodernism" was initiated in architecture and other arts as representing the repudiation of modern art forms imbued with rationalist function for more whimsical and unpredictable forms. More generally, the term came to mean "the critique of modernity as a set of assumptions about industrial and technological forms of life." The modern worldview which arose during the eighteenth century, at the expense of established religious belief, tradition, and custom, gave rise to " the notion of the freely acting, freely knowing individual whose experiments can penetrate the secrets of nature and whose work with other individuals can make a new and better world."[5] Postmodernists attacked this optimistic worldview and its idea of progress based upon objective scientific knowledge, focusing in particular upon the idea of the individual self, which they understood as nothing more than an ideological construct: "a myth perpetuated by liberal societies whose legal systems depend upon the concept of individual responsibility. By making this argument against the unified self--postmodernists call it 'the subject' to underline its lack of autonomy--they also, perhaps inadvertently, undermine the premises of multiculturalism."[6] As for the environment, postmodernist critics of science and history point to the twentieth century and its pollution, genocide, famine, world wars, and depressions as evidence of the failure of the Enlightenment and its ideals of democracy, science, and reason. They also argue that science and technology are corrupt because they necessarily are shaped by political agendas seeking power over both nature and other humans. Thus, they implicitly support Ophuls's thesis that the Enlightenment mollified two of the four great ills of civilization only by worsening the other two, i.e., overexploitation of nature and imperialistic violence against outsiders.

The postmodern critique of the purported objectivity of science and history has also extended into the realms of ecology, environmental philosophy, and environmental history. The validity of terms such as "wilderness" and "nature" has been challenged or "deconstructed" by postmodernists who maintain that all such terms are socially constructed, and are therefore ephemeral and relative since they reflect changing beliefs about nature (as well as other categories) over time under the influence of social, economic, and political forces. Thus, there are as many conceptions of nature as there are (or have been) different worldviews. This postmodern constructivism applies linguistic analysis to the task of repudiating any beliefs that epistemological foundations to knowledge transcend social-cultural-historical contexts. Knowledge can only be contingent, ephemeral, and relative. American environmental historian William Cronon was one of the earliest proselytizers of constructivism applied to environmental concepts such as "nature," "wilderness," or "ecosystem." In 1995, for example, he wrote that the removal of Native Americans during the late nineteenth century in order to create "uninhabited wilderness" in the new national parks "reminds us just how invented, just how constructed, the American wilderness really is. To return to my opening argument: there is nothing natural about the concept of wilderness. It is entirely a creation of the culture that holds it dear, a product of the very history it seeks to deny."[7] Cronon's ideas initiated a furious controversy among environmentalists, ecologists, and especially environmental historians and philosophers. His suggestion that "wilderness" as a concept has encouraged us to believe we are separate from nature, which would reinforce environmentally irresponsible behavior, was particularly aggravating to many environmental historians because it appeared to assume naivete on the part of the reader regarding the relationship between humans as hunter-gatherers and the biomes which they inhabited.[8] It is common knowledge that hominids have disturbed biomes for several million years, and that *Homo sapiens* has disrupted much of Earth for ca. 100,000- 150,000 years. However, by civilized standards, the unexplored regions of the nineteenth century world usually, but not always, inhabited by hunter-gatherers, were nevertheless "wilderness" to those who explored those regions. On the other hand, Charles C. Mann's *1491: New Revelations of the Americas Before Columbus* documents the surprising degree to which the Americas before Columbus were settled with a mix of hunter-gatherers, farmers in the incipient stages of agriculture, and full-fledged farmers who laid the foundations of Aztec, Mayan, Incan, and other lesser known civilizations. The European-induced demographic collapse of Native American societies due to virgin soil epidemics did indeed return large tracts to wildness, if not "wilderness," created out of abandoned utilized lands. Cronon's ideas, in fact, directly inspired Mann to demonstrate how much of the American "wilderness" was created by the abandonment of artifactual environments. [9]

The usefulness of Cronon's much contested thesis has been, of course, to make environmental historians and others more aware of the sociocultural context of words and concepts too often carelessly used, such as "nature" and "wilderness." However, although the historiographical skills of environmental historians have been sharpened by the debates over fundamental concepts, the counterattack against postmodern constructivism continues. As Eileen Crist writes, regarding the endangered idea and reality of wilderness:

> There is nothing intellectually or socially innocent about the timing of the disclosure that "wilderness" is a cultural concept: as wild nature sinks into the quicksand of all manner of development, the idea itself starts to feel like gossamer. What poses as a sophisticated argument--that wilderness is a construct since it has been a (non) idea amenable to historically diverse conceptions--in socio-historical context can be understood as an unsurprising ideological reverberation of the appropriation of wild nature.[10]

In short, the almost total destruction of wilderness has been followed by its deconstruction. If, following the deconstructionist paradigm, wilderness is not a tangible thing but no more than the narratives and descriptions of an unknowable world external to culture and its linguistic symbols, then nothing will be truly lost when the last so-called "old growth forest" and its inhabitants are gone. Moreover, if nature is only a social construction, what is the point of preservationism and the Environmental Movement? The *prima facie* absurdity of such questions expresses our disbelief in the postmodernist premise that reality is invented by words. The text is the only reality says Michel Foucault. He, Jacques Derrida, Richard Rorty, Jean-Francois Lyotard and other postmodern deconstructionists seem to believe "that our role as human organisms is to replace the world with webs of words, sounds, and signs that refer only to other such constructions."[11] This model of an internal cultural reality disconnected from the social constructions "nature" and "wilderness" is implemented in at least three contemporary modes: the internet; artifactual simulacra of "nature" (such as man-made wetlands in place of wetland ecosystems destroyed by development); and the global network of corporate economic development which is completing the task of replacing "natural" ecosystems with *cultural* agrarian, industrial, and urban systems.

More and more individuals within the technological society live much, if not most, of their waking lives interacting with the electronic symbols of the computer, thoroughly accustomed to interaction with artifactual machines and their messages as the dominant manner of being in the world. In place of displaced and fragmented wilderness and wild areas, Disneylandish wildlife parks, wildlife safaris, and other simulacra of the past proliferate as human contact with "real" nature brings highways, concessions, and crowds to the accessible "wilderness" of Yellowstone and Yosemite. Theme parks and national parks become almost indistinguishable. And, above all, from one corner of the globe to another, from Hilton to Hilton, exotic lands are incorporated into the global technological network. Given the acculturating power of these forces, are contemporary individuals more in contact with nature or systems of "texts" and their symbols? Denatured, technological humanity is thus deracinated from nature and programmed for interaction with a world of machines, simulacra, and their human inventors. (Spengler's observation that "the machine leads, and humanity subserviently follows" seems prescient in this context). In the "real" contemporary world of the technological society, we appear to eerily approximate the postmodern portrayal of modern man's imprisonment in language and texts.

Fortunately, however, the premises of postmodern deconstructionists regarding humanity and nature are wrong. Language and texts *are* intermediary between individuals and the natural world; nevertheless, science *has* greatly improved our understanding of the structure and history of the natural world. The causes of our separation from nature consist of far more than ambiguities in language and texts. "Mainstream Western philosophy, together with the Renaissance liberation of Art as a separate domain and its Neoclassical thesis of human eminence, were like successive cultural wedges driven between humans and nature, hyperboles of separateness, autonomy, and control. It may be time, as the voices of deconstruction say, for much of this ideological accretion to be pulled down."[12] However, out of the bitterness of the failed ideologies of existentialism and Marxism, French philosophers and literary critics have presented us with arguments for a meaningless universe, unstructured and unfathomable beyond the cultural framework of language and texts. But not all texts and interpretations are equal! First of all, deconstructing postmodernists conflate mythopoeic texts with scientific and technological texts; i.e., works of literature and imagination with works of methodological inquiry into how nature works. We hardly could have flown a spaceship and men to the moon with the high level of ambiguity and hidden motivations that constructivists attribute to the texts comprising the "Western canon," the Bible and other great mythopoeic texts of our civilization. Some postmodern critics have erred in treating scientific or veridical texts, which seek to describe nature and its functions unambiguously for a community of like-minded practitioners, as comparable to traditional works of art, poetry, and literature. Also, a postmodern philosopher abandoned in a wild landscape would find no smokescreen of language separating him from the water, fruit and berries which might sustain him. As human ecologist Paul Shepard writes: "As for 'truth,' 'origins,' or 'essentials' beyond the 'metanarratives,' the naturalist has a peculiar advantage--by attending to species which have no words and no text other than context and yet among whom there is an unspoken consensus about the contingency of life and real substructures."[13] The sense organs and body language of millions of species show them to be focused on something beyond themselves that sustains their existence. "To argue that because we interpose talk or pictures between us and this shared immanence, and that it therefore is meaningless, contradicts the testimony of life itself."[14] Humans live in the same world, interacting with the same processes, structures, and events as other organisms. Against this intractable reality of the planet and its evolutionary ecosystems, the nihilism of postmodernists appears itself fragile and evanescent.

Despite the growing commonsensical, historical, and ecological bases for refuting the arguments of postmodern deconstructionists, or perhaps because of their intersection, one brave literary critical theorist took it upon himself to write a thick volume, *Evolution and Literary Theory* (1995), designed to rebut and intellectually dismantle the community of postmodern deconstructionists once and for all. It was inevitable that the very tools of textual deconstruction would ultimately be applied to the deconstructionists themselves. The author, Joseph Carroll, exposes the sophistries lurking within the doctrines of Derrida, Foucault and others at length, but in the specific context of Darwinian evolutionary theory, which he recognizes as the only viable foundation for a modern theory of culture. Carroll argues that the cognitive and linguistic categories important to both constructivists and evolutionists have evolved in adaptive relation to the environment. "They correspond to the world not because they 'construct' the world in accordance with their own autonomous, internal principles but because their internal principles have evolved as a means of comprehending an actual world that exists independently of the categories."[15] Working from the foundation of evolutionary theory as the best or relatively "true" explanation of humanity's relationship to nature, Carroll asserts that this qualifies it as the necessary basis for any sufficient account of culture and literature. Thus, his test of any theory of culture and literature is its compatibility with the Darwinian evolutionary paradigm. If it cannot be reconciled with that paradigm it must be false. "The poststructuralist explanation of things cannot be reconciled with the Darwinian paradigm or modified and assimilated to it. Poststructuralism is an alternative, competing paradigm. It operates on principles that are radically incompatible with those of evolutionary theory. It should, consequently, be rejected"[16] Having made the case that the past several decades of work in critical theory have been a wrong turn leading to an academic dead end, Carroll nevertheless anticipates a powerful resistance to change on the part of literary critics who have committed so much energy to building the postmodernist ideology. It is his optimistic belief, however, that the same power of truth which overcame the Victorian resistance to Darwinian evolutionary theory will enable the intrusion of the Darwinian paradigm into the social sciences at the expense of postmodern theory within the next two decades. Carroll asserts that it will take longer for the Darwinian paradigm to establish itself in the humanities and in literary theory, however, because literary theory is less constrained by empirical findings than the social sciences, and also "partly because literary theory is the last refuge of mystical indeterminacy...."[17]

Carroll's rejection of deconstructive postmodernism is part of a growing counterattack against postmodernist critical theory since the mid-1990s. The implications for scientific "facts" and historical "events" such as the Nazi Holocaust deriving from postmodern deconstructionism has led to critical analysis of postmodernist thought by both scientists and historians. "The logic of their playful insistence that there were no certainties or realities, and their refusal to acknowledge the legitimacy of value-judgments, led to a free-floating relativism....."[18] Furthermore, the postmodernist critique of progress and the ideological foundations of modernity opened the door to religious revivalism amalgamated with nationalism and post-colonialism.

> Postmodernism is an enigmatic concept, whose very ambiguity reflects the confusion and uncertainty inherent in contemporary life. The term is applied in and to many diverse spheres of human life and activity. It is important for politics as it decisively reflects the end of belief in the Enlightenment project, the assumption of universal progress based on reason, and in the modern Promethean myth of humanity's mastery of its destiny and capacity for resolution of all its problems.[19]

Thus, not only has deconstructionist postmodern philosophy questioned the truth or validity of grand narratives such as the modern Western rationalization of progress, but it has done so within a cultural milieu of decreasing belief in the progressive myth, to which it has contributed in the academic world.

On the other hand, a more positive and useful aspect of postmodern philosophy is suggested by Oelschlaeger, who distinguishes deconstructive postmodernists from reconstructive postmodernists, the latter generally seeking to change the course of history by reworking existing social frameworks into new forms. The variants of postmodernism are built upon "the linguistic turn," the realm of interpretation explored by hermeneutical philosophers such as Martin Heidegger (1889-1976) and Hans-Georg Gadamer (1900-2002). Oelschlaeger traces "the linguistic turn" to changes in modern science a century ago, when the widely accepted modernist belief in the ability of science to produce knowledge of absolute certainty (apodictic knowledge) prevailed. Even today such a belief among scientists and their followers predominates. "So-called modernists typically believe that human reason--epitomized by modern science--is supreme, that it exists without limits, as it were, that the whole world lies open to disclosure by human intelligence."[20] Observing that the modern worldview has abandoned the pre-modern view of language that described nature as the words (i.e., works) of God, Oelschlaeger writes that "nature is no longer thought of as an expression of a divine semiosis but as nothing more than objectively described matter in motion."[21] The modernist understands true statements as "the mirror of nature," representing reality through scientific law and description, the appropriate form of objective knowledge.

Unfortunately for this comfortable model of modern, veridical knowledge (having succeeded pre-modern mythopoeic accounts of nature), the fabric of the mechanistic universe began to tear early in the twentieth century as quantum theory and the theory of indeterminacy transformed the Newtonian, static view of nature into one in which dynamic change and uncertainty prevailed. Once-solid atoms were deconstructed into yet smaller particles or packets of energy and their behavior seemed more random than determinate, suggesting "that science is incomprehensible apart from a culture that gives it meaning, purpose, a raison d'être."[22] Thus, even science, the modern cynosure of objectivity, was, after all, socially constructed, subject to the ambiguities of language and to the historical zeitgeists that infuse the language of science with different meanings during distinct historical periods. "To say that science itself is linguistically and historically constructed is not only permissible, but perhaps the only defensible position. A scientific account of the world is not more and no less than an explanation, proffered at a particular place and time that is judged by a particular community of researchers to be true." [23]

Intellectual historian Franklin L. Baumer writes: "After Einstein's paper on special relativity, it made no sense to speak of an absolute motionless space, or of absolute time. Einstein showed that space and time measurements varied with the motion of the observer...."[24] The early twentieth century physicists described a system of nature that could not be pictured and which was, from atom to cosmos, for all purposes ultimately mysterious and incomprehensible. Inscrutability had replaced Newton's tidy model of God's creation in the form of a mathematically calculable mechanical system. If the supposed absolutes of scientific knowledge (as apodictic) were enmeshed in linguistic differences and historicity, then what of the rest of human knowledge and truth? The hermeneutical philosophers attempted to answer this question. "If this situation is paradigmatic for all knowledge-claims, as hermeneutical thinkers are inclined to argue, then knowledge is not so much an ahistorical reconstruction of reality *sub specie aeternitatis* but a way of seeing things from the standpoint of a historically mediated set of concerns and preunderstandings, which is subject to inevitable change whenever our historically mediated standpoint shifts its focus."[25] Science offers no theory-free perspective on nature, but only varying perspectives from evolving theoretical frameworks embedded in historical settings.

One of the founders of postmodern hermeneutics is Martin Heidegger, who argued that our understanding of virtually everything is influenced by our own "type" or condition of historical existence, which he called "being-in-the-world." What Heidegger calls "preunderstanding" is an inherited "way of understanding that is so inextricably part of us that it constitutes our very being itself."[26] This definition of "preunderstanding" would also serve as a definition of "worldview" as I have applied it throughout this book. ("Worldview" is defined in Chapter Three as "the presuppositions of thought in a given historical epoch," etc.). Thus any "rational" individuals, including scientists, who are not aware of the "preunderstanding" or worldview which conditions their way of "being-in-the-world" and understanding it, are incapable of transcending their own historicity in order to grasp more precisely the meaning of "truth" in their own age. Such "truth" is always seen from a particular *perspective*, Heidegger insists. Hermeneutical philosopher Brice Wachterhauser writes: "If we are immersed in our preunderstanding, then a vicious kind of relativism must follow. These same preunderstandings must block us off from the possibility of grasping anything as it really is." [27]

Hans-Georg Gadamer has contributed to postmodern hermeneutics by further developing Heidegger's insight into human understanding as linguistically mediated. "Gadamer would say that it is only through language that we have a world. The world does not stand over against language as a pregiven intelligible entity in its own right. The world in its brute givenness is not intelligible at all."[28] Gadamer's controversial thesis that *all* understanding involves linguistically mediated prejudices is intended to challenge Enlightenment assumptions regarding the rational, autonomous self with the insight that all claims to knowledge, including those of science, are rooted in the context of a dialectical history. Regarding scientific practice: "Even practitioners of the same 'method' have need of a genuine exchange of viewpoints if they are to secure a firm grasp on the subject matter."[29] The latter statement is indicative of Gadamer's position that the natural sciences suffer less confusion due to their "preunderstandings" than the social sciences and humanities. Regarding the gap between scientific practice and the dissemination of scientific knowledge into the larger societal consciousness through education and the media, Gadamer writes:

> For the *natural* sciences, of course, this gap and the methodical alienation of research *are of less consequence than for the social sciences* (italics mine). The true natural scientist does not have to be told how very particular is the realm of knowledge of his science in relation to the whole of reality. He does not share in the deification of his science that the public would press upon him. All the more, however, the public (and the researcher who must go before the public) needs hermeneutical reflection on the presuppositions and limits of science."[30]

Gadamer's reflection on the need for scientific researchers to exercise caution in the course of conveying their findings to the general public seems far less radical than his earlier pronouncement that the world (nature) itself is unintelligible outside of a linguistic context. (On the other hand, the very fact that I regularly dream of nature, of mountains and wilderness, without reference to language suggests that the pre-linguistic interaction with nature experienced by pre-human hominids and other mammals involved direct knowledge of nature without language, or survival would not have been possible.) However, Gadamer's main point in the preceding quotation is that, properly contextualized in its historicity and linguistic preunderstandings, scientific knowledge is generally less vulnerable to error than that of the social sciences and humanities, in part due to the self-critical character of scientific method itself. Acknowledging that modern humanity has been liberated from many prejudices and disabused of many illusions by science, Gadamer nevertheless expresses his concern that scientific method has narrowed the range of questions asked about the nature of reality. "For only that which satisfies its own methods of discovering and testing truth has meaning for science. This uneasiness vis-à-vis science's claim to truth makes itself felt preeminently in religion, philosophy, and issues of worldview."[31]

Thus, Gadamer praises science for its myth-destroying methodology, while condemning it for narrowing our breadth of inquiry. But is "science" a monolithic mega-enterprise or a broadly diverse aggregate of disciplines holding in common little more than the inductive-deductive method introduced by Bacon and Descartes? Some sciences are laboratory-oriented, such as physics, chemistry, and more recently, biology. Others, such as geology and biology (especially in the 18th and 19th centuries) are predominantly field-oriented.   In terms of understanding nature, the extremes of nature, i.e., the realm of sub-atomic particles and energy studied by physicists on the one hand, and the cosmic realm of galaxies studied by astronomers and physicists on the other, have produced uncertainties which have challenged the "truths" of the mechanistic Newtonian worldview where method breaks down at the microcosmic and macrocosmic levels. In the "middle realm" of nature examined at hand and beneath our feet, geology and evolutionary biology have provided abundant evidence of past inorganic and organic processes, but the big questions of "What is Life?" or "What are the laws of evolution?" may be beyond our grasp as much as the nature of matter when reduced to sub-atomic particles and energy. We possess a detailed record of Earth's geologic and evolutionary history, but the ultimate meaning of these processes may not be accessible to the methods of science.

The limitations of science acknowledged by philosophers since Heisenberg are also related to the problem which arises when the scientific methodology of physics is transposed to the practice of biological science. The question "What is life?" may not be answerable within a scientific framework generated in order to explain the inanimate world investigated by physicists and astronomers. "The idea of a God-given, pervasive law that links the initial conditions (i.e., the state of the system) with its nature (e.g., its motion) has made mechanics the prototype of exact physical science to be emulated by all the sciences."[32] Are there, then, biological laws similar to those of physics or, alternatively, less quantifiable sets of rules whose guiding principle is natural selection? While the laws of physics generally are assumed to hold throughout the universe, the "laws of biology" if such exist, may be "either broad generalizations that describe rather elaborate chemical mechanisms that natural selection has evolved over billions of years or specific, ad hoc local instructions of a limited time scale."[33]

Despite its possible disjuncture with the methods of physics, evolutionary biological science, for all its uncertainties, has produced a modernist paradigm of science inadvertently containing a potential antidote to Francis Bacon's dream of dominion over nature (leading to the Christian Paradise restored). Baconian mechanistic science manifested an early modern preoccupation with astronomy and physics long before the establishment of modern geology and evolutionary biology. These scientific developments were completely unanticipated by mechanistic scientists whose science was permeated with Christian attitudes of human superiority and dominance over a static nature manifesting God's plan for "Lord Man." Gradually, the detachment of the geological and evolutionary biological scientific paradigms from the medieval providential framework (unlike the "Big Bang") led to the undermining of the Christian view of time to which Bacon and other mechanistic scientists adhered. This schism within the modern scientific enterprise detaches Western science from its medieval philosophical context; i.e., the relatively new geology and evolutionary science is implicitly in total contradiction with both the Baconian worldview and its philosophy of science while adhering to the Baconian methodology.

I contend, along with radical environmental philosophers, including "deep ecologists," that the new paradigms of science (involving a conception of extended natural time since the discoveries of Hutton and Darwin) have the potential to revitalize our conceptions of nature and reverse our destructiveness of nature, although the prospects for such a general transformation in worldview (metanoia) are not promising in a time of religious revivalism and retrenchment in the face of a world destabilized by the expansion of Western capitalism.

Postmodern philosophy as applied to nature has the potential to further weaken this transformation. Not even geology and evolutionary biology are exempt from the postmodern critique of claims of absolute or certain "objective" knowledge. However, the postmodern critique of truth and knowledge cannot be applied evenly to all knowledge because of the differences between mythopoeic and veridical knowledge (which is "true" enough to have put men on the moon), and the nature of scientific method, which is at least partly hermeneutically self-correcting in the long run. This process of self-correction is powerful enough to have initiated the geologically and biologically informed environmentalist critique of Bacon's mechanistic scientific paradigm.

Consequently, can postmodern critical philosophy do much more than reinforce the skepticism associated with the self-correcting activity of modern science? The majority of postmodern criticism is focused upon social science rather than natural science, for the reasons stated above. Perhaps contemporary mechanistic scientists (physicists, astronomers, chemists, and medical biologists) have not sufficiently imbibed the message of earth and evolutionary scientists working within the dynamic (rather than static) time framework, which implies that nature as a whole on planet Earth is perhaps better understood as a complex system of organisms and their physical support systems than as a machine. Geology and evolutionary biology have superseded the old, Baconian mechanistic sciences conditioned by the providential worldview, and in that sense are privileged over the mechanistic paradigm: they are free from viewing nature as a *timeless machine.* Without the dimension of cosmic and geologic time as a frame of reference, science is blind to the essential historicity of nature. A foundation for a new worldview of *becoming* rather than of *being* has been inferred from the facts and theories of modern geology, astronomy, and evolutionary biology. Isn't this gradual self-correction of the scientific understanding of nature and growing awareness of its historicity in conformance with the objectives of hermeneutical philosophy?

The practice of science today does not necessarily result in adherence to a belief in the certainty of scientific knowledge. The continuous testing and scrutiny of scientific research leads not to certainty but to novel insights into the workings of nature because "the system is fundamentally adversarial, and nature itself is the ultimate judge of who is correct."[34] But absolute certainty, i.e., apodictic knowledge, is nevertheless forever denied to scientists. Many scientific generalizations such as gravity, the laws of thermodynamics, and the role of natural selection in evolution have never been proven, but are supported by a body of evidence so convincing that scientists, for all purposes, assume them as "certain" or "true." This does not mean that these concepts cannot be overturned, but it would take a veritable scientific revolution to bring them down.

The factual basis for the theory of evolution by natural selection has gained greatly in validity since Darwin. As an evolutionary biologist recently wrote:

> Evolution by natural selection, the central concept of the life's work of Charles Darwin, is a theory. It's a theory about the origin of adaptation, complexity, and diversity among Earth's living creatures. If you are skeptical by nature, unfamiliar with the terminology of science, and unaware of the overwhelming evidence, you might even be tempted to say that it's "just" a theory. In the same sense, relativity as described by Albert Einstein is "just" a theory. The notion that Earth orbits around the sun rather than vice versa, offered by Copernicus in 1543, is a theory. Continental drift is a theory. The existence, structure, and dynamics of atoms? Atomic theory. [35]

The scientific evidence for evolutionary theory provides us with the best available knowledge about how nature, particularly organic nature, functions. Complexes of organisms and their surroundings, i.e., ecosystems, are the current manifestations of evolutionary processes, and are dynamic, historically constructed entities, as opposed to their perceived fixed condition in the older static, mechanistic view of organisms and physical systems. This is the best approximation to "truth" or certainty that we can have regarding nature. The difficulty comes when scientists or interpreters of scientific findings attempt to extrapolate from what is tentatively known about nature to something about humanity or humanity's relations to nature. We are informed from the start by Scottish philosopher David Hume that this is a bad idea which has been labeled the "naturalistic fallacy" or "is-ought dichotomy," namely that propositions based upon premises involving "fact" should not lead to conclusions regarding moral imperatives or "ought," i.e., that the terms of the argument should be commensurable.

Unfortunately, numerous scientists and environmental ethicists, formal and informal, have blundered into this realm of uncertainty, including such proselytizers of evolutionary ideas as Thomas Henry Huxley (1825-1895), Herbert Spencer (1820-1903), Julian Huxley (1887-1975), George Gaylord Simpson (1902-1984), and E.O. Wilson (b.1929). All of these men adhered to a faith in science that was tantamount to religious belief. In short, they were scientific positivists, and the positivist faith has generally led to a host of scientisms such as Freudian science and Marxian science. Confronted with such ancient questions as "What is nature?   What is man?   What is mind?," proselytizers of the Darwinian worldview have floundered under scrutiny by historians of ideas and philosophers. As intellectual historian John C. Greene writes, "from what we have seen of Darwinism as a world view it seems unlikely that evolutionary biology in and of itself will ever provide intelligible answers..."[36] to the great questions first raised by ancient Greek philosophers. Nevertheless, scientists can learn to carry on their quest for knowledge of nature in a humbler spirit by accepting that science cannot tell us about the ultimate intelligibility of things (if such a thing exists!). Also, the foundations of environmental ethics, of our perception of nature, of our assessment of its value, and of reconstructed moral guidelines in directing our behavior towards nature can be usefully informed by our knowledge of evolutionary biology and geology without conformance to a new ideology repeating the mistakes of earlier scientists. The idea of the general historicity of nature itself is generally missing from public debate and political decision-making. In hermeneutical terms, philosopher Marjorie Grene suggests the limiting linguistic and historical parameters within which evolutionary explanations of human behavior and morals should be framed:

> ...I am not now preposing to replace the principle of the primacy of historicity by a principle of the primacy of life. What we have or recognize is the place cleared within nature for the possibility of the human, that is, historical, or historicizing-historicized, nature. There have been recurrent attempts to understand human beings purely biologically, and, in particular, in terms of a theory of evolution. Whether in evolutionary ethics, evolutionary epistemology, or a functionalist theory of social organization, however, these are mistaken extrapolations from a theory of change in the genetic composition of populations to fields where some reference to human symbol systems is fundamental to the questions asked, let alone the answer given.[37]

Western Metanarratives (Philosophies of History) in the Twentieth Century

On the one hand, historians, including environmental historians, were surprised and intimidated by the onslaught of postmodern philosophy during the 1970s through 1990s; on the other, they have assimilated this challenge to their credibility and proceeded to write history with greater caution and precision than previously. Multicultural history and environmental history are thriving today, for the most part happily disassociated from the grand narratives known also as metahistory or speculative philosophy of history, which explains historical events on a grand scale as a manifestation of some architectonic principle which gives order to historical events. Leopold von Ranke established the antipathetic attitude of the traditional historical discipline toward these grand narratives when he opposed Hegel's grand historical vision--of history as the temporal realization of Spirit or the Absolute in the material world--in favor of temporally and geographically focused history solidly grounded in empirical facts. Thereafter, American "scientific" historians eschewed speculative philosophies of history, and when the metanarratives of Europeans such as Spengler and Toynbee arrived in America, they were scornfully rejected by the academic historical establishment. What traditional American historians failed to realize or admit was that "between the lines" of their own "scientific" histories they were writing, collectively, their own metanarrative of "progress," the dominant ideology bequeathed to the modern world by the Enlightenment. Postmodernists were quick to identify the implicit grand narrative of progress as an antagonist worthy of deconstruction, "progress as the relentless and also merciless march of reason through time toward a totally new stage of human existence."[38] They focused on the idea of progress articulated by the Marquis de Condorcet, in which the unfolding of reason led toward the complete emancipation of humanity. In their deconstruction of Western historical writing the postmodernists "regarded the very assertion of one encompassing and universal history as a grand and dangerous illusion and cited as proof the hegemonies, dominations, and tyrannies of the twentieth century. All grand conceptualizations of history must be rejected."[39] Avoidance of future grand narratives claiming universal validity and authoritative truth would prevent future hegemonies, oppressions, and tyrannies. Postmodernists were so focused on the metanarrative of progress that they essentially ignored the two other major metanarratives or philosophies of history of the West, cycles and the medieval Christian view of history, providence. "In their criticism and rejection of metanarratives, postmodernists have virtually ignored the cyclical paradigm. They would have found congenial its affirmation of the relativity and uniqueness of cultures in the aimless sequence-of-cultures pattern."[40]

The foregoing explanation of the interaction between postmodernist deconstructionism and Western metanarratives explains why, by the early 1970s, the study of speculative philosophy of history had all but died out in American colleges and universities. Even around mid-century anthropologists like Alfred

Kroeber and Ruth Benedict, and sociologists such as Pitirim Sorokin had moved into the academic territory abandoned by traditional American historians, who, from the perspective of a Spengler or Toynbee, could not see the forest for the trees. In 1965 intellectual historian Frank Manuel published *Shapes of Philosophical History* and in 1970 Jesuit historian John Edward Sullivan published *Prophets of the West: An Introduction to the Philosophy of History*, and afterwards, silence, except for works published by Warren W. Wagar, who actually turned to future studies due to lack of interest in the philosophy of history in America. We were content with our implicit philosophies of history underlying the "American Dream," providence and progress, and Karl Marx had given a particularly bad name to speculative philosophy of history, especially as viewed from a capitalist, Christian mass democracy. The idea of progress had been all but obliterated, however, long before its deconstruction by the angry postmodernists, by the shattering impact of two world wars and the Great Depression, the Nazi Holocaust, and Hiroshima. A hero was needed to set history back on the course of progress.

In 1963 such a historical prophet appeared with the publication of William McNeill's *The Rise of the West: A History of the Human Community*, a history so titled as to suggest that the cyclical philosophy of history in Spengler's *Decline of the West* was to be dethroned by a new progressive history appropriate to the difficult times of the twentieth century. The essential theme of *The Rise of the West* is that instead of separate, distinctive civilizations such as those portrayed by Spengler and Toynbee, world history is predominantly the history of cultural diffusion, from the Ancient Mediterranean riparian civilizations on down to the spread of Western civilization through global imperialism in recent centuries. The West is the latest, and for a long time to come, primary center of global cultural diffusion. The subtitle of *The Rise of the West* alludes to McNeill's intention, described in the book's conclusion, to give his history a purpose, namely to control the power of human civilization, or rather, of centers of power over rival centers, through development of an integrated cosmopolitan civilization which would end human warfare. McNeill perceived the power structures of the twentieth century to be hurtling towards a potentially catastrophic climax. "The globe is finite and if the rival political-social-economic power systems of our time coalesce under an overarching world sovereignty, the impetus now impelling men to develop new sources of power will largely cease."[41] Given McNeill's optimistic expectation that an empowered United Nations-like world government could come to power in the twenty-first or twenty-second century, he further prognosticates that "a stalwart, more than Chinese bureaucratic immobility would, in all probability, soon define the daily life of cosmopolitan world society."[42] This global society would also bear the imprint of the modern West. "This would be the case even if non-Westerners should happen to hold the supreme controls of world-wide political-military authority, for they could only do so by utilizing such originally Western traits as industrialism, science, and the public palliation of power through advocacy of one or other of the democratic political faiths."[43]

McNeill's vision of the future is not all that different from that of postmodernism or of Oswald Spengler in *The Decline of the West*. Postmodernist philosopher Jean Baudrillard envisioned "a basically stable world without hope for amelioration that must be accepted stoically or passively by people."[44] For Spengler, the triumphs of money and the machine point towards a spiritually and artistically bankrupt technological society.[45] Later in this chapter, however, we shall see that Spengler challenged his own philosophy of history and its implications for the future of humanity in his later, relatively unexamined writings.

William McNeill has never wavered in his hopes for a future global civilization free of the destructiveness of modern warfare. Delivering the presidential address for the American Historical Association's annual meeting in 1985, titled "Mythistory, or Truth, Myth, History, and Historians," he acknowledged that uncertainty enters historical interpretation due to the mediation of symbols between phenomena and the historian. Without paying homage to postmodernist critiques of historiography, he recognizes the inexact, relative nature of historical truth. Historical relativism was not a major problem until world civilizations came into closer contact during and after the Age of Exploration. Until then, relatively isolated cultures adhered to largely homogeneous worldviews. They were societies of true believers. McNeill observed that "the will to believe is as strong today as at any time in the past; and true believers nearly always wish to create a community of the faithful, so as to be able to live more comfortably, insulated from troublesome dissent."[46] True believers in the "scientific" history produced by American historians were shocked by the iconoclastic revisionists who initiated "multicultural" history, but the new school of historians lacked any architectonic vision of their own to take the place of the old progressive and democratic myths which ignored marginal races and classes. Throughout the early and middle twentieth century it was left to speculative philosophers of history and world historians to risk expounding some overarching principle of history from historical facts, such as the historical cycle or movement towards world government. In 1985 McNeill persisted in his belief in the latter, writing that "an intelligible world history might be expected to diminish the lethality of group encounters by cultivating a sense of individual identification with the triumphs and tribulations of humanity as a whole."[47] He lectured his fellow historians that it was a moral duty to develop such an ecumenical history, which would remind us not only of our membership in the human race, but also of our responsibilities for the larger community of life on planet Earth. McNeill also attempted to restore the value of world history in the minds of the majority of historians who wrote highly specialized, parochial histories. Looking back twenty years since McNeill introduced the term *mythistory* as "a useful instrument for piloting human groups in their encounters with one another and with the natural environment,"[48] it is clear that historians generally rejected this idea of a purposefully tendentious history. Nevertheless, a small coterie of twentieth century world historians have kept this broad approach to history alive, and, as one of them, McNeill has continued to have his say on the subject on into the twenty-first century.

World Historians, the Philosophy of History, and the Environment

World historians are, either explicitly or implicitly, also speculative philosophers of history or metahistorians. It seems to be almost impossible for world historians to describe the broad sweep of global history without drawing some conclusion regarding an intrinsic principle, direction, or goal in history. Consequently, all of Western speculative philosophy of history emphasizes patterns of cycles, providence, progress, and, as we shall see, also regress when we consider the fate of the natural world along with that of humanity. The most impressive attempt to discern and explain the speculative tendencies of world historians in recent years is Paul Costello's *World Historians and Their Goals: Twentieth-Century Answers to Modernism* (1993).[49] An intellectual historian, Costello surveys the historical writings of H.G. Wells, Oswald Spengler, Arnold Toynbee, Pitirim Sorokin, Christopher Dawson, Lewis Mumford, and William McNeill, from which he attempts to draw some conclusions regarding the usefulness and futuristic prognostications of the major twentieth century speculative world historians. He also credits several of them, notably H.G. Wells, Spengler, Toynbee, Mumford, and McNeill, for recognizing the onset of a global ecological crisis. I will briefly summarize the environmental insights of these five historians in the paragraphs that follow before explaining Costello's own opinions regarding the importance of environmental concerns for the discipline of history.

H.G. Wells, although not known primarily as a historian, published *The Outline of History* in 1920 with the hope of ameliorating the threat of humanity's increased power to destroy civilization through industrial, mechanized warfare. The West, to his mind having lost touch with the Bible as its founding text and moral preceptor, was in need of a modern common belief system, a new basis for progress through global unification. He thought that *The Outline of History* could fulfill this need. In later writings during the 1930s he deplored the effects of technological progress and unrestrained capitalism, including deforestation, desertification, and the destruction of rare and beautiful species. As Costello points out, Wells "was the first writer to consistently apply the science of organic evolution, ecology, and social Darwinism to a chronologically developed world history and as a prospectus on future development."[50] William McNeill regards Wells as a brilliant amateur whose *The Outline of History* was of great value in proselytizing the idea of a world state, concretely realized in the League of Nations and the United Nations.

Although Costello acknowledged the ecological insights of Oswald Spengler, which I have discussed at length in Chapter I, he, like most intellectual historians, was unable to discern the shift in Spengler's philosophy of history and its implications for environmental history in his later writings, many of them untranslated from the German. I will discuss this in the following section in relation to Spengler's *Man and Technics*. Arnold Toynbee took Spengler's model of the historical cycle and fused it with Christian teleology and the modern idea of progress. The degree of Christian providentialism to be found in Toynbee's historical interpretation is suggested by his reflection on the "goal" of human nature: "Its goal is to transcend the intellectual and moral limitations that its relativity imposes on it. Its intellectual goal is to see the Universe as it is in the sight of God, instead of seeing it with the distorted vision of one of God's self-centered creatures. Human Nature's moral goal is to make the self's will coincide with God's will."[51] It was teleological statements of this kind that turned most positivist, "scientific" American and European historians against Toynbee's "synoptic view of history," within which he claimed to have discovered laws directed by supernatural forces. After completing his twelve-volume *A Study of History* in 1961, he was still deeply concerned with the role of religion in civilizations, but he also addressed such contemporary issues as nuclear proliferation and global ecology. "In both these areas Toynbee saw crises of catastrophic dimensions in the making that could be avoided only through the spiritual transformation of individuals."[52]

Like Spengler, Toynbee was very much aware of the impact of nature on civilizations and their origins. Egyptian civilization was his classic paradigm of "challenge and response" because it developed in response to the challenge of post-Pleistocene, i.e., Holocene desiccation of the steppe of North Africa, and developed the large-scale environmental management projects which required concomitant large-scale social organization. Toward the end of his life Toynbee recognized the potentially devastating force of modern technology in modern history, consequently warning humanity, in *Mankind and Mother Earth*, of the human greed that threatened to destroy the biosphere. For him, as for H.G. Wells, the future "World State" had become an ecological imperative. Such an institution could manage world population through supervision of global food production and distribution. The three elements essential to the creation of a world government would include: constitutional integration of existing nation states into a cooperative world government, a functional compromise between capitalist and socialist economic systems, and the underpinning of the secular superstructure of society by a universal religious foundation.[53] Viewed from the twenty-first century, sadly, all three of these pre-requisites to international global government appear as hurdles impossible to overcome.

Like Toynbee, the maverick American intellectual Lewis Mumford in his later years envisioned a future facing catastrophe through technological warfare or destruction of the natural environment, to which he added the process of dehumanization resulting from environmental degeneration. Mumford characterized modern Western society as suffering from a "collective compulsive neurosis" which rationalized a Faustian bargain of abundant material power and goods in trade for the soul of Western humanity. As a generalist turned metahistorian, Mumford claimed: "My specialty is that of bringing the scattered specialisms together, to form an overall pattern that the expert, precisely because of his overconcentration on one small section of existence, fatally overlooks or deliberately ignores."[54] Focusing upon the role of the city in history, as well as the history of technics, particularly in Western civilization, Mumford had arrived at a sanguine expectation of a future "Biotechnic" age in his early works, an age when "new alloys, technologies of instantaneous communication, new technologies of light such as telescopes and microscopes, the industrialization of agriculture, conservation practices, birth control, collectivization, all would serve to humanize the machine, to integrate its positive benefits with human needs."[55] However, subsequently encountering Jacques Ellul's *The Technological Society* (1964), Mumford abandoned his optimism for the future of modern humanity. He was strongly influenced by Ellul's view that the rational pursuit of efficient technologies gradually transforms society in the image of those technologies. Mumford shifted his optimistic expectations of the future to that of "A world state of totalitarian efficiency and 'chromium gleam'where the media message will be a desensitizing monologue, where life would be dominated by mind-numbing therapies of adjustment and the individual would become a genetically engineered 'organization man,' a 'depersonalized servo-mechanism in the megamachine.'"[56] Mumford's antidote to this brave new technological society was resistance to the Faustian pact with technology in order to restore individual freedom and true community. His ideas were well received by the Environmental Movement, along with those of Theodore Roszak and other theorists of the counterculture.

The last of the metahistorians considered by Costello is William McNeill, whose ideas we have encountered in this and earlier chapters, and whose historical writing continues on to the present day. McNeill had the good fortune early in his career to study with and act as a teaching assistant for the great American historian Carl Becker, and to work for and form a friendship with Arnold Toynbee. By 1954 McNeill set to work on his own world history, in which he would reject the cyclical theories of Spengler and Toynbee. Proposing to turn Spengler and Toynbee on their heads, as Karl Marx claimed to have done with Hegel, McNeill believed that those two metahistorians had overstressed the independence of civilizations and their internal rhythms. Much influenced by recent work in anthropology, McNeill intended to focus on the diffusion of culture through the adoption by less developed cultures of the technologies, skills, and customs of more advanced societies. As he worked out his own theories, McNeill also rejected Toynbee's supernaturalism for a naturalistic agnosticism which explained history strictly in terms of natural processes. He "employs an overarching perspective on perpetual stimulus-response relations of human beings with disease fluctuations, agricultural technology, zootechnical advances, and resultant changes in food supplies, human habitat, and environmental control."[57] The role of epidemiology in world history was an important aspect of McNeill's diffusionism since, as in the case of the New World, discovery, conquest, and trade brought with them European diseases against which native populations had little to no natural resistance, resulting in massive die-offs in the Americas, Australia, and New Zealand, as described in Chapter Eight. McNeill developed the concept of macroparasitism by analogy with his epidemiological study of microparasitism. An invading population (macroparasite) allows the host population to survive and contribute to the colony or empire of the invading population through tribute or taxation. This macroparasitical relationship allows continuing health for the host population in contiguous societies while maximizing the spread of the invading population (i.e., macroparasitical infection). Macroparasitism was for McNeill the "hallmark of civilization," and European political history an unending fluctuation between imperial consolidation and feudal anarchy, "punctuated from time to time by epidemics of nomad invasions whenever the defenses of settled agricultural communities became insufficient to hold back armed raiders from the steppe."[58]

McNeill's views on religion are useful for understanding the contemporary religious revival in America and elsewhere. The axial religions which developed after the 6th c. B.C. were universalist, that is, they were open to all ethnic groups. McNeill asserts that, in addition to offering some form of heavenly compensation for individual suffering in the world, Christianity, Mahayana Buddhism, and Hinduism promoted an adaptive forbearance in the face of adversity while the first Eurasian ecumene occurred. He also thinks, like Voltaire, that modern society will require religious grounding in response to the rapid social change propelled by the Western tourbillon, our modern whirlwind of frenetic activity. He projects such religious revivalism into an uncertain and demoralizing future. The individualism promoted by the Enlightenment does not provide the needed solace provided by "participation in groups, where shared values and goals, cooperative behavior, and mutual aid can flourish."[59] Not at all religious himself, McNeill acknowledges the needs of *Homo religiosus* in any future society, as well as its pragmatic value in preparing individuals of different cultures for membership and participation in a world system of government. "McNeill has retained as his own the task of salvaging what can be rescued from Whig progressivism in an age whose central metaphor is ecological and whose social order is increasingly centralized and bureaucratic."[60] McNeill no longer finds it tenable to think of European history in terms of the rise of liberty. He believes strongly that, under Western leadership, humanity must develop a new mythistory, with ecological knowledge and globalism at its core, to make the complex modern world intelligible to the ordinary citizen. The social myths that we employ not only make sense of life, but also guide our actions to the point that they become self-validating (consider the rule of modern life by economic theory!).

McNeill also claims that there is a historical law of the "Conservation of Catastrophe," meaning that with every civilized adaptation to the challenges posed to humanity by the natural world, we raise the stakes of the potential to destroy one another or collapse from natural disasters such as drought, overpopulation, and disease. Modern life is further destabilized by the disappearance of traditional agrarian life and customs as the industrial global network has spread, leaving individuals bereft of moral and religious convictions in a stressful and uncertain world. McNeill is also apprehensive about the potential hazards of the modern disease pool in an overpopulated world. At the same time, he seems to be almost totally focused on *human* outcomes, with little concern expressed for the fate of other species on planet Earth. As Costello writes:

> In line with his perspective, and of course no less influenced by the dramatic requirements of our times, the ecological study of history seems likely to reign paramount in the efforts of world historical scholars into the new century. World studies of progressive patterns of environmental management, disease control, racial and ethnic integration, and the process of growth of sovereignties in history will continue to serve as ground for understanding as these processes are furthered, perhaps toward some of their logical conclusions. Gaps in McNeill's ecological paradigm, particularly in the systematic evaluation of land and water use, species loss, pollution, and consumptive waste, will presumably spark further world historical work in line with these areas as vital concerns for survival into the future.

Costello appears not to have been aware at the time of publication of *World Historians and Their Goals* in 1993 that his prognostication of the rapid development of "ecological" (environmental) history was already somewhat tardy, the journal of the American Society for Environmental History having been established twenty years before, and Clive Ponting's *A Green History of the World* having been published in 1991. What is important is that Costello had the foresight, despite the isolation of traditional academic historians from environmental concerns, to recognize the connections between world history, philosophy of history, and environmental history, a relationship which has been relatively ignored by environmental historians, although J. Donald Hughes's *An Environmental History of the World* and other works have, like Ponting's, sought to bridge the gap between world history and environmental history. However, the larger gap, or should I say "chasm," between environmental history and the speculative philosophy of history, the essential topic of this book, is what Costello has lucidly defined. Having summarized his views on H.G. Wells, Toynbee, Mumford, and McNeill, all of whom recognized more or less the inevitability of and need for a future world government in an interconnected modern world, as well as the increased potential for ecological Armageddon, I now turn to Spengler in the following pages to explain the transformation of his own philosophy of history after *The Decline of the West*, and its significance for the deeper meaning of global environmental history.

Spengler's Philosophy of History and the Destruction of Nature

From a Spenglerian point of view, the schizoid culture of the West lies at the roots of our environmental crisis. The medieval Christian stage of Western historical development gave rise to the *idea* of total control over nature. The rise of secular humanism in the modern stage of Western development paralleled the historical development of modern technics, the *tools* for the manipulation and destruction of nature. Thus, the control and domination of nature is the consistent theme of Faustian civilization even though the medieval and modern periods created the religious-secular schism of modernity. If Western, Faustian civilization has become the chosen vehicle out of all the world's great civilizations for perpetrating the *coup de grace* upon the natural world, then how can this be reconciled with the philosophy of history enunciated by Spengler in *The Decline of the West*, namely that the "Great Cultures" or individual civilizations have originated, flourished, decayed, and typically collapsed over the millennia, only to be replaced by new and future civilizations comparable to organisms in the pattern of their histories? The answer to this question is complex but accessible. It begins with the publication of *Man and Technics* in 1931 and is further illuminated by subsequent scholarly exegesis of Spengler's lesser known writings which have not been translated into English. What is most surprising, in short, is that during the period 1924-1936 Spengler developed a *new, or at least revised,* philosophy of history which subsumes the fundamental explanation of the meaning of history presented in the *Decline*.

Man and Technics is the revised and enlarged version of a lecture presented at the Deutsches Museum, a prestigious technological museum, in 1931. Upon first reading, most scholars, my self included, have long since written off this slim volume as of no great importance. However, scholar John Farrenkopf, in a remarkable work, *Prophet of Decline: Spengler on World History and Politics* (2001), has elucidated the deeper meaning and environmental implications of *Man and Technics* within the larger context of carefully researching Spengler's later writings. Farrenkopf's exegesis supersedes all earlier Spengler scholarship, particularly in regard to the meaning of Spengler's speculative philosophy of history for humanity's relationship with the environment. As he writes, a careful reading of *Man and Technics* "offers insight into the remarkable transformation of his panorama of a virtually eternal series of independent cultural cycles into a tragic, catastrophic vision of world history as a largely integrated process...."[61]

I would like to point out some of Spengler's salient references to mankind and the environment in *Man and Technics* prior to explaining Farrenkopf's insights into Spengler's generally ignored revisions to his philosophy of history. First of all, Spengler cites *Homo sapiens*, appropriately, at the apex of the genre of predatory mammals. Like Montaigne and Nietzsche, Spengler does not hold back from viewing predatory animals as superior to the herbivores and other animals. The acuity of sight and cleverness of the lion and leopard in seeking their prey has been developed to higher levels by humankind, resulting in the highest level of perception and intelligence in the animal kingdom. "The *world* is the prey, and in the last analysis it is owing to this fact that human culture has come into existence."[62] Rejecting the Rousseauian model of man as "a peaceful and virtuous creature until Culture came to ruin him...,"[63] Spengler sees our species as having, from the start, become accustomed to killing not only his prey, but also his human competitors, but with the unique advantage of the combination of large brain, hand, and tool having developed synergistically over time. "Every work of man is artificial, unnatural, from the lighting of a fire to the achievements that are specifically designated as 'artistic' in the high Cultures. *The privilege of creation has been wrested from Nature.*" In other words, Spengler saw creative man as having stepped outside of nature, and with every further act of creation becoming more and more her enemy. "*That* is his 'world-history,' the history of a steadily increasing, fateful rift between man's world and the universe--the history of a rebel that grows up to raise his hand against his mother."[64] This human tragedy inevitably grows out of the human struggle with nature, and the outcome is the defeat of every higher culture, of every civilization in the long run. Although this denouement is consistent with the Spenglerian cultural cycle described in The *Decline* (as well as with the recent discoveries of environmental history and other disciplines) , in *Man and Technics* it is the result of the universal struggle between humankind and nature.

With the undertaking of civilization and the initiation of "enterprise" (collective action under a plan) in the fifth millennium B.C., the human intellect was emancipated by speech from simple action to calculation. "Man, the preying animal, insists *consciously* on increasing his superiority far beyond the limits of his bodily powers."[65] Thereafter, sophisticated, civilized men would become the slaves of their thought, above all, manipulated by the ideas of building structures and breeding animals and plants. "Now, this verbally managed enterprise involves an immense loss of freedom--the old freedom of the beast of prey--*for the leader and the led alike.*" Out of this shift from organic to organized existence springs warfare "as an enterprise of tribe against tribe, with leaders and followers, with organized marches, surprises, and actions. Out of the annihilation of the vanquished springs the *law* that is imposed upon the vanquished."[66] Following Spengler's reflections upon the origins of civilized society, recall the natural and human tragedies of Alfred Crosby's *Ecological Imperialism* and Ophul's "four great ills of civilization," and one begins to grasp the trajectory of his argument. All history, old and new, is the history of war! The internal order of tribes and states is designed for struggle with similar groups of humans. Certainly, the history of the Maori in New Zealand supports his idea. Standing Thomas Hobbes on his head, Spengler is suggesting that societies are designed for a struggle of all against all rather than to protect the individual from a pre-civilized struggle between savage men.

... history, of old as now, is war-history. Politics is only a temporary substitute for war that uses more intellectual weapons. And the male part of a community is originally synonymous with its *host*. The character of the free beast of prey passes over, in its essential features, from the individual to the organized people, the animal with one soul and many hands. The technics of government, war, and diplomacy have all this same root and have in all ages a profound inward relationship with each other.[67]

Several prominent negative consequences follow from the passage to civilization. First, civilized man, having wrested from nature the privilege of creation, becomes the prisoner of the civilization he has created, that is, of the aggregate of artificial, self-made institutions and practices. Secondly, these accomplishments enable the expansion of human populations to the point that there are scarcely any regions on the globe free of humans. With people bordering upon people, the very idea of the frontier arouses the old instincts of predatory man to hatred and the (often repressed) desire to annihilate his alien neighbors. At the heart of the civilized cultures that press cheek to jowl across the landscape resides the artificial stone city, the citadel of "artificial living, that has become divorced from mother earth and is *completely* anti-natural--the city of rootless thought, that draws the streams of life from the land and uses them up into itself."[68] Of all these civilized cultures, Spengler recognizes the Faustian, Western European as the most tragic of them all because it is in our own culture "that the struggle between Nature and the Man whose historic destiny has made him pit himself against her is to all intents and purposes ended."[69] Certainly Spengler suggested this at the end of the *Decline*, but he places the Faustian destruction of nature in a more pessimistic perspective in *Man and Technics*. Reaching back to the fascination of medieval monks with the idea of perpetual motion, Faustian man seeks to create a small world of his own creation moving like the great world but subject to the hand of man alone. "To build a world *oneself*, to be *oneself* God--that is the Faustian inventor's dream, and from it has sprung all our designing and re-designing of machines to approximate as nearly as possible to the unattainable limit of perpetual motion."[70] The natural world itself, complete with its secret of force, is to be dragged away like prey by the ultimate predatory beast to be built into our culture. Spengler finds something devilish or demonic in all this, especially since men, at least prior to modernity, generally have regarded machines as the invention of the devil.

Writing a decade after publication of *The Decline of the West*, Spengler claimed to recognize the prelude to the culmination of the Faustian environmental tragedy, but also of civilized humanity as a whole:

Every high Culture *is* a tragedy. The history of mankind *as a whole* is tragic. But the sacrilege and the catastrophe of the Faustian are greater than all others, greater than anything Aeschylus or

Shakespeare ever imagined. The creature is rising up against its creator. As once the microcosm Man against Nature, so now the microcosm Machine is revolting against Nordic man. The lord of the World is becoming the slave of the Machine, which is forcing him--forcing us all, whether we are aware of it or not--to follow its course. The victor, crashed, is dragged to death by the team.[71]

Spengler recognized this race towards disaster to be under the leadership of a group of "Nordic" nations, including the British, Germans, French and, above all, the Americans, whose industrial strength and political power enable the rapid march towards Armageddon. The mechanization of the world has developed so suddenly that: "In a few decades most of the great forests have gone, to be turned into news-print, and climatic changes have been thereby set afoot which imperil the land-economy of whole populations. Innumerable animal species have been extinguished, or nearly so, like the bison...."[72] He concludes *Man and Technics* with the observation that machine-technics will end with the collapse of Faustian and global civilization, and that one day our railroads and skyscrapers will lie as dead as Roman roads and the Great Wall of China.[73]

Farrenkopf's assessment of Spengler's forecast of environmental apocalypse at the hand of the machine, i.e., of the mechanized world developing at a destructive pace beyond human control, takes his pessimism very seriously, including the idea that the purpose of Western man's scientific and technological activity is not to further the happiness of the greatest number, but rather "to satisfy the spiritual longing of Faustian man to conquer the infinite, to dethrone nature, and to elevate himself as a deity above its exploited, prostrate form."[74] For Farrenkopf, Spengler's second philosophy of world history provides an appropriate and useful frame of reference for grasping the enormity of the contemporary global ecological crisis. He also thinks that Spengler deserves credit as a prophet for recognizing the seriousness of this developing crisis at the end of *The Decline of the West*, a judgment with which I implicitly concur in Chapter One. Within this recently reinforced Spenglerian perspective the "problem of development" described in the preceding chapter takes on a far deeper significance than that understood by environmentalists, ecologists, environmental historians, and environmental philosophers down to the present time. As Farrenkopf writes, "Spengler's achievement was twofold. First, he sensed the gravity of the threat to the global environment posed by industrialization. Second, he grasped the centrality of the struggle between man and nature in all of world history, not merely in modern Western civilization, though it is here that this struggle has climaxed."[75]

Not all Spengler scholars would fully agree with Farrenkopf's interpretation that Spengler intended a paradigm shift in his later, "second philosophy of history." Klaus P. Fischer, for example, asserts that Spengler did not substantially change his overall philosophy of history. As Fischer observes, "Spengler assumed as a point of departure the metaphysical truth of the life cycle and its applicability to human cultures. He did see Faustian civilization as the most developed and the most predatory of all the civilizations to date. Under the darkening shadow of Nazism and the coming of another world war, which he predicted, he may well have surrendered to the worst kind of pessimism about the end of civilization as such." Nevertheless, Fischer believes that Spengler's view of decline was intended as a metaphor for "fading out," while at the same time leading to the birth, growth, development, and decline of the next culture. I agree with Fischer that this is the case, but with the emphasis that the next culture or cultures will arise in the context of a thoroughly humanized, ecologically moribund planet unless a profound revolution in global attitudes toward nature occurs (i.e., ecological metanoia). Regarding the veracity of Spengler's tragic speculative philosophy of history, Farrenkopf is neither glib nor naive. He carefully weighs the alternative perspectives regarding modern industrial-technological society: (1) the optimistic view that modern science and technology can be refined and utilized to render industrial civilization compatible with preservation and conservation of global ecosystems; and (2) Spengler's pessimistic outlook that long-term modification of industrial civilization cannot avoid ecological crisis due to the synergistic interaction of that crisis with the global population explosion and its attendant socioeconomic and political stresses. "The ongoing population explosion in the developing world, combined with the imperative to promote robust economic growth, increases the consumption of raw materials and energy resources, placing additional severe burdens upon the global environment."[76] Farrenkopf believes that, from the Spenglerian perspective, citizens of the developed countries of the northern hemisphere will strongly resist adjusting their life-styles to prevent ecological catastrophe. I tend to agree with Farrenkopf's conclusion, and much of the remainder of this chapter will be devoted to assessing this pessimistic, Spenglerian view of the future against the widely-held optimistic perspective held by the European and American elite and propagandized through the electronic media. However, whatever judgments that any of us might make regarding the relative validity of any given philosophy of history, including the entrenched Western beliefs in providence and progress, they are not likely to affect the established dominant worldview to any great extent, if at all. Worldviews are powerful and subliminal, and if they lead a culture to the edge of the abyss, even to environmental catastrophe, they are not likely to change over short periods of time.

During the past several decades, speculative philosophy of history and associated scholarship have disappeared concurrently with the rise of multicultural and environmental history, as well as of postmodern criticism of grand narratives. Certainly, the ultimate grand narratives have been the philosophies of providence, progress, and neoclassical cycles. Of these historical perspectives it was the cycles of Spengler, Toynbee, Kroeber, Sorokin, and others which attracted attention to the theme of the rise and fall of civilizations. Their various explanations of the collapse and end of civilizations, however, were generally unsatisfying to traditional historians as well as to postmodern critics.

Environmental historians such as J. Donald Hughes and Clive Ponting have partly filled the void on the subject of civilizational decline and collapse, but even their explanations are incomplete in the mind of Joseph A. Tainter, whose *The Collapse of Complex Societies* (1988) attempts to solve the mystery of the collapse of sophisticated civilizations in terms of multiple causation that includes such factors as resource shortages, but emphasizes an economic explanation of collapse as intrinsic to the process of civilizational development. According to Tainter four key concepts lead to the understanding of collapse. These include: (1) that human societies are problem-solving organizations; (2) sociopolitical systems require energy for their maintenance; (3) increased complexity carries with it increased costs per capita; and (4) investment in sociopolitical complexity as a problem-solving response often reaches a point of declining marginal returns. It has been shown that modern industrial societies are now experiencing declining marginal returns for increased expenditures in such areas as agricultural and resource production because "rationally acting human populations first make use of sources of nutrition, energy, and raw materials that are easiest to acquire, extract, process, and distribute. When such resources are no longer sufficient, exploitation shifts to ones that are costlier to acquire, extract, process, and distribute, while yielding no higher returns."[77] Tainter's argument for ultimate economic decline leading to collapse is as follows:

> Sociopolitical organizations constantly encounter problems that require increased investment merely to preserve the status quo. This investment comes in such forms as increasing size of bureaucracies, increasing specialization of bureaucracies, cumulative organizational solutions, increasing costs of legitimizing activities, and increasing costs of internal control and external defense. All of these must be borne by levying greater costs on the support population, often to no increased advantage. As the number and costliness of organizational investments increases, the proportion of a society's budget for investment in future economic growth must decline.[78]

Tainter's study of the causes of collapse of complex civilizations, much like Costello's *World Historians and Their Goals*, has partly filled the gap left by the demise of metahistory (speculative philosophy of history). Although professionally an archeologist, Tainter approaches the problem of collapse from an interdisciplinary perspective which includes history, politics, economics, environmental concerns, and archaeology. He inquires: "How could flourishing civilizations have existed in what are now such devastated circumstances? Did the people degrade their environment, did the climate change, or did civil conflict lead to collapse? Did foreign invaders put these cities to an end? Or is there some mysterious, internal dynamic to the rise and fall of civilizations?"[79] It is the latter explanation, of the kind proposed by Spengler and other speculative philosophers of history, that Tainter wishes to dispel by approaching the problem of collapse more scientifically and objectively, and less intuitively. In the process he tends to support the notion of multiple causation, with resource shortages and environmental degradation playing an important role, much as I have argued in Chapter Two, especially, but also in later chapters. Tainter eschew's Spengler for his use of the organic analogy in describing the rise, flourishing, decline, and collapse of civilizations, and Toynbee for his supernaturalism. Surprisingly, given the general thoroughness and caution of Tainter's analysis of the causes of civilizational collapse, it is clear that he misapprehends Spengler's conception of "civilization," described in my Chapters One and Two. Higher cultures possessing cities, writing, architecture, etc. are generally referred to as "civilizations." Spengler preferred to use "Great Cultures" for the major civilizations and to use "civilization" more specifically to refer to the later, petrified and uncreative winter stage of the Great Cultures. Noting that Toynbee thought of civilizations as "progressive movements," Tainter remarks that Spengler thought that "civilizations are undesirable, even evil," quoting Spengler: "They are a conclusion ... death following life, rigidity following expansion ... They are an end, irrevocable, yet by inward necessity reached again and again."[80] Spengler meant by this statement that the Great Cultures (i.e., civilizations in the typical use of the term) *became* "civilizations," by which he meant moribund civilizations.

Despite this obfuscation, Tainter's study is highly pertinent to my own interest in Spengler seen through the lens of environmental history. Tainter views collapse as a *political* process: "A society has collapsed when it displays a rapid, significant loss of an established level of sociopolitical complexity."[81] In seeking the cause or explanation of collapse, Tainter acknowledges that the gradual deterioration or depletion of a resource base such as agricultural soil or minerals is important. Nevertheless, he is ambivalent about such explanations, arguing that the major factor in cases of depletion is not the depletion of a resource but how respective societies respond to the problem. If they neglect the problem, collapse likely follows, but if they respond to the challenge the society may be spurred to renewed vigor and economic development, as in the case of the post-medieval deforestation of England leading to the use of coal, which contributed to the Industrial Revolution. On the other hand, Tainter admits that certain environmental fluctuations or deterioration may not be amenable to solutions forthcoming from existing production systems and social arrangements. To demonstrate that such situations could cause collapse would require collection of data concerning "climate, population, crop or other resource yields, yearly requirements of the population and of the sociopolitical system, *and the adaptive capabilities of the society in question. Such data have not been systematically sought in the study of collapse.*"[82] (Italics mine.)

Tainter was well aware of global warming and other environmental problems, and of the Environmental Movement as well, when he published his scholarly study of collapse, and, more recently, Jared Diamond's *Collapse* (2005) has emphasized environmental factors as causes of the collapse of civilizations. The environmental parameters that Tainter describes in the above quotation have been studied at length for decades precisely with regard to the fear of collapse of Western civilization due to overpopulation, ecological devastation, and resource shortages. *World Watch* and other publications are exploding with encyclopedic data describing overpopulation and the destruction of global ecosystems. And yet the public listens but does not really assimilate the bad news, or the concerned minority is politically impotent against the power and rigidity of the established corporate capitalist order. The foregoing quotation also mentions our lack of knowledge concerning "the adaptive capabilities of the society in question," which is exactly what this book has been seeking to understand, from our Judeo-Christian roots, through our Enlightenment origins of modernity, to the Christian-corporate capitalist structure of contemporary Western civilization. Corporate capitalism is the dominant factor determining our way of life, whether or not individual nations have become predominantly secular or have strongly maintained the Christian belief system, as in Poland, Ireland, and the United States. The latter combination of Christianity, particularly in its fundamentalist variants, and pervasive corporate capitalism probably forms the greatest obstacle to the possibility of radical change towards solving the crisis of overpopulation, ecological collapse, and increasing resource scarcity. In the remainder of this chapter I will first reassess the severity of our environmental crisis, and then consider the problem of unsustainable economic development in the light of the contemporary Western worldview. I am interested in documenting what has happened during our recent history rather than adding to the hundreds of books that have proffered solutions to the environmental crisis, virtually all of which have been ignored by the vast majority of the American public and its political establishment. I believe that, at this point in the ongoing crisis, it is more important to understand why we are incapable of meaningful reform than to add to the list of needed reforms. The American and other Western developed nations desperately need to look at themselves in the psychological mirror reflecting the rationalizations and ideals, i.e., the collective worldview which allows us to practice blatantly maladaptive, not to mention immoral, behavior in the face of an ecocatastrophe of our own making which bears down upon Earth with exponentially increasing speed.

Evaluating the Ecological Crisis

Do I exaggerate the extremity of our situation? Not at all. Peter Ward, Professor of Geological Sciences and Zoology, and curator of paleontology at the University of Washington, views the recent and ongoing rapid extinction and near extinction of mammals and other species at the hands of humankind as comparable to the two greatest mass extinctions recorded in the geological record approximately 250 million years ago and 65 million years ago, both events having been caused or compounded by devastating asteroid collisions (see Chapter Nine for a detailed discussion). We have already annihilated a large proportion of the magnificent array of mammals which evolved following the late Cretaceous mass extinction of 65 million years ago, when dinosaurs were wiped off the face of the planet, and once ubiquitous ammonites were eradicated from Earth's oceans. As population growth and economic development increase during the twenty-first century the work of the "human asteroid collision" will near completion, and evolution as we know it will come to an end, those regions untouched by humans and their machines nevertheless thrown off the course of past evolutionary processes by the ubiquitous effects of global warming, ozone depletion, and other worldwide climatic changes.

From the point of view of an evolutionary biologist, ecologist, or environmentalist this exhaustion of Earth's species appears to be a terrible disaster. From the perspective of modern man completing the task of displacing nature with a totally humanized and mechanized environment, humans have arrived at a status approaching godhood, manipulating the entire planet to the greater good of our species, in both numbers and improved quality of life. Subliminally, the ideal of a technological utopia drives us in our daily lives even as we pay homage to the wonder of nature and its disappearing species on the electronic screen.

Peter Ward suggests that the greatest hope for preservation of the world's animal and plant species depends upon "the expanding realization of how real the danger of mass extinction is."[83] He had high hopes for the ability of the 1992 environmental summit held in Rio de Janeiro to effect constructive changes. Unfortunately, as we saw in Chapter Eleven, those hopes were dashed, and ecologically unsound global economic development has jumped over any similar barriers erected to slow and redirect the existing path of development. I am convinced by the failure of recent attempts to manage growth towards sustainable practices that the present goal of environmentalism should be to open up a dialogue that focuses upon our attitudes toward nature in a last-ditch attempt to convert a large enough majority of voters to support environmentally responsible candidates for office at all levels of government. Such a process is already well underway in some European nations, who are already years ahead of the United States in energy policies and practices alone. Unhappily, the American Christian-capitalist paradigm is so deeply entrenched that only literal collapse of key resources, and continued environmental disasters may create the glimmering of self-awareness necessary to propel significant political action. Meanwhile, global warming, ecosystem destruction, and galloping resource consumption and attendant pollution of the biosphere spin off of technologically enabled unsustainable global economic development, leaving us with the neglected and perhaps unanswerable question of how might it be possible to capture the attention, interest, and commitment of a large enough proportion of humanity to enable political leadership and collective action to deflect our present path towards global destruction, certainly for many species and ecosystems, but perhaps less so for humanity itself, although terrible crises for humans can be expected as well.

Even optimistic "old warriors" of the Environmental Movement such as James Speth have become increasingly apprehensive that modernity and its present mode of economic development are leading us to the brink of disaster. Speth is presently a dean and professor of environmental policy at the Yale School of Forestry and Environmental Studies, having served over many years as founder and president of the World Resources Institute, co-founder of the Natural Resources Defense Council, environmental adviser to Presidents Carter and Clinton, and CEO of the United Nations Development Programme. His recent book, *Red Sky at Morning: America and the Crisis of the Global Environment* (2004) is once again, like hundreds of earlier books by environmentalists, a grave warning combined with a citizen's agenda for action to avert ecological and socioeconomic disaster in the decades ahead. His analysis of our failed attempts to shift towards sustainable development supports my own interpretation in Chapter Twelve of this book. Regarding the failure of the Rio objectives, he agrees that the sustainable paradigm has never been given a chance to be implemented. Instead, the established paradigm of globalization has been driven relentlessly by the global corporations of the industrialized north, under the misguided leadership of the United States, Europe, and Japan. A passage in *Red Sky at Morning* that leads us back to the underlying, inertial resistance to environmental solutions is not unlike others that I have been reading for decades: "The most fundamental transition is the transition in culture and consciousness. The change that is needed can be best put as follows: in the twentieth century we were from Mars but in the twenty-first century we must be from Venus--caring, nurturing, and sustaining."[84]

"A transition in culture and consciousness" would be nothing less than a change in worldview comparable to the shift in the Western worldview which has occurred over the past several centuries, except that the contemporary American worldview is a compound of European providentialism and the Euro-American progressivism that grew out of the Enlightenment. It is a contradictory worldview which divides religious true believers from secular humanists, and provincial, utilitarian exploiters of nature from urban worshipers of wilderness. This dichotomous worldview functions, along with corporate capitalism and the worship of technology, as a major impediment to a "change in consciousness" (metanoia) on the part of Americans and, to a lesser extent, Europeans.

The contemporary modern Western worldview that dictates so much of our socioeconomic behavior and political decision-making is constructed around two outmoded myths linking the Christian medieval heritage and the capitalist heritage of the past two hundred and fifty years. Both of these myths have been severely damaged, if not intellectually destroyed, by the accumulation of scientific knowledge. The idea of providence and its creationist view of reality has been thoroughly discredited by geological and evolutionary fact and theory. Secular humanists have acknowledged the scientific defrocking of Christian "truth" to expose it myths and absurdities simultaneously with the collapse of Catholic orthodoxy from within, as its simplistic view of human nature produces legions of twisted, broken practitioners. True it is, however, that, given the model of postlapsarian man created in the image of God, the embellishment of these ideas by such thinkers as Saint Augustine and Saint Thomas Aquinas has produced some of the greatest works of genius in the Western tradition. The idea of progress, battered by the wars and genocide of the twentieth century, persists on the basis of technopoly, the rule of society by uncontrolled and yet venerated technological change. And yet it too has been discredited by the historical evidence, the record of dead and moribund civilizations having extended their power over nature to the point of unsustainability and diminution or collapse. The providential and progressive myths are thoroughly discredited, and yet they persist as the most powerful ideals in Western civilization. Why?

Mythopoeic traditional knowledge is organized around a religious core in most complex societies prior to modernity. Religion was usually the dominant organizing principle of society. However, with the arrival of Western science upon the world stage, and the creation of secular, democratic societies, religion has become an enormous liability and obstacle to the task of maintaining and preserving the ecosystemic foundations of global civilizations.

Religion is hardly about to dry up and blow away in a postmodern world. There will be some form of religion so long as *Homo sapiens* exists, for it is a universal propensity of humankind to establish some kind of belief system that attempts to explain life in this world while, in most religions, offering some hope of continued individual existence in an afterlife, in a transcendent, supernatural realm. The Darwinian evolutionary biologist Richard Dawkins raises the question, "What use is religion?" Despite the millions of true believers who have been slaughtered or tortured over the millennia, often "for loyalty to one religion against a scarcely distinguishable alternative",[85] religion has proved its universality and its durability, but no one knows exactly why. Dawkins considers and rejects the explanation that religion is a medicine-like placebo which reduces stress, for it just as frequently seems to increase rather than decrease it. He then considers the controversial possibility that Christianity and other religions have survived by some obscure form of group-selection because they have fostered ideas of in-group loyalty and brotherly love. What was selected in our ancestors, perhaps, was not religion *per se* but some as yet unspecified psychological characteristic or genetic variation for individual survival within the larger religious group. Such an argument allows for the co-existence of true believers with non-religious, secular individuals not blessed with a genetic predilection for religious belief.

Subsuming the human need for religion is the broader set of social beliefs, values and institutions known as "traditional." Pre-modern traditional societies almost always included a powerful religious tradition as a subset and foundation for the broader parameters of traditional cultures. The rise of modernity has threatened both the old traditional ways and the mythopoeic religious explanations of the world, including nature itself, as in the creation myths. The present culture wars raging around the globe have been set in motion generally by Western colonization, imperialism, and modernization. Simultaneously, *modern* traditions, including individualism, secular humanism, and mass democracy, have transformed the worldviews of large numbers of individuals, varying in particular nations and regions, away from the older traditional values. This ferment of cultural transformation, the modern tourbillon described in Chapter Nine, has left little room for concern and deliberation about how we treat the environment, either in traditional societies or under conditions of modern economic development and destructive bulldozer technologies. Only a small minority of the population, including scientists, politicians, environmentalists, and developers, are regularly engaged in environmental controversies while the general public of the world's nations remain largely disengaged, as well as screened off from reality by adherence to traditional beliefs, including antiquated creationist and related religious beliefs which obscure our perception of the natural world and human overexploitation of it. For most people in our "global society" all appears panglossianly well with the world while modernization, alias global economic development driven by the wealthy nations of the West and Japan, marches on.

Christianity and Environmentalism: How Compatible Are They?

Given its half-modern, schizoid condition, the United States, in religious commitment, lies halfway between European secularism and Muslim fundamentalism in the intensity of its Christian belief. The United States is a social powder keg of opposites waiting to explode under conditions of environmental and social stress. Traditional Christian religion very likely will not be an important part of a postmodern solution to the environmental crisis, despite the rise of "evangelical environmentalism" in recent years. Although global warming, resource scarcity, and other environmental issues have captured the attention of evangelicals and other fundamentalists, complex ecological problems cannot be solved from a creationist perspective. Only a general knowledge of ecology understood within its evolutionary framework can lead towards lasting solutions to the environmental crisis which confronts humanity and global ecosystems. Further American retreat into fundamentalism is more likely to lead to further empire-building, war, chaos, and massive ecological degradation. A scientifically ignorant American populace will be more prone to political manipulation by conservative ideologues. The failure of secular humanism (the ultimate European legacy of the Enlightenment) in the United States has already allowed a drift towards theocracy in the view of some social critics.[86]

The claims of "evangelical environmentalists" to have learned enough about the environmental crisis to be able to offer solutions are misleading. According to Kevin Phillips, approximately 30 to 40 percent of Republicans are preoccupied with biblical prophecy and the expectation of a coming Christian millennium. Thus, the concept of Christian "stewardship" is delimited for radical

Christian believers to the "end times" leading to Armageddon, apocalypse, and the last judgment. As evangelicals have gained in importance, more corporations have begun hiring Washington lobbyists with Christian worldviews. Republican administrations have done the same, particularly "within the sections of the federal government that regulate the environment, mining, oil and petrochemicals, ranching, and logging --- in short, the principal units charged with resources stewardship (the Environmental Protection Agency and the departments of the Interior and Energy)."[87] From James Watt in the Reagan administration to Gail Norton under George W. Bush, it is the same old story of evangelical permissiveness toward regulated industries.

Phillips also points out that the Interfaith Council for Environmental Stewardship formed under the leadership of such evangelicals as Richard Land of the Southern Baptist Convention requires "sound theology" for humans to be good stewards of God's creation. Evangelical environmentalists note, under "sound theology", that the creation was designed for human use, consequently emphasizing economic development and strong property rights as central to stewardship over nature.[88] It is clear that evangelical environmentalists understand "stewardship" as the implementation of human dominion over nature or "wise use" while being recognized by the conservative media and voters as "green". If, as polls suggest, close to a majority of those who voted for George W. Bush in the elections of 2000 and 2004 believe the Bible to be literally true, the conflation of dominion over nature with stewardship is understandable. However, as Phillips points out, the consequences are deplorable. "Their biblically-viewed world is at most ten thousand years old, not the millions of years established by scientists, whose insistence on this longer time frame is said to usurp God's prerogative. In considering stem-cell research or Iraq-as-Babylon, depleting oil or melting polar ice caps, the thought processes of such true believers have at best limited openness to any national secular dialogue." [89]

Contemporary American religiosity, with its powerful core of evangelical and fundamentalist belief, has evolved over several centuries. Only 17 percent of Americans stated some religious preference in 1776, which rose to 34 percent in 1850, 45 percent in 1890, 56 percent in 1926, 62 percent in 1980, and 63 percent by 2000.[90] The proselytizing, missionary nature of American religion, particularly the biblical literalists who repeatedly broke away from mainstream churches simultaneously with westward expansion, has produced a society in which secular humanism and scientific knowledge are continually challenged by true believers. Control of all three branches of American democratic government by the Republican party's religious right has intensified the creationist-evolution controversy, and, as Kevin Phillips has observed, created the potential for military and energy crises. "Never before has a U.S. political coalition been so dominated by an array of outsider religious denominations caught up in biblical morality, distrust of science, and a global imperative of political and religious evangelicalism. These groups may represent only a quarter to a third of the U.S. population, but they are mobilized...." [91]

The United States has gradually drifted to the political right in recent years, simultaneously with the rise of evangelical Protestantism, resulting in the presidency of George W. Bush, who actually claims to receive guidance directly from God as a basis for his foreign policy and other decisions. Perhaps it is not remarkable that American citizens have not exploded with outrage at this quasi-theocratic situation because the revival of evangelical Christianity is so strong. Certainly the September 11, 2001 attacks sent lukewarm Christians in search of deeper spiritual meaning or some guarantee of individual salvation, thereby contributing to the religious revival already underway for decades. During the 2004 campaign for the presidency, *Time* perceived it to be developing into "the most religiously infused political campaign in modern history."[92] Commenting on our divinity-divided nation, the author noted that voters identifying themselves as "very religious" supported President Bush over Democrat John Kerry by a margin of 59% to 35%. Those who considered themselves "not religious" favored Kerry 69% to 22%. Of Democrats asked if a President should be guided by his faith in making policy, 70% replied no. Republicans replied 63% yes.[93] Thus, contemporary religious faith plays an enormous role in determining our leadership and governance, and since American fundamentalists have unsuccessfully launched candidates such as Pat Robertson and Pat Buchanan, they have sought to advance their moral and political goals chiefly through the Republican Party. A two-pronged attack on environmentalism and the environment itself follows this strategy, and very probably will gain strength as terrorism and resource shortages diminish the "American way of life."

First of all, if the present and near future are dominated by the Republican Party, we will witness the continued decimation of environmental legislation enacted during the environmental era. The George W. Bush administration has already made major progress towards weakening environmental laws, and, further empowered by a strongly Christian-based party for a second term, his administration is making a shambles of environmental protection (little fuss was made by the American public over continued Republican attempts to open ANWR to petroleum development in the autumn of 2005). Secondly, on the education front, fundamentalist Christians have been working relentlessly in recent decades to undermine the scientific teaching of evolutionary theory in the K-12 public schools, although direct attempts to require the teaching of Christian creationist theory as a scientifically viable alternative to Darwinian evolutionary theory has had only mixed success, largely at the state level. The American public is particularly vulnerable to this relentless movement. For example, a 1991 Gallup poll showed that only 9% of Americans adhere to a strict, naturalistic, Darwinian view of evolution. Approximately 47% of Americans believed that "God created man pretty much in his present form at one time within the last ten thousand years." Another 40% believed that "Man has developed over millions of years from less advanced forms of life, but God guided this process, including man's creation. Only 4% answered that they didn't know." In sum, 87% of Americans believed that God was involved in evolutionary processes, whereas only 1% of natural scientists would agree with them![94] So much for the efficacy of K-12 science education in the realm of evolutionary biology. Obviously, higher education did little to mitigate American ignorance in this matter.

The religious atavism of the American public is hardly surprising, given that the first wave of colonists was comprised of predominantly rigorous Protestants, with a few quiet deists maintaining a low profile in their midst. Thus, it is hardly unreasonable for modern fundamentalists to argue that the United States was founded as a *Christian* nation. One fundamentalist author disgustedly describes the influence of Darwinian evolutionary theory upon the American public as follows: "Evolution is the root of atheism, of communism, Nazism, anarchism, racism, economic imperialism, militarism, libertinism, anarchism (sic), and all manner of anti-Christian systems of belief and practice."[95] It seems a remarkable oversight that "environmentalism" was not included amongst the list of ills fomented by exposure to evolutionary theory, since the main object of the affection and preservationism fostered by the Environmental Movement is precisely the *evolutionary* ecosystems which have evolved over eons of geologic time. Likewise, the science of geology is almost equally deserving of damnation by Christian fundamentalists. Not only has evolutionary biology been under attack in our K-12 public educational system, but the paucity of geology courses taught in American high schools has indirectly contributed to the general earth science illiteracy that I have noticed during my own teaching career. In fact, in my own introductory Physical Geology course taught over a period of twenty-six years, I have had numerous students squirm and suffer through exposure to the vicissitudes of learning about geologic time in particular. The inculcation of ignorance by circumventing the teaching of geology in high school science curricula (for example, only five geology courses are offered in all of the high schools in the entire state of Oregon), combined with fundamentalist Christian brainwashing through the K-12 years appears to close a majority of American minds to the time frame introduced by the combination of geologic science and Darwinian evolutionary biology. In addition to the withdrawals and dropout engendered by exposure to "deep time," I have had students come by and tell me that they don't believe any of what I am teaching them in geology. My patent response is, "you don't have to believe it; you only have to learn it." Why so few geology courses are taught in our high schools is clearly related to the ongoing struggle of American Christians to retain the Biblical time scale and perspective. The destructive consequence for environmentalism is that there is no general appreciation in the American mind for the wonder and awe associated with a grasp of evolutionary changes in species and evolving patterns of ecosystems over the vast expanse of geologic time. As a consequence, the public in general can grasp only the utilitarian and "human welfare ecology" foundations of environmental ethics, along with preservationism as valuable for what organisms and ecosystems can provide to *humans* in terms of medicinal or aesthetic values. The grasp of geologic and evolutionary processes necessary to the appreciation of "animal rights" and the intrinsic value of species, or of "deep ecology," are never learned by the vast majority of Americans.

Apathy regarding nature, based upon a shallow approach to understanding nature, or representing it as a mechanistic system, the model that physicians, chemists, and physicists imbibe, is bad enough, but the assault upon evolutionary biology has been intensifying as the United States has reverted to both political and religious conservatism, with the No Child Left behind Act and other factors contributing to the increased numbers of students in home schooling and Christian charter schools acting as a Trojan Horse carrying the latest form of anti-evolutionary Christian creationism into the K-12 system. The newest form of creationism, "intelligent design," involves "a system of public and political strategies which operate on a very detailed plan and a set of well articulated goals named 'The Wedge' by its executors. It offers an upgraded form of the religious fundamentalist creationism long familiar in America."[96] Earlier in this book, particularly in Chapter Seven, I have discussed the idea of intelligent design in relation to Enlightenment deism, in which the Creator establishes the laws of the universe, sets them in motion, and thereafter withdraws from the material world. The new creationism or intelligent design differs radically from the Enlightenment, Newtonian model in that God remains continually involved with the material world, creating new species and destroying outmoded ones. Perhaps contemporary intelligent design is best understood in terms of its major objective, which has been to engage in a pseudo-scientific effort to denigrate and undermine current Darwinian evolutionary biology, with the long-term goal of requiring K-12 schools and even institutions of higher learning to include intelligent design theory as a "reasonable" alternative to Darwinian evolutionary theory in biology courses and other courses as well. However, Richard Dawkins and other evolutionary biologists have criticized intelligent design as unscientific because it lacks any sort of research program to support its creationist anti-evolutionary theory. In other words, intelligent design theorists do not utilize the scientific method, but, rather, search for flaws in Darwinian evolutionary biology, and then offer up the alternative explanation of divine intervention in nature. What proponents of intelligent design fail to understand is the difference between the use of the word "theory" as representing any speculation about causation with scientific theory, which requires speculation about natural causes to follow the procedures of scientific method involving extensive observation, data collection, and experimentation (where possible) prior to deducing a theory from extensive information acquired by inductive reasoning. Furthermore, tentative hypotheses and theories alike are subjected to thorough scrutiny and criticism by the community of scientific practitioners, and scientific theories are always subject to revision as new knowledge is acquired.

Barbara Forrest and Paul R. Gross, in *Creationism's Trojan Horse: the Wedge of Intelligent Design* (2004) have written an extensive history and critique of contemporary intelligent design theory and its program for introducing neo-creationism into American public schools and universities. "The attempts to insert religion into public elementary and secondary science education are unceasing, and they now include direct efforts to influence college students as well. Efforts to force it into curricula--especially those having anything at all to do with biology and the history of Earth--have been unremitting since the late nineteenth century, and they have continued into the present."[97] As battles over the content of K-12 curricula occur from state to state, first in Kansas, then in Ohio and New Mexico, pro-evolution (anti-intelligent design) victories are secure only until the next election. Contrary to public perception, creationism as a cultural presence has generally grown stronger in the recent past. The antipathy of neo-creationists towards science-based evolutionary theory is not manifested in alternative scientific practice, but by seeking out internal disagreements among evolutionary theorists (such as are common to all natural science practitioners), which "are then presented dramatically to lay audiences as evidence of the fraudulence and impending collapse of 'Darwinism'. How are such audiences to know that modern biology is *not* a house of cards, *not* founded on a 'dying theory.'[98] The epicenter of intelligent design proselytization is the Discovery Institute's Center for the Renewal of Science and Culture (recently re-titled the Center for Science and Culture) in Seattle, Washington. The Discovery Institute of neo-creationists operates under the plan of action called "The Wedge," and through a well-funded program of conferences, publication, and public appearances is working its way into the American cultural mainstream.

With this brief account of neo-creationist, intelligent design activity in mind, simple statements, such as George W. Bush's "The jury is out on evolution," take on a deeper meaning, especially in light of Republican lobbying for the No Child Left behind Act of 2001, which has begun the process of shifting large numbers of students from problematical public schools to government-funded private schools which are frequently Christian in orientation, thereby often exposing young minds to creationism and its latest manifestation as intelligent design theory. Long before the Bush administration's manipulation of the K-12 system for purposes of Christian proselytization, my own son, upon taking a job with a computer data service company, was shocked upon starting work to find that anti-scientific attitudes were common in this organization, from the top down. Having mentioned something about the Pleistocene Ice Ages during casual conversation, his new boss and an upper echelon employee, after snickering a bit, asked him how anyone could really believe in something like an Ice Age. He learned quickly to avoid issues related to natural science during conversations. Creationism had seen to that. I have no doubt that its proliferation will seriously undermine public support of environmentalism.

Beginning in the spring of 2000 The Wedge introduced a political strategy designed to cultivate and convince congressional staff of the merits of introducing intelligent design theory into the nation's K-12 system. Congressional staffers were told that polls showed a majority of Americans favoring the teaching of creationism along with evolution in public schools. A year later, an amendment couched in vague language to accomplish this objective was introduced by

Pennsylvania Senator Rick Santorum (R), and was attached to the No Child Left behind Act, initially approved by a Senate vote of 91-8. Fortunately, the amendment was eventually relegated to the legislative history of the No Child Left behind Act, but it is a warning that any further drift to the political right in the United States could have damaging consequences for the public's understanding of evolutionary theory and its ecological implications. Overall, the American public is oblivious to the challenge of proselytizing radical Christianity to both political democracy and scientific leadership. The evolutionary scientist Niles Eldridge, curator of the American Museum of Natural History, warns us of the potentially dire consequences of neglecting the challenge of contempory creationism:

The United States, amazingly yet thankfully, remains at or near the forefront of the bulk of the world's scientific research. But our efforts to remain in a leading position in science are being relentlessly hammered by creationists, who insist that they just want to give equal time to an equally valid alternative theory ("Intelligent Design" being the latest version), when in point of fact what they really want to do is get their own version of religious truth into the public school arena. Our capacity to produce more scientists is hobbled by their efforts. Worse, we are in danger of having an even more scientifically illiterate electorate than ever before – in an era when so many public, political issues (such as stem cell research) call for the considered judgment of an informed body politic. [99]

Christianity, the Philosophy of History, and Nature

The French Jesuit geologist and paleontologist Pierre Teilhard de Chardin (1881-1955) became a sensationalistically successful author in the United States as well as in France during the mid-twentieth century, unfortunately too late for him to enjoy. His major work, *The Phenomenon of Man,* was published in France in 1955, and translated into English and published in the United States in 1959. His "Christian evolutionism" created a major intellectual controversy and was repressed by the Catholic Church. His own Jesuit order forbade him to continue teaching in 1926, essentially forcing him back to his professional career as geologist and paleontologist. Chardin made important contributions to our knowledge of Asian geology and the origins of *Homo sapiens.* Against the background of a quiet controversy between spiritualistic interpretations of evolution such as Henri Bergson's *Creative Evolution* (1911) and materialistic interpretations such as paleontologist George Gaylord Simpson's *The Meaning of Evolution* (1967), Teilhard re-ignited debate over the possible religious implications of Darwinian evolutionary theory.

Early in the twentieth century the Roman Catholic Church opposed rationalism and modernism, and attempted to stem the tide of dangerous modern ideas by developing a Neo-Thomist school of theology which revived the scholastic rationalism of St. Thomas Aquinas. A revival of scholasticism was already underway in the nineteenth century, and three popes had proclaimed the primacy of St. Thomas for Christian theology, including the encyclical *Studiorum Ducem* (1923).[100] During this period of revived Thomism, Chardin was laying the foundations of his own rationalistic synthesis, eventually revised and published as *The Phenomenon of Man*. Teilhard was considerably influenced by the spiritualistic French philosopher of evolution, Henri Bergson (1859-1941), who balked at the spiritless mechanistic explanations of Darwinian evolutionary biologists, and proposed that evolution, which he personally accepted as true, was not mechanistic, but rather was driven by some vital force which he named the "elan vital." Teilhard would take this idea of a spiritually driven evolutionary process and rationalize it Thomistically by making the Catholic God and Jesus Christ the vital forces of the evolutionary process. Much of *The Phenomenon of Man* is devoted to describing the evolutionary biological processes out of which consciousness, spontaneity, and, finally, human thought emerged. Unfortunately, since his ideas are Christianity-based, unscientific preconceived notions are the premises (not fully explained but presented as what Chardin saw as obvious truths from his Christian perspective) that give rise to obscure and unscientific concepts, such as the "Without" and "Within" of things. For it was Teilhard's heartfelt desire to explain the *complete* phenomenon of man, including outward materialistic aspects on the one hand, and inward spiritualistic aspects on the other, *entirely in terms of modern science*. Modern scientific method, however, breaks down when confronted with the mysteries of the spirit. For example:

> It is impossible to deny that, deep within ourselves an 'interior' appears at the heart of beings, as it were seen through a rent. This is enough to ensure that, in one degree or another, this 'interior' should obtrude itself as existing everywhere in nature from all time. Since the stuff of the universe has an inner aspect at one point of itself, there is necessarily a *double aspect to its structure*, that is to say in every region of space and time--in the same way, for instance, as it is granular: co-extensive with their Without, there is a Within to things.[101]

Such speculations might be convincing to a Thomistic rationalist, but not to someone trained in the scientific method of reasoning. Teilhard's synthesis is a curious mix of scientific reasoning and rationalistic speculation somewhat in the mode of Hegel, who reasoned that the wonders of history and the human thought it has recorded are a realization of the Absolute, i.e. of God's intellect expressing itself through time. Applying his extensive knowledge of evolution, Chardin's thesis is that the realm of mind, which he calls the "noosphere," inevitably *had* to evolve from the increasing complexity of organic life over eons of geologic time. "The idea is that of noogenesis ascending irreversibly towards Omega through the strictly limited cycle of a geogensis."[102] Conceiving of evolution as an ascent toward consciousness, the Omega Point is for Chardin some future time when consciousness is fully realized in human individuals. "By its structure Omega, in its ultimate principle, can only be a distinct Centre radiating at the core of a system of centres; a grouping in which personalization of the All and personalization of the elements reach their maximum, simultaneously and without merging, under the influence of a supremely autonomous focus of union."[103] This autonomous, central focus is the Omega Point. Chardin sees contemporary humankind as having traveled far towards the Omega Point, but with having a considerable future of further psychic evolution before attaining it fully. Therefore, he envisions a progressive development towards complete human consciousness the future goal of humanity. If evolution is progressive in this sense, as well as being inevitable, then a theory of providential or millenarian progressivism is contained in Teilhard's comprehensive theory of evolution.

Teilhard's evolutionary theory of history is thoroughly integrated with the Christian worldview. Although a scientist, Chardin never rejected the Christian religion he absorbed as a Jesuit, but felt compelled to reconcile evolution and Christianity in a meaningful synthesis. The spread of humanity around the globe and the Neolithic Revolution were necessary for the word of Christ to be spread, thereby raising the level of human consciousness in preparation for further development before converging with the supreme consciousness of God at the Omega Point. Thus, for Teilhard de Chardin world history is also the history of a triumphant global Christianity. As he writes:

> Christianity is in the first place real by virtue of the spontaneous amplitude of the movement it has managed to create in mankind. It addresses itself to every man and to every class of man, and from the start it took its place as one of the most vigorous and fruitful currents the noosphere has ever known. Whether we adhere to it or break off from it, we are surely obliged to admit that its stamp and its enduring influence are apparent in every corner of the earth today.[104]

Teilhard de Chardin's philosophy, combined with Arnold Toynbee's *A Study of History*, together almost constitute a modern-day equivalent of the *Summa Theologica* of St. Thomas Aquinas. Whereas Aquinas built an architectonic structure of rational Christian thought which culminated the intellectual achievements of the High Middle Ages, Teilhard and Toynbee were attempting to bring the Christian religion into the modern world. As a Jesuit, Chardin hoped to reconcile evolutionary thought and the Catholic Church. Obviously, he failed to accomplish this with the Vatican, whose intellectuals may pay lip service to evolutionary theory, but without changing the rigid Catholic worldview and its growing influence over the desperately poor masses of Latin America, Africa, and parts of Asia. With a well-established European-American empire in hand, Roman Catholicism is rapidly building a new empire in the Third World,[105] with potentially dire consequences for long-term population trends and accompanying ecological devastation, compounded by a naive conception of the natural world. In contrast, many informed Catholics in Europe and North America are at least cognizant of evolutionary thought and have shifted to a more liberal viewpoint, for which Chardin is at least partly responsible.

One of the major obstacles to solving the environmental crisis is the schizoid culture of *modernity*, which is actually only *half-modern*, i.e., it is split between belief in traditional religious soteriology and associated mores, and Enlightenment secularism and rationalism, producing in the United States and other religiously conservative nations the inherent conflict implicit in a nation of two cultures in one. In the United States, the antipathy of Christian fundamentalists towards evolution (and by association, towards ecological knowledge), is the analogue of the anger of radical Muslims against modernity in its full-blown secular mode, as manifested in secularized western Europe, where the philosophical implications of the Enlightenment, i.e. a secular humanistic society, were taken seriously (at least by many intellectuals). I doubt that most Muslims are aware that the Enlightenment in America produced "Christian Democracy," a characterization lauded by nationalistic American historian Ralph Barton Perry during the mid-twentieth century in his classic, *Puritanism and Democracy*. What he extolled is actually Christian-capitalist democracy in which belief in divine providence and the wonders of the marketplace produce an outcome vaguely defined as "progress." Across the Atlantic Ocean, in contrast, predominantly secular-capitalist democracies educate populaces possessing little to no animosity towards the findings of modern science, including evolutionary and ecological ideas. Although the European continent has undergone far more ecological damage than North America (as described in Chapter Four), their relative openness to a possible postmodern evolutionary-ecological synthesis of ideas is promising for some degree of caution in tampering with natural processes in the future, as reflected by their concern over global warming compared to the apathy of the United States in this regard.

Tradition, Progress, and the Environment

What, precisely, is tradition? Loosely, tradition consists of beliefs, customs, and conventions either handed down orally or in foundational documents such as "old" traditionalism's Judeo-Christian *Bible* and the *Koran*, or the "new" traditionalism of *The Wealth of Nations* or *Das Kapital*. The modern world includes a typically incompatible mix of the old and new traditions, which is one of the causes of ecological crisis. Old traditions that rationalize high birthrates combine with new traditions that rationalize high consumption.

The Third World, trapped in traditionalism while experiencing industrialization and population growth following Western colonization, is rapidly destroying its ecosystems and beginning the long-term process of genocide in Africa, an inevitable consequence of environmental degradation as carrying capacity is overrun. Without solving these problems, Western solutions to the ecological crisis will be undermined. Given its half-modern, schizoid condition, the United States, culturally halfway between Western European secularism at the center of Europe and the traditional fundamentalism of Islam and the Third World, is a powder keg of opposites waiting to explode under future conditions of social and environmental stress. Traditional religious values are not going to disappear, but they are not likely to become part of a postmodern solution to the ecological crisis. They are more likely to become enmeshed in political and social chaos, genocide, and warfare propelled by our general neglect of environmental constraints to the modern dream of progress: the global dissemination of "the American way of life." Ralph Barton Perry wrote glowingly of "Christian democracy," but Christianity, the natural political outcome of which is *theocracy* or "crypto-theocracy", is antithetical to democracy because mass democracy is a *modern* institution informed by both implicit and explicit critiques of Christianity as a political foundation for government.[106] As a child of the Enlightenment (and its critique of historical, autocratic Christianity) democracy has been associated with social movements towards secular humanism and away from the church. This is an historical *fait accompli* in Western Europe but the transition in the U.S. has been contested in a culture war which teeters on the edge of regression to increased neo-fundamentalism and a *de facto* theocracy, as in the case of George W. Bush believing or claiming that he receives his most important decision-making advice directly from God. Europeans both titter and shudder at the anti-intellectualism, self-righteousness, and ignorance of American political leaders and the large proportion of half-literate and fundamentalist voters who project their ignorance and anti-environmentalism by enabling the election of anti-intellectual, reactionary leaders such as Ronald Reagan and the Bushes.

A hidden agenda is present in capitalism and the Republican Party, which supports capitalism in its unmitigated form to a greater extent than the Democratic Party. By maintaining socioeconomic inequality, class and racial division, as well as the Christian-secularist schism, the United States sputters along in a politics of division, rather than, with quasi-socialistic institutions in place to solve the socioeconomic problem, focusing on the most serious issues of the twenty-first century: the future of a technological society; population decrease; maintenance and expansion of ecological systems; leadership in a world headed towards energy and environmental disaster as the effects of global warming multiply; and empowerment of a United nations and United Nations Environment Program with teeth in order to mitigate against global ecological degradation and associated genocide. Along with these issues, there remains the constant threat of global and nuclear warfare in an increasingly Hobbesian world running short on resources.

The roots of our contemporary impasse regarding the destruction of nature run deep, and the modern way of life, i.e., the inexorable *tourbillon* of unsustainable economic development, is now so well established and rationalized that we seem to be doomed to ecological catastrophe in the long run. Aiding and abetting this rush towards environmental oblivion is the political incommensurability of modern problems and traditional beliefs and values. The rapid and exponentially increasing degradation of nature is a uniquely modern phenomenon that cannot be solved without modern knowledge, knowledge of science and philosophy in general, and of evolutionary biology and ecology in particular.The battle to preserve nature from thorough destruction cannot be won on the political front with a citizenry armed with pre-modern beliefs in creationism and other antediluvian anachronisms about human nature and the natural world (combined with technological utopianism). A half-ignorant citizenry has imposed a ponderous failure upon environmental education, constantly undermined by traditional values held by the right and center of the political spectrum in the United States and elsewhere. The disjuncture between modern problems and traditional values has created both a schizoid culture and the impossibility of realistic solutions to environmental problems that require collective and individual sacrifice. Former President Ronald Reagan neatly represented our rejection of such sacrifice when he stated that quality of life is more important than preservation. And so we press on in pursuit of a "quality of life" measured quantitatively (in the size of one's home, SUV, and domestic budget) and, by all historical standards, obscenely.

This "modern impasse" is not what angered postmodern philosophers of the 1960s and 1970s. Derrida, Foucault and others were enraged at the inegalitarianism, social oppression, and imperialistic bullying by Western civilization in the name of a philosophy of progress. To them, this progressive speculative philosophy of history or "metanarrative," as they preferred to call it, was an overarching historical narrative that rationalized Western oppression over the "other," i.e., the inhabitants of the colonized lands of America, Australia, and Asia. However, the postmodern philosophers were naive and anthropocentric, ultimately becoming self-entangled in a critical theory pathetically ignorant of the evolutionary ecosystems within which cultural and linguistic systems develop. As a consequence the postmodern philosophers themselves have been deconstructed, their intellectual narrowness leading them towards extinction in the struggle of ideas. Nevertheless, they have accomplished the positive, useful function of forcing all of us to use terms such as "nature" and "ecosystem" with far greater care than previously. Their wariness of certainty is relevant to the problem of traditional resistance to the environmentalist's desire to "change the world" overnight. The natural world that we hoped to save from relentless development has become epistemologically ambiguous even as it is rapidly destroyed. "Nature" and "ecosystems" may be largely gone by the time we do understand what they really are---or were.

The ambiguities of meaning revealed by postmodern philosophy undermine the idea of converting people's traditional beliefs to the environmentalist's point of view, which is presently fractured into its own chaos of competing philosophies and positions. Given such uncertainty, why should Americans or other citizens in developed countries give up or transform their traditional beliefs? Instead, religious conservatism has deepened with the growth of evangelical Christianity, "family values" is the catchphrase of all political parties, and the ecological crisis has all but dried up as a political issue. From a politically conservative point of view, all is well with the world. We have seen that American conservatism is essentially two-tiered, consisting of an older layer of traditional values clustered around Christian religion, and a younger layer of economic values, which grew out of the rationalistic thought of the Enlightenment. The Christian layer includes a mythopoeic view of reality which rejects evolution and prevents reforms related to population and birth control. The younger, capitalist layer ignores ecological constraints to economic growth, and encourages weakening or removal of environmental legislation which might deter or re-orient our "bulldozer technology" style of development. Many Americans adhere to both sets of conservative values, but either set in itself impedes serious environmental reform and regulation.

Tradition is the cement that holds civilizations together. Traditions such as the world's great religions and their core values have survived terrible vicissitudes over long periods of history. The Catholic Church has been around for 1600 years after surviving the collapse of the western Roman Empire! It survived the Dark Ages, the Black Plague of the fourteenth century, and the challenges of Renaissance, Reformation, and the Enlightenment. The ecological ravaging of Earth today appears to its intelligentsia as another bump on the road to eternity, even though many of its more liberal adherents think otherwise. The traditions of the great religions are here to stay, unfortunately carrying with them the massive ignorance which increases inertial resistance to creating something like a "sustainable society" infused with new, ecological-evolutionary values.

Modernity, whose leaders attempted to sweep aside the old religion-based values with the Enlightenment broom of rationalism, instead added its own layer of values, of "laissez-faire" economic ideology and unlimited economic growth to produce "the American Dream" of unprecedented material abundance. These two-hundred year old values merged with Protestant Christianity to create a "Christian democracy." Then, a mere forty-odd years ago, Rachel Carson and a few others sounded the environmentalist alarm which initiated the clash of values which we call the Environmental Movement, which has become somewhat obscured by the rise of multiculturalism, political correctness, and the demands of liberal secular humanists and moderate Christians for a society free of the constraints of traditional, religion-based values. The countercultural "revolution," and the sexual, multicultural, feminist, and gay movements have collectively driven holders of traditional Christian values into a corner, and they have lashed out at the new, discordant, even barbarous secular culture that assaults all and sundry from the television screen, the cinema, newspapers and magazines. Somewhere in the midst of this frenzy of social change environmentalism has been diluted and forced towards the end of the line of a host of issues, from abortion to inner city violence, drug use, and educational collapse. Today, the ecological crisis, like the nuclear crisis, has been upstaged by a host of strongly anthropocentric and therefore more immediate issues. The cacophony of the modern-traditional struggle has drowned out the warning cries of the Environmental Movement. The battle to save traditional values has taken center stage.

Religious fundamentalists reach back to the Bible or Koran to justify their behavior regarding child bearing and the threat of population control, while the modernist fundamentalists cite Adam Smith or Karl Marx to justify their overexploitation of nature. Our ability to cope with modern ecological degradation is seriously constrained by these two layers of fundamentalist belief. The paradox of the modern developed world is that our need for tradition, particularly the old religious traditions (which *Homo religiosus* is to some degree genetically programmed for through some form of social selection that benefited some aspect of tribal life), is compromised by the dynamism of science, technology, and the new tradition of economic beliefs and institutions. This has resulted in a schizoid culture, with one part of the contemporary mind seeking the stability and personal security offered by traditional religion while the other approves of "progress" in the form of technological and economic development, the very forces that propel all of the rapid change and uncertainties that undermine not only nature but also the stable milieu of traditional societies and their religious paradigms. Belief in providence and progress, and an insidiously false and ecologically destructive idea of progress at that, as I explained at length in Chapter Six, are responsible for a modern culture which at present appears locked on a course of self-destructive behavior at the cost of the Earth's ecological richness and long-term evolutionary stability. If *Homo religiosus* is a fixed aspect of human nature, however, the same cannot be said for *Homo oeconomicus*, the greedy, self-interested aspect of modern humanity, because this variation on human behavior has been around in a magnified form for a mere several centuries. Any hope for transformation of the current, insane, growth-oriented practices of the developed countries and their third world satellites must rest upon our ability to challenge the false progressive paradigm which rules and ruins the world. As I have attempted to demonstrate, there is little evidence to suggest that such a transformation is seriously underway at present, although faint stirrings of environmental responsibility exist in a small minority of individuals and environmental groups. Consequently, one must surmise that metanoia, or transformation towards an ecological worldview, will be slow and difficult, and just as likely to culminate in a pattern of resource scarcity, international friction, warfare, and genocide as in a comfortable, sustainable, ecologically sound society embracing new traditions of responsibility for Earth and her non-human inhabitants.

The mythopoeic realm of the Christian millennium lies forever in the future of the true believer, and progress towards Ecotopia remains an environmentalist fantasy. We appear to be drifting towards the unfortunate possibilities of either a technological society in full command of a totally humanized "spaceship Earth" in which all of the planet's resources satisfy *only* human needs, or, following the scenario of resource shortages, warfare, and genocide, the final acts of the cycle of civilizations, leaving behind an impoverished planet and a human remnant of Stone Age survivors such as William Miller, Jr. depicted in *A Canticle for Leibowitz,* a novel depicting the conditions of humanity following global thermonuclear war. Even barring nuclear war, humanity is dismantling Earth's terrestrial and aquatic ecosystems gradually and relentlessly by "ordinary means" even without the impacts of global warming.

Although the old traditional values associated with Western Christianity and its idea of providence operate on the limited scale of individuals, families, and parishes for the most part, the new traditional values of modernity and their associated scientific, technological, educational, business, and governmental institutions constitute the global network of economic development and transnational corporations which are destroying ecosystems at a breakneck pace. Underlying this global network of frenzied activity is the idea of progress, the core rationalizing philosophy of modernity which, even though long since condemned by historians and other academics, remains the traditional myth of the masses and of their educators, politicians, and the media. The idea of progress of the past two centuries is actually an insidiously destructive one, which assumes beneficial future changes in human institutions and the conditions of life to be the natural result of continuing innovations in science and technology, when in truth such changes have produced economically inegalitarian and ecologically destructive societies as well as an overpopulated Earth. Concurrently, partial destruction of the viability of the old, pre-Enlightenment traditions has produced social barbarism on the one hand and the extremes of religious reaction on the other. A leading critic of the Enlightenment and its distorted legacy of science and technology-based progressivism, United Kingdom philosopher Anthony O'Hear also bemoans the loss of the old religious traditions which once elevated Englishmen, producing a culture that loves the old gods no longer.

> We have no comparable vision or hope. Science has destroyed these edifying and elevating beliefs. Physics and modern cosmology tell us that we live in a universe which is self-contained and self-sufficient, governed only by the impersonal operations of chance and causal law. In this vast universe we are no more than biological survival machines on an insignificant planet. We are dominated by the genetic imperatives to survive and reproduce, and the playthings of forces, biological and social, which we cannot fully understand, let alone control.[107]

These rather grim observations of O'Hear reflect the responses of some intellectuals to modern progress, both the idea of progress and "progressive" change itself, which has destroyed the old traditional beliefs and values for most Frenchmen, Englishmen, Scandinavians, Germans, and most of the populations of developed nations in general. As we saw earlier in this chapter, postmodern philosophers generally are as pessimistic about the actual legacy of progress as the traditionalists, rejecting the idea of progress as a metanarrative which makes sweeping and unjustified claims about the direction of human history. The assault on progressivism comes from both right and left, leaving it to the masses of untutored working people to acknowledge and support this modern tradition as a shibboleth fit for political propaganda along with lip service to the gods of the old traditions. "This outlook ratifies the idea of the domination of life by large technological systems, by default if not by design. The accompanying mood varies from a sense of pleasurably self-abnegating acquiescence in the inevitable to melancholy resignation or fatalism."[108]

Cycles in Nature and History

The word "nature," in the sense of the natural world, has several key meanings: (1) the sum total of all things in time and space; i.e., the entire physical universe; (2) the power or principle that seems to regulate the physical universe, sometimes personified as "mother nature"; (3) natural scenery, including the plants and animals that are part of it; (4) the natural world as it exists without human beings and civilization; and (5) the elements of the natural world, as mountains, trees, animals, or rivers.[109] Not mentioned in these definitions is the fact that nature, as experienced by human beings on planet Earth, displays multiple cyclical aspects, some quite obvious, such as the cycle of night and day caused by the Earth's rotation, and the seasons of the year dictated by Earth's annual revolution about the sun. Other cycles such as biological life cycles (the carbon cycle, the nitrogen cycle, etc.) are adapted to the daily and annual astronomical cycles. Even the cosmos or universe itself may be cyclical, the "Big Bang" being part of an eternal process of pulsating expansion and contraction, which some contemporary astronomers admit to even though the majority prefer the model of a beginning and end to the cosmos, undoubtedly an expression of linear, providential historical thought in Western civilization. Modern science has revealed the existence of ubiquitous cycles in nature, including complex movements such as the wobble of Earth's axis, the plate tectonic cycle, glacial cycles, and meteorological cycles related to cyclical changes in ocean currents and temperatures such as El Niño and La Niña.

Beginning with the Babylonians of the ancient world, intelligentsias of early civilizations gradually took note of the regular motions of the moon, planets, and stars relative to earth. In addition to the practical applications of astronomy, the Babylonians developed the first cyclical philosophy of history. Their approach to the universe was mythopoeic, involving the personalization of the powers of nature. Observation of the daily sun cycle, probably the first idea of a recurrent cycle, understood this cycle in terms of the daily birth and death of the sun-god. Later, the cycle of the seasons was mythopoeically conceived as the death and rebirth of fertility powers associated with the death of an early leader or king. The Mesopotamians eventually conceived of a much larger cycle of 3000 years, an "age of the moon god," which consequently led to the conception of a recurrent "Great Year of the cosmos" of twelve divine months or 36,000 years. "This idea of a Divine Year cycle or moon cycle is found in many other civilizations and may have originated in Babylonia."[110] The idea of a world-year does not appear to have developed in Egypt, however. Recall that in Chapter One I mentioned Plato's quantification of the cycle of civilization as a period of 36,000 years. As philosopher Grace Cairns observed: "The cyclical view of Plato is akin to that of the Pythagoreans, for it, too, is connected with the movements of the heavenly bodies. In Plato's *Timaeus* the Great or Cosmic Year is described as the interval of Time which must elapse before the planetary bodies return to the same relative positions from which they originated."[111]   Although some scholars believe that Plato chose 10,000 years as the duration of his Great year on the basis of a multiple of ten, the perfect holy number of the Pythagoreans (the tetractys), others argue that Plato's Great Year was a period of 36,000 years on the basis of either Greek astronomical and numerological calculations or adoption of the Babylonian cycle.[112]

The major points of this discussion are: (1) that observers of nature in Babylonian, Greek, and other ancient civilizations (as well as pre-civilized cultures) took the order and regularities of nature very seriously, and with a solemnity inspired by their awe of what we today call "natural law"; (2) that the cyclical philosophy of history is a venerable tradition of pre-scientific metahistory as worthy of study by our own modern intelligentsia as the providential and progressive philosophies of history; and (3) that modern science, social science, and particularly ecological science have suggested that modern neocyclical philosophies of history such as those of Spengler and Toynbee are more correct historical models of history than those of providence and progress. The twentieth century cyclical philosophers were right, but for the *wrong* reasons; namely, that long-term ecological degradation, rather than historical predestination, was a dominant factor in determining the rise, flourishing, decline, and fall or transformation of many, if not most, of the world's civilizations. Our own Western, Faustian civilization regenerated itself out of the ecological and epidemiological turmoil of the fourteenth century (as described in Chapter Four), and, in the process, experienced the gradual shift in worldview (metanoia) that created modern, largely secular contemporary Western civilization around a dwindling core of pre-modern, essentially medieval beliefs. This transformation has left us with a schism that divides Western, and especially American culture today.

Our scientific understanding of cyclical processes in nature has been greatly increased during the past half century in the disciplines of geology, biology, archeology, and meteorology. Our understanding of the rise and fall of civilizations has been enhanced by the accumulation of detailed knowledge regarding past climates during the Pleistocene Ice Ages and the Holocene Epoch. This knowledge has allowed archaeologists like Charles Redman and Brian Fagan (see Chapter Two) to sort out the human causes of cultural decline from the causes related to natural cyclical phenomena like El Niños in explaining the decline and collapse of Mayan and Greco-Roman civilization, and the vicissitudes of Western civilization during the Late Middle Ages and early modern centuries. Our geological and meteorological understanding of the Pleistocene Ice Ages has also established a baseline for evaluating the impact of modern industrial man upon the atmosphere and hydrosphere, thereby sounding an alarm that global warming is occurring. In the arid west of the United States, wet and dry cycles, i.e., periods of above average and below average rainfall of one to several decades duration, have been recorded since the mid-nineteenth century. They show a striking trend towards shorter wet cycles and longer dry cycles over the past 150 years, which means that planning for surface and groundwater use in the western United States, which is based upon a nineteenth and twentieth century period of record, may have drastically overcommitted western water resources at a time when gradual desiccation of the region is likely to intensify due to global warming, and population is swelling for various reasons, including the frostbelt to sunbelt migration and immigration from Latin America and elsewhere. The long-term consequences for the west's ecosystems and human inhabitants could be devastating.

Applying our increasing knowledge of historic cyclical variations in climate to the history of civilizations (as well as to ecosystems, which Charles Redman and Brian Fagan have already done) will refine our knowledge of human responses to periodic fluctuations such as flood and drought, as well as our understanding of ancient civilizations and the hypotheses they generated to explain cyclical phenomena, including their cyclical philosophies of history. Unfortunately, any such new knowledge will be of interest chiefly to academic specialists, and will not likely affect the general outlook or worldview of the vast majority of human beings during the twenty-first century. We appear to be locked on a course largely determined by the Western linear belief systems of providence and progress. I have never met a non-environmental historian who was familiar with such works as Clive Ponting's *A Green History of the World* or J. Donald Hughes's *Pan's Travail: Environmental Problems of the Ancient Greeks and Romans*. The seemingly mind-boggling insights of environmental history have not penetrated the armor of the established belief system of the West. The minds of our children, as well, are filled with antiquated and outmoded providential and progressive myths designed to keep them content and optimistic while the dismantling of nature continues--not unnoticed by a few, to be sure--but definitely ignored by the vast majority who have been well trained to pursue their own interests and pleasures. Worse, when resource shortages begin to pinch the inhabitants of America and the rest of the developed world in the decades ahead, the myth of providence will most certainly overshadow the myth of progress as we begin our descent into an age of scarcity. We have lost the capacity for awe and for the deep respect for nature that undergirded the ancient belief in a cyclical cosmos. As Hughes has shown, even such awe-inspired respect for nature was not enough to prevent the ancients from destroying the Mediterranean bioregion. Today, with all such constraining beliefs swept aside, with no courage whatsoever available to moderate burgeoning human population, with economically-based progressive myths controlling governments and the mass mind of the "global marketplace," Spengler's perception of humanity as the ultimate omnivore-predator-dominator of nature and humanity itself (*homo homini lupus*) makes more than a little sense. The humanity-induced "organic asteroid collision" of Peter Ward is near to completing its mission. An ecologically moribund and totally humanized planet appears to be the likely outcome, as the possibility of ecological metanoia becomes more remote in the face of intractable human problems and beliefs. National parks and "wilderness" may continue to exist as carefully managed remnants of Earth's evolutionary ecosystems, but the evolutionary process as we know it will have ended.

Hope and the Future

Out of this ecological denouement, the odds are that populous *Homo sapiens* will survive on a planet shorn of its ecosystems and biodiversity. Presently, we have managed to damage all the components of Gaia: lithosphere (mining and soil erosion), biosphere (destruction of ecosystems and biodiversity), hydrosphere (acidification and pollution of fresh and oceanic waters, destruction of coral reefs and fish populations), and atmosphere (global warming, ozone depletion, acid rain). In time, new myths may be created to rationalize "a planet made for man," what Buckminster Fuller, the ultimate technocrat, called "spaceship Earth." On the other hand, nuclear weapons and new biological weapons could even threaten Promethean mankind itself.

In 1992, Max Oelschlaeger proposed, in *Caring for Creation*,[113] that because the majority of Americans are of Christian or other religious faiths, the churches of America possess the potential for engendering a deep concern for the future of Earth and the biosphere. Unfortunately, such a transformation faces the prevailing corporate and governmental pursuit of endless economic and material growth. American culture encourages utilitarian individualism which is incompatible with high-minded communal goals such as population control, ecologically responsible land-use planning, the setting aside of more extensive wilderness and open space, and tightening our wasteful energy use.

Although the "greening" of religion is underway, it still has a long way to go. Furthermore, the number of post- patriarchal Christians who follow the radical Christianity of reformers such as Matthew Fox ("*Eco-justice* is a necessity for planetary survival and human ethics; without it we are crucifying the Christ all over again in the form of destruction of forests, waters, air, and soil")[114] is quite small. Fox's *eco-justice* is a Christianized version of Aldo Leopold's land ethic, but even if endorsed by most churches, would it be much more than a shibboleth? The promise of individual salvation will hold center stage during the era of scarcity and discomfort which awaits us. Most Americans are Christian-capitalist utilitarians. Changing their values will be a difficult task, as the failure of much of the agenda of 1960s-1970s environmentalism suggests.

As a secular humanist, I think that the public schools and colleges also offer some hope of educating the American public to reform their ecologically destructive values. Perhaps, complemented by the environmental education of children and adults in many churches, we can transform the leadership of the corporate military-industrial complex as well. Whatever we can save of the biosphere through socio-economic changes and political action will be of inestimable value.

NOTES

[1]. Edward Hallett Carr, *What Is History?* (New York: Alfred A. Knopf, 1967), p.21.
[2]. Ibid., p.22.
[3]. Ibid., p.26.
[4]. Ibid.,p.52.
[5]. Joyce Appleby, Lynn Hunt, and Margaret Jacob, *Telling the Truth About History* (New York: W.W. Norton and Company, 1994), p.201.
[6]. Ibid., p.202.
[7]. William Cronon, "The Trouble with Wilderness; or, Getting Back to the Wrong Nature" in William Cronon, ed., *Uncommon Ground: Toward Reinventing Nature* (New York: W.W. Norton and Company, 1995), p.79.
[8]. Ibid.,p.87.
[9]. Charles C. Mann, *1491: New Revelations of the Americas Before Columbus* (New York: Alfred A. Knopf, 2005).
[10]. Eileen Crist, "Against the Social Construction of Nature and Wilderness," in *Environmental Ethics,* Spring, 2004, Vol. 26, No. 1, p.20. See also p.17 where Crist attempts to clarify some of the muddy waters created by postmodern deconstruction of the ecosystem concept. Crist writes: "A case about stable scientific knowledge can also be made regarding the understanding of ecosystems. It is well known that views about the stability versus flux of ecosystems, and the relationship between biological diversity and ecological resilience have markedly shifted; they are likely to shift again. But the general insight into--along with the innumerable concrete facts about--what Darwin called "the entangled bank" of organisms interlocked in food pyramids, relationships of symbiosis, tolerance, and competition, conversion of nutrients, waste assimilation and decomposition, and element cycling is so solid as to have become nearly prosaid: it constitutes the *ground* from which debates about the relative stability versus dynamism of ecosystems are launched. To focus on how perspectives within ecology have shifted may be intellectually stimulating, but to obscure the background of accruing ecological knowledge in relation to which scientific analysis has changed is to elide a huge portion of the spectrum that composes "scientific knowledge."
[11]. Paul Shepard, "Virtually Hunting Reality in the Forests of Simulacra" in Michael E. Soulé, Gary Lease, *Reinventing Nature: Responses to Postmodern Deconstruction* (Washington, D.C.: Island Press, 1995), p.21.
[12]. Ibid., pp.24-25.
[13]. Ibid., p.27.
[14]. Ibid.
[15]. Joseph Carroll, *Evolution and Literary Theory* (Columbia: University of Missouri Press, 1995), p.3.
[16]. Ibid., p.468.
[17]. Ibid., p.469.
[18]. Francis Wheen, *Idiot Proof: Deluded Celebrities, Irrational Power Brokers, Media Morons and the Erosion of Common Sense* (New York: Public Affairs, 2004), p.81.
[19]. Jeff Haynes, "Religion, Secularization and Politics: A Postmodern Conspectus," *Third World Quarterly,* Vol. 18, No. 4(1997), p.715, quoted in Malise Ruthven, *Fundamentalism: The Search for Meaning* (Oxford: Oxford University Press, 2004), p.197.

[20]. Max Oelschlaeger, ed., *Postmodern Environmental Ethics* (State University of New York Press, 1995), Introduction, p.2.

[21]. Ibid., p.3.

[22]. Ibid., p.3.

[23]. Ibid., p.4. For an excellent concise summary of how scientific relativism and uncertainty forced our reconceptualization of Newtonian mechanistic nature into a once again "mysterious" universe, see Franklin L. Baumer, *Modern European Thought: Continuity and Change* (New York: Macmillan Publishing Co., Inc., 1977), pp.456-474.

[24]. Franklin L. Baumer, *Modern European Thought*, p.461.

[25]. Brice R. Wachterhauser, ed., *Hermeneutics and Modern Philosophy* (Albany: State University of New York Press, 1986), p.8.

[26]. Ibid., p.22.

[27]. Ibid., p.24.

[28] .Ibid., p.30.

[29]. Ibid., p.31. Although this may be true for humans, other animals, bears and elephants for example, have direct apprehension of the world with little or no mediating language, and they understand the world well enough to make a comfortable living from it. Also, the "world" that I experience in dreams generally is not linguistically mediated.

[30]. Hans-Georg Gadamer, "On the Scope and Function of Hermeneutical Reflection," in Brice R. Wachterhauser, *Hermeneutics and Modern Philosophy* (Albany: State University of New York Press, 1986), p.296.

[31]. Hans-Georg Gadamer, "What Is Truth?" in Brice R. Wachterhauser, *Hermeneutics and Truth* (Evanston, Illinois: Northwestern University Press, 1994), p.35.

[32]. Martin Carrier, Gerald J. Massey, and Lara Reutsche, eds., *Science at Century's End: Philosophical questions on the Progress and Limits of Science* (University of Pittsburgh Press, 2000), p.296.

[33]. Ibid., p.298.

[34]. Paul R. Ehrlich and Anne R. Ehrlich, *Betrayal of Science and Reason* (Washington D.C.: Island Press, 1996), p.27.

[35]. David Quammen, "Was Darwin Wrong? No. The Evidence for Evolution is Overwhelming," *National Geographic*, Vol. 206, No. 5, Nov., 2004, pp. 2-35, p.4.

[36]. John C. Greene, *Science, Ideology, and World View* (Berkeley: University of California Press, 1981), p.188. See Ch.7, "From Huxley to Huxley: Transformations in the Darwinian Credo."

[3737]. Marjorie Grene, "The Paradoxes of Historicity" in Brice R. Wachterhauser, ed., *Hermeneutics and Modern Philosophy* (Albany: State University of New York Press, 1986), p.185.

[38]. Ernst Breisach, *On The Future of History: The Postmodernist Challenge and Its Aftermath* (Chicago: The University of Chicago Press, 2003), p.12.

[39]. Ibid., p.23.

[40]. Ibid., p.123.

[41]. William McNeill, *The Rise of the West: A History of the Human Community* (Chicago: The University of Chicago Press, 1963), p.806.

[42]. Ibid.

[43]. Ibid.

[44]. Ernst Breisach, *On the Future of History*, p. 34.

[45]. Oswald Spengler, *The Decline of the West*, Vol. II (New York: Alfred Knopf, 1928), pp. 504-507.

[46]. William McNeill, *Mythistory and Other Essays* (Chicago: University of Chicago Press, 1986), p.10.

[47]. Ibid., p.16.

[48]. Ibid., p.22.

[49]. Paul Costello, *World Historians and Their Goals: Twentieth-Century Answers to Modernism* (DeKalb: Northern Illinois University Press, 1993).

[50]. Ibid., p.42.

[51]. Ibid., Toynbee quoted by Costello, p.72, from *A Study of History*, Vol. 12, p.563.

[52]. Paul Costello, *World Historians and Their Goals*, p.77.

[53]. Ibid., p.91.

[54]. Ibid., p.156.

[55]. Ibid., p.170.

[56]. Ibid., p.179.

[57]. Ibid., p.194.

[58]. Ibid., p.202.

[59]. Ibid., p.206.

[60]. Ibid., p.208.

[61]. John Farrenkopf, *Prophet of Decline: Spengler on World History and Politics* (Baton Rouge: Louisiana State University Press, 2001), p.192.

[62]. Oswald Spengler, *Man and Technics: A Contribution to a Philosophy of Life* (Honolulu: University Press of the Pacific, 1931, 2002), p.25.

[63]. Ibid., p.3.

[64]. Ibid., p.44.

[65]. Ibid., p.58.

[66]. Ibid., p.66.

[67]. Ibid., p.67.

[68]. Ibid., p.76.

[69]. Ibid., p.78.

[70]. Ibid., pp.84-85.

[71]. Ibid., pp.90-91.

[72]. Ibid., pp.93-94.

[73]. Ibid., p.103.

[74]. John Farrenkopf, *Prophet of Decline*, p.201.

[75]. Ibid., p.205.

[76]. Ibid., p.206.

[77]. Ibid., p.194.

[78]. Ibid., p.195.

[79]. Joseph A. Tainter, *The Collapse of Complex Societies* (Cambridge; Cambridge University Press, 1988), p.1.

[80]. Ibid., p.41.

[81]. *Ibis.*, p.4.

[82]. Ibid., p.50.

[83]. Peter Ward, *The End of Evolution: On Mass Extinctions and the Preservation of Biodiversity* (New York: Bantam Books, 1994), p.276.

[84]. James Speth, *Red Sky at Morning: America and the Crisis of the Global Environment* (Yale University Press: New Haven and London, 2004), p.191.

[85]. Richard Dawkins, "What Use is Religion? : Part I," *Free Inquiry*, Vol. 24, No. 4; June-July, 2004, pp.13-14, 56.

[86]. Ronnie Dugger, "It's Time for an American Offensive against Theocracy," *Free Inquiry*, Vol. 24, No. 3, April-May, 2005, pp.16-17.

[87] Kevin Phillips, *American Theocracy: The Peril and Politics of Radical Religion, Oil, and Borrowed Money in the 21st Century* (New York: Viking, 2006, p.65.

[88] Ibid., p.66.

[89] Ibid., p.62.

[90] Ibid., p.116.

[91] Ibid., p.393.

[92]. Nancy Gibbs, "The Faith Factor," *Time*, Vol. 163, No. 25, June 21, 2004, p.33.

[93]. Ibid., pp.26-28.

[94]. Michael Shermer, *Why People Believe Weird Things: Pseudoscience, Superstition, And Other Confusions of Our Time* (New York: W.H. Freeman and Company, 1997), p.156.

[95]. Henry Morris, author of *The Remarkable Birth of Planet Earth*, Institution of Creation Science, 1986, quoted in Malise Ruthven, *Fundamentalism* (Oxford: Oxford University Press, 2004), p.19.

[96]. Barbara Forrest and Paul R. Gross, *Creationism's Trojan Horse: The Wedge of Intelligent Design* (Oxford: Oxford University Press, 2004), p.6.

[97]. Ibid..6.

[98]. Ibid., p.7.

[99] Niles Eldredge, *Darwin: Discovering the Tree of Life* (W.W. Norton & Company, Inc., 2005), p. 221

[100]. Franklin L. Baumer, *Modern European Thought*, pp. 449-450.

[101]. Teilhard de Chardin, *The Phenomenon of Man* (New York: Harper and Row, Publishers, 1959), p.56.

[102]. Ibid., p.273.

[103]. Ibid., p.262-263.

[104]. Ibid., p.295.

[105]. See Stephen R. Welch, "The Sins of the Missionaries," *Free Inquiry*, Vol. 24, No. 2, Feb.-Mar., 2004.

[106]. Ralph Barton Perry, *Puritanism and Democracy* (New York: The Vanguard Press, 1944).

[107]. Anthony O'Hear, *After Progress: Finding the Old Way Forward* (New York: Bloomsbury Publishing, 1999), p.228.

[108]. Leo Marx, "The Idea of 'Technology' and Postmodern Pessimism" in Merritt Roe Smith and Leo Marx, eds., *Does Technology Drive History?: The Dilemma of Technological Determinism* (Cambridge, Massachusetts; The MIT Press, 1994), p.257.

[109]. Michael Agnes, ed., *Webster's New World College Dictionary*, Fourth Edition (New York: Macmillan, 1999), p.960, and Stuart Berg Flexner, ed., *The Random House Dictionary of the English Language*, Second Edition, Unabridged (New York: Random House, 1987), p.1281.

[110]. Grace E. Cairns, *Philosophies of History* (New York: The Citadel Press, 1962), p. 33.

[111]. Ibid. p.207.

[112]. Ibid. pp.208-211.

[113] Max Oelschlaeger, *Caring for Creation*, pp. 216-220.

[114] Matthew Fox, *A New Reformation: Creation, Spirituality and the Transformation of Christianity* (Rochester, Vermont: Inner Traditions, 2006), p. 69. A detailed explanation of Fox's "creation spirituality," in contrast to the Augustinian fall/redemption spiritual tradition, is presented in Matthew Fox, *Original Blessing* (Santa Fe, New Mexico: Bear and Company, Inc. 1983).

LIST OF SOURCES

Agnes, Michael, editor. *Webster's New World College Dictionary*, Fourth Edition (New York: Macmillan, 1999).

Alvarez, L.W., W. Alvarez, F. Asaro, and H.V. Michel. 1980, "Extraterrestrial cause for the Cretaceous-Tertiary extinction." *Science* 208: 1095-1108.

Appleby, Joyce, Lynn Hunt, and Margaret Jacob. *Telling the Truth about History* (New York: W.W. Norton and Company, 1994).

Arnold, David. *The Problem of Nature: Environment, Culture and European Expansion* (Cambridge, Massachusetts: Blackwell Publishers, Inc., 1996).

Bacon, Francis. *Novum Organum*, translated and edited by Peter Urbach and John Gibson (Chicago and Lasalle, Illinois: Open Court, 1994).

Bacon, Francis. *The New Atlantis* in *Ideal Commonwealths,* H. Morley, ed. (New York: The Colonial Press, 1901).

Bate, Jonathan. *The Song of the Earth* (Cambridge, Massachusetts: Harvard University Press, 2000).

Baumer, Franklin L. *Modern European Thought: Continuity and Change in Ideas 1600-1950* (New York: Macmillan Publishing Co. Inc., 1979).

Beauchamp, Gorman. "The Politics of Progress," *Michigan Quarterly Review*, Vol. 21, 658-73, Fall, 1982.

Bechmann, Roland. *Trees and Man: The Forest in the Middle Ages* (New York: Paragon House, 1990).

Bengtsson, Jan, Nilsson, Sven G., Franc, Alain, Menozzi, Paolo. "Biodiversity, disturbances, ecosystem function and management of European forests," *Forest Ecology and Management* 132, 2000.

Berman, Marshall. *All That is Solid Melts Into Air: The Experience of Modernity* (New York: Penguin Books, 1988).

Berneri, Louise Marie. *Journey Through Utopia* (London: Routledge and Kegan Paul Ltd., 1950).

Biese, Alfred. *The Development of the feeling for Nature in the Middle Ages and Modern Times* (London: George Routledge and Sons, Ltd., 1905).

Blaut, J.M. *The Colonizer's Model of the World: Geographical Diffusionism and Eurocentric History* (New York: The Guilford Press, 1993).

Bowlus, Charles R. "Ecological Crisis in Fourteenth Century Europe" in *Historical Ecology: Essays on Environment and Social Change*, ed. Lester T. Bilsky (Port Washington, NY: Kennikat Press, 1980).

Bratton, Susan Power. *Christianity, Wilderness, and Wildlife: The Original Desert Solitaire* (Scranton: University of Scranton Press; London and Toronto: Associated University Press, 1993).

Breisach, Ernst. *On The Future of History: The Postmodernist Challenge and Its Aftermath* (Chicago: The University of Chicago Press, 2003).

Brinton, Crane. *The Shaping of Modern Thought* (Englewood Cliffs, N.J.: Prentice Hall, Inc., 1963).

Brownell, Baker. *The Human Community* (New York: Harper & Row, 1950).

Brumfitt, J.H. *The French Enlightenment* (London: MacMillan, 1972).

Bruun, Geoffrey. *Nineteenth Century European Civilization* (New York: Oxford University Press, 1960).

Bury, J.B. *The Idea of Progress: An Inquiry into Its Origin and Growth* (New York: Dover Publications, Inc., 1955, 1932).

Cairns, Grace E. *Philosophies of History* (New York: The Citadel Press, 1962).

Callenbach, Ernest. *Ecotopia* (New York: Bantam Books, 1975).

Callicott, J. Baird. "The Conceptual Foundations of the Land Ethic" in J. Baird Callicott, ed. *Companion to A Sand County Almanac: Interpretive and Critical Essays* (Madison: The University of Wisconsin Press, 1987).

Callicott, J. Baird. *In Defense of the Land Ethic: Essays in Environmental Philosophy* (Albany: State University of New York Press, 1989).

Callicott, J. Baird. *Beyond the Land Ethic: More Essays in Environmental Philosophy* (Albany: State University of New York Press, 1999).

Cameron, J.M. *New York Review of Books*, April 17, 1980.

Canter, Larry. *Environmental Impact Assessment*   (New York: McGraw-Hill, 1977).

Cantor, Norman E. *Civilization of the Middle Ages* (New York: Harper Collins Publishers, 1963).

Carr, Edward Hallett. *What Is History?* (New York: Alfred A. Knopf, 1967).

Carrier, Martin, Gerald J. Massey, and Lara Reutsche, eds., *Science at Century's End: Philosophical questions on the Progress and Limits of Science* (University of Pittsburgh Press, 2000).

Carroll, Joseph. *Evolution and Literary Theory* (Columbia: University of Missouri Press, 1995).

Chardin, Teilhard de. *The Phenomenon of Man* (New York: Harper and Row, Publishers, 1959).

Charlton, D.G. *New Images of the Natural in France* (Cambridge: Cambridge University Press, 1984).

Chew, Sing C. *World Ecological Degradation*: *Accumulation, Urbanization, and Deforestation* (Walnut Creek, CA: Alta Mira Press, 2001).

Chew, Sing C. *The Recurring Dark Ages: Ecological Stress, Climate Changes, and System Transformation* (Lanham, Maryland: Altamira Press, 2007).

Chua, Amy. *World on Fire: How Exporting Free Market Democracy Breeds Ethnic Hatred and Global Instability* (New York: Doubleday, 2003).

Coates, Peter. *Nature: Western Attitudes Since Ancient Times* (Berkeley: University of California Press, 1998).

Collingwood, Robin G. *The Idea of History* (London: Oxford University Press, 1946).

Costello, Paul. *World Historians and Their Goals: Twentieth Century Answers to Modernism* (DeKalb: Northern Illinois University Press, 1993).

Crawford, Peter. *The Living Isles: A Natural History of Britain and Ireland* (New York: Charles Scribner's Sons, 1985).

Crist, Eileen. "Against the Social Construction of Nature and Wilderness," *Environmental Ethics* Spring, 2004, Vol. 26, No. 1

Cronon, William. "The Trouble with Wilderness; or, Getting Back to the Wrong Nature" in William Cronon, ed. *Uncommon Ground: Toward Reinventing Nature* (New York: W.W. Norton and Company, 1995).

Crosby, Alfred. *Ecological Imperialism: The Biological Expansion of Europe, 900-1900* (Cambridge: Cambridge University Press, 1986).

Cunningham, Andrew, and Grell, Ole Peter. *The Four Horsemen of the Apocalypse: Religion, War, Famine and Death in Reformation Europe* (Cambridge: Cambridge University Press, 2000).

Darby, H.C. "The Clearing of the Woodland in Europe" in *Man's Role in Changing the Face of the Earth* (Chicago: University of Chicago Press, 1958).

Darwin, Charles. Quoted by editor Anthony Flew in introduction to *Malthus, An Essay on the Principle of Population,* p. 49.

Darwin, Charles. *The Origin of Species* (Oxford: Oxford University Press, 1996).

Darwin, Charles. *The Voyage of the Beagle* (New York: Bantam Books, 1958).

Dawkins, Richard. "What Use is Religion? : Part I," *Free Inquiry*, Vol. 24, No. 4; June-July, 2004.

Devall, Bill and George Sessions. *Deep Ecology* (Salt lake City: Gibbs M. Smith, 1985).

Diamond, Jared. *Guns, Germs, and Steel: The Fates of Human Societies* (New York: W.W. Norton and Company, 1999).

Diamond, Jared. *Collapse: How Societies Choose to Fail or Succeed* (New York: Viking, 2005).

Diggins, John Patrick. *The Rise and Fall of the American Left* (New York: W.W. Norton and Company, 1992).

Dobkowski, Michael N. and Wallimann, Isidor, editors. *The Coming Age of Scarcity: Preventing Mass Death and Genocide in the Twenty-first Century* (Syracuse, New York: Syracuse University Press, 1998).

Drew, Wayland. "Killing Wilderness," *The Trumpeter* Vol. 3, No. 1, Winter 1986, pp. 19-23.

Dubos, René. *The Wooing of Earth: New Perspectives on Man's Use of Nature* (New York: Charles Scribner's Sons, 1980).

Dugger, Ronnie. "It's Time for an American Offensive against Theocracy," *Free Inquiry*, Vol. 24, No. 3, April/May, 2005.

Eckersley, Robyn. *Environmentalism and Political Theory: Toward an Ecocentric Approach* (Albany: State University of New York Press, 1992).

Edelstein, Ludwig. *The Idea of Progress in Classical Antiquity* (Baltimore, Maryland: The Johns Hopkins Press, 1967).

Ehrlich, Paul R. and Anne R. Ehrlich. *Betrayal of Science and Reason* (Washington D.C.: Island Press, 1996).

Ehrlich, Paul. *Human Natures: Genes, Cultures, and the Human Prospect* (Washington, D.C.: Island Press, 2000).

Ellul, Jacques. *The Technological Society* (New York: Vintage Books, 1964).

Fagan, Brian. *The Journey from Eden: The Peopling of Our world* (London: Thames and Hudson, 1990).

Fagan, Brian. *Floods, Famines, and Emperors*, (New York: Basic Books, 1999).

Fagan, Brian. *The Little Ice Age: How Climate Made History, 1300-1850* (New York: Basic Books, 2000).

Farrenkopf, John. *Prophet of Decline: Spengler on World History and Politics* (Baton Rouge: Louisiana State University Press, 2001).

Fischer, Klaus P. *History and Prophecy: Oswald Spengler and the Decline of the West* (New York: Peter Lang, 1989).

Flannery, Tim. *The Weather Makers: How Man Is Changing the Climate and What It Means for Life on Earth* (New York: Atlantic Monthly Press, 2005).

Flexner, Stuart Berg. *The Random House Dictionary of the English Language*, Second Edition, Unabridged (New York: Random House, 1987).

Forrest, Barbara and Paul R. Gross. *Creationism's Trojan Horse: The Wedge of Intelligent Design* (Oxford: Oxford University Press, 2004).

Foster, John Bellemy. "Planetary Defeat: The Failure of Global Environmental Reform," *Monthly Review*, January, 2003, pp. 1-9.

Fox, Matthew. Original Blessing (Santa Fe, New Mexico: Bear and Company Publishing, 1983).

Fox, Matthew. *A New Reformation: Creation Spirituality and the Transformation of Christianity* (Rochester, Vermont: Inner Traditions, 2006).

Frank, Robert Worth Jr. "The 'Hungry Gap,' Crop Failure, and Famine: The Fourteenth Century Agricultural Crisis and *Piers Plowman*." *Agriculture in the Middle Ages: Technology, Practice, and Representation* (Philadelphia: University of Pennsylvania Press, 1995).

Frankfort, Henri. *The Birth of Civilization in the Near East* (Garden City, New York: Doubleday & Company, Inc., 1956).

Fukuyama, Francis. *The End of History and the Last Man* (New York: Avon Books, 1992).

Gadamer, Hans-Georg. "On the Scope and Function of Hermeneutical Reflection," in Brice R. Wachterhauser, *Hermeneutics and Modern Philosophy* (Albany: State University of New York Press, 1986).

Gadamer, Hans-Georg. "What Is Truth?" in Brice R. Wachterhauser, *Hermeneutics and Truth* (Evanston, Illinois: Northwestern University Press, 1994).

Gardner, Eldon J. *History of Biology* (Minneapolis, Minnesota: Burgess Publishing company, 1972).

Gastil, Raymond Duncan. *Progress: Critical Thinking about Historical Change* (Westport, Connecticut: Praeger, 1993).

Gay, Peter. *The Enlightenment: An Interpretation: The Rise of Modern Paganism* (New York: Alfred A. Knopf, 1967).

Gibbon, Edward. *The Decline and Fall of the Roman Empire* (Chicago: Encyclopedia Brittanica, Inc., 1952), Vol. I.

Gibbs, Nancy. "The Faith Factor," *Time*, Vol. 163, No. 25, June 21, 2004.

Glacken, Clarence. *Traces on the Rhodian Shore: Nature and Culture in Western Thought from Ancient Times to the End of the Eighteenth Century* (Berkeley: University of California Press, 1967).

Golley, Frank Benjamin. *A History of the Ecosystem Concept in Ecology* (New Haven and London: Yale University Press, 1993).

Gore, Al. *An Inconvenient Truth: The Planetary Emergency of Global Warming and What We Can Do about It* (Emmaus, Pennsylvania: Rodale, 2006).

Gould, Stephen J. *The Structure of Evolutionary Theory* (Cambridge, Massachusetts: The Belknap Press of Harvard University Press, 2002).

Gould, Stephen J. "Church, Humboldt, and Darwin: The Tension and Harmony of Art and Science" in Franklin Kelly, *Frederic Edwin Church* (Washington: Smithsonian Institution Press, 1989).

Grant, Michael. *The Founders of the Western World: A History of Greece and Rome* (New York: Charles Scribner's Sons, 1991).

Greene, John C. *Science, Ideology, and World View: Essays in the History of Evolutionary Ideas* (Berkeley: University of California Press, 1981).

Grene, Marjorie. "The Paradoxes of Historicity" in Brice R. Wachterhauser, ed. *Hermeneutics and Modern Philosophy* (Albany: State University of New York Press, 1986).

Guegen, J.A. Review of Robert Nisbet's *History of the Idea of Progress* in *The American Political Science Review*, Vol. 75, 176-177, March 1981.

Hardin, Garrett. "The Tragedy of the Commons." *Science,* 162: 1243-1248, 1968.

Harris, Sam. *The End of Faith: Religion, Terror, and the Future of Reason* (New York: W.W. Norton and Company, 2004).

Hay, Peter. *Main Currents in Western Environmental Thought* (Bloomington: Indiana University Press, 2002).

Haynes, Jeff. "Religion, Secularization and Politics: A Postmodern Conspectus," *Third World Quarterly*, Vol. 18, No. 4 (1997), quoted in Malise Ruthven, *Fundamentalism: The Search for Meaning* (Oxford: Oxford University Press, 2004).

Hays, Samuel P. *A History of Environmental Politics since 1945* (Pittsburgh: University of Pittsburgh Press, 2000).

Hays, Samuel P. *Conservation and the Gospel of Efficiency: The Progressive Conservation Movement, 1890-1920* (New York: Atheneum, 1969).

Heilbroner, Robert. *The Nature and Logic of Capitalism* (New York: W.W. Norton and Company, 1985).

Henderson, Hazel. *Beyond Globalization: Shaping a Sustainable Global Economy* (West Hartford, Connecticut: Kumarian Press, Inc., 1999).

Herlihy, David J. *The Black Death and the Transformation of the West* (Cambridge, Massachusetts: Harvard University Press, 1997).

Herlihy, David J. "Attitudes Toward the Environment in Medieval Society," in *Historical Ecology: Essays on Environment and Social Change*, ed. Lester J. Bilsky (Port Washington, New York: National University Publications, Kennikat Press, 1980).

Heuer, Karsten. *Walking the Big Wild: From Yellowstone to the Yukon on the Grizzly Bear's Trail* (Seattle: The Mountaineer's Books, 2004).

Highet, Gilbert. *Poets in a Landscape* (Harmondsworth, Middlesex: Penguin Books Ltd, 1959).

Himmelfarb, Gertrude. "In Defense of Progress," *Commentary*, Vol. 69, 53-60, June, 1980.

Himmelstein, Jerome. "The Two Nisbets: The Ambivalence of Contemporary Conservatism," *Social Forces*, vol. 60, 231-36, September, 1981.

Hobsbawm, Eric. *The Age of Extremes: A History of the World, 1914-1991* (New York: Pantheon Books, 1994).

Hoffer, Eric. *The True Believer* (New York: Time, Inc., 1963).

Hoffert, Robert W. "The Scarcity of Politics: Ophuls and Western Political Thought," *Environmental Ethics*, Vol. 8, No. 1 (Spring 1986), pp. 5-32.

Hollister, C. Warren. *Medieval Europe: A Short History* (New York: McGraw-Hill Publishing Company, 1990).

Hollister, C. Warren. *Roots of the Western Tradition: A Short History of the Ancient World* (New York: McGraw-Hill Inc., 1982).

Hollister, C. Warren. *Roots of the Western Tradition: A Short History of the Ancient World* Fifth Edition (New York: McGraw-Hill, Inc., 1991).

Hopkins, Keith. *A World Full of Gods: The Strange Triumph of Christianity* (New York: The Free Press, 2000).

Horowitz, Asher. *Rousseau, Nature, and History* (Toronto: University of Toronto Press, 1987).

Hughes, H. Stewart. *Oswald Spengler* (New York: Charles Scribner's Sons, 1962).

Hughes, J. Donald, editor. *The Face of the Earth: Environment and World History* (Armonk, New York: M.E. Sharpe, 2000).

Hughes, J. Donald. "Theophrastus as Ecologist," *Environmental Review*, Vol. 9, No. 4, Winter, 1985), pp. 296-306.

Hughes, J. Donald. *An Environmental History of the World: Humankind's Changing Role in the Community of Life* (London: Routledge, 2001).

Hughes, J. Donald. *Ecology in Ancient Civilizations* (Albuquerque: University of New Mexico Press, 1975).

Hughes, J. Donald. *Pans Travail: Environmental Problems in Ancient Greece and Rome* (Baltimore: Johns Hopkins University Press, 1994).

Israel, Jonathan I. *Radical Enlightenment: Philosophy and the Making of Modernity 1650-1750* (Oxford: Oxford University Press, 2001).

Jacoby, Susan. *Freethinkers: A History of American Secularism* (New York: Henry Holt and company, 2004).

Johnson, Warren. *Muddling Toward Frugality: A Blueprint for Survival in the 1980s* (Boulder, Colorado: Shambhala, 1979).

Johnstone, Ronald L. *Religion in Society: A Sociology of Religion* (Upper Saddle River, New Jersey: Prentice Hall, 6th Ed., 2001).

Jonas, Hans. "Reflections on Technology, Progress, and Utopia." *Social Research* 48 (Autumn, 1981).

Josephson, Paul R. *Industrialized Nature: Brute Force Technology and the Transformation of the Natural World* (Washington: Island Press, 2002).

Kaplan, Robert D. *The Coming Anarchy: Shattering the Dreams of the Post Cold War* (New York: Vintage Books, 2001).

Kelly, Franklin (with Stephen Jay Gould, James Anthony Ryan, Debora Rindge). *Frederic Edwin Church* (Washington: National Gallery of Art and Smithsonian Institution Press, 1989).

Kemp, Jonathan, ed. *Diderot, Interpreter of Nature: Selected Writings* (New York: International Publishers, 1963).

Klare, Michael T. *Resource Wars: The New Landscape of Global Conflict* (New York: Henry Holt and Company, 2001).

Kline, Benjamin. *First Along the River: A Brief History of the U.S. Environmental Movement* (San Francisco: Acada Books, 2000).

Knight, Frank. "Environmental Celebrations: The Ten Most Significant Environmental Events of the 20th Century," *Nature Study*, Vol. 44, Nos. 2 and 3, Feb., 1991, pp. 1-2.

Kolbert, Elizabeth. *Field Notes from a Catastrophe: Man, Nature, and Climate Change* (New York and London: Bloomsbury Publishing: 2006).

Korten, David C. "The Mythic Victory of Market Capitalism" in Jerry Mander and Edward Goldsmith, eds., *The Case Against the Global Economy*.

Kovel, Joel. *The Enemy of Nature: The End of Capitalism or the End of the World?* (Nova Scotia: Fernwood Publishing Ltd, and London: Zed Books Ltd, 2002).

Kroeber, Alfred L. *Configurations of Culture Growth* (Berkeley and Los Angeles: University of California Press, 1944).

Kroeber, Alfred L. *Style and Civilizations* (Ithaca: Cornell University Press, 1957).

Kuhn, Thomas S. *The Structure of Scientific Revolutions* (Chicago: The University of Chicago Press, 1970).

LaFreniere, Gilbert F. *Jean-Jacques Rousseau and the Idea of Progress* (Ann Arbor, Michigan University Microfilms International,1979. (Ph.D. dissertation, University of California at Santa Barbara, 1976).

LaFreniere, Gilbert F. "Rousseau's *First Discourse* and the Idea of Progress," *The Willamette Journal of the Liberal Arts*, V.1, No.1 (Fall 1983), pp.7-26.

LaFreniere, Gilbert F. "World Views and Environmental Ethics," *Environmental Review*, V. 9, No. 4 (Winter 1985), pp. 307-322.

LaFreniere, Gilbert F. "The Redefinition of Progress," *The Willamette Journal of the Liberal Arts*, Vol. 4, No.2 (Summer 1989), pp.73-93.

LaFreniere, Gilbert F. "Rousseau and the European Roots of Environmentalism," *Environmental History Review*, V.14, No. 4 (Winter 1990), pp. 41-72.

LaFreniere, Gilbert F. "Land-Use Planning and the Land Ethic," *The Trumpeter*, V. 10, No. 2, 1993, pp. 59-62.

LaFreniere, Gilbert F. "Greenline Parks in France: Les Parcs Naturels Régionaux." *Agriculture and Human Values*, V.14, No. 4 (Dec. 1997), pp. 337-352.

Lambert, David. *Wild Mammals of the Countryside* (Covent Garden, London: Concertina Publications Limited, 1979).

Lander, Jerry and Edward Goldsmith, editors. *The Case against the Global Economy; And for a Turn toward the Local* (San Francisco: Sierra Club Books, 1996).

Lasch, Christopher. *The True and Only Heaven: Progress and Its Critics* (New York: W. W. Norton & company, 1991).

Leiss, William. *The Domination of Nature* (New York: George Braziller, 1972).

Leopold, Aldo. *A Sand County Almanac, with Essays on Conservation from Round River* (New York: Ballantine Books; San Francisco: Sierra Club, 1974).

Liberati, Anna Maria and Fabio Bourbon. *Ancient Rome: History of a Civilization that Ruled the World* (New York: Steward, Tabori and Chang, 1996).

Lindberg, David C. *The Beginnings of Western Science* (Chicago: The University of Chicago Press, 1992).

Little, Charles E. *Hope for the Land* (New Brunswick: Rutgers University Press, 1992).

Locke, John. *Two Treatises of Government* (New York: Cambridge University Press, 1960)

Lovelock, James. The *Revenge of Gaia: Earth's Climate Crisis and the Fate of Humanity*
(New York: Basic Books, 2006).

Lucretius, *On Nature* (Indianapolis: The Bobbs-Merrill Company, 1965).

Malthus, Thomas. Edited by Anthony Flew. *An Essay on the Principle of Population* (Harmondsworth, Middlesex, England: Penguin Books Ltd., 1970).

Mann, Charles C. "1491," The Atlantic Online, http://www.theatlantic.com. (*The Atlantic*, March 2002).

Mannheim, Karl. *Ideology and Utopia* (New York: Harcourt, Brace & World, Inc., 1936).

Manuel, Frank E. "Men of Ideas," *New York Times*, March 16, 1980, VII.

Marsak, Leonard M. "Bernard de Fontenelle: In Defense of Science," in *The Rise of Science in Relation to Society*, Leonard M. Marsak, ed. (London: Collier-Macmillan Limited, 1964).

Marsak, Leonard. "Bernard de Fontenelle: The Idea of Science in the French Enlightenment." *Transactions of the American Philosophical Society* 49, Part 7 (December 1959), pp.1-64.

Marsak, Leonard. *The Nature of Historical Inquiry* (New York: Holt, Rinehart and Winston, 1970).

Marsh, George Perkins. *Man and Nature*, David Lowenthal, ed. (Cambridge, Massachusetts: The Belknap Press of Harvard University Press, 1985).

Martin, Luther H. *Hellenistic Religions: An Introduction* (New York: Oxford University Press, 1987).

Marx, Leo. "Does Improved Technology Mean Progress?" *Technology Review* (January 1987), pp. 32-41.

Marx, Leo. *The Pilot and the Passenger: Essays on Literature, Technology, and Culture in the United States* (Oxford: Oxford University Press, 1988).

Marx, Leo. "The Idea of 'Technology' and Postmodern Pessimism" in Merritt Roe Smith and Leo Marx, editors. *Does Technology Drive History? : The Dilemma of Technological Determinism* (Cambridge, Massachusetts; The MIT Press, 1994).

Marx, Leo and Bruce Mazlish, eds. *Progress: Fact or Illusion?* (Ann Arbor: The University of Michigan Press, 1996).

Mason, Stephen F. *A History of the Sciences* (New York: Collier Books, 1962).

McNeill, J. R. and William McNeill. *The Human Web: A Bird's-Eye View of World History* (New York: W.W. Norton and Company, 2003).

McNeill, J.R. *Something New Under the Sun: An Environmental History of the Twentieth-Century World* (New York: W.W. Norton and Company, 2000).

McNeill, William. *The Rise of the West: A History of the Human Community* (Chicago: The University of Chicago Press, 1963).

McNeill, William. *Plagues and People* (Garden City: Anchor books, 1976).

McNeill, William H. *Mythistory and Other Essays* (Chicago: University of Chicago Press, 1986).

Merchant, Carolyn. *The Death of Nature: Women, Ecology and the Scientific Revolution* (San Francisco: Harper and Row, Publishers, 1980).

Merchant, Carolyn. "Reinventing Eden: Western Culture as a Recovery Narrative" in *Uncommon Ground: Toward Reinventing Nature*, William Cronon, ed. (New York: W.W. Norton and Company), pp. 132-159.

Merchant, Carolyn. *Reinventing Nature: The Fate of Nature in Western Culture* (New York: Routledge, 2003).

Meyer, Stephen. *The End of the Wild* (Cambridge, Massachusetts: The MIT Press, 2006.

Middleton, Nick. *The Global Casino: An Introduction to Environmental Issues*, 3rd Edition (New York: Oxford University Press Inc., 2003).

Miller, Char, editor. *The Atlas of U.S. and Canadian Environmental History* (New York: Routledge, 2003).

Miller, Char. *Gifford Pinchot and the Making of Modern Environmentalism* (Washington: Island Press, 2002).

Moore, John A. *Science As A Way of Knowing: The Foundations of Modern Biology* (Cambridge, Massachusetts: Harvard University Press, 1993).

Mornet, Daniel. *French Thought in the Eighteenth Century* (New York: Prentice-Hall, Inc. 1929).

Mornet, Daniel. *Le Sentiment de La Nature en France, de J.J. Rousseau à Bernardin de Saint-Pierre* (Genève-Paris: Slatkine Reprints 1907, 1980).

Morris, Henry. *The Remarkable Birth of Planet Earth*, Institution of Creation Science, Seattle, 1986, quoted in Malise Ruthven, *Fundamentalism* (Oxford: Oxford University Press, 2004).

Moyers, Bill. "There is no Tomorrow," *Free Inquiry*, Vol. 25, No. 4, June-July 2005, pp. 21-23.

Mumford, Lewis. *The Condition of Man* (New York: Harcourt, Brace, 1944).

Mumford, Lewis. *The Story of Utopias* (New York: The Viking Press, 1922, 1962).

Mumford, Lewis. *Interpretations and Forecasts* (New York: Harcourt, Brace, Jovanovich, Inc., 1973).

Murray, Gilbert. *Five Stages of Greek Religion (Garden City, New York: Doubleday and Company, Inc., 1955.*

Nash, Roderick. *Wilderness and the American Mind*, Third Ed. (New Haven: Yale University Press, 1982).

National Geographic Society. *Wild Animals of North America* (Washington, D.C.: National Geographic Society, 1960).

Nicolson, Marjorie Hope. *Mountain Gloom and Mountain Glory: The Development of the Aesthetics of the Infinite* (New York: W.W. Norton & Company, Inc., 1963).

Nisbet, Robert. *History of the Idea of Progress* (New York: Basic Books, 1980).

Norgaard, Richard B. *Development Betrayed: The End of Progress and A Coevolultionary Revisioning of the Future* (London: Routledge, 1994).

O'Hear, Anthony. *After Progress: Finding the Old Way Forward* (New York: Bloomsbury Publishing, 1999).

Oelschlaeger, Max. *The Idea of Wilderness* (New Haven: Yale University Press, 1991).

Oelschlaeger, Max. *Caring for Creation: An Ecumenical Approach to the Environmental Crisis* (New Haven: Yale University Press, 1994).

Oelschlaeger, Max, ed. *Postmodern Environmental Ethics* (State University of New York Press, 1995).

Oelschlaeger, Max. Professor of Philosophy and Religion, Northern Arizona University, Personal communication, 2006.

Olafson, Frederick A. "The Idea of Progress: An Ethical Appraisal." In *Progress and its Discontents*, Gabriel A. Almond, Marvin Chodurow, and Roy Harvey Peace, editors. (Berkeley: University of California Press, 1982).

Olson, Theodore. *Millennialism, Utopianism, and Progress* (Toronto: University of Toronto Press, 1982).

Ophuls, William. *Ecology and the Politics of Scarcity: Prologue to a Political theory of the Steady State* (San Francisco: W.H. Freeman and Company, 1977). Revised as *Ecology and the Politics of Scarcity Revisited: The Unraveling of the American Dream*, with A. Stephen Boyan, Jr. (New York: W.H. Freeman and Company, 1992).

Ophuls, William. *Requiem for Modern Politics: The Tragedy of the Enlightenment and the Challenge of the New Millennium* (Boulder, Colorado: Westview Press, 1997).

Opie, John. *Nature's Nation: An Environmental History of the United States* (Fort Worth: Harcourt Brace College Publishers, 1998).

Opie, John. "Renaissance Origins of the Environmental Crisis." *Environmental Review*. Vol. 11. No. 1, Spring, 1987, pp. 3-17.

Oregonian, The. "Bones Yield Signs of Malaria Hastening the Fall of Rome" (Portland, Oregon), Feb. 21, 2001, pp. 2-17.

Osborn, Fairfield. *Our Plundered Planet* (New York: Pyramid Books: 1970). Little, Brown edition published March, 1949.

Osborn, Matthew. " 'The Weirdest of all Undertakings': The Land and the Early Industrial Revolution in Oldham, England," *Environmental History Vol.* 8 (April, 2003), pp.246-249.

Oursel, Raymond. *Living Architecture: Romanesque* (New York: Grosset and Dunlap, 1967).

Outram, Dorinda. *The Enlightenment* (Cambridge: Cambridge University Press, 1995).

Pakev, Sybil P., ed. *Grzimek's Encyclopedia of Mammals* (New York: McGraw-Hill Publishing company, 1990, Vol. 5).

Palmer, R.R. , Joel Colton, and Lloyd Kramer. *A History of the Modern World*, 9th ed. (New York: Alfred A. Knopf, 2002).

Perlin, John. *A Forest Journey: The Role of Wood in the Development of Civilization* (Cambridge, Massachusetts: Harvard University Press. 1989).

Perry, Ralph Barton. *Puritanism and Democracy* (New York: The Vanguard Press, 1944).

Petulla, Joseph. *American Environmental History* (Columbus, Ohio: Merrill Publishing Company, 2nd Edition., 1988).

Phillips, Kevin. *Wealth and Democracy*. (New York: Broadway Books, 2002).

Pinker, Steven. *The Blank Slate: The Modern Denial of Human Nature* (New York: Viking, 2002).

Plato, *Dialogues*, Edith Hamilton and Huntington Cairns, eds., translated by A.E. Taylor (Princeton, New Jersey: Princeton University Press for Bollingen Foundation, 1961).

Plattner, Marc F. *Rousseau's State of Nature* (DeKalb: Northern Illinois University Press, 1979).

Polanyi, Karl. *The Great Transformation: The Political and Economic Origins of Our Time* (Boston: Beacon, 1957).

Ponting, Clive. *A Green History of the World: The Environment and the Collapse of Great Civilizations* (New York: Penguin Books, 1991).

Pyne, Stephen J. *Vestal Fire: An Environmental History, Told Through Fire, of Europe and Europe's Encounter with the World* (Seattle: University of Washington Press, 1997).

Quammen, David. "Was Darwin Wrong? No. The Evidence for Evolution is Overwhelming," *National Geographic*, Vol. 206, No. 5, Nov., 2004, pp. 4-35.

Quinn, Daniel. *Ishmael: An Adventure of the Mind and Spirit* (New York: Bantam/Turner, 1992).

The Random House Dictionary of the English Language, Second Edition, Unabridged (New York: Random House, 1987).

Redman, Charles L. *Human Impact on Ancient Environments* (Tucson: The University of Arizona Press, 1999).

Roderick Nash. *Wilderness and the American Mind* (New Haven: Yale University Press, 1967, 1982).

Roszak, Theodore. *Where the Wasteland Ends: Politics and Transcendence in Postindustrial Society* (New York: Doubleday and Company, 1972).

Roszak, Theodore. *The Making of a Counterculture: Reflections on the Technocratic Society and its Youthful Opposition* (Garden City, New York: Anchor Books, Doubleday and Company, 1969).

Roszak, Theodore. *The Voice of the Earth* (New York: Simon and Schuster, 1992).

Rothman, Hal K. *Saving the Planet: The American Response to the Environment in the Twentieth Century* (Chicago: Ivan R. Dee, 2000).

Rothman, Hal. "A Decade in the Saddle: Confessions of a Recalcitrant Editor," *Environmental History*, 7:1 (January, 2002), pp. 9-21.

Rothschild, Emma. *Economic Sentiments: Adam Smith, Condorcet, and the Enlightenment* (Cambridge, Massachusetts: Harvard University Press, 2001).

Rousseau, Jean-Jacques. *Oeuvres Complètes, Tome III, Du contrat Social, Écrits Politiques* (Paris: Éditions Gallimard, 1964).

Rousseau, Jean-Jacques. *Julie, ou La Nouvelle Hèloïse*, edited and translated by Judith H. McDowell (University Park: The Pennsylvania State University Press, 1968).

Rousseau, Jean-Jacques. *The Social Contract* and *Discourses*, translated by G.D.H. Cole (New York: E.P. Dutton and Company, Inc., 1950).

Rousseau, Jean-Jacques. *Emile,* translated by Barbara Foxley (London: Dent, 1969).

Rousseau, Jean-Jacques. *Reveries of the Solitary Walker* (Harmondsworth, Middlesex, England: Penguin Books, Ltd., 1979).

Ruddiman, William F. *Plows, Plagues, and Petroleum: How Humans Took Control of Climate* (Princeton and Oxford: Princeton University Press, 2005).

Runte, Alfred. *National Parks: The American Experience* (Second Edition, Revised). (Lincoln: University of Nebraska Press, 1987).

Saggs, H.W.F. *Civilization Before Greece and Rome* (New Haven: Yale University Press, 1989).

Sale, Kirkpatrick. *The Green Revolution: The American Environmental Movement 1962- 1992* (New York: Hill and Wang, 1993).

Sale, Kirkpatrick. T*he Conquest of Paradise: Christopher Columbus and the Columbian Legacy* (New York: Alfred A. Knopf, 1990).

Sale, Kirkpatrick. *After Eden: The Evolution of Human Domination* (Durham and London: Duke University Press, 2006).

Schama, Simon. *Landscape and Memory* (New York: Alfred A. Knopf, 1995).

Schiavone, Aldo. *The End of the Past: Ancient Rome and the Modern West* (Cambridge, Massachusetts: Harvard University Press, 2000).

Seidel, Peter. *Invisible Walls: Why We Ignore the Damage We Inflict on the Planet ... and Ourselves* (Amherst, New York: Prometheus Books, 2001).

Shellenberger, Michael and Ted Nordhaus, "The Death of Environmentalism: Global warming politics in a post-environmental world" *Grist Magazine*, 3/22/05, http://www.grist.org/cgi-bin.

Shepard, Paul. "Virtually Hunting Reality in the Forests of Simulacra" in Michael E. Soulé, Gary Lease. *Reinventing Nature: Responses to Postmodern Deconstruction* (Washington, D.C.: Island Press, 1995).

Shermer, Michael. *Why People Believe Weird Things: Pseudoscience, Superstition, And Other Confusions of Our Time* (New York: W.H. Freeman and Company, 1997).

Simmons, I.B. *Changing the Face of the Earth: Culture, Environment, History* (Oxford: Basil Blackwell, 1989).

Soper, Kate. *What is Nature? Culture, Politics and the Non-Human* (Oxford and Cambridge, Mass.: Blackwell, 1995).

Sorokin, Piterim A. *Social and Cultural Dynamics* (1937-1941; reprint (New York: Bedminster, 1962).

Spengler, Oswald. *Man and Technics: A Contribution to a Philosophy of Life* (Honolulu: University Press of the Pacific, 1931, 2002 reprint).

Spengler, Oswald. *The Decline of the West* (New York: Alfred A. Knopf, Vol. I, 1926; Vol. II, 1928). Original German publication, 1918, 1922.

Speth, James. *Red Sky at Morning: America and the Crisis of the Global Environment* (Yale University Press: New Haven and London, 2004).

Starobinski, Jean. *Jean-Jacques Rousseau: Transparence and Obstruction* (Chicago: The University of Chicago Press, 1971, 1988).

Stierlen, Henri. *Living Architecture: Mayan* (New York: Grosset and Dunlap, 1964).

Stiglitz, Joseph E. *Globalization and its Discontents* (New York: W.W. Norton & Company, 2002).

Stoddard, Whitney S. *Monastery and Cathedral in France* (Middletown, Connecticut: Wesleyan University Press, 1966).

Strahler, Arthur N. and Alan H. Strahler. *Geography and Man's Environment* (New York: John Wiley and Sons, 1977).

Suckling, Norman. "The Enlightenment and the Idea of Progress," *Studies in Voltaire and the Eighteenth Century*, No. 58, 1967, pp. 1461-1480.

Sutter, Paul S. *Driven Wild: How the Fight Against Automobiles Launched the Modern Wilderness Movement* (Seattle: University of Washington Press, 2002).

Tacitus, *On Britain and Germany* (Harmondsworth: Penguin Books Ltd., 1948).

Tainter, Joseph A. *The Collapse of Complex Societies* (Cambridge; Cambridge University Press, 1988).

Tarbuck, Edward J. and Frederick K. Lutgens. *Earth: An Introduction to Physical Geology* (Upper Saddle River, New Jersey: Prentice Hall, 2002).

Tarnas, Richard. *The Passion of the Western Mind: Understanding the Ideas That Have Shaped Our World View* (New York: Ballantine Books, 1991).

Teggart, Frederic J. *Theory and Processes of History* (Berkeley: University of California Press, 1962.

Terborgh, John. Review of *The Eternal Frontier: An Ecological History of North America and its Peoples* (by Tim Flannery) (*The New York Review of Books*, Vol. 48, No. 14, Sept. 20, 2001).

Thacker, Christopher. *The Wildness Pleases: The Origins of Romanticism* *(London: Croom Helm, 1983).*

Toulmin, Stephen and June Goodfield, *The Discovery of Time* (New York: Harper & Row, Publishers, Incorporated, 1965).

Toynbee, Arnold J. *A Study of History* (New York & London: Oxford University Press, 1947. Abridgement of Volumes I-VI by D.C. Somervell).

Truettner, William H. and Alan Wallach, eds. *Thomas Cole: Landscape into History* (New Haven: Yale University Press, and Washington D.C.: National Museum of American Art, Smithsonian Institution, 1994).

Tuchman, Barbara W. *A Distant Mirror: The Calamitous Fourteenth Century* (New York: Alfred A. Knopf, Inc., 1978).

Turner, Frederick. *Beyond Geography: The Western Spirit Against the Wilderness* (New Brunswick, New Jersey: Rutgers University Press, 1983).

Tuveson, Ernest Lee. *Millennium and Utopia: A Study in the Background of the Idea of Progress* (New York: Harper & Row, 1964).

Van Doren, Charles. *The Idea of Progress* (New York: Frederick A. Praeger, 1967).

Vyverberg, Henry. *Historical Pessimism in the French Enlightenment* (Cambridge: Harvard University Press, 1958).

Wachterhauser, Brice R., ed. *Hermeneutics and Modern Philosophy* (Albany: State University of New York Press, 1986).

Wagar, W. Warren. *American Historical Review*, (Review of Robert Nisbet's History of the Idea of Progress), 568, June 1981.

Wagar, W. Warren, editor. *The Idea of Progress since the Renaissance* (New York: John Wiley and Sons, 1964).

Wagar, W. Warren. *Good Tidings: The Belief in Progress from Darwin to Marcuse* (Bloomington: Indiana University Press, 1972).

Wagar, Warren E. *Worldviews: A Study in Comparative History* (Hinsdale Illinois: The Dryden Press, 1977).

Wagar, W. Warren. *The Three Futures: Paradigms of Things to Come* (New York: Praeger, 1991).

Wagar, Warren, Editor. *The Secular Mind* (New York: Holmes J. Meier Publishers, Inc., 1982)

Ward, Peter. *The End of Evolution: On Mass Extinctions and the Preservation of Biodiversity* (New York: Bantam Books, 1994).

Ward, Peter. *Future Evolution* (New York: Henry Holt and Company, 2001).

Webb, Walter Prescott. *The Great Frontier* (Austin: University of Texas Press, 1952).

Welch, Stephen R. "The Sins of the Missionaries," Free *Inquiry*, Vol. 24, No. 2, Feb.-Mar., 2004.

Wells, H.G. *The Outline of History* (Garden City: Garden City Publishing, 1920).

Wertheim, Margaret. *Pythagoras' Trousers* (New York: Random House, 1995).

Westcott, Roger W. "The Enumeration of Civilizations," *History and Theory*, Vol. IX, No. 1 pp.59-85, 1970).

Wheen, Francis. *Idiot Proof: Deluded Celebrities, Irrational Power Brokers, Media Morons and the Erosion of Common Sense* (New York: Public Affairs, 2004).

White, Gilbert. *The Natural History of Selborne* (London: Cassell and Company, 1908).

White, Lynn Jr. "The Historical Roots of Our Ecologic Crisis." *Science*, V. 155, No. 3767, 10 March 1967.

Whitehead, Alfred North. Quoted in Henri Frankfort, *The Birth of Civilization in the Near East* (Garden City, New York: Doubleday and Company, Inc., 1956).

Wilderness Society, The (Ben Beach, Bart Koehler, Leslie Jones, and Jay Watson, eds.) *The Wilderness Act Handbook* (40th Anniversary Edition) (Washington, D.C.: The Wilderness Society, 2004).

Whitty, Julia. "Gone: Half of Earth's Species Face Extinction this Century." *Mother Jones*, V. 32, No. 3, May-June, 2007, pp. 36-45, 88-90.

Wilson, Edward O. The Future of Life (New York: Alfred Knopf, 2002).

Worster, Donald, editor. *The Ends of the Earth* (Cambridge: Cambridge University Press, 1988).

Worster, Donald. *Nature's Economy: A History of Ecological Ideas* (Cambridge: Cambridge University Press, (1977, 1985).

Worster, Donald. *The Wealth of Nature: Environmental History and the Ecological Imagination* (New York: Oxford University Press, 1993).

Wright, Patricia. "Forest Primeval: Life and Death among the ancient trees of Poland's Bialowieza Forest", *Massachusetts,* Massachusetts Alumni Quarterly, Fall 1991.

Wright, Richard T. *Environmental Science: Toward a Sustainable Future* (Upper Saddle River, New Jersey: Pearson-Prentice Hall, 2005).

Yalden, Derek. *The History of British Mammals* (London: T & AD Poyser Ltd., 1999).

Zimmerman, Michael E. *Contesting Earth's Future: Radical Ecology and Postmodernity* (Berkeley: University of California Press, 1994).

INDEX